This Book Belongs To

Dr. Michael R. Chernick

Member of
the
American Statistical Association

AMERICAN·STATISTICAL·ASSOCIATION·FOUNDED·1839

Statistical Quality Assurance Methods for Engineers

STEPHEN VARDEMAN
Iowa State University

J. MARCUS JOBE
Miami University

John Wiley & Sons, Inc.
New York • Chichester • Weinheim • Brisbane • Singapore • Toronto

ACQUISITIONS EDITOR Wayne Anderson
MARKETING MANAGER Katherine Hepburn
DESIGNER Michael Jung
FREELANCE PRODUCTION MANAGEMENT Hermitage Publishing Services
COVER ART Michael Jung
ILLUSTRATION EDITOR Jaime Perea

This book was set in Times Roman by Publication Services and printed and bound by R. R. Donnelley & Sons. The cover was printed by Lehigh Press.

This book is printed on acid-free paper. ♾

The paper in this book was manufactured by a mill whose forest management programs include sustained yield harvesting of its timberlands. Sustained yield harvesting principles ensure that the numbers of trees cut each year does not exceed the amount of new growth.

Library of Congress Cataloging in Publication Data:

ISBN 0-471-15937-3

Printed in the United States of America

10 9 8 7 6 5 4 3 2 1

Soli Deo Gloria

Preface

This is a book primarily about statistical methods. It is intended for an advanced undergraduate audience that has the background of a first course in probability and statistics, and needs to know how tools from that course can be extended and applied in the production of quality goods and services. The outline of the book and the material in it have evolved over some 15 plus years of teaching statistical methods for quality assurance to undergraduate engineering and statistics majors, and have been significantly influenced by experiences teaching industrial short courses and by interactions with working engineers and industrial statisticians. Although the title of the book identifies engineering students as our primary target audience, we hope that properly prepared readers from other disciplines (such as business administration) will also find it accessible and useful.

The book contains somewhat more material than can be comfortably covered in a standard (three semester hour) college/university course, and so allows instructors some flexibility in topic choice. The course at ISU that provided the motivation for beginning writing on the book currently covers Chapters 1 through 4, Sections 5.1 through 5.3, Chapter 6, Sections 7.1 and 7.2, Sections 8.1, 8.3, and 8.5, and Chapter 9. Most of the remaining sections cover matters that sometimes arise in student projects in the ISU course, and so represent material that we feel also needs to be available to students in a modern quality assurance text.

Novel Features of the Text

Some of the features of the text that we expect instructors to find novel and effective include:

1. a thorough treatment of methods for assessing the adequacy of measurement precision (including gage R&R and LLD methodologies) (see Section 2.2),

2. a discussion of the differences between (and proper roles of) engineering control and statistical process monitoring, including an introduction to the ideas of PID control schemes and a real application of them (see Section 3.6),
3. consistent use of the ARL notion in the design and analysis of process monitoring schemes (including EWMA, CUSUM and X/MR schemes in Chapter 4),
4. an estimation-oriented treatment of the analysis of data from factorial and fractional factorial studies that doesn't rely on ANOVA methods or software and can handle unbalanced data (see Chapters 6 and 7),
5. an introduction to the design and analysis of mixture studies (see Section 7.3),
6. an elementary discussion of important qualitative considerations in experimental design for quality improvement, including ideas given emphasis by G. Taguchi (see Section 7.4),
7. provision of elementary design tools for sampling inspection (see Sections 8.1 and 8.2),
8. clear discussion of what can and cannot be accomplished by product-oriented sampling inspection (see Section 8.5), and
9. a dispassionate description of the main elements of TQM and analysis of their legitimate realms of application and limitations (in Chapter 9).

Web Site

Students and instructors will find current supporting materials for the text at the Web site: www.wiley.com/college/vardeman. Check this site for electronic versions of the book's data sets, complete solutions to selected exercises, practice exams, help with Minitab and Excel calculations for text examples, known errata, and model instructions for an outside-class quality assurance project together with "best in class" final reports from the ISU course where this assignment is used.

Many of the scenarios and data sets used in this book come from student projects at ISU. (Those examples with named contributors but no journal or book references derive from such projects.) We gratefully acknowledge the excellent work of the students who carried these out and their important contribution to the text. A number of students also aided our proofreading efforts on early versions of the manuscript, notably Dewi Rahardja and Mike Eraas.

We wish to thank Mr. Gary Norman of GE Aircraft Engines for his interest in this project and kind help in securing permission to make use of company materials in a number of exercises. We also appreciate the help of Mr. Doug Hart and the cooperation of the Miami University Paper Science Lab in developing the PID control example in Section 3.6 and the related exercises.

We are grateful to the Iowa State University Statistics Department and Ms. Sharon Shepard for the conversion of the early EXP version of the manuscript of this book to an incredibly clean LaTEX version that served both as class notes for several ISU and Miami U courses, and eventually also as a production manuscript. Kevin Dodd was a lifesaver in setting up the computing resources needed to support the conversion. Jeanne R. Jones of Miami University provided significant typing assistance on the first draft and Sara Udstuen of Miami produced initial versions of many of the book's figures. Bob Evans of Oxford, Ohio gave us helpful advice on potential design matters. We thank all of these people.

People at Wiley including Charity Robey, Wayne Anderson, Susanne Dwyer, Penny Perrotto, Charlotte Hyland and others have been very professional, helpful, and easy to work with. So too have Larry Meyer and Hermitage Publishing Services. These organizations and people have our gratitude.

Bobby Mee, Bill Woodall, Jayant Rajgopal, Diane Schaub, Jacek Dmochowski, Nasser Fard and Ahmad Scifoddini all provided helpful reviews of early versions of the manuscript and made many good suggestions. We've adopted many of them and appreciate the careful work of these people. Jimmy Wright and Andy Chiang did an excellent job of checking this book's solutions manual. We appreciate their thoroughness.

Finally and most importantly, we are grateful for the support of wonderful families, both throughout the preparation of this book and in all the endeavors of life. Thanks are due to John and Caryl Jobe, Jo Ellen, Micah, Andrew, Bruce, and Helen Vardeman for filling our lives with love, sanity, opportunity, and industry.

We hope that this book proves useful to college and university instructors and students, and to practitioners in industry. We would be glad to hear suggestions and comments from its users. Our current e-mail addresses are:

vardeman@iastate.edu
jobejm@muohio.edu

March 1998
Ames, Iowa
Oxford, Ohio

Contents

CHAPTER 1

Introduction

This brief chapter introduces the subject of quality assurance and lays out the connection between it and the subject of statistics. Then some fairly standard emphases in modern quality assurance are introduced and a six-step process-oriented quality assurance cycle is put forward as a framework for approaching problems in this field.

1.1 THE NATURE OF QUALITY AND THE ROLE OF STATISTICS

This book's title, *Statistical Quality Assurance Methods for Engineers,* raises at least two basic questions, namely "What is 'quality'?" and "What does 'statistics' have to do with assuring it?" It is thus sensible to begin by trying to provide at least preliminary answers to these basic questions. So consider first the matter of defining "quality." What does it mean to say that a particular good is a quality product? And what does it mean to call a particular service provided by a company a quality service?

In the case of manufactured goods (for instance, automobiles and dishwashers), issues of reliability (the ability to function consistently and effectively across time), appropriateness of configuration, fit and finish of parts, and so on, typically come to mind when one thinks about quality. In the realm of services provided (for example, telecommunications and transportation services) one thinks of consistency of availability and performance, esthetics, convenience, and so forth. And in evaluating the "quality" of both goods and services, there is typically at least an implicit understanding that these issues will be balanced against corresponding costs to determine overall "value." A popular definition of quality that reflects some of these notions is next.

Definition 1.1 **Quality** in a good or service is fitness for use. That fitness includes aspects of both product design and conformance to the (ideal) design.

Quality of design has to do with appropriateness, the choice and configuration of features that define what a good or service is supposed to be like and is supposed to do. In many cases it is essentially a matter of matching product "species" to an arena of use. One needs different things in a vehicle that is supposed to drive the dirt roads of the Baja Peninsula than in one that is to drive the German Autobahn. Vehicle quality of design has to do with providing the "right" features for the environment at an appropriate price. With this understanding, there is no necessary contradiction between thinking of both a Rolls Royce and a Ford economy car as quality vehicles. Similarly, there is not necessarily any contradiction between thinking of both a particular fast food outlet and a particular four star restaurant as quality eateries.

Quality of conformance has to do with living up to the specifications laid down in the design of a product. It is concerned with small variation from what is specified or expected. It is a fact of life that variation tends to make goods and services undesirable. Mechanical devices whose parts vary substantially from their ideal/design dimensions tend to be noisy, inefficient, prone to breakdown, difficult to service, and so on. They simply don't work well. In the service sector, variation from what is promised/expected is the principal source of customer dissatisfaction. A city bus system that runs on schedule every day that it is supposed to run can be seen as a quality transportation system. One that fails to do so cannot. And an otherwise elegant hotel that fails to see that its rooms have the spotless bathrooms its customers expect will soon be without those customers.

This book is concerned primarily with tools for assuring quality of conformance (although some of the tools discussed could also find application in the realms of product and process design). This is not in any way to say that quality of design is of secondary importance. Designing effective goods and services is a highly creative and important engineering activity, but it is just not the primary topic of this text.

Then what does the subject of statistics have to do with the assurance of quality of conformance? To answer this question, it is helpful to have clearly in mind a definition of statistics.

Definition 1.2 **Statistics** is the study of how best to

1. collect data,
2. summarize or describe data, and
3. draw conclusions or inferences based on data,

all in a framework that recognizes the reality and omnipresence of variation.

If quality of conformance has to do with small variation and one wishes to assure it, it will be necessary to measure, monitor, find sources of, and seek ways to reduce variation. All of these require data, information on what is happening in a system producing a product, and therefore the tool of statistics. The intellectual framework of the subject of statistics, emphasizing as it does the notion of variation, makes it a natural for application in the world of quality assurance. We will see that both the very simple and somewhat more advanced methods of statistics have their uses in the quest to produce quality goods and services.

1.2 MODERN QUALITY PHILOSOPHY AND A SIX-STEP PROCESS-ORIENTED QUALITY ASSURANCE CYCLE

Modern quality assurance methods and philosophy are focused not (primarily) on products, but rather on the **processes** used to produce them. The basic notion is that if one gets processes to work effectively, the resulting products will automatically be good. On the other hand, if one only focuses on screening out or reworking bad product, one is unlikely to ever get down to root causes of quality problems and make the changes necessary to improve quality. The importance of this process orientation can be illustrated by means of an example.

Example 1.1 **Process Improvement in a Clean Room.** One of the authors of this text was once given a tour of a so-called "clean room" at a division of a large electronics manufacturer. In this clean room, integrated circuit (IC) chips critical to the production of the division's most important product were being manufactured. The clean room was, in fact, the bottleneck of the whole production process for that product. Initial experience with that (very expensive) facility included 14% yields of good IC chips, with over 80 people working hard in the room, trying to produce the precious chips.

Early efforts at quality assurance of these important chips centered on testing the product and sorting good chips from bad. But it was soon clear that those efforts alone would never be adequate to bring yields up to a level adequate to supply the numbers of chips needed for the end product. So, rather than simply concentrating on the product, a project team went to work on improving the production process. They found that by carefully controlling the quality of some incoming raw materials, adjusting some process variables, and making measurements on wafers of chips early in the process (aimed at identifying and culling ones that would almost certainly in the end produce primarily bad chips) they could make the process much more efficient. At the time your author was shown the room, process improvement efforts had raised yields to 65% (that's like more than quadrupling production capacity with no capital expenditure!), drastically reduced material waste, and cut the staff necessary to run the facility from the original 80 to only eight technicians. The point is that process-oriented efforts are what enabled this success story. No amount of attention to the yield of the process as it was originally running would have produced these important results.

It is important to note that while process-oriented quality improvement efforts have center stage these days, product-oriented methods and efforts still have their place. Consider again the clean room of Example 1.1. At the time that your author toured the facility, process improvement efforts had in no way eliminated the need for end-of-the-line testing of the IC chips made in the room. Bad chips still needed to be identified and culled. Product-oriented inspection was still necessary, but it alone was not sufficient to produce important quality improvements.

A second important emphasis of modern quality philosophy is its **customer orientation**. This customer orientation has two faces. In the first place, the final or end user of a good or service is viewed as being supremely important. Much effort is expended by modern corporations in seeing that the "voice of the customer" (the will of the end user) is heard and carefully considered in all decisions involved in the design and production of a product. There are many currently popular communication and decision-making techniques (such as "quality function deployment") that are used to see that this happens.

But the customer orientation in modern quality philosophy extends beyond concentration on an end user. All workers are taught to view their efforts in terms of processes that have both "vendors" from whom they receive input and "customers" to whom they pass work. One's most immediate customer need not be the end user of a company product. But (so the modern philosophy goes) it is still important to do one's work in a way that those who handle one's personal "products" are able to do so efficiently. It is important to produce work that does not cause problems for the next person in the production chain.

A third major emphasis in modern quality assurance is that of **continual improvement.** The thinking here is that what is state of art today will be woefully inadequate tomorrow. Consumers are expecting (and getting!) ever more effective computers, cars, home entertainment equipment, package delivery services, and communications options. And modern quality philosophy says that this kind of improvement must and will continue. This is both a statement of what "ought" to be and a recognition that in a competitive world, if an organization does not continually improve what it does and makes, it will not be long before aggressive competition drives it from the marketplace.

This text presents a wide array of (almost exclusively statistical) tools for quality assurance. But even students armed with these tools do not always seem to know where to begin a quality assurance/improvement project. So, it is useful to present an outline for approaching modern quality problems that places (at least a majority of) the methods of this book into context, and is consistent with the elements of modern quality philosophy discussed thus far in this section. Table 1.1 presents this six-step process-oriented quality assurance cycle and indicates corresponding tools discussed in this book.

A sensible first step in any quality improvement project is to attempt to thoroughly understand the current and ideal configurations of the processes

Table 1.1 A Six-Step Process-Oriented Quality Assurance Cycle (and Corresponding Tools)

Step	Tools
1. Attempt a logical analysis of how a process works (or should work) and where potential trouble spots, sources of variation, and data needs are located.	• Flowcharts (§2.1) • Ishikawa/fishbone/cause-and-effect diagrams (§2.1)
2. Formulate appropriate (customer-oriented) measures of process performance and develop corresponding measurement systems.	• Basic concepts of measurement/metrology (§2.2) • Gage repeatability and reproducibility studies (§2.2)
3. Habitually collect and summarize process data.	• Simple quality assurance data collection principles (§2.3) • Simple statistical graphics (§2.4)
4. Assess and work toward process stability.	• Control charts (Ch. 3, Ch. 4)
5. Characterize current process and product performance.	• Statistical graphics for process characterization (§5.1) • Measures of process capability and performance and their estimation (§5.2, §5.3) • Probabilistic tolerancing and propagation of error (§5.4) • Estimation of variance components (§5.5)
6. Work to improve those processes that are unsatisfactory.	• Design and analysis of experiments (Ch. 6, Ch. 7)

involved. This matter of *process mapping* can be aided by very simple tools like the flowcharts and Ishikawa diagrams discussed in Section 2.1.

The business of proper measurement is foundational to engineering efforts to improve processes and products. If one cannot effectively measure important characteristics of what is being done to produce a good or service, there is clearly no way to tell whether design requirements are being met and customer needs genuinely addressed. Section 2.2 introduces some basic concepts of metrology and some methodology for assessing the extent to which a measurement system is adequate for a given purpose.

When adequate measurement systems are in place, one can begin to collect data on process performance. But there are pitfalls to be avoided in this collection, and if data are to be genuinely helpful in addressing quality assurance issues, they typically need to be summarized and presented effectively. So Sections 2.3 and 2.4 contain discussions of some elementary principles of quality assurance data collection and effective presentation of such data.

Once one recognizes uniformity as essentially synonymous with quality of conformance (and variation as synonymous with "unquality"), it is clear that one would like processes to be perfectly consistent in their output. That, of course, is too much to hope for in the real world. Variation is a fact of life, and the most that one can hope is that a process be at least consistent in its pattern of variation, that it be describable as physically stable. Control charts are tools for monitoring processes and issuing warnings when there is evidence in process data of physical instability. These essential tools of quality assurance are thoroughly discussed in Chapters 3 and 4.

Even those processes that can be called physically stable need not be adequate for current or future needs. (Indeed modern quality philosophy would suggest that one should view *all* processes as inadequate and in need of improvement!) So it is important to be able to characterize in fairly precise terms what a process is currently doing and to have tools for finding ways of improving it. Chapter 5 of this text discusses a number of methods for quantifying current process and product performance, while Chapters 6 and 7 deal with methods of experimental design and analysis especially helpful in process improvement efforts.

The steps outlined in Table 1.1 are a useful framework for approaching most process-related quality assurance projects. They are presented here not only as something of a road map for the first seven chapters of this book, but also as a list of steps to follow for students wishing to get started on a class project in process-oriented quality improvement.

1.3 CHAPTER SUMMARY

Modern quality assurance is concerned with quality of design and quality of conformance. Statistical methods, dealing as they do with data and variation, are essential tools for producing quality of conformance. Most of the tools presented in this text are useful in the process-oriented approach to assuring quality of conformance that is outlined in Table 1.1.

1.4 CHAPTER 1 EXERCISES

1.1. An engineer observes several responses for a variable of interest. The average of these recorded measurements is exactly what the engineer desires for any single response. Why should the engineer be concerned about variability in this context? How does the engineer's concern relate to product quality?

1.2. What is the difference between quality of conformance and quality of performance?

1.3. Suppose 100% of all brake systems produced by an auto manufacturer have been inspected and meet safety standards. What type of quality is this? Why?

1.4. Describe how a production process could be characterized as exhibiting quality of conformance but potential customers are wisely purchasing a competitor's version of the product made using the process in question.

1.5. In Example 1.1, initial experience at an electronics facility involved 14% yields of good IC chips.

a. Explain how this number (14%) may have been obtained.

b. Describe how the three parts of Definition 1.2 are manifested in your answer for part (a).

1.6. The improved yield discussed in Example 1.1 came as a result of improving the chip production process. Material waste was reduced and the staff necessary to run the facility was reduced from the original 80 to only eight technicians. What motivation do engineers have to improve processes if the improvement can lead to their own layoff? Discuss.

1.7. Suppose an engineer must choose among vendors 1, 2, and 3 to supply tubing for a product he is developing. Vendor 1 charges $20 per tube, vendor 2 charges $19 per tube, and vendor 3 charges $18 per tube. Vendor 1 has implemented the Six-Step Process-Oriented Quality Assurance Cycle (and corresponding tools) in Table 1.1. As a result, only 1 tube in a million from vendor 1 is judged to be nonconforming. Vendor 2 has just begun implementation of the six steps and is producing 10% nonconforming tubes. Vendor 3 does not apply quality assurance methodology and has no idea what percent of its tubing is nonconforming. What is the price per *conforming* item for vendors 1, 2, and 3?

1.8. The following matrix is due to Dr. Brian Joiner and can be used to classify production outcomes. Good "result of production" means there is a large proportion of product meeting engineering specifications. (Bad result of production means there is a low proportion of product meeting requirements.)

Good "method of production" means that quality variables are consistently near their targets. (Bad method of production means there is considerable variability about target values.)

		Result of Production: Good	Result of Production: Bad
Method of Production	Good	1	2
	Bad	3	4

Describe product characteristics for items produced under circumstances described by each of cells 1, 2, 3, and 4.

1.9. **Plastic Packaging.** Hsiao, Linse, and McKay investigated the production of some plastic bags, specifically hole positions on the bags. Production of these bags is done on a model 308 poly bag machine using preprinted, prefolded plastic film delivered on a roll. The plastic film is drawn through a series of rollers to punches that make holes in the bag lip. An electronic eye scans the film after it is punched and triggers heated sills which form the seals on the bags. A conveyor transports the bags to a machine operator who counts and puts them onto wickets (by placing the holes of the bags over 6-inch metal rods) and then places them in boxes. Discuss how this process and its output might variously be classified in cell 1, 2, 3, or 4 introduced in problem (1.8).

1.10. In the **Plastic Packaging** case described in problem (1.9):

a. Who is the immediate customer of the hole-punching process?

b. Is it possible for the hole-punching process to produce hole locations with small variation and yet still produce a poor quality bag? Why or why not?

c. After observing that 100 out of 100 sampled bags fit over the two 6-inch wickets, an analyst might conclude that the hole-punching process needs no improvement. Is this thinking correct? Why or why not?

d. Hsiao, Linse, and McKay used statistical methodologies consistent with steps 1, 2, 3, and 4 of the Six-Step Process-Oriented Quality Assurance Cycle and detected unacceptable variation in hole location. Would it be advisable to pursue step 6 in Table 1.1 in an attempt to improve the hole-punching process? Why or why not?

1.11. **Hose Skiving.** Siegler, Heches, Hoppenworth, and Wilson applied the Six-Step Process-Oriented Quality Assurance Cycle to a skiving operation. Skiving consists of taking rubber off the ends of steel-reinforced hydraulic hose so that couplings may be placed on these ends. A crimping machine tightens the couplings onto the hose. If the skived length or diameter are not as designed, the crimping process can produce an unacceptable finished hose.

a. What two variables did the investigators identify as directly related to product quality?
b. Which step in the Six-Step Cycle probably identified these two variables as important in the skiving process?
c. The analysts applied steps 3 and 4 of the Six-Step Cycle and found that for a particular production line, aim and variation in skive length were consistent. (Unfortunately, outside diameter data were not available, so study of the outside diameter variable was not possible.) In keeping with the doctrine of continual improvement, steps 5 and 6 were considered by the analysts. Was this a good idea? Why or why not?

1.12. Engineers at GE Aircraft Engines have identified several "givens" regarding cost of quality problems. Two of these are "Making it right the first time is always cheaper than doing it over" and "Fixing a problem at the source is always cheaper than fixing it later." Describe how the Six-Step Process-Oriented Quality Assurance Cycle in Table 1.1 relates to the two givens.

1.13. A common rule of thumb used by engineers when evaluating the cost of quality is the rule of 10. This rule can be characterized as follows (in terms of the dollar cost required to fix a nonconforming item):

Design $1	→	Production $10	→	Assembly/Test $100	→	Field $1000

The cost history of nonconforming parts for GE Aircraft Engines has been roughly as follows:

Nonconforming Item Found	Cost to Find/Fix
At Production Test	$200
At Final Inspection	$260
At GE Rotor Assembly	$20,000
At GE Assembly Teardown	$60,000
In Customer's Airplane	$200,000
At Unscheduled Engine Removal	$1,200,000

a. Calculate the ratios of costs to fix to the cost to fix at production.
b. How do these ratios compare to the rule of 10 illustrated in the four-box schematic?

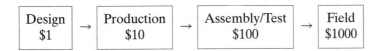

c. What does your response to (b) suggest about implementation of step 3 in the Six-Step Cycle of Table 1.1?

1.14. The following quotes are representative of some engineering attitudes toward quality assurance efforts. "Quality control is just a police force." "The quality control people are the ones who come in and shoot the wounded." "Those machinists will do what's easiest for them, so we'll start out with really tight engineering specifications on that part dimension."

a. How might an engineer develop such attitudes?

b. How can quality engineers avoid these perspectives as personal attitudes and work to change them in others?

1.15. Brush Ferrules. Adams, Harrington, Heemstra, and Snyder did a quality improvement project concerned with the manufacture of some paint brushes. Bristle fibers are attached to a brush handle with a so-called "ferrule." If the ferrule is too thin, bristles fall out. If the ferrule is too thick, brush handles are damaged and can fall apart. At the beginning of the study there was some evidence that bristle fibers were falling out. Crank position, slider position, and dwell time are three process variables that may affect ferrule thickness.

a. What variable should the analysts measure on a given brush?

b. Suggest how an engineer might evaluate whether the quality problem is due to poor conformance or to poor design.

c. From the limited information given above, what seems to have motivated the investigation?

d. The analysts considered plotting the variable identified in (a) versus the time at which the corresponding brush was produced. One of the analysts suggested first sorting the brushes according to the different crank position, slider position, and dwell time combinations, then plotting the variable chosen in (a) versus time of production *on separate graphs.* The others argued that no insight into the problem would be gained by having separate graphs for each combination. What point of view do you support? Defend your answer.

1.16. Window Frames. Christenson, Hutchinson, Mechem, and Theis worked with a manufacturing engineering department in an effort to identify cause(s) of variation and possibly reduce the amount of offset in window frame corner joints. (Excessive offset had previously been identified as the most frequently reported type of window nonconformity.)

a. Suggest how the company might have come to know that excessive offset in corner joints was the most frequently occurring type of problem.

b. What step in the Six-Step Cycle corresponds to your answer in (a)?

c. The team considered the following six categories of factors potentially contributing to unacceptable offset: (1) measurements, (2) materials, (3) men, (4) environment, (5) methods, (6) machines. Suggest at least one possible cause for each of these categories.

d. Which step in the Six-Step Cycle of Table 1.1 is most clearly related to the kind of categorization of factors alluded to in part (c)?

1.17. Machined Steel Slugs. Harris, Murray, and Spear contacted a plant that manufactures steel slugs used to seal a hole in a certain type of casting. The group's first task was to develop and write up a standard operating procedure for data collection on several critical dimensions of these slugs. The slugs are turned on a South Bend Turrett Lathe using 1018 cold rolled steel bar stock. The entire manufacturing process is automated by means of a CNC (computer numerical control) program and only requires an operator to reload the lathe with new bar stock. The group of analysts attempted to learn about the CNC lathe program. They discovered it was possible for the operator to change the finished part dimensions by adjusting the offset on the lathe.

a. Briefly, what benefit is there to having a standard data collection procedure in this context?

b. Why was it important for the group to learn about the CNC lathe program? Which step of the Six-Step Cycle is directly affected by their knowledge of the lathe program? Briefly discuss.

1.18. Cut-Off Machine. Wade, Keller, Sharp, and Takes studied factors affecting tool life for carbide cutting inserts. The group discovered that feed rate and stop delay were two factors known by a production staff to affect tool life.

a. What steps might the group have taken to independently verify that feed rate and stop delay impact tool life for carbide cutting inserts?

b. What is the important response variable in this problem?

c. How would you suggest that the variable in (b) be measured?

d. Suggest how increasing tool life might be attractive to customers using the inserts.

1.19. Potentiometers. Chamdani, Davis, and Kusumaptra worked with personnel from a potentiometer assembly plant to improve the quality of finished trimming potentiometers. The fourteen wire springs fastened to the potentiometer rotor assemblies (produced elsewhere) were causing short circuits and open circuits in the final potentiometers. Engineers suspected that the primary cause of the problems was a lack of symmetry on metal strips upon which these springs were fastened. Of concern was the distance from one edge of the metal strip to the first spring and the corresponding distance from the last spring to the other end of the strip.

a. Suggest how the assembly plant might have discovered the short and open circuits.

b. Suggest how the plant producing the rotor assemblies became aware of the short and open circuits. How does your response relate to the Six-Step Cycle in Table 1.1?

c. If "lack of symmetry" is the true cause of quality problems, what should henceforth be recorded for each metal strip sampled?

d. What kind of measurement corresponds to perfect symmetry?

1.20. "Empowerment" is a term frequently heard in today's organizations in relation to quality improvement. Empowerment concerns moving decision-making authority in an organization down to the lowest appropriate levels.

Unfortunately, the concept is sometimes employed only until a mistake is made, then a severe reprimand occurs and/or the decision-making privilege is moved back up to a higher level.

a. Name two things that are lacking in this approach to quality improvement.

b. Where does real (and consistent) empowerment logically fit in the Six-Step Quality Improvement Cycle?

CHAPTER 2

Simple Quality Assurance Tools

This chapter introduces some of the simplest of all quality assurance tools. As indicated in Chapter 1, they are useful early in the Six-Step Process-Oriented Quality Assurance Cycle. They also are both powerful and accessible enough that they deserve very wide dissemination and application. Included in the discussion in this chapter (without being specifically identified as such) are six out of most people's list of "the seven Japanese QC tools" widely advertised as things that all modern corporate citizens should know and use. (The "Japanese" tool not covered in this chapter is the Shewhart control chart discussed extensively in Chapter 3.) The chapter begins with a brief discussion of process mapping/analysis. Section 2.2 considers basic issues in metrology and assessing the adequacy of a measurement system. Then follow discussions of some simple principles of quality assurance data collection in Section 2.3 and the usefulness of simple statistical graphics in Section 2.4.

2.1 LOGICAL PROCESS IDENTIFICATION AND ANALYSIS

Often, simply comparing "what is" in terms of process structure to "what is supposed to be" or to "what would make sense" is enough to identify real opportunities for improvement. Particularly in service industry contexts, the mapping of a process and identification of redundant and unnecessary steps can lead very quickly to huge reductions in cycle times and corresponding improvements in customer satisfaction. But even in cases where how to make such easy improvements is not immediately obvious, a process identification exercise is often invaluable in locating potential process trouble spots, possibly important sources of process variation, and data collection needs.

The simple **flowchart** is one effective tool in process identification. Figure 2.1 is a flowchart for a printing process similar to one prepared by students (Drake, Lach, and Shadle) in a quality control course. The figure gives a fairly high-level view of the work flow in a particular print shop. Nearly any one of the boxes on the chart could be expanded to provide more detailed information about the printing process if such detail were needed.

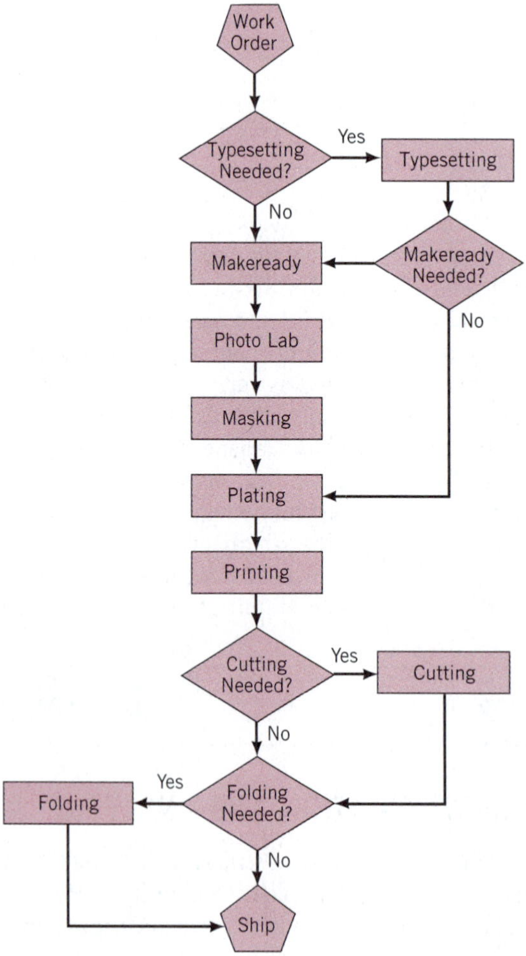

Figure 2.1
Flowchart of a printing process.

There are a variety of ways that people have suggested increasing the amount of information carried by a flowchart. One possibility is the use of different shapes for the boxes on the chart, according to some kind of a classification scheme for the activities being portrayed. Figure 2.1 uses only three different shapes, one each for input/output, decisions, and all else. An interesting discussion on pages 205–213 of Kolarik's *Creating Quality: Concepts, Systems, Strategies and Tools* suggests the use of seven different symbols for flowcharting industrial processes, corresponding to operations, transportation, delays, storage, source inspection, SPC charting, and sorting inspection. Of course, many other schemes are possible and potentially useful in different circumstances.

A second notion that has been suggested for enhancing the analytical value of the flowchart is to make good use of both spatial dimensions on the chart. Typically, top to bottom on a chart corresponds at least roughly to time order of activities. That leaves the possibility of using left-to-right positioning on a flowchart to indicate some other important variable. For example, a flowchart might be segmented into several "columns" left to right, each one indicating a different physical location where the indicated activities are to take place. Or the columns might indicate different departmental spheres of responsibility. Such positioning is an effective way of further organizing one's thinking about a process.

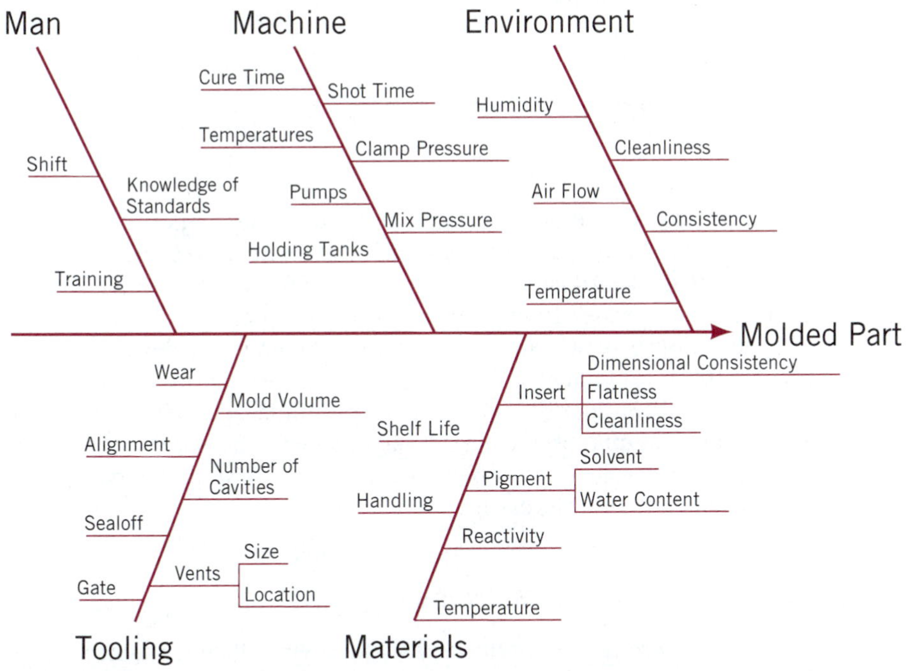

Figure 2.2
Cause-and-effect diagram for an injection molding process.

Another simple but very popular device for use in process identification/mapping activities is that of the **Ishikawa diagram** (otherwise known as the **fishbone diagram** or **cause-and-effect diagram**). Suppose one has a desired outcome or (conversely) a quality problem in mind and wishes to lay out the various possible contributors to the outcome or problem. It is often helpful to organize thinking about these factors by placing them on a tree-like structure, where the further one moves into the tree, the more specific or basic the contributor becomes. For example, if one were interested in quality of an airline flight, general contributors might include on-time performance, baggage handling, in-flight comfort, and so on. In-flight comfort might be further amplified as involving seating, air quality, cabin service, and so on. Cabin service could be broken down into components like flight attendant availability and behavior, food quality, entertainment, and so on.

Figure 2.2 on page 15 is part of an Ishikawa diagram made by an industrial team analyzing an injection molding process. It seems obvious that without some kind of organized method of putting down the various contributors to the quality of the molded parts, anything like an exhaustive listing of potentially important factors would be nearly impossible. The cause-and-effect diagram format provides an easily made and effective organization tool. It is an especially helpful device in group brainstorming sessions, where people are offering suggestions from many different perspectives, and some kind of structure needs to be provided for suggestions, pretty much on the fly.

2.2 MEASUREMENT

Good measurement is absolutely fundamental to quality assurance efforts. That which cannot be effectively measured cannot be guaranteed to a customer. If a customer wants a Brinell hardness of 220 for certain castings and one has no means of reliably measuring hardness, there is clearly no way to proceed to provide such castings. It is thus not surprising that most successful companies devote substantial resources to the development and maintenance of good measurement systems. In this section we will consider some basic concepts of measurement, discuss statistical studies aimed at quantifying the effectiveness of a measurement method, and then briefly consider the issue of how measurement variability impacts one's ability to detect changes or differences.

2.2.1 Basic Concepts in Metrology

Metrology is the science of measurement. Measurement of some physical quantities (like lengths on the order of inches to miles and weights on the order of ounces to tons) is so commonplace that we think little about basic issues involved in metrology. But often engineers are forced by circumstances to leave the world of off-the-shelf measurement technology and devise their own instruments. And frequently because of externally imposed quality requirements for a product, one is led to ask "Can we even measure that?" In such circumstances, the fundamental issues of validity, precision, and accuracy start to come into focus.

Definition 2.1 A measurement or measuring method is said to be **valid** if it usefully or appropriately represents the feature of the measured object or phenomenon that is of interest.

Definition 2.2 A measurement system is said to be **precise** if it produces small variation in repeated measurement of the same object or phenomenon.

Definition 2.3 A measurement system is said to be **accurate** (or sometimes unbiased) if on average it produces the true or correct values of quantities of interest.

Validity is the most fundamental matter to be faced when developing a measurement method. Without it, there is no point in proceeding to consider precision or accuracy. The issue in establishing validity is whether or not a method of measurement will faithfully portray the quantity of interest. When developing a new pH meter, one wants a device that will react to changes in acidity, not to changes in temperature of the solution being tested or to changes in the amount of light incident on the container holding the solution. And when looking for a measure of customer satisfaction with a new model of automobile, one needs to consider those things that are important to customers, not those important to some intermediate party like the dealer that sells the vehicle. For example, number of warranty service calls per vehicle is probably a more valid measure of customer satisfaction (or aggravation) with a new car than warranty dollars spent per vehicle by the manufacturer.

Precision of measurement has to do with getting essentially the same value every time a particular measurement is done. A bathroom scale that can produce any number between 150 lbs and 160 lbs when one gets on it repeatedly is really not very useful. Once one has established that a measurement system produces valid measurements, one needs consistency of those measurements. Figure 2.3 portrays some hardness measurements made by a group of students (Blad, Sobotka, and Zaug) on a single metal specimen with three different hardness testers. It is clear from the figure that by most standards the Dial Rockwell tester produced the most consistent results and would therefore be termed the most precise.

Precision is largely an intrinsic property of a measurement method or device. There is not really any way to "adjust" for poor precision or to remedy it except to overhaul or replace measurement technology or to average multiple

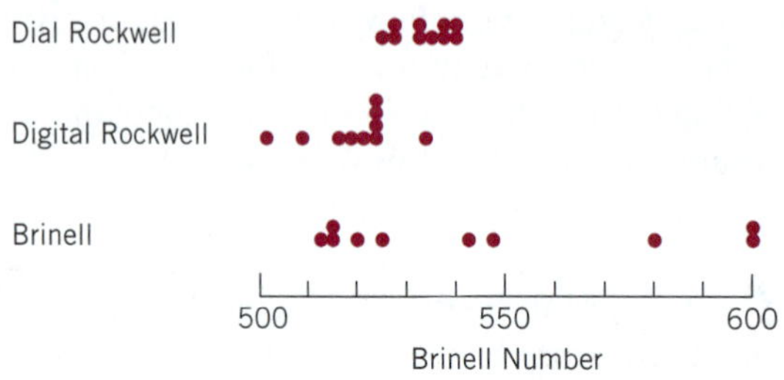

Figure 2.3
Brinell hardness measurements made on three different machines.

measurements. In this latter regard, the reader should be familiar with the fact from elementary statistics that if $x_1, x_2, \ldots, x_n$ can be thought of as independent measurements of the same quantity, each with some mean μ (the true value of interest) and standard deviation σ, then the sample mean, $\overline{x}$, has expected or average value μ and standard deviation $\sigma/\sqrt{n}$. So people sometimes rely on multiple measurements and averaging to reduce an unacceptable precision of individual measurement (quantified by σ) to an acceptable precision of average measurement (quantified by $\sigma\sqrt{n}$).

Of course, even validity and precision don't tell the whole story regarding the usefulness of real-world measurements. This can be illustrated very clearly by again considering Figure 2.3. It is true that the Dial Rockwell tester is apparently the most precise of the three testers. But it is *not* obvious from the figure what the truth is about "the" Brinell hardness of the specimen. That is, the issue of accuracy remains. Whether any of the three testers produces essentially the "right" hardness value on average is not clear. In order to assess that, one needs to reference the testers to an accepted standard of hardness measurement.

The business of comparing a measurement method or device to a standard one and, if necessary, working out conversions that will allow the method of interest to produce "correct" (converted) values on average is the business of **calibration.** In the United States, it is the National Institute of Standards and Technology (NIST) that is responsible for maintaining and disseminating consistent standards for calibrating measurement equipment. One could imagine (if the problem were important enough) sending the students' specimen to NIST for an authoritative hardness evaluation and using the result to calibrate the testers in question. Or more likely, one might test some other specimens supplied by NIST as having known hardnesses, and use those to assess the accuracy of the testers in question (and guide any recalibration that might be needed).

An analogy that is sometimes helpful in remembering the difference between accuracy and precision of measurement is that of target shooting. Accuracy in target shooting has to do with producing a pattern centered on the bull's eye (the ideal). Precision has to do with producing a tight pattern (consistency). Figure 2.4 illustrates four possibilities for accuracy and precision in target shooting.

It is useful at this point to briefly consider the elementary probability modeling of the effects of measurement error. If a fixed quantity x is to be measured with error (as all real-world quantities are!) one might represent what is actually

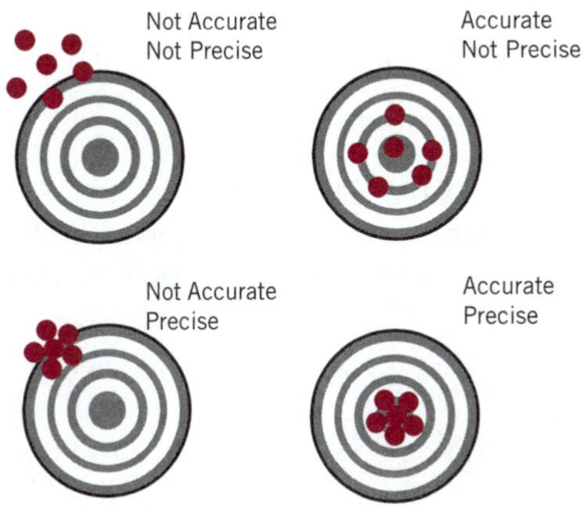

Figure 2.4
Measurement/target-shooting analogy.

observed as

$$y = x + \epsilon \tag{2.1}$$

where ϵ is a random variable, say with mean β and standard deviation $\sigma_{\text{measurement}}$. Model (2.1) says that the mean of what is observed is

$$\mu_y = x + \beta.$$

If $\beta = 0$, the measurement of x is accurate or unbiased. If β is not 0, it is called the **measurement bias.** The standard deviation of y is (for fixed x) the standard deviation of ϵ, $\sigma_{\text{measurement}}$. So $\sigma_{\text{measurement}}$ quantifies measurement precision in model (2.1).

Thinking in terms of model (2.1) is especially helpful when x itself is subject to variation. For instance, when parts produced on a machine have varying diameters x, one might think of model (2.1) as applying separately to each individual part diameter. But then in view of the reality of manufacturing variation, it makes sense to think of diameters as random, say with mean μ_x and standard deviation σ_x, independent of the measurement errors. This combination of assumptions then implies that the mean of what is observed is

$$\mu_y = \mu_x + \beta,$$

and the standard deviation of what is observed is

Standard Deviation of Observations Subject to Measurement Error

$$\sigma_y = \sqrt{\sigma_x^2 + \sigma_{\text{measurement}}^2}\,. \tag{2.2}$$

A nonzero β is still a measurement bias, but now observed variation across parts is seen to include a component due to variation in x and another due to measurement error.

Notice that equation (2.2) implies that

$$\sigma_x = \sqrt{\sigma_y^2 - \sigma_{\text{measurement}}^2}\,.$$

This suggests a means of estimating σ_x alone. If one has (single) measurements y for several parts that produce a sample standard deviation s_y, and several measurements on a single part that produce a sample standard deviation s, then a plausible estimate of σ_x is

Estimator of Process or Part Variation Excluding Measurement Error

$$\hat{\sigma}_x = \sqrt{\max\left(0, s_y^2 - s^2\right)}\,. \tag{2.3}$$

Whether or not one goes so far as to use formula (2.3), it is important to keep relationship (2.2) in mind. Variation *observed* in a variable is typically due both to real variation in the actual value x and to measurement variation.

2.2.2 Gage R&R Studies

In quality assurance contexts, it is very important to be able to quantify the precision of gages and other measurement systems. This usually needs to be done in a framework where multiple operators or technicians will be called upon to use a gage. In such cases there is at least the potential that different operators will use the measurement equipment in slightly different ways, leading to systematic differences in measurement between operators. So-called "Gage R&R" studies are used to provide an assessment of measurement precision in this kind of circumstance.

In a typical gage R&R study, each of J operators uses the same gage (or other measurement device) to measure each of I parts (common to all operators) a total of m different times. In rough terms, variation typical of that seen in the m measurements for a particular operator on a particular part is called the **repeatability** of the gage (hence the first "R" in "R&R"). Similarly, variation similar to that which can be attributed to differences between the J operators is called **reproducibility** of the measurement system (hence the second "R" in "R&R").

Example 2.1 Gage R&R for a 1-Inch Micrometer Caliper. Heyde, Kuebrick, and Swanson conducted a gage R&R study on a certain micrometer caliper as part of a class project for a quality control course. Table 2.1 shows the data that the $J = 3$ (student) operators obtained, each making $m = 3$ measurements of the heights of $I = 10$ steel punches.

Notice that even for a given punch/student combination, measured heights are not exactly the same. Further, it is possible to verify that averaging the 30 measurements made by student 1, a mean of about 498.53 is obtained, while

Table 2.1 Measured Heights of 10 Steel Punches in 10^{-3} Inch

Punch	Student 1	Student 2	Student 3
1	496, 496, 499	497, 499, 497	497, 498, 496
2	498, 497, 499	498, 496, 499	497, 499, 500
3	498, 498, 498	497, 498, 497	496, 498, 497
4	497, 497, 498	496, 496, 499	498, 497, 497
5	499, 501, 500	499, 499, 499	499, 499, 500
6	499, 498, 499	500, 499, 497	498, 498, 498
7	503, 499, 502	498, 499, 499	500, 499, 502
8	500, 499, 499	501, 498, 499	500, 501, 499
9	499, 500, 499	500, 500, 498	500, 499, 500
10	497, 496, 496	500, 494, 496	496, 498, 496

corresponding means for students 2 and 3 are respectively about 498.13 and 498.40. Student 1 may tend to measure slightly higher than students 2 and 3. That is, by these rough "eyeball" standards, there is some hint in these data of both repeatability and reproducibility components in the overall gage imprecision.

We proceed to make these ideas more precise and describe calculations based on gage R&R data like those in Table 2.1 that can be used to judge the sizes of repeatability and reproducibility components for a given gage. To begin with, it will be very helpful to have a model for how gage R&R data are generated. The most common and tractable model used in this context is the so-called "two-way random effects model" that can be found in many intermediate-level statistical methods texts. (See, for example, Section 8.4 of Vardeman's *Statistics for Engineering Problem Solving*.) Let

$$y_{ijk} = \text{the } k\text{th measurement made by operator } j \text{ on part } i.$$

Then, the model is that

$$y_{ijk} = \mu + \alpha_i + \beta_j + \alpha\beta_{ij} + \epsilon_{ijk}, \qquad (2.4)$$

where μ is an (unknown) constant, the α_i are normal random variables with mean 0 and variance σ_α^2, the β_j are normal random variables with mean 0 and variance σ_β^2, the $\alpha\beta_{ij}$ are normal random variables with mean 0 and variance $\sigma_{\alpha\beta}^2$, the ϵ_{ijk} are normal random variables with mean 0 and variance σ^2, and all of the α's, β's, $\alpha\beta$'s, and ϵ's are independent. In this model, the unknown constant μ is an average (over all possible operators and all possible parts) measurement, the α's are (random) effects of different parts, the β's are (random) effects of different operators, the $\alpha\beta$'s are (random) joint effects peculiar to particular part/operator combinations, and the ϵ's are (random) measurement errors. The variances σ_α^2,

Table 2.2 Measurements in a Hypothetical Gage R&R Study

	Operator 1	Operator 2
Part 1	$y_{111} = \mu + \alpha_1 + \beta_1 + \alpha\beta_{11} + \epsilon_{111}$	$y_{121} = \mu + \alpha_1 + \beta_2 + \alpha\beta_{12} + \epsilon_{121}$
	$y_{112} = \mu + \alpha_1 + \beta_1 + \alpha\beta_{11} + \epsilon_{112}$	$y_{122} = \mu + \alpha_1 + \beta_2 + \alpha\beta_{12} + \epsilon_{122}$
Part 2	$y_{211} = \mu + \alpha_2 + \beta_1 + \alpha\beta_{21} + \epsilon_{211}$	$y_{221} = \mu + \alpha_2 + \beta_2 + \alpha\beta_{22} + \epsilon_{221}$
	$y_{212} = \mu + \alpha_2 + \beta_1 + \alpha\beta_{21} + \epsilon_{212}$	$y_{222} = \mu + \alpha_2 + \beta_2 + \alpha\beta_{22} + \epsilon_{222}$

σ_β^2, $\sigma_{\alpha\beta}^2$, and σ^2 are called "variance components" and their sizes govern how much variability is seen in the measurements y_{ijk}.

To perhaps make the model equation (2.4) slightly more understandable, consider a hypothetical case with $I = 2$, $J = 2$, and $m = 2$. Model (2.4) says that there is a normal distribution with mean 0 and variance σ_α^2 from which α_1 and α_2 are drawn. And there is a normal distribution with mean 0 and variance σ_β^2 from which β_1 and β_2 are drawn. And there is a normal distribution with mean 0 and variance $\sigma_{\alpha\beta}^2$ from which $\alpha\beta_{11}$, $\alpha\beta_{12}$, $\alpha\beta_{21}$, and $\alpha\beta_{22}$ are drawn. And there is a normal distribution with mean 0 and variance σ^2 from which eight ϵ's are drawn. Then these realized values of the random effects are added to produce the eight measurements as indicated in Table 2.2.

Either directly from equation (2.4) or as illustrated in Table 2.2, it should be clear that according to the two-way random effects model, the only differences between measurements for a fixed part/operator combination are the measurement errors ϵ. And the variability of these is governed by the parameter σ. That is, σ is a measure of repeatability in this model, and one objective of an analysis of gage R&R data is to estimate it.

There are other functions of the model parameters that are also of interest. Consider the quantity

$$\sigma_{\text{reproducibility}} = \sqrt{\sigma_\beta^2 + \sigma_{\alpha\beta}^2}\,. \tag{2.5}$$

According to the model, this is the standard deviation that would be experienced by many operators making a single measurement on the same part *assuming that there is no repeatability component to the overall variation*. (Another way to say the same thing is to recognize this quantity as the standard deviation that would be experienced computing with long-run average measurements for many operators on the same part.) That is, the quantity (2.5) is a natural measure of reproducibility in this model.

Further, the quantity

$$\sigma_{\text{overall}} = \sqrt{\sigma_\beta^2 + \sigma_{\alpha\beta}^2 + \sigma^2} = \sqrt{\sigma_{\text{reproducibility}}^2 + \sigma^2} \tag{2.6}$$

is the standard deviation implied by the model (2.4) for many operators each making a single measurement on the same part. That is, quantity (2.6) is a measure of the combined imprecision in measurement attributable to *both* repeatability and reproducibility sources. And one might think of

$$\frac{\sigma^2}{\sigma^2_{\text{overall}}} = \frac{\sigma^2}{\sigma^2_\beta + \sigma^2_{\alpha\beta} + \sigma^2} \quad \text{and} \quad \frac{\sigma^2_{\text{reproducibility}}}{\sigma^2_{\text{overall}}} = \frac{\sigma^2_\beta + \sigma^2_{\alpha\beta}}{\sigma^2_\beta + \sigma^2_{\alpha\beta} + \sigma^2} \tag{2.7}$$

as the fractions of total measurement variance due respectively to repeatability and reproducibility. Pretty clearly, if one can produce estimates of σ and $\sigma_{\text{reproducibility}}$, estimates of these quantities (2.6) and (2.7) will follow in straightforward fashion.

So consider first the estimation of σ. Restricting attention to any particular part/operator combination, say part i and operator j, the model says that observations obtained for that combination differ only by independent normal random measurement error with mean 0 and variance σ^2. That suggests that a measure of variability for the ij sample might be used as the basis of an estimator of σ. We will here primarily consider using sample ranges in the estimation process. (Other methods are possible, essentially built upon the use of sample standard deviations, and these will be presented very briefly at the end of this subsection.)

So let R_{ij} be the range of the m measurements on part i by operator j. It is a mathematical fact that the expected value of the range of a sample from a normal distribution is a constant (depending upon m) times the standard deviation of the distribution being sampled. The constants are well known and called d_2. (We will write $d_2(m)$ to emphasize their dependence upon m and note that values of $d_2(m)$ are given in Table A.1.) It then follows that

$$\text{E}R_{ij} = d_2(m)\sigma,$$

which in turn suggests that the ratio

$$\frac{R_{ij}}{d_2(m)}$$

is a plausible estimator of σ. Or better yet, one might average these over all $I \times J$ part/operator combinations to produce the range-based estimator of σ,

$$\hat{\sigma}_{\text{repeatability}} = \frac{\overline{R}}{d_2(m)}. \tag{2.8}$$

Range-Based Estimator for Repeatability Standard Deviation

Example 2.1 continued. Subtracting the smallest measurement for each part/operator combination in Table 2.1 from the largest for that combination, one obtains the ranges in Table 2.3.

Table 2.3 Ranges of 30 Part/Operator Samples of Measured Punch Heights

Punch	Student 1	Student 2	Student 3
1	3	2	2
2	2	3	3
3	0	1	2
4	1	3	1
5	2	0	1
6	1	3	0
7	4	1	3
8	1	3	2
9	1	2	1
10	1	6	2

It is then easy to verify that the 30 ranges in Table 2.3 have mean $\overline{R} = 1.9$. From Table A.1, $d_2(3) = 1.693$. So using expression (2.8) an estimate of σ, the repeatability standard deviation for the caliper studied by the students, is

$$\hat{\sigma}_{\text{repeatability}} = \frac{\overline{R}}{d_2(3)} = \frac{1.9}{1.693} = 1.12 \times 10^{-3} \text{ inch}.$$

Again, this is an estimate of the (long-run) standard deviation that would be experienced by any particular student measuring any particular punch many times.

Consider now the standard deviation (2.5) representing the reproducibility portion of the gage imprecision. In order to discuss estimation of this quantity, it will be convenient to have some additional notation. Let

$$\overline{y}_{ij} = \text{the (sample) mean measurement made on part } i \text{ by operator } j$$

and

$$\Delta_i = \max_j \overline{y}_{ij} - \min_j \overline{y}_{ij} = \text{the range of the mean measurements made on part } i.$$

Notice that with the obvious notation for the sample average of the measurement errors ϵ, according to model (2.4)

$$\overline{y}_{ij} = \mu + \alpha_i + \beta_j + \alpha\beta_{ij} + \overline{\epsilon}_{ij}.$$

Thus, for a fixed part i these means $\overline{y}_{ij}$ differ only by independent normal random variables $\beta_j + \alpha\beta_{ij} + \overline{\epsilon}_{ij}$. And these have mean 0 and variance $\sigma_\beta^2 + \sigma_{\alpha\beta}^2 + \sigma^2/m$. Thus, it should not be surprising that their range, Δ_i, has mean

$$\mathrm{E}\Delta_i = d_2(J)\sqrt{\sigma_\beta^2 + \sigma_{\alpha\beta}^2 + \sigma^2/m}\,.$$

This suggests $\Delta_i/d_2(J)$, or better yet, the average of these over all parts i, $\overline{\Delta}/d_2(J)$, as an estimator of $\sqrt{\sigma_\beta^2 + \sigma_{\alpha\beta}^2 + \sigma^2/m}$. This in turn suggests that one can estimate $\sigma_\beta^2 + \sigma_{\alpha\beta}^2 + \sigma^2/m$ with $(\overline{\Delta}/d_2(J))^2$. Then remembering that $\overline{R}/d_2(m) = \hat{\sigma}_{\text{repeatability}}$ is an estimator of σ, an obvious estimator of $\sigma_\beta^2 + \sigma_{\alpha\beta}^2$ becomes

$$\left(\frac{\overline{\Delta}}{d_2(J)}\right)^2 - \frac{1}{m}\left(\frac{\overline{R}}{d_2(m)}\right)^2. \tag{2.9}$$

The quantity (2.9) is meant to approximate $\sigma_\beta^2 + \sigma_{\alpha\beta}^2$, which is obviously non-negative. But the estimator (2.9) has the unfortunate property that it can on occasion turn out to give negative values. When this happens, it is sensible to replace the negative value by 0, and thus expression (2.9) by

$$\max\left(0, \left(\frac{\overline{\Delta}}{d_2(J)}\right)^2 - \frac{1}{m}\left(\frac{\overline{R}}{d_2(m)}\right)^2\right). \tag{2.10}$$

So finally, an estimator of the reproducibility standard deviation can be had by taking the square root of expression (2.10). That is, one may estimate the quantity (2.5) with

Range-Based Estimator for Reproducibility Standard Deviation

$$\hat{\sigma}_{\text{reproducibility}} = \sqrt{\max\left(0, \left(\frac{\overline{\Delta}}{d_2(J)}\right)^2 - \frac{1}{m}\left(\frac{\overline{R}}{d_2(m)}\right)^2\right)}. \tag{2.11}$$

Example 2.1 continued. Table 2.4 organizes $\bar{y}_{ij}$ and Δ_i values for the punch height measurements of Table 2.1.

Then $\overline{\Delta} = 8.67/10 = .867$, and since $J = 3$, $d_2(J) = d_2(3) = 1.693$. So using equation (2.11) one has

$$\begin{aligned}
\hat{\sigma}_{\text{reproducibility}} &= \sqrt{\max\left(0, \left(\frac{.867}{1.693}\right)^2 - \frac{1}{3}\left(\frac{1.9}{1.693}\right)^2\right)}, \\
&= \sqrt{\max(0, -.158)}, \\
&= 0.
\end{aligned}$$

This is a case where (relatively speaking) σ appears to be so large that the reproducibility standard deviation cannot be seen above the "noise" in measurement

Table 2.4 Part/Operator Means and Ranges of Such Means for the Punch Height Data

Punch (i)	$\bar{y}_{i1}$	$\bar{y}_{i2}$	$\bar{y}_{i3}$	Δ_i
1	497.00	497.67	497.00	.67
2	498.00	497.67	498.67	1.00
3	498.00	497.33	497.00	1.00
4	497.33	497.00	497.33	.33
5	500.00	499.00	499.33	1.00
6	498.67	498.67	498.00	.67
7	501.33	498.67	500.33	2.67
8	499.33	499.33	500.00	.67
9	499.33	499.33	499.67	.33
10	496.33	496.67	496.67	.33

due to the repeatability component of variation. Fairly obviously, in this particular problem, estimates of the ratios (2.7) based on $\hat{\sigma}_{\text{repeatability}}$ and $\hat{\sigma}_{\text{reproducibility}}$ would attribute fractions 1 and 0 of the overall variance in measurement to respectively repeatability and reproducibility.

Remembering that the quantity σ_{overall} given in display (2.6) is the standard deviation for many operators each making a single measurement on the same part, it is common to treat some multiple of it (often the multiplier is six) as a kind of uncertainty associated with a measurement made using the gage or measurement system in question. And when a gage is being used to check conformance of a part dimension or other measured characteristic to engineering specifications (say, some lower specification L and some upper specification U) this multiple is compared to the spread in specifications with the hope that it is at least an order of magnitude smaller than the spread in specifications. (**Specifications** U and L are numbers set by product design engineers that are supposed to delineate what is required of a measured dimension in order that the item in question be functional.) Some organizations go so far as to call the quantity

Engineering Specifications

$$GCR = \frac{6\sigma_{\text{overall}}}{U - L} = \frac{6\sqrt{\sigma_\beta^2 + \sigma_{\alpha\beta}^2 + \sigma^2}}{U - L} \tag{2.12}$$

a **gage capability ratio,** estimate it by

Estimator for a Gage Capability Ratio

$$\widehat{GCR} = \frac{6\sqrt{\hat{\sigma}^2_{\text{repeatability}} + \hat{\sigma}^2_{\text{reproducibility}}}}{U - L},$$

and require that it be no larger than .1 (and preferably as small as .01) before using the gage for checking conformance to such specifications.

Some brief observations regarding the planning of a gage R&R study are in order at this point. The precisions with which one can estimate σ and $\sigma_{\text{reproducibility}}$ obviously depend upon I, J, and m. Roughly speaking, precision of estimation of σ is governed by the product $(m-1)IJ$, so increasing any of the "dimensions" of the data array will improve estimation of repeatability. However, it is primarily J that governs the precision with which $\sqrt{\sigma_\beta^2 + \sigma_{\alpha\beta}^2}$ can be estimated. Only by increasing the number of operators in a gage R&R study can one substantially improve the estimation of reproducibility.

While this fact about the estimation of reproducibility is perfectly plausible, its implications are not always fully appreciated (or at least not kept clearly in mind) by quality assurance practitioners. For example, many standard gage R&R data collection forms allow for at most $J = 3$ operators. But three is a *very* small sample size when it comes to estimating a variance or standard deviation. So although the data in Table 2.1 are perhaps more or less typical of many R&R data sets, the small ($J = 3$) number of operators evident there should *not* be thought of as in any way ideal. To get a really good handle on the size of reproducibility, many more operators would be needed.

As a final matter concerning gage R&R studies, it needs to be said that other (actually theoretically superior) means of estimating the parameters of the model (2.4) (and then the repeatability and reproducibility measures discussed here) are available. Estimators can, for example, be based on the so-called "two-way ANOVA" discussed in many places (including Section 8.4 of Vardeman's *Statistics for Engineering Problem Solving*). The ones presented here have the virtues of being computationally simple and related to ones fairly common in quality assurance circles. But for reference purposes (for those readers familiar with two-way Analysis of Variance) we also record the forms of the standard ANOVA estimators. These are

ANOVA-Based Estimators for a Gage R&R Study

$$\hat{\sigma}^2_{\text{repeatability}} = \hat{\sigma}^2 = MSE,$$

$$\hat{\sigma}^2_\beta = \max\left(0, \frac{1}{mI}(MSB - MSAB)\right),$$

$$\hat{\sigma}^2_{\alpha\beta} = \max\left(0, \frac{1}{m}(MSAB - MSE)\right),$$

and

$$\hat{\sigma}^2_{\text{reproducibility}} = \max\left(0, \frac{1}{mI}(MSB + (I-1)MSAB) - \frac{1}{m}MSE\right).$$

2.2.3 Measurement Precision and the Ability to Detect a Change or Difference

Estimating the repeatability and reproducibility of a gage or measurement system amounts to finding a sensible standard deviation (or "sigma") to associate with it. (Where only one operator is involved, the repeatability measure, σ, is appropriate. Where multiple operators are going to be used, the measure σ_{overall} defined in display (2.6) is more germane.) After an appropriate sigma has been identified, it can form the basis of judgments concerning the adequacy of the gage or system for specific purposes. For example, looking at measures like the gage capability ratio (2.12) is a way of judging (based on an appropriate sigma) the adequacy of a gage for the purpose of checking conformance to a set of engineering specifications. As a final topic in this section, we consider the slightly different problem of judging the adequacy of a gage or measurement system for the purpose of detecting a change or difference.

The problem of determining whether "something has changed"/"there is a difference" is fundamental in engineering and technology. For example, in the context of process monitoring, engineers need to know whether process parameters (e.g., the mean widget diameter being produced by a particular lathe) are at standard values or have changed. And, in evaluating whether two machines are producing similar output, one needs to assess whether product characteristics from the two machines are the same or are consistently different. And, for example, when using hazardous materials in manufacturing, engineers need to compare chemical analyses for current environmental samples to analyses for "blank" samples, looking for evidence that important quantities of toxic materials have escaped a production process and thus increased their ambient level from some "background" level.

For the remainder of this section, let $\sigma_{\text{measurement}}$ stand for an appropriate standard deviation for describing the precision of some measurement system. (Depending upon the context this could be σ or σ_{overall} from a gage study.) We will consider the impact of $\sigma_{\text{measurement}}$ on one's ability to detect change or difference through consideration of the distribution of the statistic

$$\bar{y}_{\text{new}} - \bar{y}_{\text{old}}, \tag{2.13}$$

where $\bar{y}_{\text{new}}$ is the sample mean of n_{new} independent measurements taken on a particular "new" object and $\bar{y}_{\text{old}}$ is the sample mean of n_{old} independent measurements taken on a particular "old" object. (The issue of how to think about situations where particular *objects* being measured are themselves of interest only to the extent that they represent "new" and "old" *conditions* that produced them will be considered briefly at the end of this subsection.) This discussion will allow the possibility that the information on the "old" object is strong enough that n_{old} can be thought of as being essentially infinite, and thus $\bar{y}_{\text{old}}$ essentially equal to the (long-run) mean of old observations, μ_{old}. If μ_{new} is the long-run mean (or expected value) of the new observations and the new and old measurements are

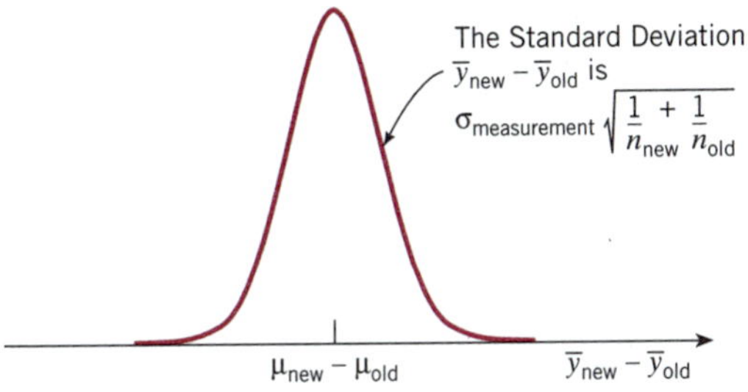

Figure 2.5
The probability distribution of $\bar{y}_{new} - \bar{y}_{old}$.

independent, the random variable in display (2.13) has

$$E(\bar{y}_{new} - \bar{y}_{old}) = \mu_{new} - \mu_{old}, \tag{2.14}$$

Mean for the Random Variable $\bar{y}_{new} - \bar{y}_{old}$

and

$$\text{Var}(\bar{y}_{new} - \bar{y}_{old}) = \sigma^2_{measurement}\left(\frac{1}{n_{new}} + \frac{1}{n_{old}}\right). \tag{2.15}$$

Variance for the Random Variable $\bar{y}_{new} - \bar{y}_{old}$

In the case where the information on the old object is strong enough to consider n_{old} to be essentially infinite, expression (2.15) reduces to

$$\text{Var}(\bar{y}_{new} - \mu_{old}) = \sigma^2_{measurement}\left(\frac{1}{n_{new}}\right). \tag{2.16}$$

In the event that it is sensible to think of $\bar{y}_{new} - \bar{y}_{old}$ as normally distributed (for example, either because repeat measurements are themselves approximately normally distributed or both n_{new} and n_{old} are large) one then has the picture of the distribution of $\bar{y}_{new} - \bar{y}_{old}$ given in Figure 2.5. If $\mu_{new} = \mu_{old}$ then the normal curve in the figure is centered at 0. If $\mu_{new} \neq \mu_{old}$ then the normal curve in the figure is centered at the nonzero difference between the new and old means.

Example 2.2 **Chemical Analysis for Benzene.** An appropriate standard deviation for characterizing an industrial laboratory's precision in a particular analysis for the benzene content of samples of a particular type is $\sigma_{measurement} = .03\mu g/l$. Suppose that in order to determine whether the amount of benzene in a particular environmental sample exceeds that in a particular similar "blank" sample (supposedly containing only background levels of the substance), the environmental sample will be analyzed $n_{new} = 1$ time and its measured content compared to the mean measured content from $n_{old} = 5$ analyses of a single blank sample. Then from equations (2.14) and (2.15) the random variable $\bar{y}_{new} - \bar{y}_{old}$ has mean

$$E(\bar{y}_{new} - \bar{y}_{old}) = \mu_{new} - \mu_{old},$$

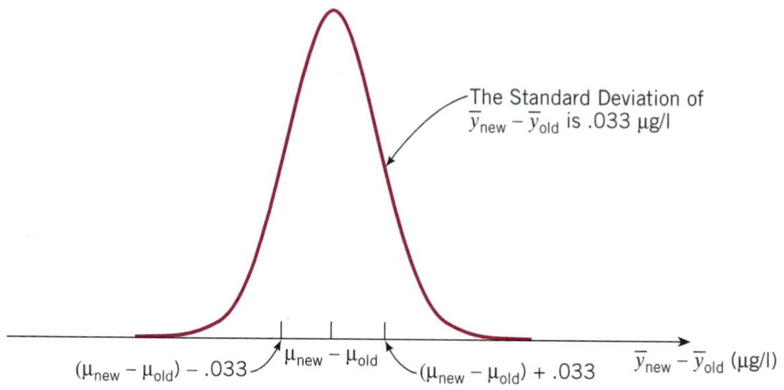

Figure 2.6
The probability distribution of $\bar{y}_{\text{new}} - \bar{y}_{\text{old}}$ in the benzene example.

and standard deviation

$$\sqrt{\text{Var}(\bar{y}_{\text{new}} - \bar{y}_{\text{old}})} = \sigma_{\text{measurement}}\sqrt{\frac{1}{1} + \frac{1}{5}} = .03\sqrt{\frac{6}{5}} = .033\ \mu\text{g/l}.$$

Figure 2.6 shows a corresponding probability distribution for the difference $\bar{y}_{\text{new}} - \bar{y}_{\text{old}}$ assuming that repeat measurements of a given sample are normally distributed.

Note, by the way, that if the information on μ_{old} was essentially perfect and equation (2.16) was relevant, the standard deviation associated with $\bar{y}_{\text{new}} - \mu_{\text{old}}$ would be .030 μg/l, not much smaller than the .033 μg/l value found here.

The picture of $\bar{y}_{\text{new}} - \bar{y}_{\text{old}}$ given in equations (2.14) and (2.15) and Figure 2.5 forms the basis for several common ways of evaluating the adequacy of a measurement technique to characterize a change or difference. One simple-minded rule of thumb often employed by analytical chemists is that a difference or change in mean, $\mu_{\text{new}} - \mu_{\text{old}}$, needs to be on the order of 10 times the standard deviation of $\bar{y}_{\text{new}} - \bar{y}_{\text{old}}$ before it can be adequately characterized by a measurement process with standard deviation $\sigma_{\text{measurement}}$ using sample sizes n_{new} and n_{old}. For example, in the benzene analyses of Example 2.2, with sample sizes $n_{\text{new}} = 1$ and $n_{\text{old}} = 5$, this rule of thumb says that only increases in real benzene content on the order of at least

$$10 \times .033 = .33\ \mu\text{g/l}$$

can be reliably characterized. This somewhat ad hoc (but nevertheless popular) guideline amounts to a requirement that one's "signal-to-noise ratio" for determination of a difference (ratio of mean to standard deviation) be at least 10 before being comfortable with the resulting precision.

A second approach to using the picture of $\bar{y}_{\text{new}} - \bar{y}_{\text{old}}$ given in equations (2.14) and (2.15) and Figure 2.5 to describe one's ability to detect a difference between objects involves some ideas from the elementary theory of hypothesis testing usually studied in beginning statistics courses. In interpreting an observed value of $\bar{y}_{\text{new}} - \bar{y}_{\text{old}}$, one might require that it be of a certain minimum magnitude before declaring that there is clearly a difference between new and old objects. For the sake of concreteness, suppose for the rest of this section that one is concerned about detecting an increase in response. That is, suppose one is interested in detecting the possibility that $\mu_{\text{new}} - \mu_{\text{old}} > 0$. It then makes sense to set some **critical limit, L_c**, and to only declare that there has been a change (that there is a difference) if

$$\bar{y}_{\text{new}} - \bar{y}_{\text{old}} > L_c. \tag{2.17}$$

If one wishes to limit the probability of a "false positive" (i.e., a type I error) L_c should be chosen large enough that the eventuality (2.17) occurs rarely when in fact $\mu_{\text{new}} = \mu_{\text{old}}$. For example, if one may assume that $\bar{y}_{\text{new}} - \bar{y}_{\text{old}}$ is essentially normally distributed, it is possible to use the fact that when $\mu_{\text{new}} = \mu_{\text{old}}$

$$\frac{\bar{y}_{\text{new}} - \bar{y}_{\text{old}}}{\sigma_{\text{measurement}}\sqrt{\dfrac{1}{n_{\text{new}}} + \dfrac{1}{n_{\text{old}}}}}$$

is standard normal to set L_c. From the normal table given in this book as Table A.2, one may pick a number z_1 so that for a standard normal variable Z, $P[Z > z_1] = \alpha$, for α any small number of one's choosing. Then with

Critical Value for Normally Distributed $\bar{y}_{\text{new}} - \bar{y}_{\text{old}}$ Where an Increase in Response Is to Be Detected

$$L_c = z_1 \sigma_{\text{measurement}}\sqrt{\frac{1}{n_{\text{new}}} + \frac{1}{n_{\text{old}}}}, \tag{2.18}$$

the probability of a false positive is no more than α.

Once one has established a critical value L_c (using equation (2.18) or otherwise) it is then reasonable to ask what is the probability of detecting a change of a given size, or equivalently what size change can be reliably detected. Again, assuming that $\bar{y}_{\text{new}} - \bar{y}_{\text{old}}$ is essentially normally distributed as in Figure 2.5, it is possible to answer this question. That is, under the normal model, for

z-score for the Critical Value L_c

$$z_2 = \frac{L_c - (\mu_{\text{new}} - \mu_{\text{old}})}{\sigma_{\text{measurement}}\sqrt{\dfrac{1}{n_{\text{new}}} + \dfrac{1}{n_{\text{old}}}}}, \tag{2.19}$$

the probability (depending upon $\mu_{\text{new}} - \mu_{\text{old}}$) of declaring that there has been a change (that there is a difference) is

Probability of Declaring That Response Has Increased (for Normal $\bar{y}_{\text{new}} - \bar{y}_{\text{old}}$)

$$\gamma = P[Z > z_2], \tag{2.20}$$

(for Z standard normal). Or by rewriting equation (2.19), for z_2 chosen so that γ in display (2.20) is large, one can solve for the size of change required to produce a large (at least γ) probability of detection, namely

Change in Mean Response with γ Probability of Detection (for Normal $\bar{y}_{\text{new}} - \bar{y}_{\text{old}}$)

$$\mu_{\text{new}} - \mu_{\text{old}} = L_c - z_2\sigma_{\text{measurement}}\sqrt{\frac{1}{n_{\text{new}}} + \frac{1}{n_{\text{old}}}}. \tag{2.21}$$

Notice that in display (2.21) z_2 is typically negative so that $\mu_{\text{new}} - \mu_{\text{old}}$ is then larger than L_c.

In analytical chemistry, the value of $\mu_{\text{new}} - \mu_{\text{old}}$ required to produce a large probability of detecting an increase in mean is given a special name.

Definition 2.4 For a given standard deviation of measurement $\sigma_{\text{measurement}}$, sample sizes n_{new} and n_{old}, critical value L_c and desired (large) probability γ, the **lower limit of detection, L_d,** of a measurement protocol is the smallest difference in means $\mu_{\text{new}} - \mu_{\text{old}}$ producing

$$P[\bar{y}_{\text{new}} - \bar{y}_{\text{old}} > L_c] \geq \gamma.$$

When $\bar{y}_{\text{new}} - \bar{y}_{\text{old}}$ is normal and z_2 is chosen according to display (2.20), equation (2.21) implies that

Lower Limit of Detection (for Normal $\bar{y}_{\text{new}} - \bar{y}_{\text{old}}$)

$$L_d = L_c - z_2\sigma_{\text{measurement}}\sqrt{\frac{1}{n_{\text{new}}} + \frac{1}{n_{\text{old}}}}. \tag{2.22}$$

Example 2.2 continued. Consider again the benzene analysis example. Suppose that it is desirable to limit the probability of producing a "false positive" (a declaration that the environmental sample clearly contains more benzene than the blank sample when in fact there is no real difference in the two) to no more than $\alpha = .10$. Then consulting Table A.2, it can be seen that for Z standard normal, $P[Z > 1.282] = .10$. So, using equation (2.18), an appropriate critical value is

$$L_c = 1.282(.030)\sqrt{\frac{1}{1} + \frac{1}{5}} = .042.$$

Should one wish to evaluate the probability of detecting an increase in real benzene content of size $\mu_{\text{new}} - \mu_{\text{old}} = .02\ \mu\text{g/l}$ using this critical value, equations (2.19) and (2.20) show that with

$$z_2 = \frac{.042 - .02}{.030\sqrt{\frac{1}{1} + \frac{1}{5}}} = .67,$$

the probability is only about

$$P[Z > z_2] = P[Z > .67] = .2514.$$

There is a substantial (75%) chance of failing to identify a .02 μg/l increase in benzene content beyond that resident in the blank sample. This unpleasant fact motivates the question "How big does the increase in benzene concentration need to be in order to have a large (say 95%) chance of seeing it above the measurement noise?" Since consultation of the standard normal table shows that

$$P[Z > -1.645] = .95,$$

equations (2.20) and (2.22) imply that for $\gamma = .95$

$$L_{\text{d}} = \mu_{\text{new}} - \mu_{\text{old}} = .042 - (-1.645)(.030)\sqrt{\frac{1}{1} + \frac{1}{5}} = .096\ \mu\text{g/l}$$

is (in the language of Definition 2.4) the lower limit of detection in this situation. A real increase in benzene content must be of at least this size for there to be a large (95%) chance of "seeing" it through the measurement noise.

The particular example used in the foregoing discussion of the implications of equations (2.14), (2.15), and Figure 2.5 is from analytical chemistry. But the basic method illustrated is perfectly general and could, for example, be equally well applied to the consideration of the implications of measurement precision on one's ability to detect a difference in diameters of two particular parts turned on a lathe.

It is, however, an extremely important distinction that the discussion thus far in this section has been phrased in terms of detecting a difference between two particular objects and *not* between processes or populations standing behind those objects. Example 2.2 concerns comparison of a particular environmental sample and a particular blank sample. It does *not* directly address the issue of how the population of environmental samples from a site of interest compares to a population of blanks. Similarly, in a manufacturing context, comparisons based on the foregoing material would concern two particular measured parts, not the process conditions operative when those parts were made. The point here is that only measurement variation has been taken into account, and not object-to-object variation for processes or populations the measured objects might represent. And unless there is no "within-population" or "process" variation, detection of a difference between an old and a new object is not the same as detection of (for example) a difference between old and new population or process means.

There is, however, one obvious way to reinterpret the formulas and language here in order to allow application to the problem of detecting changes in a process or population mean. If "measurement" for a process is taken to mean "(random) selection of an object from the process or population and measurement of that object," "$\sigma_{\text{measurement}}$" can be taken as a measure of variation including *both* process/population *and* measurement variation. Of course, under such a scenario, "n_{new}" and "n_{old}" refer to the numbers of objects selected and measured (not to counts of how many times one item is measured). (If the object in a scenario like that of Example 2.2 is to compare not a particular field sample to a particular blank, but a site mean to a mean of blanks, what governs "precision" is the number of field samples and blanks studied.) And constant "$\sigma_{\text{measurement}}$" means that the combination of process/population variation and real measurement variation remains unchanged from old to new conditions.

2.3 ELEMENTARY PRINCIPLES OF QUALITY ASSURANCE DATA COLLECTION

Good (practically useful) data do not collect themselves. Neither do they magically appear on the desk of a quality engineer, ready for analysis and lending insight into how to improve processes. In the abstract, these points are quite obvious. But it sometimes seems that little is said in statistical quality control texts about data collection. And in practice, quality engineers sometimes lose track of the fact that no amount of clever analysis will make up for lack of intrinsic information content in poorly collected data. Often, wisely and purposefully collected data will carry such a clear message that they essentially "analyze themselves." So it seems wise to make some early comments in this book about general considerations in quality assurance data collection.

A first observation regarding the collection of useful data is closely related to the discussion of measurement in the previous section. That is that if data are to be at all helpful, there must be a consistent understanding about exactly how they are to be collected. This involves having **operational definitions** for quantities to be observed and personnel who have been **well trained** in using the definitions and any relevant measurement equipment. Consider, for example, the apparently fairly "simple" problem of measuring "the" diameters of (supposedly circular) steel rods. Handed a gage and told to measure diameters, one would not really know where to begin. Should the diameter be measured at one end of the rods, in the center, or where? Should the first diameter seen for each rod be recorded, or should perhaps the rods be rolled in the gage to get maximum diameters (for those cases where rods are not perfectly circular in cross section)?

Or consider a case where one is to collect qualitative data on, say, defects in molded glass automobile windshields. Exactly what constitutes a "defect"? Surely a bubble one inch in diameter directly in front of the driver's head position is a defect. But would a 10^{-4}-inch diameter flaw in the same position be a problem? Or what about a one-inch diameter flaw at the very edge of the windshield that would be completely covered by trim molding? Should such a flaw be called a defect? Clearly, if useful data are to be collected in a situation like this, very careful operational definitions need to be developed and personnel need to be taught to use them.

This point about the importance of consistency of observation/measurement in quality assurance data collection cannot be overemphasized. When, for example, different technicians use measurement equipment in substantially different ways, what looks in process monitoring data like a big process change can in fact be nothing more than a change in the person doing the measurement. If gage reproducibility variation is of the same magnitude as important physical effects and multiple technicians are going to make measurements, it must be reduced through proper training and practice before there is reason to put much faith in data that are collected.

A second important point in the collection of quality assurance data has to do with **when and where** they are gathered. The closer in time and space that data are taken to an operation whose performance they are supposed to portray, the better. The ideal here is probably for well-trained workers actually doing the work or running the equipment in question to do their own data collection. There are several reasons for this. For one thing, it is such people who are in a position (after being trained in the proper interpretation of quality assurance data and given the authority to act on them) to react quickly and address any process ills portrayed by the data that they collect. (Quick reaction to process information can prevent process difficulties from affecting additional product and producing unnecessary waste.) For another, it is simply a fact of life that data collected far away in time and space from a process rarely lead to important insights into "what is going on." Over the years, your authors have seen many student groups (against good advice) take on company projects of the variety "Here are some data we've been collecting for the past three years. Tell us what they mean." These essentially synthetic postmortem examinations are yet to produce anything helpful for the companies involved. Even if an interesting pattern is found in such data, it is very rare that root causes can be identified completely after the fact.

If one accepts that much of the most important quality assurance data collection will be done by people whose primary job is not data collection but rather working in or on a production process, a third general point comes into focus. That is that routine data collection should be made as **convenient** as possible and where at all feasible, the methods used should make the data **immediately useful.** These days, quality assurance data are often entered as they are collected into

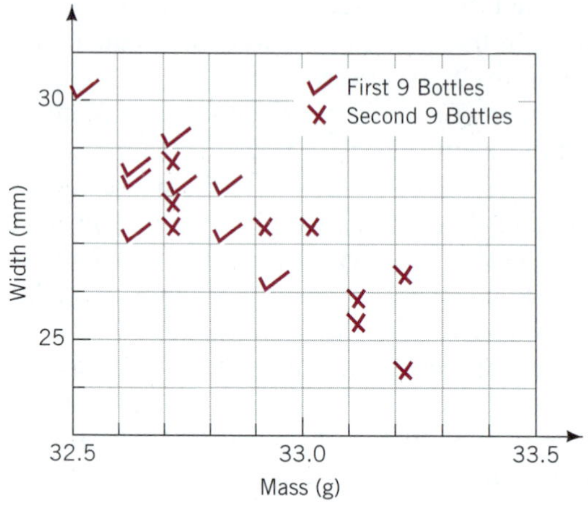

Figure 2.7
Check sheet for bottle mass and width of bottom piece for 18 PVC bottles.

computer systems that produce real-time displays intended to show those who gathered them their most important features.

Where pencil and paper data recording methods are used, thought needs to go into the making of the forms employed so that there is no need, for example, for a transfer to another form or medium before using data. Figure 2.7 is a so-called two-variable "check sheet." Rather than making a list of (x, y) pairs and later transferring them to a piece of graph paper or a computer program for making a scatterplot, use of a form like this allows immediate display of any relationship between x and y. (Note that the use of different symbols or even colors can carry information on variables besides x and y, like time order of observation.) The point here is that if one's goal is quality improvement, data are for using, not for archiving away. Data collection is a tool, not an end in itself, and it needs to be designed to be practically effective.

A fourth general principle of quality assurance data collection regards proper **documentation.** One typically collects process data hoping to locate (and subsequently eliminate) possible sources of variation. If this is to be done, care needs to be taken to keep track of conditions associated with each data point. One needs to know not only that a measured widget diameter was 1.503 mm, but also the machine on which it was made, who was running the machine, what raw material lot was used, when it was made, what gage was used to do the measuring, who did the measuring, and so on. Without such information there is, for example, no way to ever discover consistent differences between two machines that contribute significantly to overall variation in widget diameters. Simply having a sheet full of numbers without the histories of those numbers is of little help in quality assurance.

Several additional important general points about the collection of quality assurance data have to do with the **volume** of information one is to handle. In the first place, a small or moderate amount of carefully collected (and immediately used) data will typically be worth much more than even a huge amount that is haphazardly collected (or never used). It has already been said in this section that student project groups usually find the autopsy of a large industrial data set to be a fruitless exercise. One is almost always better off trying to learn about a process based on a small data set collected with specific purposes and questions in mind than when rummaging through a large "general purpose" database assembled without the benefit of such focus.

Further, when trying to answer the perennial question "How much data do I need to...?" one needs at least a basic qualitative understanding (hopefully gained in a first course in statistics) of what things govern the information content of a sample. For one thing (even in cases where one is gathering data from a particular finite lot of objects rather than from a process) it is the absolute (and not relative) size of a sample that governs its information content. So blanket rules like "Take a 10% sample" are not wise. They will sometimes call for the collection of too much data and other times fail to collect enough data.

Rather than seeking to choose sample sizes in terms of some fraction of a universe of interest, one should think instead in terms of the size of the unavoidable background variation and of the size of an effect that is of practical importance. If there is no variation at all in a quantity of interest, a sample of size $n = 1$ will characterize it completely! On the other hand, if substantial variation is inevitable and small overall changes are of practical importance, huge sample sizes will be needed to illuminate important process behavior.

A final general observation about quality assurance data collection is that one must take careful account of **human nature, psychology, and politics** when assigning data collection tasks. If one wants useful information, he or she had better see that those who are going to collect data are convinced that doing so will genuinely aid (and *not* threaten) them and that accuracy is more desirable than "good numbers" or "favorable results." People who have seen data collected by themselves or colleagues used in ways that they perceive as harmful (for instance, identifying one of their colleagues as a candidate for termination) will simply not cooperate. Nor will people who see nothing coming of their honest efforts at data collection. People who are to collect data need sufficient motivation to believe that data can help them do a better job and help their organization be more successful.

2.4 SIMPLE STATISTICAL GRAPHICS AND QUALITY ASSURANCE

The old saying "a picture is worth a thousand words" is especially true in the realm of statistical quality assurance. Simple graphical devices that have the potential to be applied effectively by essentially all workers have a huge potential impact that is sometimes obscured by their almost trivial appearance. In this section, the usefulness of simple histograms, Pareto charts, scatterplots, and run charts in quality assurance efforts is discussed. This is done with the hope that readers will see the

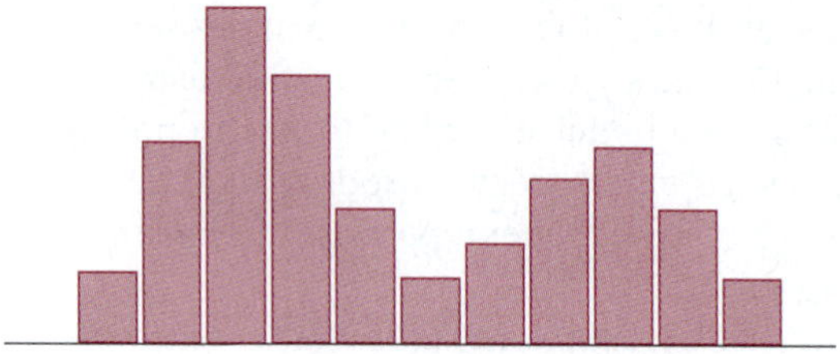

Figure 2.8
A bimodal distribution.

value of routinely using these as the important data organizing and communication tools that they are.

Essentially every elementary statistics book ever written has a discussion of the making of a **histogram** from a sample of measurements. Most even provide some terminology for describing various histogram shapes. That background will not be repeated here. Instead this book's discussion will center on the interpretation of various patterns seen on histograms in quality assurance contexts, and on how they can be of use in quality improvement efforts.

Figure 2.8 is a bimodal histogram of widget diameters. Observing that the histogram has two distinct "humps" is not in and of itself particularly helpful. But asking the question "Why is the data set bimodal?" begins to be more to the point. Bimodality (or multimodality) in a quality assurance data set is a strong hint that there are two (or more) effectively different versions of something at work in a process. Bimodality might be produced by two different workers doing the same job in measurably different ways, two parallel machines that are adjusted somewhat differently, and so on. The systematic differences between such versions of the same element of a process produce variation that often can and should be eliminated, thereby improving quality. On a plot like Figure 2.8, one can hope to identify and eliminate the physical source of the bimodality and effectively be able to "slide the two humps together" so that they coincide, greatly reducing the overall variation.

As a side comment, it is worth noting that the modern trend toward reducing the size of supplier bases and even "single sourcing" has its origin in the kind of phenomenon pictured in Figure 2.8. Different suppliers of a good or service will inevitably do some things slightly differently. As a result, what they supply will differ somewhat between them. Reducing a company's number of vendors then has two effects. Variation in the products that it makes from components or raw materials supplied by others is reduced and the costs (in terms of lost time and waste) often associated with switchovers between different material sources are also reduced.

Other shapes on histograms can also give strong clues as to what is going on in a process (and help guide quality improvement efforts). For example, sorting operations often produce distinctive truncated shapes. Figure 2.9 shows two different histograms for the net contents of some containers of a liquid. The first portrays a distribution that is almost certainly generated by culling those containers (filled by an imprecise filling process) that are below label contents. The second looks as

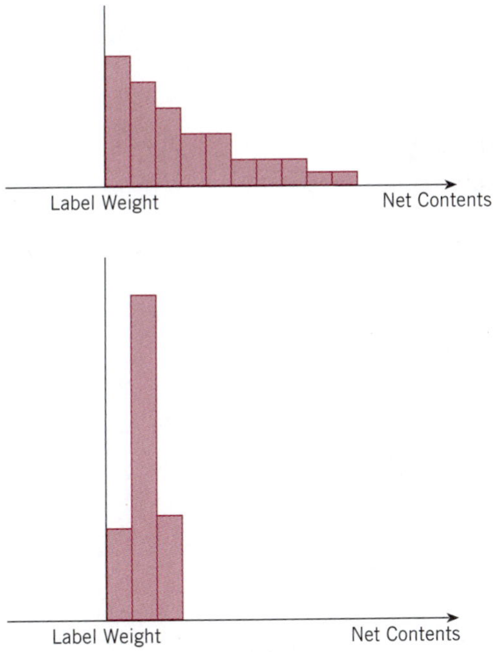

Figure 2.9
Two distributions of bottle contents.

if it might be generated by a very precise filling process aimed only slightly above the labeled contents. The histograms give both hints at how the guaranteed minimum contents are achieved in the two cases, and also a pictorial representation of the waste produced by imprecision in filling. A manufacturer supplying a distribution of net contents like that in the first histogram must both deal with the rework necessitated by the part of the first distribution that has been "cut off" and also suffer the "give away cost" associated with the fact that much of the truncated distribution is quite a bit above the label value.

Figure 2.10 is a histogram for a very interesting set of data from *Engineering Statistics and Quality Control* by I. W. Burr. The very strange shape of the data set portrayed there almost certainly also arose from a sorting operation. But in this case, it appears that the center part of the distribution is missing. In all probability, one large production run was made to satisfy several orders for parts of the same

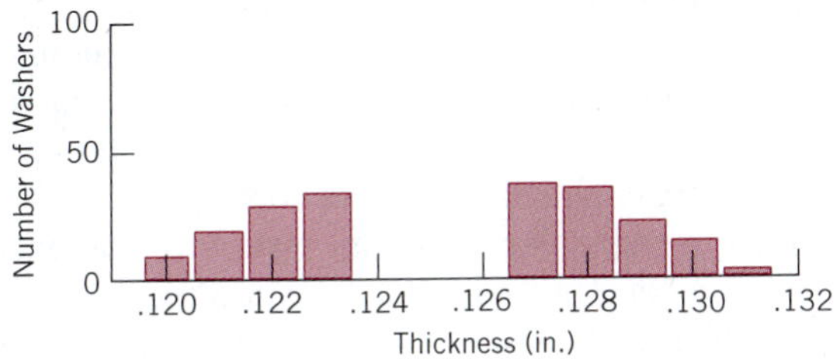

Figure 2.10
Thicknesses of 200 mica washers (specifications .125 ± .005 in.).

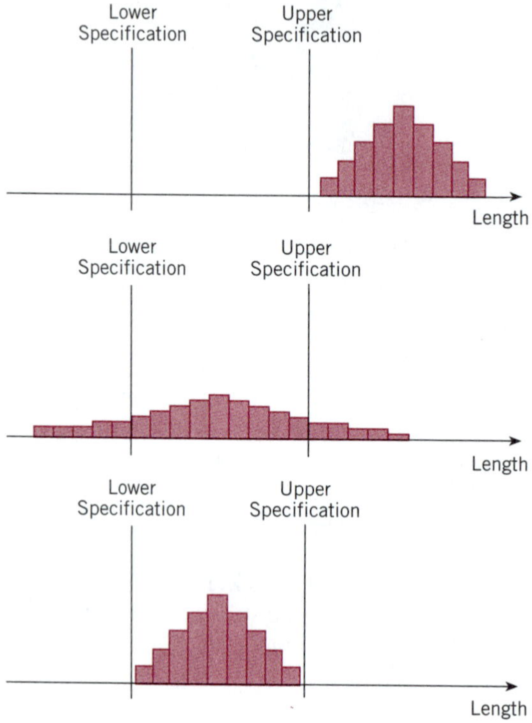

Figure 2.11
Three distributions of a critical machined dimension.

nominal type. Then a sorting operation graded those parts into classes depending upon how close actual measurements were to nominal. Customers placing orders with tight specifications probably got (perhaps at a premium price) parts from the center of the original distribution, while others with looser specifications likely received shipments with distributions like the one in Figure 2.10.

Marking engineering specifications on a histogram is a very effective way of communicating to even very nonquantitative people what is needed in the way of process improvements. Figure 2.11 shows a series of three histograms with specifications for a part dimension marked on them. In the first of those three histograms, the production process seems quite "capable" of meeting specifications for the dimension in question (in the sense of having adequate intrinsic precision), but clearly needs to be "reaimed" so that the mean measurement is lower. The second histogram portrays the output of a process that is properly aimed, but incapable of meeting specifications. The intrinsic precision is not good enough to fit the distribution between the engineering specifications for the dimension in question. The third histogram represents data from a process that is both properly aimed *and* completely capable of meeting engineering specifications.

Another kind of bar chart that is quite popular in quality assurance contexts is the so-called **Pareto diagram**. This tool is especially useful as a political device for getting people to prioritize their efforts and focus first on the biggest quality

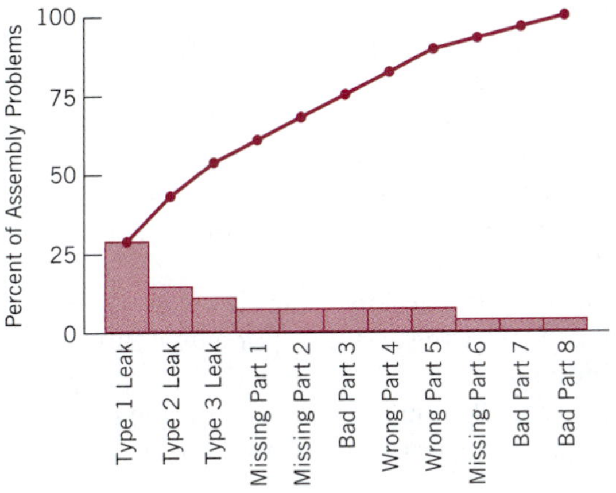

Figure 2.12
Pareto chart of assembly problems.

problems an organization faces. One makes a bar chart where problems are listed in decreasing order of frequency, dollar impact, or some other measure of importance. Often, a broken line graph indicating the cumulative importance of the various problem categories is also added to the display. Figure 2.12 shows a Pareto diagram of assembly problems identified on a production run of 100 pneumatic hand tools. Clearly, by the measure of frequency of occurrence the most important quality problem to be addressed is that of leaks.

The name "Pareto" is that of a mathematician who studied wealth distributions and concluded that most of the money in Italy belonged to a relatively few people. His name has become associated with the so-called "Pareto principle" or "80–20 principle." This states that "most" of anything (like quality problems or beer consumption) is traceable to a relatively few sources (like root causes of quality problems or big beer drinkers). Conventional wisdom in modern quality assurance is that attention to the relatively few major causes of problems will result in huge gains in efficiency and quality.

Discovering *relationships* between variables is often important in discovering means of process improvement. An elementary but most important place to start in looking for such relationships is the making of simple **scatterplots** (plots of (x, y) pairs). Consider, for example, Figure 2.13. This figure consists of two scatterplots

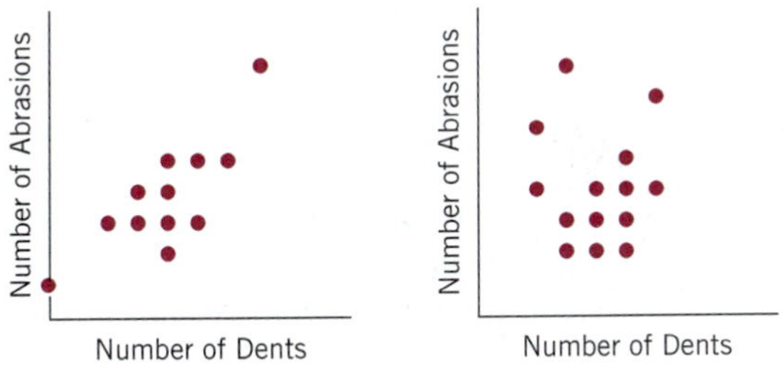

Figure 2.13
Two scatterplots of numbers of occurrences of manufacturing defects.

of the numbers of occurrences of two different quality problems in lots of widgets. The stories told by the two scatterplots are quite different. In the first, there seems to be a positive correlation between the numbers of problems of the two types, while in the second no such relationship seems evident. The first scatterplot suggests that a single root cause may be responsible for both types of problems and that in looking for it, one can limit attention to those causes that could possibly produce both effects. The second scatterplot suggests that two different causes are at work and one will need to look for them separately.

It is true, of course, that one can use numerical measures (like the sample correlation) to investigate the extent to which two variables are related. But a simple scatterplot can be understood and used even by people with little quantitative background. Besides, there are occasionally things that can be seen in plots (like, for example, nonlinear relationships) that will be missed by looking only at numerical summary measures.

The habit of plotting and looking at data is one of the best habits a quality engineer can develop. And one of the most important ways of plotting data is in a scatterplot against time order of observation. Where there is only a single measurement associated with each time period and one connects consecutive plotted points with line segments, it is common to call the resulting plot a **run chart**. Figure 2.14 is a run chart of some data studied by a student project group (Williams and Markowski). Pictured are 30 consecutive outer diameters of metal parts turned on a lathe.

Investigation of the somewhat strange pattern on the plot led to a better understanding of how the turning process worked (and could have led to appropriate compensations to eliminate much of the variation in diameters seen on the plot). In general the first 15 diameters seem to decrease, then there is a big jump in diameter, after which diameters again decrease. Checking production records, the students found that the lathe in question had been shut down and allowed to cool

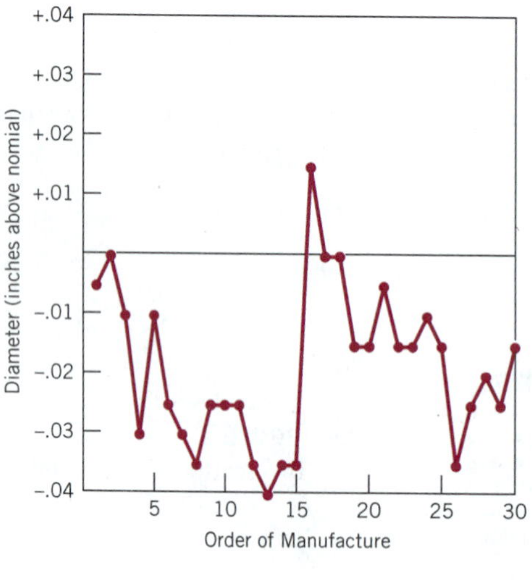

Figure 2.14
A run chart for 30 consecutive outer diameters turned on a lathe.

off between parts 15 and 16. It is quite likely that the pattern seen on the plot is related to the dynamics of the lathe hydraulics. When cold, the hydraulics didn't do as good a job pushing the cutting tool into the workpiece as when they were warm. Hence the diameters tended to decrease as the lathe warmed up. (Apparently the data collection in question did not cover a long enough period to see the effects of tool wear, which would have tended to increase part diameters as the length of the cutting tool decreased.) Note that if one knows that this kind of phenomenon exists, it is possible to compensate for it (and increase part uniformity) by setting artificial target diameters for parts made during a warmup period below those for parts made after the lathe is warmed up.

2.5 CHAPTER SUMMARY

Steps 1 through 3 of the Six-Step Cycle presented in Table 1.1 can often be accomplished using fairly simple/basic tools. This chapter has introduced some of these concepts and methods. Section 2.1 considered elementary tools for use in process mapping. Basic issues in metrology, including validity, precision, accuracy, gage repeatability, gage reproducibility, gage capability, and the adequacy of a measurement system to detect change were discussed in Section 2.2. Important qualitative principles of engineering and quality assurance data collection were presented in Section 2.3. And Section 2.4 demonstrated how effective the simplest methods of statistical graphics can be when wisely used in quality improvement efforts.

2.6 CHAPTER 2 EXERCISES

2.1. Pellet Densification. Crocfer, Downey, Rixner, and Thomas studied the densification of Nd_2O_3. Pellets of this material were fired at 1400°C for various lengths of time and the resulting densities measured, giving the values below. (Densities are in g/cc and times are in minutes.) Before firing, pellet sizes were similar, but clearly not identical. The effect of firing time (at 1400°C) on the final pellet density was of interest.

Density, y	6.521	6.662	4.907	5.102	5.380	5.657	5.895	6.207	6.325
Time, x	150	270	1	3	7	15	30	60	90

a. There is a glaring weakness in the data collection effort. What is it?

b. Make a scatterplot of these data and comment on "how linear" the relation between y and x appears to be. As a second attempt to find a simple relationship between y and x, take the natural logarithms of the densities and times. Plot $y' = \ln(y)$ versus $x' = \ln(x)$. Does this relationship appear to be approximately linear? What relationship between y and x is implied by a linear relationship between y' and x'?

2.2. Laser Metal Cutting. Davis, Martin, and Poppinga used a Ytterbium Argon gas laser to make some cuts in 316 stainless steel. Using 95 mJ/pulse and 20 Hz settings on the laser and a 15.5 mm distance to the steel specimens (set at a 45° angle to the laser beam) the students made cuts in specimens using 100, 500, and 1000 pulses. The measured depths of cut (in standard units) are given below.

100 Pulses	500 Pulses	1000 Pulses
7.4, 8.6, 5.6, 8.0	24.2, 29.5, 26.5, 23.8	33.4, 37.5, 35.9, 34.8

a. What is the response variable in this problem?
b. Give the sample average values for the 100, 500, and 1000 Pulse levels.
c. Calculate the sample range for the data at each pulse level. Give estimates of the standard deviation of depth of cut for each level of the pulse variable, first based on the sample range and then using the sample standard deviation. (You will have two estimates for each of the three population standard deviations.)
d. Assuming variability is the same for all three pulse levels, give an estimate of the common standard deviation based on the three sample ranges.
e. The concepts of measurement validity, precision, and accuracy are discussed in Section 2.2. The analysts here decided to report the average values for the different numbers of pulses. This averaging can be thought of in terms of improving which of (1) validity, (2) precision, or (3) accuracy (over the use of any single measurement)? The concept of calibration is most closely associated with which of the three?

2.3. Fiber Angle. Grunig, Hamdorf, Herman, and Potthoff studied a carpet-like product. Specifically they measured the angle at which fibers were glued to a sheet of base material. A portion of finished product was obtained and cut into five sections. Each of the four team members measured the fiber angle eight times for each section. The results of their measuring are given below in degrees (above an undisclosed reference value).

a. Say what each term in equation (2.4) means in this problem (including the i, j, and k).
b. Give the (different) estimates of the standard deviation quantifying repeatability variation derived from (Analyst 1, Section 2), (Analyst 2, Section 3), and (Analyst 4, Section 4).
c. Give the "pooled" estimate of the repeatability standard deviation. What (in the context of this particular problem) must be true so that your pooled estimate is meaningful?
d. Consider Section 4 of the product. Give the four angle averages for the different analysts. Give the range of these four averages.

Section	Analyst 1	Analyst 2	Analyst 3	Analyst 4
1	10, 20, 20, 23 20, 20, 20, 15	20, 25, 17, 22 23, 15, 23, 20	20, 19, 15, 16 20, 19, 12, 14	10, 10, 10, 5 5, 5, 5, 5
2	15, 17, 20, 20 10, 15, 15, 15	15, 13, 5, 10 8, 8, 10, 12	15, 20, 14, 16 13, 20, 15, 15	10, 10, 10, 10 10, 15, 15, 10
3	23, 20, 22, 20 25, 22, 20, 23	20, 23, 20, 20 23, 23, 22, 20	15, 20, 22, 18 15, 20, 16, 20	10, 10, 10, 15 15, 10, 10, 10
4	15, 16, 22, 15 15, 15, 22, 17	20, 22, 18, 23 23, 23, 24, 20	13, 13, 15, 20 11, 20, 13, 15	5, 10, 10, 10 10, 10, 10, 10
5	20, 20, 22, 20 27, 17, 20, 15	18, 20, 18, 23 20, 20, 18, 15	10, 14, 17, 12 11, 10, 15, 10	5, 10, 10, 10 10, 10, 10, 10

e. Using only the value requested in (d), give an estimate of the variance of the angle averages for Section 4. Continuing, give an estimate of the variance of the sample average angle for Section 4 and Analyst 2 based on your answer to (c).

f. Using only your answers to (e), give an estimate of the standard deviation that would be experienced by many analysts making a single measurement on the same section assuming there is no repeatability component to the overall variation.

g. Using all the data, give an estimate of the same quantity estimated in (f).

h. Specifications on the fiber angle are known to be $\pm 5°$. Estimate the gage capability ratio. Does it appear this measurement method is adequate to check conformance to such specifications? Why or why not?

2.4. Refer to the **Fiber Angle** case in problem (2.3).

a. Is it preferable to have eight measurements on a given section by each analyst as opposed to, say, two measurements? Why or why not?

b. For a given number of measurements per analyst/section combination, is it preferable to have 4 analysts instead of 2, 6, or 8? Why or why not?

c. When making angle measurements for a given section, does it matter if the same location on the piece is repeatedly measured or is it acceptable (or even preferable?) for each analyst to measure angles at 8 different locations on the section? Discuss.

d. Continuing with (c), does it matter that, say, the location(s) measured on a given section varied analyst to analyst? Why or why not?

2.5. **Bolt Shanks.** A 1-inch micrometer is used by GE Aircraft Engines to measure the diameter of a body-bound bolt shank. Specification limits on this dimension have been set with a spread of .002 inches. Three operators and ten body-bound bolt shanks were used in a gage repeatability and reproducibility study. Each bolt shank was measured twice by each operator to produce the data below (in inches).

Part	Operator A	Operator B	Operator C
1	.3473	.3467	.3472
	.3473	.3465	.3471
2	.3471	.3465	.3471
	.3471	.3464	.3471
3	.3472	.3467	.3471
	.3472	.3464	.3471
4	.3474	.3470	.3473
	.3475	.3470	.3474
5	.3474	.3470	.3473
	.3474	.3470	.3473
6	.3472	.3463	.3471
	.3472	.3464	.3471
7	.3473	.3465	.3472
	.3473	.3469	.3471
8	.3474	.3470	.3473
	.3473	.3470	.3473
9	.3472	.3465	.3472
	.3472	.3466	.3471
10	.3474	.3470	.3474
	.3474	.3470	.3473

a. Plot the bolt shank diameter measurements versus part number using a different symbol for each operator. (You may wish to also plot and connect with line segments consecutive mean measurements for a given operator.) Discuss what your plot reveals about the measurement system.

b. Give the estimated standard deviation for repeatability.

c. Give the estimated standard deviation for measurement variation assuming there is no repeatability component to the overall variation.

d. Using your answers to (b) and (c), estimate the percent of total measurement variance due to repeatability.

e. Using your answers to (b) and (c), estimate the percent of total measurement variance due to reproducibility.

f. Discuss the relationship of your plot in (a) to your answers to (b) through (e).

g. Give an estimate of the gage capability ratio. Is the measurement process acceptable? Why or why not?

2.6. Refer to the **Bolt Shanks** case in exercise (2.5). The following data are from three new operators with a different set of ten body-bound bolt shanks.

Part	Operator D	Operator E	Operator F
11	.3694	.3693	.3693
	.3694	.3693	.3693
12	.3693	.3693	.3692
	.3693	.3692	.3692
13	.3698	.3697	.3697
	.3697	.3697	.3697
14	.3697	.3698	.3697
	.3696	.3697	.3697
15	.3694	.3695	.3695
	.3693	.3695	.3694
16	.3692	.3692	.3692
	.3693	.3692	.3691
17	.3696	.3695	.3695
	.3696	.3695	.3695
18	.3697	.3696	.3696
	.3696	.3696	.3696
19	.3697	.3696	.3695
	.3696	.3695	.3696
20	.3697	.3697	.3698
	.3697	.3698	.3697

a. Answer (a) through (g) from problem (2.5) for this new data set.

b. Are your answers to (a) through (g) qualitatively different than those for problem (2.5)? Why or why not? If your answer is yes, what might be the cause of the difference?

c. Do conclusions from the new R&R study suggest a more consistent measurement process for body-bound bolt shanks than those in exercise (2.5)? Why or why not?

2.7. Transmission Gear Measurement. Cummins, Rosario, and Vanek studied two gages used to measure ring gear height and bevel gear height in the production of transmission differentials. Ring gear height and bevel gear height determine the milling points for the customized transmission housings, creating the horizontal location in the housing and the "tightness" of the casing against the differential. A test stand (hydraulically) puts a 1000 pound force on the differential. This force is used to keep the differential from free spinning while allowing spin with some force applied. A 3″ Mitoya digital depth micrometer and a 6″ Mitoya digital depth micrometer were used to make the measurements. Vanek used the 3″ micrometer and took two ring gear height measurements on differential 8D4. Using the same 3″ Mitoya micrometer, Cummins made two ring gear height measurements on the same part. Vanek then took two bevel gear height measurements with the 6″ Mitoya micrometer on the same differential. Cummins followed with the same 6″ micrometer and took two bevel gear height measurements on differential 8D4. This protocol was repeated two more times for the differential 8D4. The whole procedure was then applied to differential 31D4. The data follow:

	Ring Gear Heights (inches) (3″ Mitoya Micrometer)			Bevel Gear Heights (inches) (6″ Mitoya Micrometer)	
	Vanek	**Cummins**		**Vanek**	**Cummins**
8D4	1.88515	1.88470	8D4	5.49950	5.49850
	1.88515	1.88470		5.49985	5.49945
	1.88540	1.88380		5.49975	5.49945
	1.88530	1.88510		5.50000	5.50005
	1.88485	1.88435		5.49930	5.50070
	1.88490	1.88450		5.49945	5.49945
31D4	1.88365	1.88270	31D4	5.49785	5.49700
	1.88370	1.88295		5.49775	5.49710
	1.88330	1.88235		5.49765	5.49615
	1.88325	1.88235		5.49750	5.49615
	1.88270	1.88280		5.49670	5.49595
	1.88265	1.88260		5.49680	5.49620

a. Consider the ring gear heights measured with the 3″ Mitoya micrometer. Give the numerical values of m, I, and J.

b. In the context of the ring gear height measurements, what do m, I, and J represent?

c. Give the estimated repeatability standard deviation for the ring gear height measuring procedure.

d. Give the estimated reproducibility standard deviation for the ring gear height measuring procedure.

e. The upper and lower specifications for ring gear heights are 1.92 and 1.88 inch. If the company requires the gage capability ratio to be no larger than .05, does the 3″ Mitoya micrometer, as currently used, seem to meet this requirement? Why or why not?

f. Repeat (a) through (e) for the bevel gear heights measured with the 6″ Mitoya micrometer. Lower and upper specifications are 5.50 and 5.53 inches for the bevel gear heights.

2.8. Computer Locks. The Engineering Research Institute (ERI) provides services and produces parts to support computer usage on the Iowa State University campus. One of the parts machined by ERI is a computer safety lock. The lock is used to prevent computer and keyboard theft. Three critical dimensions for these locks are: the length of the tip's lower neck (this must accommodate the thickness of the item being secured), tip length (if the lock is too short, the slots on the tip will not expand enough to secure the item), and tip diameter (the lock tip needs to fit snugly into the hole drilled on the item). Cheng, Lourits, Hugraha, and Sarief decided to study tip diameter. The team began its work with an evaluation of measurement precision for tip diameter. The following data are in inches and represent two diameter measurements for each of two analysts made on all 25 locks machined on one day.

Part	Lourits	Cheng
1	.375, .375	.374, .374
2	.375, .375	.377, .376
3	.375, .373	.374, .375
4	.375, .373	.375, .374
5	.374, .374	.374, .374
6	.374, .374	.374, .375
7	.374, .375	.375, .376
8	.374, .375	.374, .373
9	.374, .374	.375, .375
10	.374, .374	.374, .374
11	.375, .373	.374, .374
12	.375, .374	.376, .374
13	.376, .373	.373, .374
14	.373, .373	.379, .374
15	.372, .373	.374, .373
16	.373, .373	.374, .374
17	.373, .373	.374, .373
18	.373, .373	.373, .373
19	.373, .373	.376, .373
20	.373, .373	.373, .373
21	.374, .374	.374, .375
22	.375, .375	.374, .377
23	.375, .375	.376, .377
24	.376, .375	.376, .374
25	.374, .374	.374, .375

a. Many companies have established their own guidelines for interpreting the results of gage R&R studies. One such set of guidelines is below. $6\hat{\sigma}_{\text{repeatability}}$ divided by the difference in the upper and lower specifications corresponds to the "% Gage" for repeatability. $6\hat{\sigma}_{\text{reproducibility}}$ divided by the difference in the upper and lower specifications is the "% Gage" for reproducibility.

% Gage	Rating
33%	unacceptable
20%	marginal
10%	acceptable
2%	good
1%	excellent

Suppose that specifications for the lock diameters are .375±.002 in. According to the company guidelines above, how does the ERI diameter measuring process rate? Why?

b. Find expressions for $\bar{y}_{\text{operator1}}$ and $\bar{y}_{\text{operator2}}$ as functions of the model terms used in display (2.4).

c. Continuing with (b) and applying logic consistent with that used to develop equation (2.9), what does $|\bar{y}_{\text{operator1}} - \bar{y}_{\text{operator2}}|/d_2(2)$ estimate in terms of σ^2_α, σ^2_β, $\sigma^2_{\alpha\beta}$, and σ^2?

2.9. Refer to the **Computer Locks** case in problem (2.8). Consider the measurements made by Lourits. The sample average tip diameter for the ith lock measured by Lourits can be written (holding Lourits fixed) as

$$\bar{y}_{i\,\text{Lourits}} = \mu + \alpha_i + \beta_{\text{Lourits}} + \alpha\beta_{i\,\text{Lourits}} + \bar{\epsilon}_{i\,\text{Lourits}}.$$

a. Give the random portion of $\bar{y}_{i\,\text{Lourits}}$.

b. In terms of σ^2, σ^2_α, σ^2_β, and $\sigma^2_{\alpha\beta}$, give the variance of your answer to part (a).

c. Letting R be the range of the 25 variables $\bar{y}_{i\,\text{Lourits}}$, what does $R/d_2(25)$ estimate?

d. Give the numerical value for $R/d_2(25)$ considered in part (c).

e. In terms of σ^2, σ^2_α, σ^2_β, and $\sigma^2_{\alpha\beta}$, what is the variance of different lock tip diameters as measured by a single operator (say Lourits) assuming there is no repeatability variation?

f. In terms of σ^2, σ^2_α, σ^2_β, and $\sigma^2_{\alpha\beta}$, what is the variance of single diameter measurements made on different lock tips made by the same operator (say Lourits)? (Hint: This is the sum of your answer to (e) and the repeatability variance, σ^2.)

g. Using the data obtained by Lourits, estimate your answer to (e).

h. Using the data obtained by Lourits, estimate your answer to (f).

2.10. Implement Hardness. Olsen, Hegstrom, and Casterton worked with a farm implement manufacturer on the hardness of a certain steel part. Before process monitoring and experimental design methodology were considered, the consistency of relevant hardness measurement was evaluated. Nine parts were obtained from a production line and three operators agreed to take part in the measuring process evaluation. Each operator made two readings on the same part (so two hardness values were recorded by each operator for each of the nine parts). The data below are in mm.

Part	Operator A	Operator B	Operator C
1	3.30	3.25	3.30
	3.30	3.30	3.30
2	3.20	3.20	3.15
	3.25	3.30	3.30
3	3.20	3.20	3.25
	3.30	3.20	3.20
4	3.25	3.20	3.20
	3.30	3.25	3.20
5	3.25	3.10	3.20
	3.30	3.10	3.15
6	3.30	3.30	3.25
	3.30	3.20	3.20
7	3.15	3.10	3.15
	3.20	3.20	3.20
8	3.25	3.20	3.20
	3.20	3.20	3.25
9	3.25	3.20	3.30
	3.30	3.30	3.40

a. Is it important to begin quality improvement efforts with an evaluation of measuring processes? Why or why not?

b. Suppose management tells engineers charged with a quality improvement project "We did a gage R&R study last year and the estimated gage capability ratio was .005. You don't need to redo the gage study." How should the engineers respond and why?

c. In the context of this problem, say what each term in equation (2.4) means.

d. What are the values for I, J, and m in this study?

e. Estimate the repeatability standard deviation, σ.

f. Estimate the reproducibility standard deviation, $\sigma_{\text{reproducibility}}$.

g. Estimate the gage capability ratio if specifications on the hardness of this part are $\pm$ .10 mm.

h. Using the corporate gage rating table given in problem (2.8), rate the repeatability and the reproducibility of the hardness measurement method.

i. Does it appear the current measuring process is adequate to check conformance to $\pm$ .10 mm hardness specifications? Why or why not?

2.11. Refer to the **Implement Hardness** case in problem (2.10).

a. Suppose each operator used a different gage to measure hardness. How would this affect the interpretation of your calculations in exercise (2.10)?

b. If it were known that the measuring of part hardness actually alters the part hardness in the vicinity of the point tested, how should this be addressed in a gage R&R study?

c. When an operator measures the same part two times in a row, it is likely the second measurement is "influenced" by the first in the sense that there is psychological pressure to produce a second measurement like the initial one. How might this affect results in a gage R&R study? How could this problem be addressed/eliminated?

2.12. Paper Weight. Everingham, Hart, Hartong, Spears, and Jobe studied the top loading balance used by the Paper Science Department at Miami University, Oxford, Ohio. Two 20 cm × 20 cm pieces of 20-lb bond paper were cut from several hundred feet of paper made in a departmental laboratory. Weights of the pieces obtained using the balance are given below in grams. The numbers in parentheses specify the order in which the measurements were made. (Piece 1 was measured 15 times, 3 times by each operator. That is, piece 1 was measured 1st by Spears, 2nd by Spears, 3rd by Hart, ... 14th by Hartong, and lastly by Jobe.) Different orders were used for pieces 1 and 2, and both were determined using a random number generator. Usually, the upper specification minus the lower specification $(U - L)$ is about 4 g/m^2 for this type of paper.

Piece	Hartong	Hart	Spears	Everingham	Jobe
1	(14) 3.481	(3) 3.448	(1) 3.485	(13) 3.475	(10) 3.472
	(12) 3.477	(9) 3.472	(2) 3.464	(4) 3.472	(5) 3.470
	(7) 3.470	(6) 3.470	(11) 3.477	(8) 3.473	(15) 3.474
2	(1) 3.258	(13) 3.245	(7) 3.256	(6) 3.249	(11) 3.241
	(2) 3.254	(12) 3.247	(5) 3.257	(15) 3.238	(8) 3.250
	(3) 3.258	(9) 3.239	(10) 3.245	(14) 3.240	(4) 3.254

a. What purpose is potentially served by randomizing the order of measurement as was done in this study?
b. Give the table of operator/piece ranges, R_{ij}.
c. Give the table of operator/piece averages, $\bar{y}_{ij}$.
d. Give the ranges of the operator/piece means, Δ_i.
e. Express the observed weight range determined by Spears for piece 2 in g/m^2.
f. Find a gage repeatability rating. (See problem (2.8a).)
g. Find a gage reproducibility rating. (See problem (2.8a).)
h. Calculate the estimated gage capability ratio.
i. What minimum value for $(U - L)$ would guarantee an estimated gage capability ratio of at most .1?

2.13. Paper Thickness. Everingham, Hart, Hartong, Spears, and Jobe continued their evaluation of the measuring devices in the Paper Science Lab at Miami University by investigating the repeatability and reproducibility of the TMI automatic micrometer routinely used to measure paper thickness. The same two 20 cm × 20 cm pieces of 20-lb bond paper referred to in problem (2.12) were used in this study. But unlike measuring weight, measuring thickness alters the properties of the portion of the paper tested (by compressing it and thus changing the thickness). So, an 8 × 8 grid was marked on each piece of paper. The corresponding squares were labeled 1, 2, . . . , 64 left to right, top to bottom. Ten squares were randomly allocated to each operator. Because so many measurements were to be made, only the "turn" for each analyst was determined randomly, and each operator made all 10 of his measurements on a given piece at once. A second randomization and corresponding order of measurement was made for piece 2.

Hartong measured 3rd on piece 1 and 5th on piece 2, Hart was 1st on piece 1 and 3rd on piece 2, Spears was 5th and 4th, Everingham was 2nd and 2nd, and Jobe was 4th and 1st. The data follow (in mm). The numbers in parenthesis identify the squares measured.

Piece	Hartong	Hart	Spears	Everingham	Jobe
1	(14) .201	(51) .195	(48) .192	(33) .183	(43) .185
	(25) .190	(54) .210	(58) .191	(38) .189	(40) .204
	(17) .190	(18) .200	(15) .198	(36) .196	(49) .194
	(21) .194	(63) .203	(55) .197	(3) .195	(12) .199
	(53) .212	(20) .196	(44) .207	(59) .192	(29) .192
	(16) .209	(50) .189	(23) .202	(45) .195	(13) .193
	(47) .208	(31) .205	(64) .196	(41) .185	(56) .190
	(42) .192	(37) .203	(57) .188	(9) .193	(2) .195
	(22) .198	(34) .195	(26) .201	(62) .194	(8) .199
	(35) .191	(7) .186	(1) .181	(5) .194	(6) .197
2	(5) .188	(14) .186	(55) .177	(43) .179	(9) .191
	(16) .173	(24) .171	(51) .174	(21) .194	(3) .180
	(11) .188	(62) .178	(36) .184	(18) .187	(42) .194
	(47) .180	(34) .175	(12) .180	(39) .175	(50) .183
	(25) .178	(29) .183	(38) .179	(6) .173	(53) .181
	(15) .188	(10) .185	(41) .186	(7) .179	(17) .188
	(56) .166	(30) .190	(63) .183	(64) .171	(33) .188
	(26) .173	(40) .177	(45) .172	(54) .184	(23) .173
	(8) .175	(58) .184	(31) .174	(59) .181	(60) .180
	(52) .183	(13) .186	(2) .178	(57) .187	(22) .176

Thus, for piece 1, Hart began the measurement procedure by recording thicknesses for squares 51, 54, 18, 63, . . . , 7, then Everingham measured squares 33, 38, . . . , 5, etc. After the data for piece 1 were obtained, measurement on piece 2 began. Jobe measured squares 9, 3, . . . , 22 then Everingham measured squares 43, 21, . . . , 57, etc.

a. In the context of this problem, say what each term in equation (2.4) means.
b. How is this study different from a "garden variety" gage R&R study?
c. Will the nonstandard feature of this study tend to increase, decrease, or have no effect on the estimate of the repeatability standard deviation? Why?
d. Will the nonstandard feature of this study tend to increase, decrease, or have no effect on the estimated standard deviation of measurements from a given piece across many operators? Why?
e. Give the estimated standard deviation of paper thickness measurements for a fixed piece/operator combination.
f. Give the estimated standard deviation of thicknesses measured on a fixed piece across many operators. (The quantity being estimated should include but not be limited to variability for a fixed piece/operator combination.)
g. What percent of the overall measurement variance is due to repeatability? What part is due to reproducibility?

2.14. Paper Burst Strength. An important property of finished paper is the force (lbs/in^2) required to burst or break through it. Everingham, Hart, Hartong, Spears, and Jobe investigated the repeatability and reproducibility of existing measurement technology for this paper property. A Mullen tester in the Miami University Paper Science Department was studied. Since the same two 20 cm $\times$ 20 cm pieces of paper referred to in problems (2.12) and (2.13) were available, the team used them in its gage R&R study for burst strength measurement. The burst test destroys the portion of paper tested, so repeat measurement of exactly the same paper specimen is not possible. Hence, a grid of 10 approximately equal-sized rectangles, 10 cm $\times$ 4 cm (each large enough for the burst tester), was marked on each large paper piece. Each of the analysts was assigned to measure burst strength on two randomly selected rectangles from each piece. The measurement order was also randomized among the five operators for each paper piece. The data obtained are below. The ordered pairs (a, b) specify the rectangle measured and the order of measurement. (For example, the ordered pair (2,9) in the top half of the table indicates that 8.8 lbs/in^2 was obtained from rectangle number 2 and was the 9th rectangle measured from piece 1.)

Piece	Hartong	Hart	Spears	Everingham	Jobe
1	(9,2) 13.5	(6,6) 10.5	(4,8) 12.9	(2,9) 8.8	(3,10) 12.4
	(7,5) 14.8	(5,1) 11.7	(1,4) 12.0	(8,3) 13.5	(10,7) 16.0
2	(3,9) 11.3	(1,8) 14.0	(5,6) 13.0	(6,7) 12.6	(2,1) 11.0
	(8,10) 12.0	(7,5) 12.5	(9,3) 13.1	(4,2) 12.7	(10,4) 10.6

The following is the ANOVA table for this study corresponding to the model (2.4).

ANOVA Table

Source	*SS*	*df*	*MS*
Piece	.5445	1	.5445
Operator	2.692	4	.6730
Piece × Operator	24.498	4	6.1245
Error	20.955	10	2.0955
Total	48.6895	19	

a. To what set of operators can the conclusions of this study apply?
b. To what set of paper pieces can the conclusions of this study correctly apply?
c. What are the numerical values of I, J, and m in this study?
d. Using the ANOVA approach briefly summarized at the end of Section 2.2.2, give an estimate of the repeatability standard deviation, σ.
e. Using the range-based approach fully developed in Section 2.2.2, give another estimate of the repeatability standard deviation, σ.
f. Using the ANOVA approach, find an estimate of $\sigma_{\text{reproducibility}}$.
g. Using the range-based approach, find another estimate of $\sigma_{\text{reproducibility}}$.
h. Using the ANOVA approach, estimate the standard deviation of burst measurements on a fixed piece of paper made by many operators, σ_{overall}.
i. Using the range-based approach, estimate the standard deviation of burst measurements on a fixed piece of paper made by many operators, σ_{overall}.

2.15. Paper Tensile Strength. The final measurement method studied by Everingham, Hart, Hartong, Spears, and Jobe in the Paper Science Lab at Miami University was that for paper tensile strength. Since the burst tests discussed in problem (2.14) destroyed the two 20 cm × 20 cm pieces of 20-lb bond paper referred to in exercises (2.12) and (2.13), two new 20 cm × 20 cm pieces of paper were selected from the same run of paper. Ten 15 mm × 20 cm strips were cut from each 20 cm × 20 cm piece. Each set of ten strips was randomly allocated to the five operators (2 strips per operator for each set of ten). The order of testing was randomized for the ten strips from each piece and the same Thwing-Albert Intellect 500 tensile tester was used by each operator to measure the load required to pull apart the strips. The data appear below in kg. (Consider, for example, the data given for piece 1, Hartong, (9,2) 4.95. A 4.95 kg load was required to tear strip number 9 from piece 1 and the measurement was taken second in order among the ten strips measured for piece 1.)

Piece	Everingham	Hart	Hartong	Spears	Jobe
1	(2,8) 4.34	(1,5) 4.34	(9,2) 4.95	(6,6) 4.03	(10,4) 4.51
	(8,10) 4.71	(4,3) 4.61	(7,7) 4.53	(3,9) 3.62	(5,1) 4.56
2	(4,7) 5.65	(6,6) 4.80	(1,1) 4.38	(2,2) 4.65	(9,5) 4.30
	(8,9) 4.51	(10,8) 4.75	(3,3) 3.89	(5,4) 5.06	(7,10) 3.87

a. Make a table of load averages, $\bar{y}_{ij}$, for all 10 operator/piece combinations.

b. Using information from all operator/piece combinations, estimate the standard deviation of the sample average load from a fixed operator/piece combination based on, say, four measurements. (The quantity to be estimated is $\sigma/\sqrt{4}$.)

c. Suppose the target tensile strength for strips of 20-lb bond paper is 4.8 kg. Typically, upper and lower specifications are set 5% above and below the target. Estimate the gage capability ratio under these conditions.

d. If upper and lower specifications for tensile strength of 20-lb bond paper are equal distances above and below a target of 4.8 kg, find the upper and lower limits such that the estimated gage capability ratio is .01.

e. Redo part (d) for an estimated gage capability ratio of .1.

f. Is it easier to make a gage capability ratio better (smaller) by increasing its denominator or decreasing its numerator? Will your answer lead to a more consistent final product? Why or why not?

2.16. Thorium Detection. In the article "Limits for Qualitative Detection and Quantitative Determination," which appeared in *Analytical Chemistry* in 1968, L. Currie reported some experimental observations in the spectrophotometric determination of thorium using thorin. The response variability from measurements on "blank" material was observed to be essentially the same as that from any (fixed) sample of interest and extensive analysis of blank material produced a standard deviation of measurement ($\sigma_{\text{measurement}}$) of around .002. The absorbance response for a field sample is typically expressed as the measured response minus that of a blank, and in these terms a response for a particular sample of interest was .006.

a. If an analyst is willing to tolerate a 5% risk of incorrectly concluding that a sample contains more thorium than is present in blank material, give the critical limit L_c. For the sample mentioned above, what conclusion does one reach using this critical limit? (Here, both n_{new} and n_{old} are 1.)

b. Suppose one can tolerate only a 1% chance of incorrectly concluding there is more thorium in a sample than in a blank. Give the corresponding critical limit and say what conclusion would be reached about a sample with a response of $y_{\text{new}} = y_{\text{old}} + .006$.

c. What risk level corresponds to a critical limit of .006?

d. What model assumptions must be made in order to answer (a) through (c)?

e. Suppose one can tolerate a 5% risk of incorrectly concluding a sample contains more thorium than a blank and the critical value from (a) will be employed. What is the value A (expressed in terms of an excess over the mean value for a blank) such that the chance of not detecting a mean absorbance of at least A (and corresponding thorium content above that of the blank material) is only 5%?

f. In the vocabulary of Section 2.2.3 what is A in part (e)?

2.17. Refer to the **Thorium Detection** case in problem (2.16).

a. Consider part (e) in the previous problem. Another analyst can tolerate only a 1% risk for both types of possible errors. Find A for this analyst.

b. Currie stated that the calibration factor used to translate absorbance readings to concentration values for thorium is about $k = 58.2$ l/g. (An absorbance value divided by k gives a corresponding concentration.) Express your value for part (e) of problem (2.16) in μg/l. (Note that 1 μg is 10^{-6} g.)

c. Express your answer to part (a) of this problem in μg/l.

d. Find an increase in thorium concentration (over that in a blank) that is 10 times the standard deviation of the difference in a field sample reading and a blank reading (that is, $10\sigma_{\text{measurement}}\sqrt{1+1}$). Express your answer in absorbance units and then in μg/l.

2.18. Carbon Atmospheric Blank. Currie, et al. presented the paper "Impact of the Chemical and Isotopic Blank on the Interpretation of Environmental Radiocarbon Results" at the International Radiocarbon Conference in Glasgow, Scotland, August 1994. Part of their presentation included discussion of six carbon content measurements made on a single urban atmospheric aerosol field filter blank. These six responses (in μg) were (in the order produced)

95.6, 73.1, 56.9, 114.4, 42.3 and 35.6.

a. Give an estimate of $\sigma_{\text{measurement}}$. (You may compute either a range-based estimate or use a sample standard deviation.) Use this estimate in the balance of this problem as if it were exactly $\sigma_{\text{measurement}}$.

b. Assuming measured carbon follows a normal distribution, find a critical limit, L_C, for determining whether a single measurement from a field sample indicates a carbon content in excess of that in a (new) blank (that will also be measured once). Use 2.5% for the probability of incorrectly deciding that there is more carbon in the field sample than in the blank.

c. What additional assumption (beyond normality) must hold for your answer in (b) to be valid?

d. Suppose now that the amount of carbon in blanks is known, i.e., every blank has the same, known, amount of carbon. Using your answer to (a) as the standard deviation of a single field sample measurement, find a critical limit for deciding if the average of two measurements on a field sample indicates carbon in excess of that in any blank. Use 1% as the largest risk of incorrectly deciding that the sample contains excess carbon that can be tolerated. (The intention here is that $n_{\text{new}} = 2$ and $n_{\text{old}} = \infty$.)

e. Find a lower limit of detection based on your estimate in (a). In making this calculation use a 5% risk of incorrectly concluding a field sample contains more carbon than a new blank (of unknown content). Also use 5% for the probability that a field sample at the lower limit of detection will fail to produce a difference exceeding the critical value. (Suppose that both the field sample and the blank will be measured once.)

f. Let the two risks in (e) be 2.5%. Find a lower limit of detection.

g. As the two risk levels decrease, what happens to the lower limit of detection? Explain in the context of this problem.

2.19. Refer to the **Carbon Atmospheric Blank** case in problem (2.18). As in problem (2.18), assume that the estimate from (2.18(a)) is in fact exactly equal to $\sigma_{\text{measurement}}$.

a. Find the increase in carbon content that is 10 times the (estimated) standard deviation of the difference between the measured carbon content for a single field sample and that from a single (new) blank. (That is, find $10\sigma_{\text{measurement}}\sqrt{1+1}$.)

b. Suppose many measurements from the same blank are available. Answer part (a) under these new conditions, if the difference of interest is between a single field sample measurement and the average of the large number of measurements on a single blank. (That is, find $10\sigma_{\text{measurement}}\sqrt{1+\frac{1}{\infty}}$.)

c. Does knowing the true carbon content of a blank affect the lower limit of detection? Why or why not? (Hint: Consider a single field sample measurement and a single blank measurement. Find the lower limit of detection using, say, 5% risks. Compute lower limits of detection both when the blank content is considered known and then when it is unknown.)

2.20. Dimethyl Phenanthrene (DMP4) Atmospheric Blank. The presentation referred to in problem (2.18) also included a discussion of six measurements of the DMP4 content of an atmospheric field filter blank. (DMP4 is a product of softwood pyrolysis.) The six responses in nanograms (10^{-9} g) were

8.25, 7.30, 7.27, 6.54, 6.75 and 7.32.

a. Suppose one analyst measured the DMP4 field filter blank all six times with the same instrument. Find the sample standard deviation of these values. In terms of the gage R&R material in Section 2.2.2, what does this sample standard deviation estimate? In parts (b) and (c) of this problem, use this estimate as if it were the exact value of $\sigma_{\text{measurement}}$.

b. Find a critical limit, L_c, if the probability of a false positive is to be 10% and both a field sample and a (new) blank are to be measured once.

c. Find a lower limit of detection where the probability of a false positive detection is 10% and the probability of a false negative is to be 5% for a true content at the lower limit of detection. (Suppose as in (b) that both a field sample and a new blank are to be measured once.)

d. Continuing as in (a) to presume that a single individual made all six measurements, what must be true about the measurement method if the sample standard deviation in (a) is to correctly be taken as an estimate of $\sigma_{\text{measurement}}$? (Consider the ideas of Section 2.2.2.)

e. Suppose the first two measurements above were taken by operator 1, the second two by operator 2, and the third two readings by operator 3, but the same blank and measuring instrument were involved in each of the three pairs of measurements. Find an appropriate estimate of $\sigma_{\text{measurement}} = \sigma_{\text{overall}}$ and in the rest of this problem use this estimate as if it were perfect. Find a critical limit if the probability of a false positive is to be 10% and a randomly selected analyst is to measure a single response on both a field sample and a new blank.

f. Continuing with the scenario in (e), find a lower limit of detection where the probability of a false positive is 10% and the probability of a false negative is 5% for true content at the lower limit of detection.

2.21. In the context of Section 2.2.3, consider a situation where the probabilities of both a false positive and a false negative (when the new mean is at the lower limit of detection) are set at .05.

a. Make a 2×2 table giving formulas for critical limits in column 1 and formulas for lower limits of detection in column 2, for the cases of one measurement from a field sample and one from a blank in row 1, and one measurement from a field sample and a large number from a single blank in row 2.

b. Add an additional column to the table in (a). Fill in this column with expressions for 10 times the standard deviation of $\overline{y}_{\text{field}} - \overline{y}_{\text{blank}}$.

2.22. Lab Carbon Blank. The following data were provided by L. A. Currie at the National Institute of Standards and Technology (NIST). The data are preliminary and exploratory but real. The unit of measure is "instrument response" and is approximately equal to one microgram of carbon. That is, 5.18 corresponds to 5.18 instrument units of carbon or 5.18 micrograms of carbon. The responses come from blank material generated in the lab.

Measured Carbon	5.18	1.91	6.66	1.12	2.79	3.91	2.87	4.72	3.68	3.54	2.15	2.82	4.38	1.64
Order of Measurement	1	2	3	4	5	6	7	8	9	10	11	12	13	14

a. Plot measured carbon content versus order of measurement.

b. The data are ordered in time, but intervals between measurements are not equal and an appropriate plan for obtaining data was not necessarily in place. What feature of the plot in (a) might still have meaning?

c. If one treats the measurement of lab-generated blank material as repeated measurements of a single blank, what does a trend on a plot like that in (a) suggest regarding $\sigma_{\text{repeatability}}$? (Assume the plot is made from data equally spaced in time and collected by a single individual.)

d. Make a frequency histogram of these data with categories 1.00–1.99, 2.00–2.99, etc.

e. What could be missed if only a histogram was constructed (and one didn't make a plot like that in (a)) for data like these?

CHAPTER 3

Process Monitoring Part 1: Basics

This chapter begins discussion of the important topic of process monitoring using so-called "control charts." These are devices for the routine and organized plotting of process performance measures, with the intention of identifying process changes so that those running the process can either intervene and set things aright (if the change is detrimental) or identify the source of an unexpected process improvement.

The treatment begins with some generalities of basic control charting philosophy. Then the standard Shewhart control charts for both measurements/"variables data" and counts/"attributes data" are presented in consecutive sections. A fourth section discusses qualitative interpretation and practical implications of patterns sometimes seen on Shewhart charts, and some sets of rules often applied to check for such patterns. Then there is a discussion of the so-called Average Run Length concept that can be used to quantify what a particular process monitoring scheme can be expected to provide. Finally, the chapter closes with a section aimed at clarifying the relationship between "statistical process control" and "engineering control" and presenting some basic concepts of so-called PID engineering control schemes.

3.1 GENERALITIES ABOUT SHEWHART CONTROL CHARTING

The discussion in Section 1.2 introduced the notion of process stability as a kind of consistency over time in the pattern of process variation. Walter Shewhart, working at Bell Labs in the late 1920s and early 1930s, developed an extremely powerful (and deceptively simple) tool for use in investigating whether a process can be sensibly thought of as stable. He called the tool a "control chart." Nearly 70 years after the fact, your authors would prefer (for reasons that are laid out in Section 3.6) that Shewhart had chosen instead the more appropriate term "monitoring chart." Nevertheless, this book will use Shewhart's terminology and the monitoring chart terminology interchangeably and now begins an exposition of Shewhart's basic thinking.

Shewhart's fundamental conceptualization was that while some variation is inevitable in any real process, the variation seen in data taken on a process can be thought of as decomposable as

$$\begin{aligned}\text{overall observed variation} = {} & \text{baseline variation} \\ & + \text{variation that can be eliminated.}\end{aligned} \tag{3.1}$$

Shewhart conceived of baseline variation as that variability in production and measurement which will remain even under the most careful process monitoring and appropriate physical intervention. It is an inherent property of a particular combination of system configuration and measurement methodology that cannot be reduced without basic changes in the physical process or how it is run or observed. This variation is sometimes called variation due to "system" or "common" (universal) causes. Other names for it that will be used in this book are "random" or "short-term" variation. It is the kind of variation that one should expect to experience under the best of circumstances, measuring item to consecutive item produced on a production line. It is variation that comes from many small, unnameable, and unrecognized physical causes. When only this kind of variation is present, it is reasonable to call a process "stable" and model observations as independent random draws from a fixed population or universe.

The second component of overall variability portrayed in equation (3.1) is that which can potentially be eliminated by careful process monitoring and wise physical intervention (when such is warranted). This has variously been called "special cause" or "assignable" cause variation, "nonrandom" and "long-term" variation. It is the kind of systematic, persistent change that accompanies real (typically unintended) physical alteration of a process (or the measurement system used to observe it). It is change that is large enough that one can potentially track down and eliminate its root cause, leaving behind a stable process.

If one accepts Shewhart's conceptualization (3.1), the problem then becomes one of detecting the presence of the second kind of variation so that appropriate steps can be taken to eliminate it. The point of Shewhart control charting is to provide a detection tool.

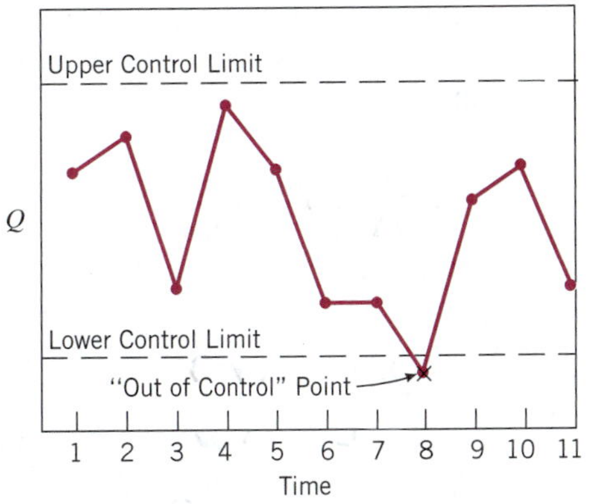

Figure 3.1
Generic Shewhart control chart (for a statistic Q).

The basic working of Shewhart's charting method is this. One periodically takes samples from the process of interest (more will be said later about the timing and nature of these samples) and computes a statistic meant to summarize process behavior at the period in question. These values are plotted against time order of observation and compared to so-called **control limits** drawn on the chart. These in some sense separate values of the statistic that are plausible if the process is in fact stable, from those that are rare or implausible under this scenario. As long as the plotted points remain inside the control limits, one presumes that all is well (the process is stable) and does not intervene in its workings. (This is an oversimplification of how these charts are often used that will be corrected in Section 3.4. But for the time being this simplified picture will suffice.) When a point plots outside control limits, one has an indication that a physical change has probably taken place and that intervention is appropriate. Figure 3.1 is a plot of a generic Shewhart control chart where the plotted statistic is Q, upper and lower control limits are UCL_Q and LCL_Q respectively, and there is one "out of control" point.

Control Limits

There are many different kinds of Shewhart charts, corresponding to various choices of the plotted statistic, Q. Some of these specific chart types will be discussed in the next two sections. But before moving to discussion of specific charts, a number of generalities remain to be considered. For one thing, there is the question of how one sets the control limits, UCL_Q and LCL_Q.

Shewhart's suggestion for setting control limits was essentially the following. If one can model the process output under stable conditions (i.e., if one can specify a sensible probability distribution for individual observations made on the process) then probability theory can often be invoked to produce a distribution for Q under stable conditions. Then small upper and lower percentage points for this distribution can be taken as providing the necessary control limits. The thinking is that only rarely will values outside these be seen under stable process conditions. Further, rather than working explicitly with probability tables or formulas for a distribution of Q, one often simply makes use of the fact that for many probability distributions most of the probability is within three standard deviations of the mean. So, if μ_Q and σ_Q are respectively a stable process mean and standard deviation for Q, then common control limits are

Generic 3-sigma Control Limits

$$UCL_Q = \mu_Q + 3\sigma_Q \quad \text{and} \quad LCL_Q = \mu_Q - 3\sigma_Q. \tag{3.2}$$

Further, it is common to draw in a "center line" on a Shewhart control chart at

$$CL_Q = \mu_Q. \tag{3.3}$$

To make this discussion slightly more concrete, consider briefly the situation where the plotted statistic is the sample mean of n individual measurements, $Q = \overline{x}$. If the process output can be modeled as independent selections from a distribution with mean μ and standard deviation σ, facts from elementary probability can be invoked to conclude that the statistic $\overline{x}$ has a distribution with mean $\mu_Q = \mu_{\overline{x}} = \mu$ and standard deviation $\sigma_Q = \sigma_{\overline{x}} = \sigma/\sqrt{n}$. Then applying relationships (3.2) and (3.3) it follows that typical control limits for $\overline{x}$ are

$$UCL_{\overline{x}} = \mu + 3\frac{\sigma}{\sqrt{n}} \text{ and } LCL_{\overline{x}} = \mu - 3\frac{\sigma}{\sqrt{n}} \tag{3.4}$$

with a center line drawn at μ.

The nature of display (3.4) helps bring into focus another general issue regarding Shewhart control charting. That is the fact that very often process parameters appear in formulas for control limits, and values for them must come from somewhere if one is going to apply a control chart. There are two possibilities in this regard. Sometimes past experience with a process, engineering standards, or other considerations made prior to a particular application specify what values should be used. This kind of situation is commonly known as a **standards given** scenario. In other circumstances, one has no information on a process outside a series of samples that are presented along with the question "Is it plausible that the process was physically stable over the period represented by these data?" In such a case, all that one can do is tentatively assume that in fact the process was stable, make provisional estimates of process parameters and plug them into formulas for control limits, and apply those limits to the data in hand as a means of criticizing the tentative assumption of stability. This kind of situation is sometimes called an **as past data** scenario and will often be referred to in this text as a **retrospective** scenario.

Standards Given Context

Retrospective Context

The difference between what is possible in standards given and retrospective contexts can be thought of in terms of two different questions addressed in the two situations. In a standards given context, one is in a position with each new sample to face the question

"Are process parameters currently at their standard values?"

In a retrospective context, one can only wait until a number of samples have been collected (often, using data from a minimum of 20–25 time periods is recommended) and then looking back over the data ask the question

"Are these data consistent with any fixed set of process parameters?"

Having introduced the notion of control limits, it is important to make an initial attempt to warn readers of a very unfortunate and persistent type of error in logic often related to their use. That is the confusion that students (and even practicing engineers) often have regarding the *much different* concepts of control limits and engineering specifications. Control limits have to do with assessing process stability. They refer to a statistic Q. They are usually derived from what a process has done in the past or is currently doing. On the other hand, engineering specifications have to do with assessing product acceptability. They almost always refer to individual measurements. They are usually derived from product performance requirements and may have little or nothing to do with the inherent capability of a process to produce a product meeting those requirements.

Control Limits versus Engineering Specifications

Despite these huge differences in meaning, it is all too common to see people confuse these concepts (for example, applying specifications to sample means as if they were control limits, or arguing that since a mean or individual is inside control limits for $\bar{x}$, the product being monitored is acceptable). It is vital that these concepts be kept separate and applied in their proper contexts. (Notice that a process that is stable and producing Q's inside appropriate control limits need *not* be producing mostly acceptable product. And conversely, a process may produce product that is acceptable by current engineering standards, but nevertheless be very unstable!)

Another issue that needs some general discussion before turning to the details of making specific Shewhart control charts is the matter of sampling, how one should go about gathering the data to be used in control charting. This matter is sometimes referred to as that of **rational subgrouping** or rational sampling. When one is collecting process-monitoring data, it is important that anything one intends to call a single "sample" be collected over a short enough time span that there is little question that the process was physically stable during the data collection period. It must be clear that an "independent draws from a single population/universe" model is appropriate for describing data in a given sample. This is because the variation within such a sample essentially specifies the level of background noise against which one looks for process changes. If what one calls "samples" often contain data from genuinely different process conditions, the apparent level of background noise will be so large that it will be hard to see important process changes. In high-volume manufacturing applications of control charts, single samples (rational subgroups) typically consist of n consecutive items taken from a production line. On the other hand, in extremely low-volume operations, where one unit might take many hours to produce and there is significant opportunity for real process change between consecutive units, the only natural samples may be of size $n = 1$.

Once one has determined to group only observations close together in time into samples or subgroups, there is still the matter of how often these samples should be taken. When monitoring a machine that turns out 1000 parts per hour, where samples are going to consist of $n = 5$ consecutive parts produced on the machine, does one sample once every minute, once every hour, once every day, or what? An answer to this kind of question of course depends upon what one expects in terms of process performance, and the consequences of process changes. If the consequences of a process change are disastrous, one is pushed toward frequent samples. The same is true if significant process upsets are a fairly frequent occurrence. On the other hand, if a process rarely experiences changes and even when those occur only a moderate loss is incurred should it take a while to discover them, long intervals between samples are sensible. Various operations-research type attempts have been made to provide quantitative guidelines in answer to this sampling frequency question, but these have proved largely unsatisfactory for practice. Nevertheless, the qualitative matters noted here clearly need to be the major considerations as one tries to find an economically feasible and effective sampling frequency for process monitoring.

As a final matter in this introductory discussion of Shewhart charting philosophy, there should probably come some reflection on what control charting can and cannot reasonably be expected to provide. It can signal the need for process intervention and can keep one from ill-advised and detrimental overadjustment of a process that is behaving in a stable fashion. But in doing so, what is achieved is simply reducing variation to the minimum possible for a given system configuration (in terms of equipment, methods of operation, methods of measurement, etc.). Once that minimum has been reached, all that is accomplished is maintaining a *status quo* best possible process performance. In today's global economy, standing still is never good enough for very long. Achieving process stability provides a solid background against which to evaluate possible innovations and fundamental/order-of-magnitude improvements in production methods. But it does not itself guide their discovery. Of the tools discussed in this book, it is the methods of experimental design and analysis covered in Chapters 6 and 7 that have the most to say about aiding fundamental innovations.

3.2 SHEWHART CHARTS FOR MEASUREMENTS/"VARIABLES DATA"

This section considers the problem of process monitoring when the data available for monitoring purposes are measurements. (Sometimes the terminology "variables data" is used in this context.) In such situations, it is common to make charts for both the process location and also for the process spread (size of the process short-term variability). So this section will consider the making of $\bar{x}$ and median ($\tilde{x}$) charts for location, and R and s charts for spread.

3.2.1 Charts for Process Location

Without doubt the most common of all Shewhart control charts is that for means of samples of n measurements, the case where $Q = \bar{x}$. As was discussed in the previous section (and portrayed in display (3.4)), the fact that sampling from a distribution with mean μ and standard deviation σ produces sample averages with expected value $\mu_{\bar{x}} = \mu$ and standard deviation $\sigma_{\bar{x}} = \sigma/\sqrt{n}$ suggests **standards given** Shewhart control limits for $\bar{x}$

$$UCL_{\bar{x}} = \mu + 3\frac{\sigma}{\sqrt{n}} \quad \text{and} \quad LCL_{\bar{x}} = \mu - 3\frac{\sigma}{\sqrt{n}}, \tag{3.5}$$

Standards Given $\bar{x}$ Chart Control Limits

and center line at

$$CL_{\bar{x}} = \mu.$$

Standards Given $\bar{x}$ Chart Center Line

Example 3.1 Monitoring the Surface Roughness of Reamed Holes. Dohm, Hong, Hugget, and Knoot worked with a manufacturer on a project involving roughness measurement after the reaming of preformed holes in a particular metal part. Table 3.1 contains some summary statistics (the sample mean $\bar{x}$, the sample median $\tilde{x}$, the sample range R, and the sample standard deviation s) for 20 samples (taken over a period of 10 days) of $n = 5$ consecutive reamed holes.

Suppose for the time being that standards (established on the basis of previous experience with this reaming process) for surface roughness after reaming these holes are $\mu = 30$ and $\sigma = 4$. Then, standards given control limits for the $\bar{x}$ values in Table 3.1 are

$$UCL_{\bar{x}} = 30 + 3\frac{4}{\sqrt{5}} = 35.37$$

and

$$LCL_{\bar{x}} = 30 - 3\frac{4}{\sqrt{5}} = 24.63.$$

Figure 3.2 is a standards given $\bar{x}$ chart for the surface roughness measurements. It is clear from this chart that as early as the second sample one would detect the fact that the reaming process is not stable at standard process parameters. Several of the sample means fall outside control limits, and had the control limits been applied to the data as they were collected, the need for physical intervention would have been signaled.

Table 3.1 Summary Statistics for 20 Samples of 5 Surface Roughness Measurements on Reamed Holes (μ-inches)

Sample	$\bar{x}$	$\tilde{x}$	R	s
1	34.6	35	9	3.4
2	46.8	45	23	8.8
3	32.6	34	12	4.6
4	42.6	41	6	2.7
5	26.6	28	5	2.4
6	29.6	30	2	0.9
7	33.6	31	13	6.0
8	28.2	30	5	2.5
9	25.8	26	9	3.2
10	32.6	30	15	7.5
11	34.0	30	22	9.1
12	34.8	35	5	1.9
13	36.2	36	3	1.3
14	27.4	23	24	9.6
15	27.2	28	3	1.3
16	32.8	32	5	2.2
17	31.0	30	6	2.5
18	33.8	32	6	2.7
19	30.8	30	4	1.6
20	21.0	21	2	1.0

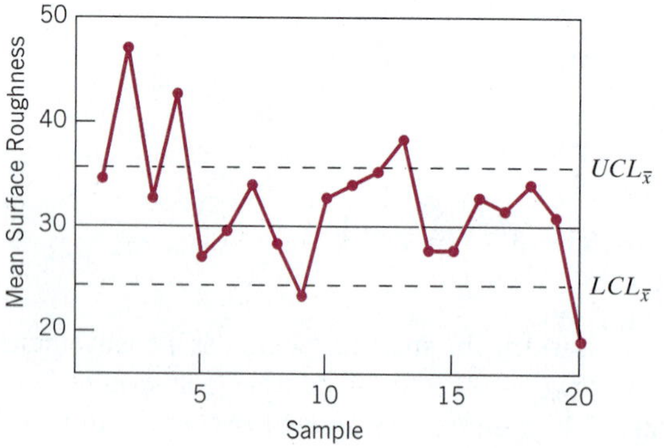

Figure 3.2
Standards given $\bar{x}$ chart for surface roughness.

In order to make a **retrospective** $\bar{x}$ chart one must derive estimates of the process parameters μ and σ from data in hand (temporarily assuming process stability) and plug them into the formulas (3.5). There are many possible ways of doing this, each leading to slightly different retrospective control limits. Here only the most common ones will be considered and we begin with the matter of estimating μ.

Let r stand for the number of samples available in a retrospective $\bar{x}$ chart analysis. A most natural (at least if all the sample sizes are the same) way of estimating a supposedly common process mean for the r periods is to simply average the r sample means. That is, standard control charting practice is to use

$$\bar{\bar{x}} = \frac{1}{r}\sum_{i=1}^{r} \bar{x}_i$$

Average Sample Mean

as an estimator of μ in making retrospective control limits for $\bar{x}$.

An answer to the question of how to estimate σ is perhaps not quite so obvious. The estimator of σ with the best theoretical properties is obtained by pooling the r sample variances to obtain (in the constant sample size case)

$$s^2_{\text{pooled}} = \frac{1}{r}\sum_{i=1}^{r} s_i^2,$$

and then taking the square root. However, this method is rarely used in practice (due at least in part to historical precedent). Instead, common practice is to use estimators based on the average sample range or the average sample standard deviation.

Consider first the estimation of σ based on

$$\bar{R} = \frac{1}{r}\sum_{i=1}^{r} R_i.$$

Average Sample Range

As in the discussion of gage R&R in Section 2.2, it is the case that if process output is normally distributed at time period i,

$$\mathrm{E}R_i = d_2\sigma$$

and thus

$$\mathrm{E}\left(\frac{R_i}{d_2}\right) = \sigma.$$

(The dependence of d_2 on n is not being displayed here, since there is no chance of confusion regarding which "sample size" is under discussion.) So assuming the process is stable over all r periods, all sample sizes are n, and that a normal distribution governs the data generation process,

$$\frac{\overline{R}}{d_2}$$

is a sensible estimator of σ. Plugging this and $\overline{\overline{x}}$ into the standards given control limits for $\overline{x}$ provided in display (3.5) one obtains retrospective Shewhart control limits for $\overline{x}$,

$$UCL_{\overline{x}} = \overline{\overline{x}} + 3\frac{\overline{R}}{d_2\sqrt{n}} \quad \text{and} \quad LCL_{\overline{x}} = \overline{\overline{x}} - 3\frac{\overline{R}}{d_2\sqrt{n}}. \tag{3.6}$$

Further, one can define a constant A_2 (depending upon n) by

$$A_2 = \frac{3}{d_2\sqrt{n}},$$

and rewrite display (3.6) more compactly as

Retrospective Control Limits for $\overline{x}$ Based on the Average Range

$$UCL_{\overline{x}} = \overline{\overline{x}} + A_2\overline{R} \quad \text{and} \quad LCL_{\overline{x}} = \overline{\overline{x}} - A_2\overline{R}. \tag{3.7}$$

Values of A_2 can be found in the table of control chart constants, Table A.1.

As an alternative to estimating σ on the basis of sample ranges, consider next estimating σ based on the average sample standard deviation,

Average Sample Standard Deviation

$$\overline{s} = \frac{1}{r}\sum_{i=1}^{r} s_i.$$

It turns out that when sampling from a normal distribution with standard deviation σ, the sample standard deviation, s, has a mean that is not quite σ. The ratio of the mean of s to σ is commonly called c_4. (c_4 depends upon the sample size, but it will not be necessary or helpful to display the dependence of c_4 on n.) Thus, if one assumes that process output is normally distributed at period i,

$$\mathrm{E}\left(\frac{s_i}{c_4}\right) = \sigma.$$

So assuming the process is stable over all r periods, all sample sizes are n, and that a normal distribution governs the data generation process,

$$\frac{\bar{s}}{c_4}$$

is a sensible estimator of σ. Plugging this and $\bar{\bar{x}}$ into the standards given control limits for $\bar{x}$ provided in display (3.5) one obtains *retrospective* Shewhart control limits for $\bar{x}$,

$$UCL_{\bar{x}} = \bar{\bar{x}} + 3\frac{\bar{s}}{c_4\sqrt{n}} \quad \text{and} \quad LCL_{\bar{x}} = \bar{\bar{x}} - 3\frac{\bar{s}}{c_4\sqrt{n}}. \tag{3.8}$$

Further, one can define another constant A_3 (depending upon n) by

$$A_3 = \frac{3}{c_4\sqrt{n}},$$

and rewrite display (3.8) more compactly as

$$UCL_{\bar{x}} = \bar{\bar{x}} + A_3\bar{s} \quad \text{and} \quad LCL_{\bar{x}} = \bar{\bar{x}} - A_3\bar{s}. \tag{3.9}$$

Retrospective Control Limits for $\bar{x}$ Based on the Average Standard Deviation

Values of A_3 can also be found in the table of control chart constants, Table A.1.

Example 3.1 continued. Returning to the reaming study, it is straightforward to verify from Table 3.1 that

$$\bar{\bar{x}} = 32.1, \qquad \bar{R} = 8.95, \qquad \text{and} \qquad \bar{s} = 3.76.$$

Further, for $n = 5$ (which was the sample size used in the study) Table A.1 shows that $A_2 = .577$ and $A_3 = 1.427$. Thus, from formulas (3.7), retrospective control limits for $\bar{x}$ based on $\bar{R}$ are

$$UCL_{\bar{x}} = 32.1 + .577(8.95) = 37.26 \quad \text{and} \quad LCL_{\bar{x}} = 32.1 - .577(8.95) = 26.94.$$

And from formulas (3.9), retrospective control limits for $\bar{x}$ based on $\bar{s}$ are

$$UCL_{\bar{x}} = 32.1 + 1.427(3.76) = 37.47 \text{ and } LCL_{\bar{x}} = 32.1 - 1.427(3.76) = 26.73.$$

Figure 3.3 shows the retrospective $\bar{x}$ control chart with control limits based on $\bar{R}$. It is clear from this chart (as it would be using the limits based on $\bar{s}$) that

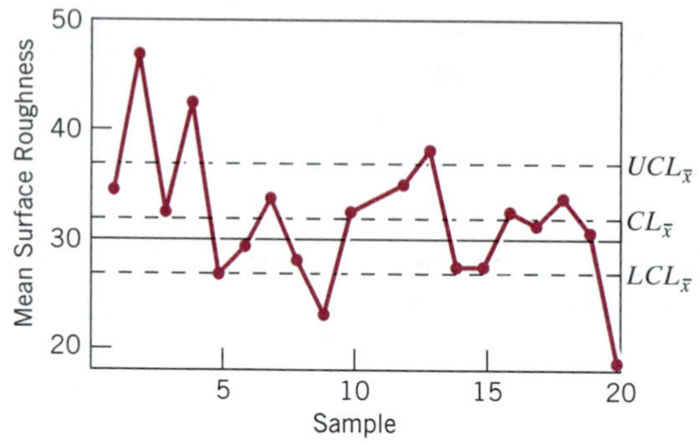

Figure 3.3
Retrospective $\bar{x}$ chart for surface roughness.

the reaming process was not stable over the period of the study. The mean measured roughness fluctuated far more than one would expect under a stable process model.

$\bar{x}$ charts are by far the most common charts for monitoring process location, but there is an alternative worth mentioning. That is to use **sample medians** in place of sample means ($\tilde{x}$ in place of $\bar{x}$). This alternative has the advantage of requiring less in the way of computational skills from those who must compute the values to be plotted, but has the drawback of generally being somewhat less sensitive to changes in process location than the $\bar{x}$ chart.

The basic probability facts that lead to control limits for $\tilde{x}$ concern sampling from a normal distribution. For a sample of size n from a normal distribution with mean μ and standard deviation σ, the random variable $\tilde{x}$ has mean $\mu_{\tilde{x}} = \mu$ and standard deviation $\sigma_{\tilde{x}} = \kappa\sigma_{\bar{x}} = \kappa\sigma/\sqrt{n}$ for a constant κ (depending upon n). Table 3.2 gives a few values of κ.

Applying these basic facts about the probability distribution of $\tilde{x}$ under a normal process model and the generic Shewhart control limits given in display (3.2), one is led to standards given control limits for $\tilde{x}$

Standards Given Control Limits for Medians

$$UCL_{\tilde{x}} = \mu + 3\kappa\frac{\sigma}{\sqrt{n}} \quad \text{and} \quad LCL_{\tilde{x}} = \mu - 3\kappa\frac{\sigma}{\sqrt{n}}. \tag{3.10}$$

And retrospective limits can be made by replacing μ and σ with any sensible estimates.

Table 3.2 Ratios κ Between $\sigma_{\tilde{x}}$ and $\sigma_{\bar{x}}$ When Sampling from a Normal Distribution

n	3	5	7	9	11	∞
κ	1.160	1.197	1.214	1.223	1.229	$\sqrt{\pi/2}$

Example 3.1 continued. Returning to the reaming study, suppose once more that process standards are $\mu = 30$ and $\sigma = 4$. Then for samples of size $n = 5$, like those used in the students' project, control limits for sample medians are

$$UCL_{\tilde{x}} = 30 + 3(1.197)\frac{4}{\sqrt{5}} = 36.42$$

and

$$LCL_{\tilde{x}} = 30 - 3(1.197)\frac{4}{\sqrt{5}} = 23.58.$$

Comparing the sample medians in Table 3.1 to these limits, it is again clear that had these limits been applied to the data as they were collected, the need for physical intervention would have been signaled as early as the second sample.

3.2.2 Charts for Process Spread

This book's exposition of control charts for measurements began with the $\bar{x}$ chart for location because it is surely the single most commonly used process monitoring tool, and because facts from elementary probability can be invoked to quickly motivate the notion of control limits for $\bar{x}$. However, in practice it is often important to deal *first* with the issue of consistency of process spread before going on to consider consistency of process location. After all, such consistency of spread (constancy of σ) is already implicitly assumed when one sets about to compute control limits for $\bar{x}$. So it is important to now consider charts intended to monitor this aspect of process behavior. The discussion here will center on charts for ranges and standard deviations, beginning with the **range chart**.

In deriving $\bar{R}/d_2$ as an estimator of σ for use in making retrospective $\bar{x}$ charts, the fact that when sampling from a normal universe with mean μ and standard deviation σ,

$$\mathrm{E}R = \mu_R = d_2\sigma \tag{3.11}$$

has already been employed. The same kind of mathematics that stands behind relationship (3.11) can be used to also derive a standard deviation to associate with R based on a sample from a normal population. (This is a measure of spread for the probability distribution of R, which is itself a measure of spread of the sample.) It turns out that the standard deviation of R is proportional to σ. The constant of proportionality is called d_3 and is tabled in Table A.1. (Again, d_3 depends on n, but it will not be useful to display that dependence here.) That is,

$$\sigma_R = d_3\sigma. \tag{3.12}$$

Now the relationships (3.11) and (3.12) together with the generic formula for Shewhart control limits given in display (3.2) and center line given in display (3.3) quickly imply that standards given control limits for R are

$$UCL_R = (d_2 + 3d_3)\sigma \quad \text{and} \quad LCL_R = (d_2 - 3d_3)\sigma \tag{3.13}$$

with a center line at

Standards Given R Chart Center Line

$$CL_R = d_2\sigma. \tag{3.14}$$

Further, if one adopts the notations $D_2 = d_2 + 3d_3$ and $D_1 = d_2 - 3d_3$ the relationships (3.13) can be written somewhat more compactly as

Standards Given R Chart Control Limits

$$UCL_R = D_2\sigma \quad \text{and} \quad LCL_R = D_1\sigma. \tag{3.15}$$

Values of the constants D_1 and D_2 may again be found in the table of control chart constants, Table A.1.

It is instructive to contemplate the tabled values of D_1. Notice that there are no tabled values for sample sizes $n \leq 6$. The reason is that for such sample sizes, the difference $d_2 - 3d_3$ turns out to be negative. Since ranges are nonnegative, a negative lower control limit would make no sense. So standard practice for $n \leq 6$ is to use no lower control limit for R.

Consider also the implications of the fact that for $n > 6$, one typically employs a positive lower control limit for R. This means that it is possible for an R chart to signal an "out of control" situation because R is *too small*. This fact sometimes causes students consternation. After all, isn't the goal to produce *small* variation? Then why signal an alarm when R is small? The answer to this conundrum lies in remembering precisely what a control chart is meant to detect, namely *process instability/change*. It is possible for unintended causes to occasionally act on a process to reduce variability. A lower control limit on an R chart simply allows one to detect such happy events. If one can detect such a change and identify its physical source, there is the possibility of making that assignable cause part of standard practice and the accompanying decrease in σ permanent. So, the practice of using positive lower control limits for R when n is sufficiently big is one that makes perfectly good practical sense.

Example 3.1 continued. Consider once more the reaming example of Dohm, Hong, Hugget, and Knoot from a standards given perspective with $\sigma = 4$. For samples of size $n = 5$, Table A.1 provides the values $d_2 = 2.326$ and $D_2 = 4.918$. So using formulas (3.14) and (3.15), standards given control chart values for R are

$$UCL_R = 4.918(4) = 19.7 \quad \text{and} \quad CL_R = 2.326(4) = 9.3.$$

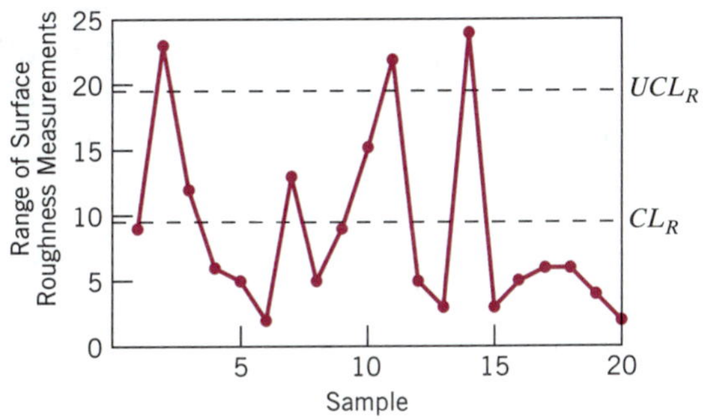

Figure 3.4
Standards given R chart for surface roughness.

Figure 3.4 is the corresponding standards given control chart for the students' ranges. There are three out-of-control points on the chart, the first coming as early as the second sample. The reaming process did not behave in a manner consistent with the $\sigma = 4$ standard over the period of the study. Samples 2, 11, and 14 simply have too much internal variability to make consistency of σ at the value 4 believable. One wonders if perhaps the reamer was changed in the middle of these samples, with the effect that some holes were very rough while others were very smooth.

Retrospective control limits for R come about by plugging an estimate for σ derived from samples in hand into the formulas (3.14) and (3.15). A particularly natural choice for an estimator of σ in this context is $\overline{R}/d_2$. Substituting this into relationship (3.14), one gets the perfectly obvious retrospective center line for an R chart,

$$CL_R = \overline{R}. \tag{3.16}$$

Retrospective R Chart Center Line

Further, substituting $\overline{R}/d_2$ into equations (3.15) for σ, one gets retrospective control limits for R

$$UCL_R = D_2\left(\frac{\overline{R}}{d_2}\right) \quad \text{and} \quad LCL_R = D_1\left(\frac{\overline{R}}{d_2}\right). \tag{3.17}$$

And if one adopts the notations $D_4 = D_2/d_2$ and $D_3 = D_1/d_2$, it is possible to write the relationships (3.17) more compactly as

$$UCL_R = D_4\overline{R} \quad \text{and} \quad LCL_R = D_3\overline{R}. \tag{3.18}$$

Retrospective R Chart Control Limits

As is by now to be expected, the constants D_3 and D_4 are tabled in Table A.1. And, of course, the table contains no values of D_3 for $n \leq 6$.

Example 3.1 continued. Recall that the 20 samples in Table 3.1 have $\overline{R} = 8.95$ and note that for $n = 5$ one has $D_4 = 2.115$. So from displays (3.16) and (3.18) a retrospective control chart for the ranges (based on $\overline{R}/d_2$ as an estimator of σ) has center line at

$$CL_R = \overline{R} = 8.95$$

and upper control limit

$$UCL_R = D_4\overline{R} = 2.115(8.95) = 18.9.$$

A plot of this retrospective R chart would look very little different from Figure 3.4. The same three ranges plot outside control limits. The conclusion one reaches is that not only is a "σ constant at 4" view of the students' data not plausible, but neither is a "σ constant at some value" view. There is solid evidence of reaming process instability in the ranges of Table 3.1. The short-term process variability changes over time.

The R chart is the most common Shewhart control chart for monitoring process spread. It has the virtue of requiring very little from its user in the way of calculations, is based on a statistic that is very easy to understand, and is firmly entrenched in quality assurance practice dating from the days of Shewhart himself. There is, however, an alternative to the R chart that tends to detect changes in process spread more quickly, at the price of increased computational complexity. Where the quantitative sophistication of a user is high and calculations are not a problem, the ***s* chart** should be considered as a viable competitor for the R chart.

The fact that when sampling from a normal distribution the mean of s is c_4 times the process standard deviation, that is,

$$\mathrm{E}s = c_4\sigma, \tag{3.19}$$

has already proved useful when making retrospective control limits for $\overline{x}$ based on $\overline{s}$. As it turns out, the same kind of mathematics that leads to relationship (3.19) can be used to find the standard deviation of s based on a sample of size n from a normal universe. (This is a measure of spread for the probability distribution of the random variable s, that is itself a measure of spread of the sample.) It happens that this standard deviation is a multiple of σ. The multiplier is called c_5 and it turns out that $c_5 = \sqrt{1 - c_4^2}$. That is,

$$\sigma_s = \sigma\sqrt{1 - c_4^2} = c_5\sigma. \tag{3.20}$$

Now relationships (3.19) and (3.20) together with the generic Shewhart control limits and center line specified in displays (3.2) and (3.3) lead immediately to standards given control limits and center line for an s chart. That is,

$$UCL_s = (c_4 + 3c_5)\sigma \quad \text{and} \quad LCL_s = (c_4 - 3c_5)\sigma \tag{3.21}$$

and

$$CL_s = c_4\sigma. \tag{3.22}$$

Standards Given s Chart Center Line

Further, if one adopts the notations $B_6 = c_4 + 3c_5$ and $B_5 = c_4 - 3c_5$, the relationships (3.21) can be written as

$$UCL_s = B_6\sigma \quad \text{and} \quad LCL_s = B_5\sigma. \tag{3.23}$$

Standards Given s Chart Control Limits

Values of the constants B_5 and B_6 may again be found in the table of control chart constants, Table A.1. Note that for $n \leq 5$ there are no values of B_5 given in Table A.1. This is because for such sample sizes $c_4 - 3c_5$ is negative. For $n > 5$, B_5 is positive, allowing the s chart to provide for detection of a decrease in σ (just as is possible with an R chart and $n > 6$).

Retrospective control limits for s can be made by substituting any sensible estimate of σ into the standards given formulas (3.22) and (3.23). A particularly natural choice for an estimator of σ in this context is $\bar{s}/c_4$. Substituting this into relationship (3.22), one gets the obvious retrospective center line for an s chart

$$CL_s = \bar{s}. \tag{3.24}$$

Retrospective s Chart Center Line

Further, substituting $\bar{s}/c_4$ into equations (3.23) for σ, one gets retrospective control limits for s

$$UCL_s = B_6\left(\frac{\bar{s}}{c_4}\right) \quad \text{and} \quad LCL_s = B_5\left(\frac{\bar{s}}{c_4}\right). \tag{3.25}$$

And if one adopts the notations $B_4 = B_6/c_4$ and $B_3 = B_5/c_4$, it is possible to write the relationships (3.25) more compactly as

$$UCL_s = B_4\bar{s} \quad \text{and} \quad LCL_s = B_3\bar{s}. \tag{3.26}$$

Retrospective s Chart Control Limits

As usual, the constants B_3 and B_4 are tabled in Table A.1. And, of course, the table contains no values of B_3 for $n \leq 5$.

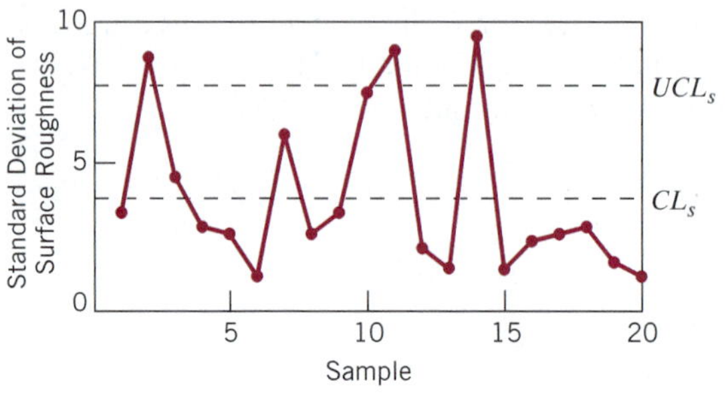

Figure 3.5
Retrospective s chart for surface roughness.

Example 3.1 continued. The 20 samples in Table 3.1 have $\bar{s} = 3.76$. Note that for $n = 5$ one has $B_4 = 2.089$. So from displays (3.24) and (3.26) a retrospective control chart for the standard deviations (based on $\bar{s}/c_4$ as an estimator of σ) has center line at

$$CL_s = \bar{s} = 3.76$$

and upper control limit

$$UCL_s = B_4\bar{s} = 2.089(3.76) = 7.85.$$

Figure 3.5 is a retrospective s chart for the sample standard deviations of Table 3.1. It carries the same message as does a retrospective analysis of the sample ranges for this example. Not only is a "σ constant at 4" view of the students' data not plausible, neither is a "σ constant at some value" view. There is solid evidence of reaming process instability in the standard deviations of Table 3.1.

3.3 SHEWHART CHARTS FOR COUNTS/"ATTRIBUTES DATA"

The control charts for measurements introduced in Section 3.2 are the most important of the Shewhart control charts. Where it is at all possible to make measurements, they will almost always provide more information on process behavior than will a corresponding number of qualitative observations. However, there are occasions where only attributes data can be collected, so this section presents Shewhart control charting methods for such cases. That is, the section considers charting *counts* and corresponding *rates* of occurrence for nonconforming items (or defectives) and for nonconformities (or defects). The case of so-called np charts and p charts for "percent nonconforming" (or percent defective) contexts is treated first. Then follows a discussion of c and u charts for "nonconformities per unit" (or defects per unit) situations.

3.3.1 Charts for Fraction Nonconforming

Consider now a situation where a process is turning out a (final or intermediate) product and one periodically samples n items and (making careful use of operational definitions) classifies each one as "nonconforming" or "conforming." (The old terminology for these outcomes is "defective" and "nondefective." The newer terminology is used in recognition of the fact that some kinds of failures to meet inspection criteria do not render a product functionally deficient. There is probably also some reluctance in today's litigious society to ever admit that anything made by an organization could possibly be "defective.")

Then let

$$X = \text{the number of nonconforming items in a sample of } n \text{ items} \tag{3.27}$$

and

$$\hat{p} = \frac{X}{n} = \text{the fraction of a sample of } n \text{ items that is nonconforming.} \tag{3.28}$$

Shewhart ***np* charts** are for the plotting of $Q = X$, and ***p* charts** are for the monitoring of $Q = \hat{p}$. Both are based on the same probability model for the variable X. (The fact that $\hat{p}$ is simply X divided by n makes it pretty clear that control limits for $\hat{p}$ should simply be those for X, divided by n.) Under stable process conditions for the creation of the n items in a sample (under the assumption that the sample in question is a rational subgroup) it is reasonable to model the variable X with a binomial distribution for n "trials" and "success probability," p, equal to the process propensity for producing nonconforming items.

Elementary properties of the binomial distribution can be invoked to conclude that

$$\mu_X = \mathrm{E}X = np \quad \text{and} \quad \sigma_X = \sqrt{\mathrm{Var}\, X} = \sqrt{np(1-p)}. \tag{3.29}$$

Then the mean and standard deviation in display (3.29) and the generic Shewhart control limits and center line specified in displays (3.2) and (3.3) lead immediately to standards given control limits for both X and $\hat{p}$. That is,

$$CL_X = np \tag{3.30}$$

Standards Given *np* Chart Center Line

while

$$UCL_X = np + 3\sqrt{np(1-p)} \quad \text{and} \quad LCL_X = np - 3\sqrt{np(1-p)}. \tag{3.31}$$

Standards Given *np* Chart Control Limits

And dividing the expressions (3.30) and (3.31) through by n, one arrives at standards given values for $\hat{p}$,

Standards Given p Chart Center Line

$$CL_{\hat{p}} = p, \tag{3.32}$$

Standards Given p Chart Control Limits

$$UCL_{\hat{p}} = p + 3\sqrt{\frac{p(1-p)}{n}} \quad \text{and} \quad LCL_{\hat{p}} = p - 3\sqrt{\frac{p(1-p)}{n}}. \tag{3.33}$$

Example 3.2 Monitoring the Fraction Nonconforming in a Pelletizing Process. Kaminiski, Rasavaghn, Smith, and Weitekamper worked with a manufacturer of hexamine pellets. Their work covered a time period of several days of production. Early efforts with the pelletizing machine (using shop standard operating procedures) produced a standard fraction nonconforming of approximately $p = .60$. On the final day of the study, after adjusting the "mix" of the powder being fed into the machine, the counts and proportions of nonconforming pellets in samples of size $n = 30$ portrayed in Table 3.3 were collected.

From equations (3.31), standards given control limits for the numbers of nonconforming pellets in the samples represented by Table 3.3 are

$$UCL_X = 30(.6) + 3\sqrt{30(.6)(.4)} = 26.05$$

and

$$LCL_X = 30(.6) - 3\sqrt{30(.6)(.4)} = 9.95,$$

Table 3.3 Counts and Fractions of Nonconforming Pellets in Samples of Size 30

Sample	X	$\hat{p}$	Sample	X	$\hat{p}$
1	14	.47	14	9	.30
2	20	.67	15	16	.53
3	17	.57	16	16	.53
4	13	.43	17	15	.50
5	12	.40	18	11	.37
6	12	.40	19	17	.57
7	14	.47	20	8	.27
8	15	.50	21	16	.53
9	19	.63	22	13	.43
10	21	.70	23	16	.53
11	18	.60	24	15	.50
12	14	.47	25	13	.43
13	13	.43			

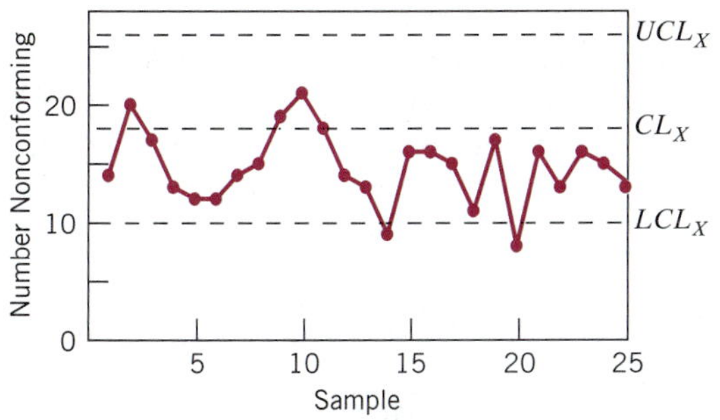

Figure 3.6
Standards given *np* chart for counts of nonconforming pellets.

and from display (3.30) a center line at

$$CL_X = 30(.6) = 18$$

is in order. Figure 3.6 is the standards given *np* control chart for the data of Table 3.3.

It is evident from Figure 3.6 that the pelletizing process was not stable at the standard value of $p = .60$ on the final day of the students' study. Notice that there are two out-of-control points on the chart (and most of the plotted points run below the center line established on the basis of the standard value of p). The message that was delivered at samples 14 and 20 (if not even before, on the basis of the plotted values running consistently below 18) was one of clear process improvement, presumably traceable to the change in powder mix.

Example 3.2 nicely illustrates the fact that a lower control limit on an *np* chart or on a *p* chart makes perfectly good sense, in terms of allowing identification of unexpectedly good process output. Remember that the objective of Shewhart charting is to detect process instability/change. On occasion, that change can be for the good.

Retrospective control limits for X or $\hat{p}$ require that one take the data in hand and produce a provisional estimate of (a supposedly constant) p for plugging into formulas (3.30) through (3.33) in place of p. If samples (of possibly different sizes) are available from r different periods, then a most natural estimator of a common p is the pooled sample fraction nonconforming

Pooled Fraction Nonconforming

$$\hat{p}_{\text{pooled}} = \frac{\sum_{i=1}^{r} n_i \hat{p}_i}{\sum_{i=1}^{r} n_i} = \frac{\sum_{i=1}^{r} X_i}{\sum_{i=1}^{r} n_i} = \frac{\text{total nonconforming}}{\text{total of the sample sizes}}. \quad (3.34)$$

Example 3.2 continued. Returning again to the pelletizing example, it is straightforward to verify that the counts of nonconforming pellets in Table 3.3 total to 367. There were 30(25) = 750 pellets inspected, so from relationship (3.34) $\hat{p}_{\text{pooled}} = 367/750 = .4893$. Substituting this into equations (3.30) and (3.31) in place of p, one arrives at retrospective values

$$CL_X = 30(.4893) = 14.68,$$

$$UCL_X = 30(.4893) + 3\sqrt{30(.4893)(.5107)} = 22.89$$

and

$$LCL_X = 30(.4893) - 3\sqrt{30(.4893)(.5107)} = 6.47.$$

Figure 3.7 is a retrospective *np* chart made using these values and the data of Table 3.3. It is clear from the figure that although it is not plausible that the pelletizing process was stable at the standard value of p (.60) on the final day of the students' study, it *is* plausible that the process was stable at *some* value of p, and .4893 is a reasonable guess at that value.

A few words need to be said about cases where sample sizes vary in a fraction nonconforming context. In such situations, it makes much more sense to plot consecutive $\hat{p}$ values than it does to plot consecutive X's based on differing sample sizes. Then at least, one has a constant center line (given by expression (3.32)). Of course, the control limits represented in display (3.33) will vary with the sample size, and many largely unsatisfactory methods have been suggested for trying to avoid this fact. In the opinion of your authors, the only sensible thing to do when sample sizes vary is to recompute control limits for every new value of n. In a day of widely available computing power this does not seem like a huge burden, and it is the only logically consistent way of treating the problem.

Notice too, that when sample sizes vary, equations (3.33) show that the larger the sample size, the tighter will be the control limits about the central value p. This

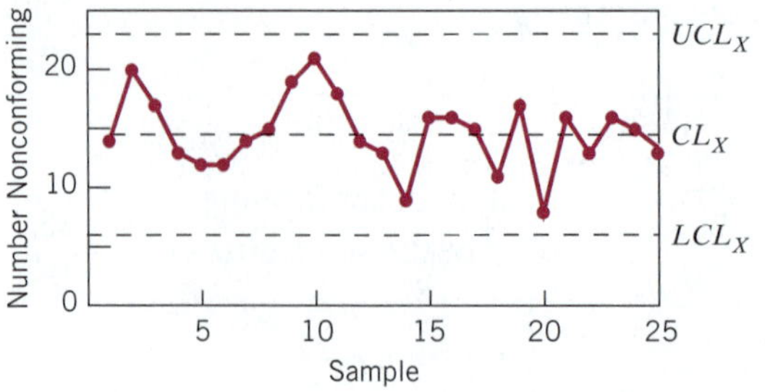

Figure 3.7
Retrospective *np* chart for counts of nonconforming pellets.

is perfectly sensible. The larger the sample, the more information about the current process propensity for producing nonconforming items, and the *less* variation one should allow from the central value before declaring that there is evidence of process instability.

3.3.2 Charts for Mean Nonconformities per Unit

Another kind of situation leading to count and rate data that is fundamentally different from the fraction nonconforming scenario is the so-called "mean nonconformances/nonconformities per unit" ("or mean defects per unit") situation. In such a context, one periodically selects k inspection units from a process output and counts

$$X = \text{the total number of nonconformities on the } k \text{ units} \tag{3.35}$$

(older terminology for nonconformities is "defects" or "flaws"). In cases where k is always equal to 1, the count X itself is plotted and the resulting chart is called a ***c* chart.** Where k varies and/or is not equal to 1, it is common to plot instead

$$\hat{u} = \frac{X}{k} = \text{the mean nonconformities per unit} \tag{3.36}$$

and the resulting chart is called a ***u* chart.**

Control limits for c and u charts are based on the Poisson process model. If one assumes that under stable process conditions the generation of nonconformities can be described by a Poisson process with (constant) rate parameter λ, the number of defects on one inspection unit has a Poisson distribution with mean λ. And X, the number of defects on k inspection units, is (the sum of k independent Poisson random variables each with mean λ and therefore itself) a Poisson random variable with mean $k\lambda$. Thus, under stable process conditions

$$\mu_X = \mathrm{E}X = k\lambda \quad \text{and} \quad \sigma_X = \sqrt{\mathrm{Var}\, X} = \sqrt{k\lambda}. \tag{3.37}$$

So using facts (3.37) and the generic Shewhart control limits and center line specified in displays (3.2) and (3.3), in the c chart situation ($k \equiv 1$) standards given values are

$$CL_X = \lambda, \tag{3.38}$$

Standards Given c Chart Center Line

and

$$UCL_X = \lambda + 3\sqrt{\lambda} \quad \text{and} \quad LCL_X = \lambda - 3\sqrt{\lambda}. \tag{3.39}$$

Standards Given c Chart Control Limits

It follows from the definition of $\hat{u}$ in display (3.36) and relationships (3.37) that

$$\mu_{\hat{u}} = \mathrm{E}\hat{u} = \lambda \quad \text{and} \quad \sigma_{\hat{u}} = \sqrt{\mathrm{Var}\,\hat{u}} = \sqrt{\frac{\lambda}{k}}. \tag{3.40}$$

Then using again the generic Shewhart control limits and center line and applying the facts (3.40) one has standards given values for a u chart

Standards Given u Chart Center Line

$$CL_{\hat{u}} = \lambda, \tag{3.41}$$

and

Standards Given u Chart Control Limits

$$UCL_{\hat{u}} = \lambda + 3\sqrt{\frac{\lambda}{k}} \quad \text{and} \quad LCL_{\hat{u}} = \lambda - 3\sqrt{\frac{\lambda}{k}}. \tag{3.42}$$

Notice that in the case $k = 1$, the u chart control limits reduce (as they should) to the c chart limits.

Retrospective control limits for X or $\hat{u}$ require that one take the data in hand and produce a provisional estimate of (a supposedly constant) λ for plugging into formulas (3.38), (3.39), (3.41), and (3.42) in place of λ. If data from r different periods are available, then a most natural estimator of a common λ is the pooled mean nonconformities per unit

Pooled Mean Nonconformities Per Unit

$$\hat{\lambda}_{\text{pooled}} = \frac{\sum_{i=1}^{r} k_i \hat{u}_i}{\sum_{i=1}^{r} k_i} = \frac{\sum_{i=1}^{r} X_i}{\sum_{i=1}^{r} k_i} = \frac{\text{total nonconformities}}{\text{total units inspected}}. \tag{3.43}$$

Example 3.3 Monitoring the Number of Leaks in Assembled Radiators. The article "Quality Control Proves Itself in Assembly," by Wilbur Burns (reprinted from *Industrial Quality Control*) in Volume 2, Number 1 of *Quality Engineering*, contains a classic set of data on the numbers of leaks found in samples of auto radiators at final assembly. These are reproduced here in Table 3.4. (Remember $\hat{u}$ is defined in equation (3.36).)

This is a nonconformities per unit situation. Each unit (each radiator) presents the opportunity for the occurrence of any number of leaks, several units are being inspected and the total number of leaks on those units is being counted. The leaks per radiator are calculated as in display (3.36) and if one wishes to investigate the statistical evidence for process instability, a u chart is in order.

Table 3.4 Counts and Occurrence Rates of Outlet Leaks Found in 18 Daily Samples of Radiators

Day	X (leaks)	k (radiators)	$\hat{u}$ (leaks/radiator)
1	14	39	.36
2	4	45	.09
3	5	46	.11
4	13	48	.27
5	6	40	.15
6	2	58	.03
7	4	50	.08
8	11	50	.22
9	8	50	.16
10	10	50	.20
11	3	32	.09
12	11	50	.22
13	1	33	.03
14	3	50	.06
15	6	50	.12
16	8	50	.16
17	5	50	.10
18	2	50	.04

The article gives no shop standard value for λ, so consider a retrospective analysis of the data in Table 3.4. It is straightforward to verify that there are 116 total leaks represented in Table 3.4, and 841 radiators were tested. So from relationship (3.43)

$$\hat{\lambda}_{\text{pooled}} = \frac{116}{841} = .138,$$

and a center line for a retrospective u chart for these data can be drawn at this value. Clearly, from equations (3.42) (using .138 in place of λ) the control limits change with k, larger k leading to tighter limits about the center line. As an example of using equations (3.42), note that for those $\hat{u}$ values based on tests of $k = 50$ radiators,

$$UCL_{\hat{u}} = .138 + 3\sqrt{\frac{.138}{50}} = .296.$$

On the other hand, since the formula (3.42) for $LCL_{\hat{u}}$ produces a negative value for the intrinsically nonnegative $\hat{u}$, no lower control limit would be used for $\hat{u}$ based on 50 radiators. (As a matter of fact, no k in Table 3.4 is large enough to lead to the use of a lower control limit.)

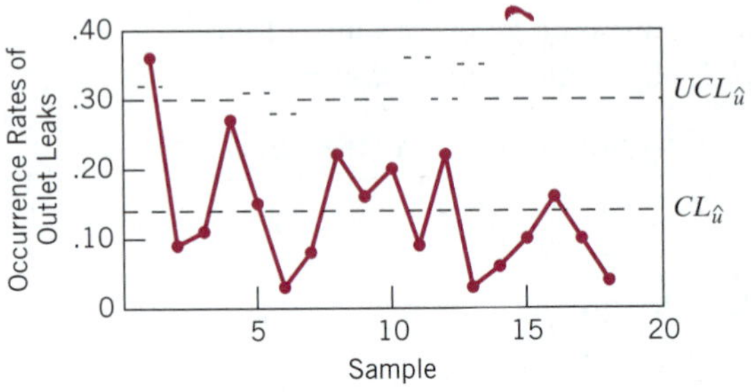

Figure 3.8
Retrospective u chart for rates of radiator outlet leaks.

Figure 3.8 is a retrospective u chart for the radiator leak data. It shows that the initial day's experience does not "fit" with the subsequent 17 days. There is evidence of process change/instability, and appearances are that things improved in the radiator assembly process after the first day.

This section opened with the disclaimer that where it is possible, the charts for measurements introduced in the previous section should be used in preference to the ones presented here. That advice bears repeating. The two examples presented in this section are reasonably convincing, but they are so in part because the relevant fraction nonconforming and mean nonconformities per unit are fairly large. Modern business pressures make standard defect rates in the "parts per million" range fairly common. And there is really no way to effectively monitor processes that are supposed to have such performance with attributes control charts (sample sizes in the millions would be required for effective detection of even doubling of defect rates!).

3.4 PATTERNS ON SHEWHART CHARTS AND SPECIAL ALARM RULES

To this point, this text's discussion has largely treated Shewhart control charts as if all one ever does with them is to compare plotted values of Q to control limits. If that were so, there would be little reason to actually make the plots. All that would really be needed would be simple numerical comparisons. But the fact is that the plots offer the possibility of *seeing* other important things in process monitoring data besides when points plot outside control limits. And it is relatively standard control charting practice to examine Shewhart control charts for these other kinds of indications of process change. The purpose of this section is to discuss some types of revealing patterns that occasionally show up on control charts (providing both jargon for naming them and discussion of the kinds of physical phenomena that can stand behind them) and some sets of rules that can be applied to identify them.

Under stable process conditions in Shewhart control charting applications (leading to Q's that can be modeled as independent and identically distributed), one expects to see a sequence of plotted values that

1. are without obvious pattern or trend,
2. only on rare occasions fall outside control limits,
3. tend to cluster about the center line, about equally often above and below it, but
4. on occasion approach the control limits.

(The tacit assumption in most applications is that the stable process distribution of Q is reasonably "mound shaped" and centered at the chart's center line.) When something other than this kind of "random scatter" picture shows up on a control chart, it can be possible to get clues as to what kinds of physical causes are acting on the process, that can in turn be used in process improvement efforts.

On occasion, when viewing a Shewhart control chart one notices **systematic variation/cycles**, regular "up then back down again" patterns. This suggests that there are important variables acting on the process whose effects are periodic. Identification of the period of variation can give one strong hints where to start looking for physical causes. Examples of the kind of factors that can produce cycles on a Shewhart chart are seasonal and diurnal variables like ambient temperature. And sometimes regular rotation of fixtures or gages or shift changes in operators running equipment or making measurements can stand behind systematic variation.

While systematic variation is variation of the "second kind" on the right side of equation (3.1), it may not always be economically feasible to eliminate it. For example, in some applications it may be preferable to live with effects of ambient temperature rather than try to control the environment in which a process operates. But in any case, recognition of its presence at least allows one to intelligently consider options regarding remedial measures, and to mentally remove that kind of variation from the baseline against which one looks for the effects of other special causes.

Instability is a word that has traditionally been applied to patterns on control charts where many points plot near or beyond control limits. This text has used (and will continue to use) the word to refer to physical changes in a process that lead to individual points plotting outside of control limits. But this second usage refers more to a pattern on the chart, and specifically to one where points outside of control limits are very frequent. Standing behind such a pattern can be more or less erratic and unexpected causes, like different lots of raw material with different physical properties mixed as process input.

Another important possible cause of many points at or beyond control limits is that of unwise operator overadjustment of equipment. Control charting is useful both because it signals the existence of conditions that deserve physical intervention, and because it tells one to leave equipment untouched when it seems to be operating as consistently as possible. When that "hands-off" advice is not followed and humans tinker with physically stable processes, reacting to every small appearance of variation, the end result is not to decrease process variation, but rather to increase it. And such fiddling can turn a process that would otherwise be turning out plotted values inside control limits into one that is regularly producing Q's near or beyond control limits.

Changes in level are sometimes seen on control charts, where the average plotted value seems to move decisively up or down. The change can be sudden and traceable to some basic change at the time of the shift. The introduction of new equipment or a clear change in the quality of a raw material can produce such a sudden change in level. Or the change can be more gradual and attributable to an important cause starting to act at the beginning of the change in level, but so to speak "gathering steam" as time goes on until its full effect is felt. For example, effective worker training in machine operation and measuring techniques could well begin a gradual decrease in level on an R chart, that over a period and with practice will reach its full potential for reducing observed variation.

Where a gradual change in level does not end with stabilization around a new mean, but would go on unabated in the absence of physical intervention, it is traditional to say that there is a **trend** on a control chart. Many physical causes acting on manufacturing processes will produce trends if they remain unaddressed. An example is tool wear in machining processes. As a cutting tool wears, the parts being machined will tend to grow larger. If adjustments are not made and the tool is not periodically changed, machined dimensions of parts will eventually be so far from ideal as to make the parts practically unusable.

There is another phenomenon that occasionally produces strange-looking patterns on Shewhart control charts. This is something the early users of control charts called the occurrence of **mixtures.** These are the combination of two or more distinct patterns of variation (in either a plotted statistic Q, or in an underlying distribution of individual observations leading to Q) that get put together on a single control chart. In "stable" mixtures, the proportions of the component patterns remain relatively constant over time, while in "unstable" versions the proportions vary with time.

Where an underlying distribution of observations has two or more radically different components, depending upon the circumstances a plotted statistic Q can be either unexpectedly variable or surprisingly consistent. Consider first the phenomenon of unexpectedly large variation in Q traceable to a mixture phenomenon. Where blunders like incomplete or omitted manufacturing operations or equipment malfunctions lead to occasional wild individual observations and correspondingly wild values of Q, the terminology **freaks** is often used to describe the impact of the mixture of normal and aberrant observations. Where individual observations or values of Q of a given magnitude tend to occur together in time, the terminology **grouping** or **bunching** is common. Different work methods employed by different operators or changes in the calibration of a measurement instrument can be responsible for grouping or bunching. So, how mixture phenomena sometimes lead to unexpectedly large variation on a control chart is fairly obvious.

How a mixture can lead to unexpectedly small variation in a plotted statistic is more subtle, but very important. It involves a phenomenon sometimes known in quality assurance circles as **stratification.** If an underlying distribution of observations has radically different components, each with small associated variation, and these components are (wittingly or unwittingly) sampled in a systematic fashion, a series of plotted values Q with unbelievably small variation can result. One might, for example, be sampling different raw material streams or the output of different machines and unthinkingly calling the resulting values a single "sample" (in violation, by the way, of the notion of rational subgrouping).

To see how stratification can lead to surprisingly small variation in Q, consider the case of a p chart and a hypothetical situation where a 10-head machine has one completely bad head and 9 perfect ones. If the items from this machine are taken off the heads in sequence and placed into a production stream, "samples" of 10 consecutive items will have fractions defective that are *absolutely constant* at $\hat{p} = .10$. A p chart for the process will look unbelievably stable about a center line at .10. (A similar hypothetical example involving $\bar{x}$ and R charts can be invented by thinking of 9 of the 10 heads as turning out widget diameters of essentially exactly 5.000, while the 10th turns out widget diameters of essentially exactly 7.000. Ranges of "samples" of 10 consecutive parts will be unbelievably stable at 2.000 and means will be unbelievably stable at 5.200.)

So, too much consistency on a control chart is not cause for rejoicing and relaxation. When plotted points hug a center line and never approach control limits something is not as it should be. There may be a simple blunder in the computation of the control limits, or the intrinsic variation in the process may be grossly overestimated. (For example, an excessive standard value for σ produces $\bar{x}$ and R chart control limits that are too wide and plotted points that never approach them under stable conditions.) And on occasion stratification may be present. When it is and it goes unrecognized, one will never be in a position to discover and eliminate the cause(s) of the differences in the components of the underlying distribution of observations. In the 10-head machine example, someone naively happy with the "$\hat{p}$ constant at .10" phenomenon will never be in a position to discover that the one head is defective and remedy it. So, a chart that looks too good to be true is as much a cause for physical investigation as is one producing points outside control limits.

Once one begins to recognize the possibility of looking for patterns on a Shewhart control chart, questions begin to arise as to exactly what ought to be considered an occurrence of a pattern. This is important for two reasons. In the first place, there is the matter of consistency within an organization. If control charts are going to be used by more than one person, those people need a common set of ground rules for interpreting the charts that they together use. Further, there is the matter that without a fair amount of theoretical experience in probability and/or practical experience in using control charts, people tend to want to "see"

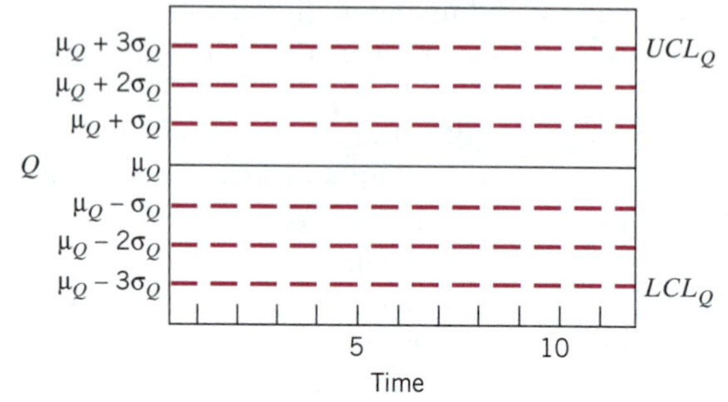

Figure 3.9
Generic Shewhart control chart with "zones" marked on it.

patterns that are in actuality very easily produced by a stable process. (This is especially true in light of the fact that process-monitoring data are ongoing and one often looks at many observations from a process.)

Since the "one point outside control limits" rule is blind to the interesting kinds of patterns discussed here and there is a need for some standardization of the criteria used to judge whether a pattern has occurred, many organizations have undertaken to develop sets of "special checks for unnatural patterns" for application to Shewhart control charts. These are usually based on segmenting the set of possible Q's into various zones defined in terms of multiples of σ_Q above and below the central value μ_Q. Figure 3.9 shows a generic Shewhart chart with typical zones marked on it.

By far the most famous set of special checks is the set of "Western Electric Alarm Rules." These are given in Table 3.5. They are discussed extensively in the *Statistical Quality Control Handbook* published originally by Western Electric and later by AT&T. Two other possible sets of rules, one taken from A.J. Duncan's excellent *Quality Control and Industrial Statistics* and the other published by Lloyd Nelson in the *Journal of Quality Technology* in 1984, are given in Tables 3.6 and Table 3.7 respectively.

Table 3.5 Western Electric Alarm Rules

- A single point outside 3 sigma control limits
- 2 out of any 3 consecutive points outside 2 sigma limits on one side of center
- 4 out of any 5 consecutive points outside 1 sigma limits on one side of center
- 8 consecutive points on one side of center

Table 3.6 Alarm Rules from Duncan's *Quality Control and Engineering Statistics*

- A single point outside 3 sigma control limits
- A run of 7 consecutive points up, down or on one side of center
- 2 consecutive points outside 2 sigma limits
- 4 consecutive points outside 1 sigma limits
- "Obvious" cycles up and down

Table 3.7 Nelson's Alarm Rules from the *Journal of Quality Technology*

- A single point outside 3 sigma control limits
- 9 consecutive points on one side of center
- 6 consecutive points increasing or decreasing
- 14 consecutive points alternating up and down
- 2 out of any 3 consecutive points outside 2 sigma limits on one side of center
- 4 out of any 5 consecutive points outside 1 sigma limits on one side of center
- 15 consecutive points inside 1 sigma limits
- 8 consecutive points with none inside 1 sigma limits

The reader should be able to see in these sets of rules attempts to provide operational definitions for the kinds of patterns discussed in this section. It is not at all obvious which set should be considered best, or even what are rational criteria for comparing them and the many other sets that have been suggested. But the motivation behind them should be clear.

3.5 THE AVERAGE RUN LENGTH CONCEPT

Once one begins to contemplate alternative schemes for issuing out-of-control signals based on process-monitoring data, the need quickly arises to quantify what a given scheme might be expected to do. For example, if one is to choose intelligently between the Western Electric and Nelson sets of alarm rules introduced in the previous section, one needs some means of predicting behavior of the two kinds of monitoring schemes. The most effective means known for making this kind of prediction is the "Average Run Length" (ARL) notion. This section introduces the concept and illustrates its use in some very simple situations.

Consider a situation where based on values of Q plotted at periods $1, 2, 3, \ldots$ one will monitor a process until an out-of-control signal is issued. It is useful to adopt the notation

$$T = \text{the period at which the process-monitoring scheme first signals.} \tag{3.44}$$

T is a random variable and is called the **run length** for the scheme. The probability distribution of T is called the **run length distribution**, and the mean or average value of this distribution is called the **Average Run Length** (ARL) for the process-monitoring scheme. That is,

$$ARL = \mathrm{E}T = \mu_T. \tag{3.45}$$

When one is setting up or choosing a process monitoring scheme, it is desirable that it produce a large ARL when the process is stable at standard values for process parameters and small ARLs under other conditions.

Finding formulas and numerical values for ARLs is not usually elementary, as some nontrivial probability and numerical analysis are often required. But there is one kind of circumstance where an explicit formula for ARLs is possible and one can illustrate the meaning and usefulness of the ARL concept in elementary terms. That is the situation where

1. the process-monitoring scheme employs only the single alarm rule "signal the first time that a point Q plots outside control limits," and
2. it is sensible to think of the process as physically stable (though perhaps not at standard values for process parameters).

Under condition 2, the values $Q_1, Q_2, Q_3, \ldots$ can be modeled as independent random variables with the same individual distribution, and the notation

Probability of an Immediate Alarm

$$q = P[Q_1 \text{ plots outside control limits}] \tag{3.46}$$

will prove useful.

In this simplest of cases, it is fairly straightforward to see that the random variable T has a geometric distribution with probability function

$$f(t) = \begin{cases} q(1-q)^{t-1} & \text{for } t = 1, 2, 3, \ldots \\ 0 & \text{otherwise.} \end{cases}$$

It then follows from the properties of the geometric distribution (or from the definition of expected value and calculations with the probability function f) and relationship (3.45) that

ARL for a "One Point Outside Control Limits" Scheme

$$ARL = \text{E}T = \frac{1}{q}. \tag{3.47}$$

Example 3.4 **Some ARLs for Shewhart $\bar{x}$ Charts.** To illustrate the meaning of relationship (3.47) consider finding ARLs for a standards given Shewhart $\bar{x}$ chart based on samples of size $n = 5$. Note that if standard values for the process mean and standard deviation are respectively μ and σ, the relevant control limits are

$$UCL_{\bar{x}} = \mu + 3\frac{\sigma}{\sqrt{5}} \quad \text{and} \quad LCL_{\bar{x}} = \mu - 3\frac{\sigma}{\sqrt{5}}.$$

Thus, from equation (3.46)

$$q = P\left[\bar{x} < \mu - 3\frac{\sigma}{\sqrt{5}} \quad \text{or} \quad \bar{x} > \mu + 3\frac{\sigma}{\sqrt{5}}\right].$$

First suppose that all is well and the process is stable at standard values of the process parameters. Then elementary probability shows that $\mu_{\overline{x}} = \mu$ and $\sigma_{\overline{x}} = \sigma/\sqrt{5}$ and if the process output is normal, so also is the random variable $\overline{x}$. Thus

$$q = 1 - P\left[\mu - 3\frac{\sigma}{\sqrt{5}} < \overline{x} < \mu + 3\frac{\sigma}{\sqrt{5}}\right] = 1 - P\left[-3 < \frac{\overline{x} - \mu}{\sigma/\sqrt{5}} < 3\right].$$

can be evaluated using the fact that

$$Z = \frac{\overline{x} - \mu}{\sigma/\sqrt{5}}$$

is a standard normal random variable. Using a normal table with an additional significant digit beyond the one in this text (Table A.2) it is possible to establish that

$$q = 1 - P[-3 < Z < 3] = .0027$$

to 4 digits. Therefore, from relationship (3.47) it follows that

$$ARL = \frac{1}{.0027} = 370.$$

The interpretation of this is that when all is OK (i.e., the process is stable and parameters are at their standard values), the $\overline{x}$ chart will issue (false alarm) signals on average only once every 370 plotted points.

In contrast to the situation where process parameters are at their standard values, consider next the possibility that the process standard deviation is at its standard value but the process mean is one standard deviation above its standard value. In these circumstances one still has $\sigma_{\overline{x}} = \sigma/\sqrt{5}$, but now $\mu_{\overline{x}} = \mu + \sigma$ (μ and σ are still the standard values of respectively the process mean and standard deviation). Then,

$$\begin{aligned} q &= 1 - P\left[\mu - 3\frac{\sigma}{\sqrt{5}} < \overline{x} < \mu + 3\frac{\sigma}{\sqrt{5}}\right], \\ &= 1 - P\left[\frac{\mu - 3\sigma/\sqrt{5} - (\mu + \sigma)}{\sigma/\sqrt{5}} < \frac{\overline{x} - (\mu + \sigma)}{\sigma/\sqrt{5}} < \frac{\mu + 3\sigma/\sqrt{5} - (\mu + \sigma)}{\sigma/\sqrt{5}}\right], \\ &= 1 - P[-5.24 < Z < .76], \\ &= .2236. \end{aligned}$$

Figure 3.10 illustrates the calculation being done here and shows the roughly 22% chance that under these circumstances the sample mean will plot outside $\overline{x}$ chart control limits.

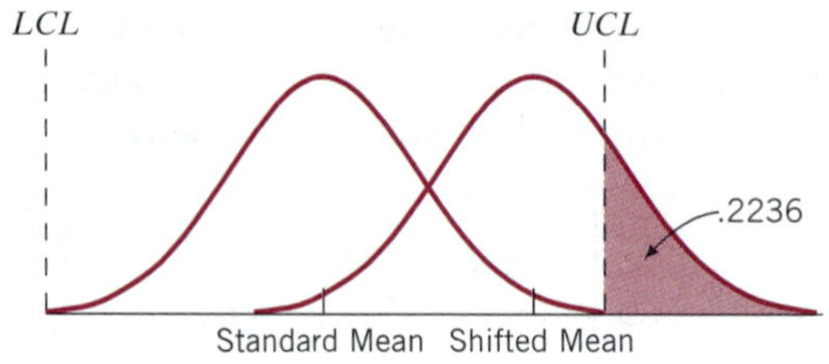

Figure 3.10
Two distributions for $\overline{x}$ and standards given control limits.

Finally, using relationship (3.47), one has

$$ARL = \frac{1}{.2236} = 4.5.$$

That is, if the process mean is off target by as much as one process standard deviation, then it will take on average only 4.5 samples of size $n = 5$ to detect this kind of misadjustment.

Example 3.4 should be entirely comforting and agree completely with the reader's intuition about "how things should be." It says that when a process is on target, one can expect long periods between signals from an $\overline{x}$ chart. On the other hand, should the process mean shift off target by a substantial amount, there will typically be quick detection of that change.

Example 3.5 **Some ARLs for Shewhart *c* Charts.** As a second example of the meaning of equation (3.47), consider finding some ARLs for two different versions of a Shewhart c chart when the standard rate of nonconformities is 1.5 nonconformities per unit. To begin with, suppose that only one unit is inspected each period. Using relationships (3.39) with $\lambda = 1.5$, it follows that since $1.5 - 3\sqrt{1.5} <$ 0 no lower control limit is used for the number of nonconformities found on an inspection unit, and

$$UCL_X = 1.5 + 3\sqrt{1.5} = 5.2.$$

So, for this situation

$$q = P[X > 5.2] = 1 - P[X \leq 5].$$

Consider evaluating q both when the nonconformity rate is at its standard value (of $\lambda = 1.5$ nonconformities per unit) and when it is at three times its standard value (i.e., is 4.5 nonconformities per unit). When the rate is standard, one uses a Poisson distribution with mean 1.5 for X and finds

$$q = 1 - P[X \le 5] = .005 \qquad \text{and} \qquad ARL = \frac{1}{.005} = 200.$$

When the rate is three times standard, one uses a Poisson distribution with mean 4.5 for X and finds

$$q = 1 - P[X \le 5] = .298 \qquad \text{and} \qquad ARL = \frac{1}{.298} = 3.4.$$

That is, completely in accord with intuition, the mean waiting time until an alarm is much smaller when quality deteriorates than when the process defect rate is standard.

Now suppose that one will inspect two units each period. One can then either use a u chart, or equivalently simply apply a c chart where the standard value of λ is 3.0 nonconformities per two units. Applying this second way of thinking and relationships (3.39) with $\lambda = 3.0$, it follows that since $3.0 - 3\sqrt{3.0} < 0$ no lower control limit is used for the number of nonconformities found on two inspection units, and

$$UCL_X = 3.0 + 3\sqrt{3.0} = 8.2.$$

So, for this situation

$$q = P[X > 8.2] = 1 - P[X \le 8].$$

Consider again the ARLs both where the nonconformity rate is at its standard value (of $\lambda = 3.0$ nonconformities per two units) and where it is at three times its standard value (i.e., is 9.0 nonconformities per two units). When the rate is standard, one uses a Poisson distribution with mean 3.0 for X and finds

$$q = 1 - P[X \le 8] = .004 \qquad \text{and} \qquad ARL = \frac{1}{.004} = 250.$$

When the rate is three times standard, one uses a Poisson distribution with mean 9.0 for X and finds

$$q = 1 - P[X \le 8] = .545 \qquad \text{and} \qquad ARL = \frac{1}{.545} = 1.8.$$

Table 3.8 summarizes the calculations of this example. It shows the superiority in terms of ARLs of the monitoring scheme based on two units rather than one unit per period. The two-unit-per-period monitoring scheme has both a larger ARL when quality is standard and a smaller ARL when the nonconformity rate degrades by a factor of 3.0 than the one-unit-per-period scheme. This, of course,

Table 3.8 ARLs for Two c Chart Monitoring Schemes for a Standard Nonconformity Rate of 1.5 Defects per Unit

	Standard Defect Rate	Three Times Standard Defect Rate
1 Unit Inspected	200	3.4
2 Units Inspected	250	1.8

does not come without a price. One must do twice as much inspection for the second plan as for the first.

Examples 3.4 and 3.5 illustrate the ARL concept in very simple contexts that are covered by an elementary formula for ARLs. Where the rules used to convert observed values $Q_1, Q_2, Q_3, \ldots$ into out-of-control signals and the probability model for these variables are more complicated than the combination of the "one point outside control limits" rule and the stable process model, explicit formulas and elementary computations are rare. This text (in Chapter 4) makes use of the ARL idea for some more complicated monitoring schemes. But the methods described there are not based on explicit formulas. Instead ARL tables prepared by researchers who have done the more advanced calculations necessary to find ARLs in these cases will be used. It is not necessary to understand the nature of the numerical analysis needed to produce the tables in order to use them and to appreciate what an ARL tells one about a monitoring scheme.

One problem whose solution has been worked out and should be mentioned here before closing this introduction to the ARL concept, is that of determining ARLs for control charts when some of the special checks for patterns discussed in the previous section are used. An important paper by Champ and Woodall appeared in *Technometrics* in 1987 and considered ARL computations for monitoring schemes using various combinations of the four Western Electric alarm rules. The reader is referred to that paper and a companion computer program published by these authors in the *Journal of Quality Technology* in 1990 for details, but as an example of the kind of conclusions that were reached, consider ARLs for an $\bar{x}$ chart. Calculations like those in Example 3.4 show the "all OK" ARL for a scheme using only the "one point outside $3\sigma_{\bar{x}}$ control limits" rule to be about 370. When all four Western Electric rules are employed simultaneously, Champ and Woodall found that the "all OK" ARL is much less than 370 (and what naive users of the rules might expect), namely approximately 92. The reduction from 370 to 92 shows the effects (in terms of increased frequency of false alarms) of allowing for other signs of process change in addition to individual points outside control limits.

3.6 STATISTICAL PROCESS MONITORING AND ENGINEERING CONTROL

Every effort has been made in this text to emphasize the fact that "Statistical Process *Control*" is really better called "Statistical Process *Monitoring*." "Engineering *Control*" is a tremendously important subject that is largely distinct from the considerations laid out thus far in this text. Unfortunately, there has been a fair amount of confusion about exactly what the two methodologies offer, how they differ, and what are their proper roles in the running of industrial processes. This section is intended to help readers better understand the relationship between them. It begins with an elementary introduction to one simple kind of engineering control, called PID control. It then proceeds to a number of general comments comparing and contrasting statistical process monitoring and engineering control.

3.6.1 Discrete Time PID Control

Engineering control has to do with guiding processes by the deliberate manipulation of appropriate process parameters. For example, in a chemical process, a temperature in a reaction vessel might be kept constant by appropriate manipulation of the position of an inlet steam valve. A very common version of engineering control in industry can be represented in terms of a feedback control diagram like that in Figure 3.11.

In Figure 3.11, a process outputs a value of a variable Y, which is fed into a control algorithm along with a value of a target T for the next output, resulting in a value for some manipulated process variable X, which together with (unavoidable) noise (somehow) produces a subsequent value of Y, and so on. Depending upon what one is willing to assume is known about the various elements in

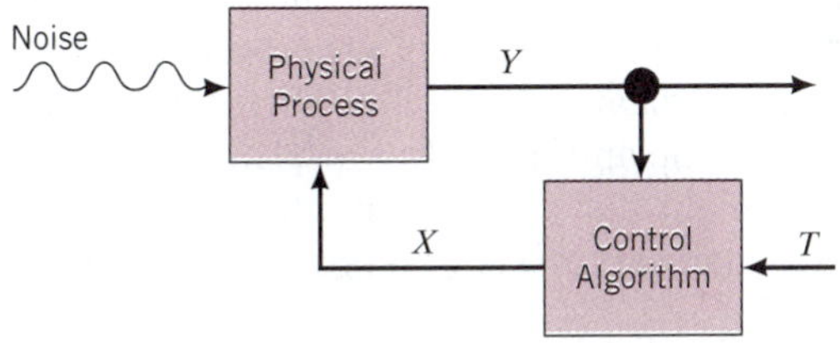

Figure 3.11
Schematic of an engineering feedback control system.

Figure 3.11, different means of choosing a control algorithm can be applied. A method that requires very little in the way of detailed knowledge about how X or the noise impact Y is that of **Proportional-Integral-Derivative (PID) control**.

The discussion here will treat the discrete time version of PID control. So consider discrete integer times $t = 1, 2, 3, \ldots$ (typically evenly spaced in real time) and as in Figure 3.11, suppose that

$$\begin{aligned} Y(t) &= \text{the value of the controlled or output variable at time } t, \\ T(t) &= \text{the value of a target for } Y \text{ at time } t, \text{ and} \\ X(t) &= \text{the value of a (manipulated) process variable that is chosen after} \\ &\quad \text{observing } Y(t). \end{aligned}$$

A control algorithm converts knowledge of $Y(1), Y(2), \ldots, Y(t)$ and $T(s)$ for all s into a choice of $X(t)$. For example, $Y(t)$ could be a measured widget diameter, $T(t)$ a target diameter, and $X(t)$ a cutting tool position. A control algorithm orders a tool position in light of all past and present diameters and all targets for past, present, and future diameters.

The practice of PID control does not typically invest much effort in modeling exactly how changes in X get reflected in Y. (Note that if the goal of a study *was* to understand that relationship, tools of regression analysis might well be helpful.) Nevertheless, in understanding the goals of engineering control, it is useful to consider two kinds of process behavior with which engineering control algorithms must sometimes deal.

For one thing, some physical processes react to changes in manipulated variables only gradually. One behavior predicted by many models of physical science is that when initially at "steady state" at time t_0, a change of ΔX in a manipulated variable introduces a change in the output at time $t > t_0$ of the form

$$\Delta Y(t) = Y(t) - Y(t_0) = G\Delta X\left(1 - \exp\left(\frac{-(t - t_0)}{\tau}\right)\right), \tag{3.48}$$

for process-dependent constants G and τ. Figure 3.12 shows a plot of ΔY in display (3.48) as a function of time. In cases where relationship (3.48) holds, G is the limit of the ratio $\Delta Y/\Delta X$ and is called the **control gain**. τ governs how quickly the limiting change in Y is reached (τ is the time required to reach a fraction $1 - e^{-1} \approx .63$ of the limiting change in Y). It is called the **time constant** for a system obeying relationship (3.48).

Another kind of phenomenon that is sometimes part of the environment in which engineering control systems must operate is that of **dead time** or **delay** between when a change is made in X and when any effect of the change begins to be seen in Y. If there are δ units of dead time and thereafter a relationship similar to that in equation (3.48) holds, one might see a pattern like that shown in Figure 3.13 following an adjustment ΔX made at time t_0 on a process at steady state at that time.

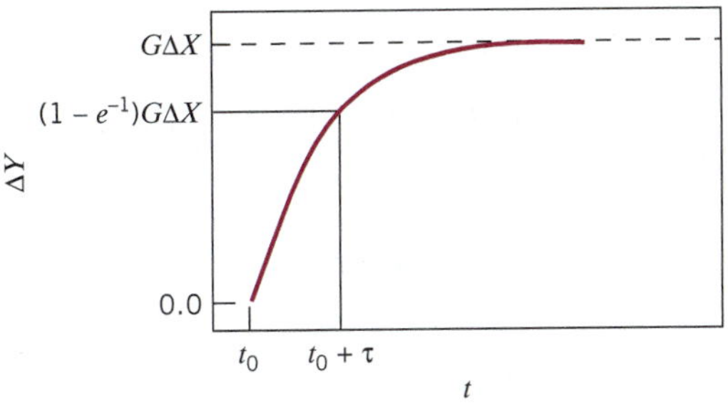

Figure 3.12
Change in the output Y (initially at steady state) in response to a ΔX change in the manipulated variable X.

Of course, not all physical systems involve the kind of gradual impact of process changes illustrated in Figure 3.12, nor do they necessarily involve dead time. (For example, real-time feedback control of machine tools will typically involve changes in tool positions that take their full effect "immediately" after being ordered.) But where these phenomena are present, they increase the difficulty of finding effective control algorithms, the dead time problem being particularly troublesome where δ is large.

To now finally get to the point of introducing the general PID control algorithm, consider a situation where roughly speaking it is sensible to expect that increasing X will tend to increase Y. Define the observed "error" at time t,

$$E(t) = T(t) - Y(t),$$

Error at Time t

and the first and second differences of errors

$$\Delta E(t) = E(t) - E(t-1)$$

First Difference in Errors at Time t

and

$$\Delta^2 E(t) = \Delta(\Delta E(t)) = \Delta E(t) - \Delta E(t-1).$$

Second Difference in Errors at Time t

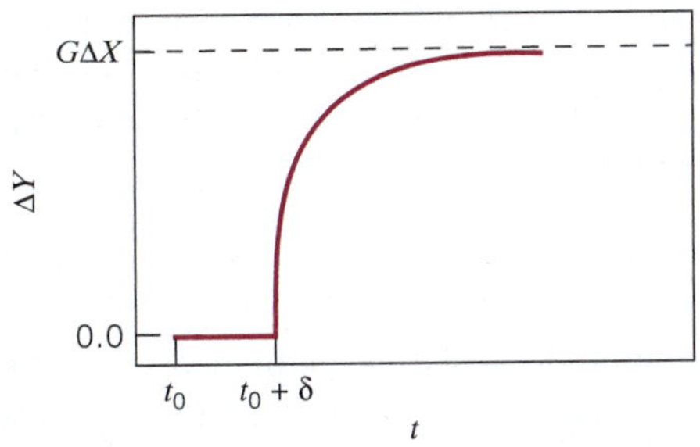

Figure 3.13
Change in the output Y in response to a change ΔX in the manipulated variable at time t_0 if there are δ units of dead time.

With some additional algebra

$$\begin{aligned}\Delta^2 E(t) &= (E(t) - E(t-1)) - (E(t-1) - E(t-2)) \\ &= E(t) - 2E(t-1) + E(t-2).\end{aligned}$$

Then, for constants κ_1, κ_2, and κ_3, a PID control algorithm sets

PID Control Algorithm

$$\Delta X(t) = \kappa_1 \Delta E(t) + \kappa_2 E(t) + \kappa_3 \Delta^2 E(t). \quad (3.49)$$

(In cases where Y tends to increase with X, the constants κ_1, κ_2, and κ_3 are typically nonnegative.) The three terms on the right of equation (3.49) are respectively the **proportional, integral,** and **derivative** parts of the control algorithm.

Example 3.6 **PID Control of Final Dry Weight of 20-lb Bond Paper.** Through the kind cooperation of the Miami University Paper Science Laboratory and Mr. Doug Hart, Research Associate at the lab, one of your authors was able to help implement a PID controller on a 13-inch Fourdrinier paper-making machine. This machine produces paper in a long continuous sheet beginning with vats of pulp mix. The final dry weight of paper is measured as the paper leaves the machine and can be controlled by the rate at which a Masterflex peristaltic pump delivers pulp mix to the machine. A manual knob is used to vary the pump speed and can be adjusted in "ticks." (Each 1-tick change corresponds approximately to a change of pump speed equal to .2% of its maximum capacity.) Past experience with the machine indicated that for 20 lb bond pulp mixture, a 1-tick increase in pump speed produces approximately a .3 g/m^2 increase in paper dry weight. But unavoidable variations in the process, including the "thickness" of the mix available to the pump, produce variation in the paper dry weight and need to be compensated for by varying the pump speed.

Since there is over a 4-minute lag between when a pump speed change is made and when paper affected by the speed change reaches the scanner that measures dry weight at the end of the machine, it was decided that measurements and corresponding adjustments to pump speed would be made only once every 5 minutes. (This choice eliminates the effect of dead time on the control algorithm, which would be a concern if measurements and adjustments were made closer together.) Some experimentation with the machine led to the conclusion that a sensible PID control algorithm for the machine (using the 5-minute intervals and measuring the control variable changes in terms of ticks) has

$$\kappa_1 = .83, \quad \kappa_2 = 1.66, \quad \text{and} \quad \kappa_3 = .83$$

in formula (3.49). Table 3.9 shows an actual series of dry weight measurements and PID controller calculations made using these constants. (Since it was impossible

Table 3.9 PID Control Calculations for the Control of Paper Dry Weight ($T, Y, E, \Delta E$ and $\Delta^2 E$ in g/m^2 and ΔX in ticks)

Period, t	$T(t)$	$Y(t)$	$E(t)$	$\Delta E(t)$	$\Delta^2 E(t)$	$\Delta X(t) = .83\Delta E(t) + 1.66E(t) + .83\Delta^2 E(t)$
1	70.0	65.0	5.0			
2	70.0	67.0	3.0	−2.0		
3	70.0	68.6	1.4	−1.6	.4	1.328
4	70.0	68.0	2.0	.6	2.2	5.644
5	70.0	67.8	2.2	.2	−.4	3.486
6	70.0	69.2	.8	−1.4	−1.6	−1.162
7	70.0	70.6	−.6	−1.4	0	−2.158
8	70.0	69.5	.5	1.1	2.5	3.818
9	70.0	70.3	−.3	−.8	−1.9	−2.739
10	70.0	70.7	−.7	−.4	.4	−1.162
11	70.0	70.1	−.1	.6	1.0	1.162

to move the pump speed knob in fractions of a tick, the actual adjustments applied were those in the table rounded off to the nearest tick.) The production run was begun with the knob (X) in the standard or default position for the production of 20-lb bond paper.

For example, for $t = 3$,

$$E(3) = T(3) - Y(3) = 70.0 - 68.6 = 1.4,$$

$$\Delta E(3) = E(3) - E(2) = 1.4 - 3.0 = -1.6,$$

$$\Delta^2 E(3) = \Delta E(3) - \Delta E(2) = -1.6 - (-2.0) = .4,$$

and so the indicated adjustment (increment) on the pump speed knob is $\Delta X(3) = .83\Delta E(3) + 1.66E(3) + .83\Delta^2 E(3) = .83(-1.6) + 1.66(1.4) + .83(.4) = 1.328$ ticks. (As actually implemented, this led to a 1-tick increase in the knob position after measurement 3.)

It is useful to consider the proportional, integral, and derivative parts of algorithm (3.49) one at a time, beginning with the integral part. Note that with $\kappa_2 > 0$, this part of the algorithm increases X when E is positive and thus $T > Y$. It is this part of a PID control algorithm that reacts to (attempts to cancel) **deviations from target**. Its function is to try to move Y in the direction of T.

To grasp why $\kappa_2 E(t)$ might be called the "integral" part of the control algorithm, consider a case where both $\kappa_1 = 0$ and $\kappa_3 = 0$ so that one has an "integral only" controller. In this case (supposing that $Y(t)$'s and $T(t)$'s with $t < 1$ are available so that one can begin using relationship (3.49) at time $t = 1$), note that

$$\sum_{s=1}^{t} \Delta X(s) = \kappa_2 \sum_{s=1}^{t} E(s). \tag{3.50}$$

But the sum on the left of equation (3.50) telescopes to $X(t) - X(0)$ so that one has

$$X(t) = X(0) + \kappa_2 \sum_{s=1}^{t} E(s).$$

That is, the value of the manipulated variable is $X(0)$ plus a sum or "integral" of the error.

"Integral only" control (especially in the presence of a large time constant and/or large dead time) often tends to overshoot target values and set up oscillations in the variable Y. The proportional and derivative parts of a PID algorithm are meant to reduce overshoot and damp oscillations. Consider next the proportional term from equation (3.49), namely $\kappa_1 \Delta E(t)$.

The proportional part of a PID control algorithm reacts to **changes in the error**. In graphical terms, it reacts to a nonzero slope on a plot of $E(t)$ versus t. Where $\kappa_1 > 0$, this part of the algorithm increases X if the error increases and decreases X if E decreases. In some sense, this part of the algorithm works to hold the error constant (whether at 0 or otherwise).

It is easy to see that when κ_1 and κ_2 have the same sign, the proportional part of a PID control algorithm augments the integral part when E is moving away from 0 and "brakes" or cancels part of the integral part when E is moving toward 0. Figure 3.14 pictures two plots of $Y(t)$ versus t for cases where the target T is constant. In the first plot, Y is approaching T from below. $E(t) > 0$ while $\Delta E(t) < 0$. This is a case where the proportional part of the algorithm brakes the integral part. In the second plot, Y is above T and diverging from it. There, both $E(t) < 0$ and $\Delta E(t) < 0$, and the proportional part of the algorithm augments the integral part. The braking behavior of the proportional part of a PID algorithm helps to resist the kind of oscillation/overshoot problem produced by "integral only" control.

To see why $\kappa_1 \Delta E(t)$ might be called the "proportional" part of the control algorithm, consider a case where both $\kappa_2 = 0$ and $\kappa_3 = 0$ so that one has a "proportional only" controller. In this case (supposing that $Y(t)$'s and $T(t)$'s with $t < 1$ are available so that one can begin using relationship (3.49) at time $t = 1$), note that

$$\sum_{s=1}^{t} \Delta X(s) = \kappa_1 \sum_{s=1}^{t} \Delta E(s). \tag{3.51}$$

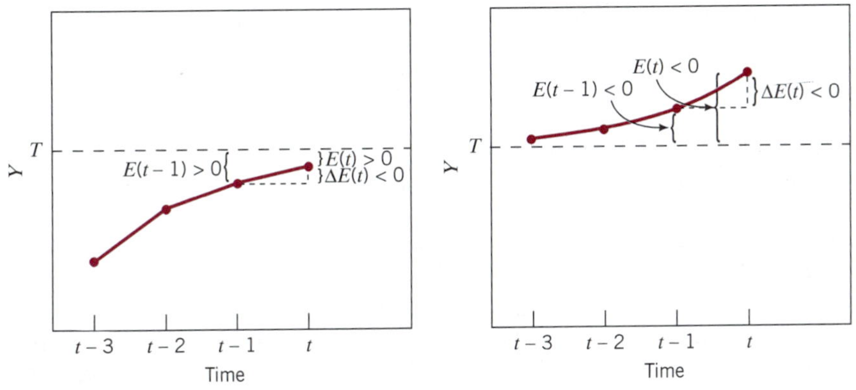

Figure 3.14
Two plots of Y against time.

But the sums on both sides of equation (3.51) telescope, so that one has

$$X(t) = X(0) - \kappa_1 E(0) + \kappa_1 E(t).$$

That is, the value of the manipulated variable is $X(0) - \kappa_1 E(0)$ plus a term "proportional" to the error.

Finally, consider the derivative part of the algorithm (3.49), namely $\kappa_3 \Delta^2 E(t)$. This part of the algorithm reacts to curvature or **changes in slope on a plot of $E(t)$ versus t**. That is, it reacts to changes in $\Delta E(t)$. If a plot of errors versus t is linear ($\Delta E(t)$ is constant), this part of the algorithm does nothing to change X. If $\kappa_3 > 0$ and a plot of errors versus t is concave up, the derivative part of algorithm (3.49) will increase X (and thus Y, decreasing E), while if the plot is concave down it will decrease X. For constant target T, this will tend to "straighten out" a plot of $E(t)$ or $Y(t)$ versus t (presumably then allowing the proportional part of the algorithm to reduce the slope to 0 and the integral part to put the process on target). Once again, since "integral only" control often produces unwanted oscillations of Y about a target, and it is impossible to oscillate without local curvature in a plot of E or Y versus t, the derivative part of the algorithm can be considered as corrective to a deficiency in the naive "integral only" idea.

The rationale for calling $\kappa_3 \Delta^2 E(t)$ the "derivative" part of the PID algorithm (3.49) is similar to the arguments made about the other two parts. Namely, if κ_1 and κ_2 are both 0 (so that one has "derivative only" control),

$$\sum_{s=1}^{t} \Delta X(s) = \kappa_3 \sum_{s=1}^{t} \Delta^2 E(s). \tag{3.52}$$

Telescoping both sides of equation (3.52) one then has that

$$X(t) = X(0) - \kappa_3 \Delta E(0) + \kappa_3 \Delta E(t),$$

and the value of the manipulated variable is $X(0) - \kappa_3 \Delta E(0)$ plus a term proportional to the change in (or "derivative" of) the error.

The primary practical problem associated with the use of PID controllers is the matter of choosing the constants κ_1, κ_2, and κ_3, sometimes called respectively the **proportional, integral, and derivative gains** for the control algorithm. In simple situations where engineers have good mathematical models for the physical system involved, those can sometimes provide at least starting values for searches to find good values of these constants. Where such models are lacking, various rules of thumb aid searches for workable values of κ_1, κ_2, and κ_3. For instance, one such rule is to initially set κ_1 and κ_3 to zero, increase κ_2 till oscillations occur, then halve that value of κ_2 and begin searching over κ_1 and κ_3. And it is pretty clear that in systems where a relationship like (3.48) holds, the gains κ_1, κ_2, and κ_3 should be inversely proportional to G. Further, conventional wisdom also says that in systems where there is dead time $\delta > 0$, the control gains should decrease (exponentially?) in δ. (The intuition for this last point is clear. One should not be changing a manipulated variable wildly if there's to be a long delay before one gets to measure the impact of those changes and to begin to correct any unfortunate effects one sees.)

In any case, the matter of finding good values for the gains κ_1, κ_2, and κ_3 is typically largely a problem of empirical optimization. Section 7.2 of this book discusses some experimental strategies in process optimization. These can be applied to the problem of finding good constants κ_1, κ_2, and κ_3 in the following way. For given choices of the constants, one may run the process in question using the PID controller (3.49) for some number of periods, say m. Then a sensible figure of merit for that particular set of constants is the random variable

$$S = \frac{1}{m}\sum_{t=1}^{m}(E(t))^2,$$

the average squared error. The empirical optimization strategies of Section 7.2 may then be applied in an attempt to find a set of values for κ_1, κ_2, and κ_3 with minimum associated mean for S, μ_S. Problems (3.38) through (3.44) describe how the average squared error idea was used to arrive at the control algorithm of Example 3.6.

3.6.2 Comparisons and Contrasts

The PID ideas just discussed are by no means the only ones used to produce engineering control algorithms. For example, where good models are available for both uncontrolled process behavior and for the impact of control actions on process outputs, mathematically optimal control algorithms (that need not be of the PID type) can sometimes be derived. And the introduction just given completely ignores real issues like the multivariate nature of most industrial applications. (The Y and X just considered are one-dimensional, while real process outputs and possible manipulated variables are often multidimensional.) But the foregoing brief discussion is intended only to give the reader enough of an idea of how engineering control operates to allow the following comments on the proper roles of engineering control and statistical process monitoring to make sense.

Since the resurgence of statistical quality control methods in the early 1980s, the relative merits of the two methodologies when applied in production contexts have been hotly debated by their proponents. On some occasions, zealots on one side or the other of the debate have essentially claimed that their methods are universally applicable and those of the other side are either without merit or are simply a weak version of their own. (Followers of the late W.E. Deming have been especially rancorous in this regard, referring in general terms to cases where poorly thought out applications of engineering control have degraded performance of industrial processes. Based on this, they have in many cases quite wrongly inferred that engineering control is without value in production.)

The truth is that the methods of statistical process monitoring and engineering control are not competitors. They are in fact, completely complementary, each having its own purposes and appropriate areas of application. When applied to the running of industrial processes, both are aimed at the reduction of unwanted variability. In many applications, they can and should be used *together* in an effort to reduce process variation and improve quality, engineering control helping to create stable conditions that are monitored using statistical process monitoring methods.

It should almost go without saying that in cases where a process is already physically stable about a target value, statistical process monitoring tools should only infrequently (and wrongly) signal the need for intervention, and engineering control *is of no help in reducing variation*. That is, in the classical stable process situation, tweaking process parameters can only make variation worse, not better. On the other hand, if successive observations on a process look as if they are dependent, or if they have means (either constant or moving) different from a target, engineering control may be able to improve process performance (uniformity of output) essentially by canceling predictable misadjustments of the process. Statistical process monitoring will then protect one from unexpected process changes.

Table 3.10 puts side by side a number of pairs of statements that should help the reader keep clear in his or her mind the basic differences between engineering control and statistical process monitoring as they are applied to industrial processes. Dr. Bill Tucker is fond of saying "You can't steer a car with statistical process control and you can't fix a car with engineering control." His apt and easily remembered analogy brings into focus the differences in intent of the two methodologies. In today's industry it should not be only one or the other, but both, in their proper places.

Table 3.10 Contrasts Between Engineering Control and Statistical Process Control for Industrial Processes

Engineering Control	Statistical Process Control
• In a typical application, there is a sensor on a process and an electromechanical adjustment mechanism that responds to orders (for change of some process parameter) sent by a computer "brain" based on signals from the sensor.	• This is either manual or automatic plotting of process performance statistics to warn of process changes.
• This is an adjustment/compensation methodology. Formulas prescribe explicit reactions to deviations from target.	• This is a detection methodology. Corrective measures for process changes that are detected are not specified.
• This is a methodology for ongoing small process adjustments.	• There is a tacit assumption here that wise intervention following detection of a process change will set things perfectly aright (for an extended period).
• There is an explicit expectation of process instability/drift in this methodology.	• There is a tacit assumption here of process stability over long periods.
• This is typically computer (or at least mechanically) controlled.	• There is typically a human agent involved in monitoring and interventions.
• The ultimate effect is to keep a process optimally adjusted.	• The ultimate effect is to warn of the presence of sources of special cause variation, to help identify them, and to lead to their permanent removal.
• This is often "tactical" and applied to process parameters.	• This is often "strategic" and applied to final quality variables.
• In its "optimal stochastic control" version, this is what one does *within* a particular probability model (for process behavior) to best exploit the probabilistic predictability of a process.	• This is what one does to monitor for "the unexpected" (departures from a stable process model of expected behavior).

3.7 CHAPTER SUMMARY

Shewhart control charts are an engineer's most widely applicable and easily understood process-monitoring tools. The first four sections of this chapter have introduced these charts for both variables data and attributes data, considered

Table 3.11 Formulas for Shewhart Control Charting

Chart	Q	μ_Q	σ_Q	Standards Given UCL_Q	Standards Given LCL_Q	Retrospective UCL_Q	Retrospective LCL_Q
$\bar{x}$	$\bar{x}$	μ	$\sigma/\sqrt{n}$	$\mu+3\sigma/\sqrt{n}$	$\mu-3\sigma/\sqrt{n}$	$\bar{\bar{x}}+A_2\bar{R}$ $\bar{\bar{x}}+A_3\bar{s}$	$\bar{\bar{x}}-A_2\bar{R}$ $\bar{\bar{x}}-A_3\bar{s}$
Median	$\tilde{x}$	μ	$\kappa\sigma/\sqrt{n}$	$\mu+3\kappa\sigma/\sqrt{n}$	$\mu-3\kappa\sigma/\sqrt{n}$		
R	R	$d_2\sigma$	$d_3\sigma$	$D_2\sigma$	$D_1\sigma$	$D_4\bar{R}$	$D_3\bar{R}$
s	s	$c_4\sigma$	$c_5\sigma$	$B_6\sigma$	$B_5\sigma$	$B_4\bar{s}$	$B_3\bar{s}$
np	X	np	$\sqrt{np(1-p)}$	$np+3\sqrt{np(1-p)}$	$np-3\sqrt{np(1-p)}$	(use $\hat{p}_{\text{pooled}}$ for p)	
p	$\hat{p}$	p	$\sqrt{\frac{p(1-p)}{n}}$	$p+3\sqrt{\frac{p(1-p)}{n}}$	$p-3\sqrt{\frac{p(1-p)}{n}}$	(use $\hat{p}_{\text{pooled}}$ for p)	
c	X	λ	$\sqrt{\lambda}$	$\lambda+3\sqrt{\lambda}$	$\lambda-3\sqrt{\lambda}$	(use $\hat{\lambda}_{\text{pooled}}$ for λ)	
u	$\hat{u}$	λ	$\sqrt{\frac{\lambda}{k}}$	$\lambda+3\sqrt{\frac{\lambda}{k}}$	$\lambda-3\sqrt{\frac{\lambda}{k}}$	(use $\hat{\lambda}_{\text{pooled}}$ for λ)	

their use in both standards given and retrospective contexts, and discussed their qualitative interpretation and supplementation with sets of "extra alarm rules." Table 3.11 summarizes many of the standard formulas used in the making of elementary Shewhart charts.

The final two sections of the chapter have provided context and perspective for the study of Shewhart charts and other process-monitoring tools. Section 3.5 introduced the ARL concept as a means of quantifying the likely performance of a monitoring scheme. Section 3.6 contrasted methods and goals of "engineering control" with those of process monitoring when they are both applied in production.

3.8 CHAPTER 3 EXERCISES

3.1. State the purpose(s) of control charting. What is suggested by out-of-control signals?

3.2. Although statistical quality control tools have historically been applied primarily in manufacturing, current conventional wisdom says that they also have important applications in service industries. Which step in the quality assurance cycle presented in Chapter 1 seems most difficult in service contexts? Explain.

3.3. Why is it essential to have a clear understanding of what constitutes a nonconformance if a Shewhart c or u chart is to be made?

3.4. Distinguish between "quality of design" and "quality of conformance." With which of these is control charting most directly concerned?

3.5. Distinguish between "control limits" and "specification limits" for variables data.

3.6. Explain the difference between "control limits" and "specification limits" in an attributes data setting.

3.7. Explain the ARL concept in terms that a person with no statistical training could understand.

3.8. When designing a control chart, what kinds of ARL values are desirable for an on-target process? For an off-target process? Explain in economic terms why your answers are correct.

3.9. State why statistical methodology is an unavoidable part of quality assurance practice. (Review Chapter 1.)

3.10. Sometimes the plotted statistics appearing on a Shewhart control chart hug (or have little scatter around) the center line. Explain why this is not necessarily a good sign.

3.11. Uninformed engineers sometimes draw in lines on Shewhart $\bar{x}$ charts at engineering specifications for individual measurements. Why is that bad practice?

3.12. It is common to hear people imply that the job of control charts is to warn one of degeneration in product quality. Do you agree with that? Why or why not?

3.13. What is the purpose of sets of "extra alarm rules" like the Western Electric, Nelson, etc. rules presented in Section 3.4?

3.14. What (relevant to quality improvement efforts) does a multimodal shape of a histogram for a part dimension suggest? (Review Chapter 2.)

3.15. At first glance, the term "control chart" perhaps suggests some type of graph depicting continuous regulatory efforts. Is this perception accurate? Why or why not? Suggest a better term than "control chart."

3.16. **Journal Diameters.** Below are some summary statistics (means and standard deviations) for journal diameters of tractor axles as the axles come off an automatic grinding machine. The statistics are based on subgroups of size $n = 4$ pieces taken once per hour. The values listed are in millimeters. Specifications on the journal diameter are from 44.975 mm to 44.990 mm.

Subgroup	$\bar{x}$	s	Subgroup	$\bar{x}$	s
1	44.9875	.0029	5	44.9783	.0039
2	44.9813	.0025	6	44.9795	.0033
3	44.9808	.0030	7	44.9828	.0021
4	44.9750	.0000	8	44.9820	.0024

(continued)

Subgroup	$\bar{x}$	s	Subgroup	$\bar{x}$	s
9	44.9770	.0024	15	44.9748	.0024
10	44.9795	.0010	16	44.9725	.0029
11	44.9815	.0017	17	44.9778	.0021
12	44.9815	.0017	18	44.9790	.0034
13	44.9810	.0024	19	44.9785	.0010
14	44.9778	.0021	20	44.9795	.0010

Note that $\sum \bar{x} = 899.5876$ and $\sum s = .0442$.

a. Are the above attributes data or variables data? Why?

b. What do these values indicate (in retrospect) about the stability of the grinding process?

c. If one judged the process to be stable based on the 20 subgroups summarized above, what could be used as an estimate of the fraction of journal diameters that currently meet specifications? (Give a number based on a normal distribution assumption for diameter measurements.)

d. Make an $\bar{x}$ chart, now using the mid-specification as the center line. What does this chart show?

Suppose that henceforth this process is to be monitored using subgroups of size $n = 5$.

e. Give control limits for a median chart for monitoring process location based on the mid-specification given above and your estimated process standard deviation.

f. Give control limits for future monitoring of sample ranges.

3.17. Refer to the **Journal Diameter** case introduced in problem (3.16). Sometimes subgroup size is not constant. When using standard deviations from subgroups of varying sizes $n_1, n_2, ..., n_r$ to estimate σ, there are several possibilities. Of commonly used ones, the one with the best theoretical properties is

$$s_{\text{pooled}} = \sqrt{\frac{(n_1 - 1)s_1^2 + (n_2 - 1)s_2^2 + \cdots + (n_r - 1)s_r^2}{(n_1 - 1) + (n_2 - 1) + \cdots + (n_r - 1)}}.$$

Another possibility is

$$\hat{\sigma} = \frac{\dfrac{(n_1 - 1)s_1}{c_4(n_1)} + \dfrac{(n_2 - 1)s_2}{c_4(n_2)} + \cdots + \dfrac{(n_r - 1)s_r}{c_4(n_r)}}{n_1 + n_2 + \cdots + n_r - r}.$$

(Since sample sizes vary in this development, we are displaying the dependence of c_4 on sample size here.) The most appropriate estimator of a common mean, μ, when sample sizes vary is

$$\bar{x}_{\text{pooled}} = \frac{n_1\bar{x}_1 + n_2\bar{x}_2 + \cdots + n_r\bar{x}_r}{n_1 + n_2 + \cdots + n_r}.$$

Consider the means and standard deviations given in problem (3.16). For sake of illustration, however, now suppose subgroups were of size $n = 4$ except for the ones indicated in the following table.

Subgroup	Sample Size
1	2
8	2
10	8
15	5
18	3
19	3
20	9

a. Find values for s_{pooled}, $\hat{\sigma}$, and $\bar{x}_{\text{pooled}}$.

b. Give two estimates of (1) the standard deviation of each subgroup mean and (2) the standard deviation of each subgroup standard deviation. (Hint: $\text{Var}\,\bar{x}_i = \sigma^2/n_i$ and $\text{Var}\,s_i = \sigma^2(1 - c_4^2(n_i))$.)

c. With the new subgroup sizes, make two retrospective control charts. One chart should be appropriate to assessing the constancy of the variability in axle journal diameters. The other chart should monitor for change in average axle journal diameter. Are the control limits constant across time? Why or why not? (See (a) and (b).)

d. Are the center lines for the two charts in (c) the same for all subgroups? Why or why not?

e. Under the new sample size configuration, if the process were judged to be stable, what would you use as an estimate of the fraction of journal diameters meeting specifications? (Assume normality of diameter measurements and that $\sigma^2 \approx s^2_{\text{pooled}}$.)

3.18. Rolled Paper. Shervheim and Snider did a project with a company that cuts rolled paper into sheets. The students periodically sampled $n = 5$ consecutive sheets as they were cut and measured their lengths. Data from 20 subgroups are summarized below. (The measurements were in 64ths of an inch above nominal.)

a. What do these values indicate (in retrospect) about the stability of the cutting process?

b. Use the values and estimate the process standard deviation.

c. If one judges the process to be stable and sheet length to be normally distributed, it is possible to estimate the fraction of sheets below nominal in length. Do this. (Show some work.)

Subgroup	$\bar{x}$	s
1	12.2	.84
2	11.2	1.64
3	10.6	2.07
4	12.2	2.49
5	11.2	.84
6	12.6	1.82
7	12.2	2.95
8	13.6	1.67
9	12.2	1.30
10	10.4	1.52
11	10.4	1.95
12	10.6	1.67
13	10.4	1.67
14	12.0	2.91
15	11.2	.84
16	10.6	1.82
17	10.4	1.14
18	9.8	2.17
19	9.6	2.07
20	10.6	1.95
	224.0	35.33

d. Each .25 inch that the cutting process mean is above nominal represents a \$100,000/year loss to the company from product "given away." On the other hand, the company wants to be sure that essentially no sheets are produced with below-nominal lengths. With this in mind, what adjustment in mean length do you suggest, and what yearly savings or additional costs do you project assuming this adjustment is made?

e. Suppose that the adjustment you recommend in (d) is made and henceforth the cutting process is to be monitored based on samples of size $n = 3$. Give control limits for future monitoring of $\bar{x}$ and s.

f. Suppose that while using your $\bar{x}$ chart from (e) the process mean suddenly drops to the point where 1% of the sheets produced are below nominal in length. On average, how many samples will be required to detect this? How does this compare in terms of quickness of detection to a scheme (essentially a p chart) that signals the first time a sample of $n = 3$ contains a sheet with below-nominal length?

3.19. Refer to the **Rolled Paper** case in problem (3.18). Again use the subgroup means and standard deviations given there, but suppose that the number of sheets per subgroup was not constant. Instead, suppose subgroups contained 5 sheets except for the ones indicated in the following table:

Subgroup	Subgroup Size
3	7
6	7
10	2
14	4
17	3
19	2
20	6

a. Give $\overline{x}_{\text{pooled}}$ and two credible estimates of σ. (See problem (3.17).)

b. Give two estimates of (1) the standard deviation of each subgroup mean and (2) the standard deviation of each subgroup standard deviation. (Hint: $\text{Var}\,\overline{x}_i = \sigma^2/n_i$ and $\text{Var}\,s_i = \sigma^2(1 - c_4^2(n_i))$.)

c. Under the new sample size configuration, construct two retrospective control charts. One chart should be appropriate for evaluating constancy of sheet length variability. The other chart should monitor for change in average sheet length. Are the control limits constant across subgroups? Why or why not?

d. Are the center lines for the two charts in (c) constant across time? Why or why not?

e. Under the new sample size configuration, if the process is judged to be stable, what would you use as an estimate of the fraction of paper sheets whose lengths are below nominal? (Assume normality of sheet lengths and that $\sigma^2 \approx s^2_{\text{pooled}}$.)

3.20. U-bolt Threads. A manufacturer of U-bolts for the auto industry measures and records thread lengths on bolts that it produces. Eighteen subgroups, each of $n = 5$ consecutive bolts, were obtained and thread lengths were measured. These and some summary statistics are indicated below (the units are .001 inch above nominal).

a. Estimate the supposedly common subgroup standard deviation, σ, using (1) the subgroup ranges (R_i) and (2) the subgroup standard deviations (s_i).

b. Find control limits for the subgroup ranges. (Use the estimate of σ based on the s_i.)

c. Find control limits for the subgroup standard deviations. (Use the estimate of σ based on the s_i.)

d. Plot the subgroup ranges and standard deviations on Shewhart charts using the retrospective limits from (b) and (c). Is it plausible that variability of thread length was constant from sampling period to sampling period? Why or why not?

e. Find retrospective control limits for the subgroup means. Use your estimate of σ based on the s_i. Plot the subgroup means on a Shewhart chart with these limits.

Subgroup	Thread Lengths	$\tilde{x}$	s	$\bar{x}$	R
1	11, 14, 14, 10, 8	11	2.61	11.4	6
2	14, 10, 11, 10, 11	11	1.64	11.2	4
3	8, 13, 14, 13, 10	13	2.51	11.6	6
4	11, 8, 13, 11, 13	11	2.05	11.2	5
5	13, 10, 11, 11, 11	11	1.10	11.2	3
6	11, 10, 10, 11, 13	11	1.22	11.0	3
7	8, 6, 11, 11, 11	11	2.30	9.4	5
8	10, 11, 10, 14, 10	10	1.73	11.0	4
9	11, 8, 11, 8, 10	10	1.52	9.6	3
10	6, 6, 11, 13, 11	11	3.21	9.4	7
11	11, 14, 13, 8, 11	11	2.30	11.4	6
12	8, 11, 10, 11, 14	11	2.17	10.8	6
13	11, 11, 13, 8, 13	11	2.05	11.2	5
14	11, 8, 11, 11, 11	11	1.34	10.4	3
15	11, 11, 13, 11, 11	11	.89	11.4	2
16	14, 13, 13, 13, 14	13	.55	13.4	1
17	14, 13, 14, 13, 11	13	1.22	13.0	3
18	13, 11, 11, 11, 13	11	1.10	11.8	2
		202	31.51	200.4	74

f. Setting the center line at $\bar{\bar{x}}$, find upper and lower control limits for the subgroup medians. Use your estimate of σ based on the s_i. Plot the subgroup medians on a Shewhart chart with these limits.

g. What do the charts in (e) and (f) suggest about the threading process?

h. A U-bolt customer requires that essentially all U-bolt thread lengths are within .011 inches of nominal. Assuming the bolt manufacturing process continues as reflected by the data given, will the customer be satisfied with current production? Why or why not? Give a quantitative defense of your answer. Assume normality of thread length.

3.21. Refer to the **U-bolt Threads** case in problem (3.20). Problem 3.17 presented ways of estimating σ when r subgroups are of varying size n_i. The formulas there are based on subgroup sample standard deviations s_i. Another expression sometimes used to estimate the process standard deviation is based on ranges, namely

$$\frac{\dfrac{(n_1-1)R_1}{d_2(n_1)} + \dfrac{(n_2-1)R_2}{d_2(n_2)} + \cdots + \dfrac{(n_r-1)R_r}{d_2(n_r)}}{n_1 + n_2 + \cdots + n_r - r}.$$

Consider the means and ranges given in problem (3.20), but now suppose that subgroups consisted of five bolts except for the ones indicated in the following table:

Subgroup	Subgroup Size
2	8
5	4
6	6
7	2
11	3
14	7
15	2
18	2

a. Give $\overline{x}_{\text{pooled}}$ and three estimates of σ. Two of the estimates of σ should be based on the subgroup standard deviations and the other on the subgroup ranges.

b. Find three estimates of the standard deviation of each subgroup mean. Two of the estimates should be based on subgroup standard deviations and one on the subgroup ranges. (Hint: $\text{Var}\,\overline{x}_i = \sigma^2/n_i$.)

c. Find three estimates of the standard deviation of each subgroup sample standard deviation. Two of these estimates should be based on subgroup standard deviations and one on the subgroup ranges. (Hint: $\text{Var}\,s_i = \sigma^2(1 - c_4^2(n_i))$.)

d. Find an estimate of the standard deviation of each subgroup range. Base these estimates on the subgroup ranges. (Hint: $\text{Var}\,R_i = d_3^2(n_i)\sigma^2$.)

e. Make retrospective $\overline{x}$ and R charts using the new configuration of subgroup sizes. Base the control limits on an estimate of σ computed from subgroup ranges.

f. For the charts in (e), are the control limits constant across subgroups? Why or why not?

g. Are the center lines for the charts in (e) constant across subgroups? Why or why not?

3.22. Turning. Allan, Robbins, and Wycoff worked with a machine shop that employs a CNC (computer numerically controlled) lathe in the machining of a part for a heavy equipment manufacturer. Some summary statistics for a particular diameter on the part obtained from 25 subgroups of $n = 4$ parts turned on the lathe are given below. The units are inches.

a. Find retrospective control limits for the values (both means and ranges). What do the $\overline{x}$ and R values indicate about the stability of the turning process?

b. Suppose that one wishes to apply the four "Western Electric Alarm Rules" to the $\overline{x}$ values. Specify the different zones to be used for the mean diameters. Are any of the rules violated in the first 10 samples? (If you find any violations, say which rule is violated for the first time where.)

Subgroup	$\bar{x}$	R
1	1.18093	.0001
2	1.18085	.0002
3	1.18095	.0002
4	1.18063	.0008
5	1.18053	.0007
6	1.18053	.0005
7	1.18058	.0005
8	1.18195	.0001
9	1.18100	.0003
10	1.18095	.0001
11	1.18095	.0006
12	1.18098	.0001
13	1.18123	.0009
14	1.18128	.0002
15	1.18145	.0007
16	1.18080	.0003
17	1.18100	.0000
18	1.18103	.0001
19	1.18088	.0003
20	1.18100	.0000
21	1.18108	.0002
22	1.18120	.0004
23	1.18088	.0002
24	1.18055	.0022
25	1.18100	.0004
	29.52421	.0101

c. Give an estimate of the process short-term standard deviation derived from the ranges and the assumption that σ is constant over the study period.

d. Engineering specifications on the diameter in question were in fact $1.1809 \pm .005$ inches. Suppose that over short production runs, diameters can be described as normally distributed and that your estimate of σ from (c) is an appropriate description of the variation seen in short runs. Give an estimate of the best possible fraction of diameters meeting specifications available using this particular lathe.

e. Make further use of your estimate of σ from (c), and set up control limits that could be used in the future monitoring of the process standard deviation via Shewhart charting of s based on samples of size $n = 5$.

f. Again use your estimate of σ from (c) and consider future monitoring of $\bar{x}$ based on samples of size $n = 4$ using "3 sigma" limits and a center line at the target diameter, 1.1809. On average, how many subgroups would be required to detect a change in mean diameter, μ, from 1.1809 to 1.1810?

3.23. Refer to the **Turning** case in problem (3.22). Problem (3.21) presented a method for estimating σ based on ranges of subgroups of varying size. Use that method in this problem. Use again the sample means and ranges given in problem 3.22, but now suppose all subgroups were of size $n = 4$ parts except for the ones indicated in the following table:

Subgroup	Subgroup Size
1	2
4	3
5	6
9	2
11	7
13	5
15	3
16	2
17	8
18	2

a. Give $\overline{x}_{\text{pooled}}$ and an estimate of σ.
b. Find an estimate of the standard deviation of each subgroup mean, $\overline{x}_i$.
c. Find an estimate of the standard deviation of each subgroup range, R_i.
d. Make retrospective $\overline{x}$ and R charts using the new configuration of subgroup sizes.
e. Are the control limits for the two charts in (d) constant across subgroups? Why or why not?
f. Are the center lines for the charts in (d) constant across subgroups? Why or why not?

3.24. Package Sorting. Budworth, Heimbuch, and Kennedy analyzed a company's package sorting system. As packages arrive at the sorting system, they are placed onto trays and the bar codes affixed to the packages are scanned (in an operation much like the scanning process at a grocery store checkout). Bar code identification numbers begin with the zip code of the package destination. This permits packages to be sorted into 40 bins, each of which represents a different bulk mail center (BMC) or auxiliary service facility (ASF). All packages in a given bin are shipped by truck to the same mail center. The bulk transportation of these packages is much cheaper than if they were mailed directly by the nearest U.S. Post Office. The large number of BMC packages handled daily by the company produces tremendous cost savings.

Initially, the three analysts tackled the so-called "no chute open" problem. This occurs when one of the BMC bins is full and packages destined for that bin cannot be dropped into it. They end up in a "no chute open" bin. This eventuality produced many inefficiencies and even shutdowns of the entire system. In fact, the system was shut down about 10 min/day on average because of this problem. This lost time cost the company the ability to process about 400 packages/day, and accumulated over a year, this represents a serious loss. The team decided to document the number of packages per shift dumped in the "no chute open" bin. The data they collected are below.

Date	Shift	Number in "No Chute Open" Bin
10/16	1	1510
10/17	3	622
10/18	1	2132
10/18	2	1549
10/19	1	1203
10/19	2	2752
10/19	3	1531
10/20	1	1314
10/20	2	2061
10/20	3	981
10/21	1	1636
10/21	2	2559
10/21	3	1212
10/22	1	2016
10/22	2	2765
10/22	3	574

a. Is this an attributes data problem or a variables data problem? Why?

b. What constitutes a "subgroup" in the context of this problem?

c. What probability model is a possible description of the number of packages routed to the "no chute open" bin during a given shift?

d. Estimate the average number of packages routed to the "no chute open" bin during a particular shift. Estimate the standard deviation of the number of packages in the "no chute open" bin. These estimates should be consistent with your answer to (c).

e. Was the number of packages in the "no chute open" bin apparently constant except for random fluctuation? Why or why not? Defend your answer using a control chart.

3.25. Refer to the **Package Sorting** case in problem (3.24). Budworth, Heimbuch, and Kennedy were told that the sorting system was set up to let a package circle on the conveyor belt for 10 cycles (once each cycle the package would fall into the correct chute if that chute was not occupied). If after 10 cycles the correct chute was always occupied, a package would be consigned to the inefficient "no chute open" bin. Upon observing the system in operation, the team immediately recognized packages dropping into the "no chute open" bin after only 2 or 3 cycles. Management was notified and the system was corrected. The team took the data below after implementation of the correction.

Date	Shift	Number in "No Chute Open" Bin
10/23	1	124
10/24	3	550
10/25	1	0
10/25	2	68
10/25	3	543
10/26	1	383
10/26	2	82
10/26	3	118

a. Extend the control limits from your chart in problem (3.24e). Plot the data above on the same chart. Does it appear the system change was effective? Why or why not?

b. Make a chart to assess stability of the number of packages in the "no chute open" bin using only the data above. Does it appear the system was stable? Why or why not?

c. Has the team solved the problem of a large number of packages in the "no chute open" bin? Defend your answer.

3.26. Refer to the **Package Sorting** case of problems (3.24) and (3.25). Budworth, Heimbuch, and Kennedy also investigated the performance of the package scanning equipment. Just as items at a cashier's scanner often are not read on the first scan, so too were bar codes on packages not necessarily read on the first or second scan. Label damage and incorrect orientation, erroneous codes, and some simply unexplained failures all produced "no read" packages. If a package was not read on the first pass, it continued on the carousel until reaching a second scanner at the end of the carousel. Failure to read at this second scanner resulted in the package being dropped into a "no read" bin and scanned manually with a substantial loss in efficiency. The team took data on the following variables over 30 consecutive one-minute periods:

n = the number of packages entering the system during a one-minute period,

X_1 = the number of those packages failing the first scan, and

X_2 = the number of those packages failing both scans.

The values they recorded follow.

Minute	n	X_1	X_2	Minute	n	X_1	X_2
1	54	10	2	16	66	17	0
2	10	3	2	17	56	11	3
3	55	22	3	18	26	6	1
4	60	18	5	19	30	6	0
5	60	12	1	20	69	14	1
6	60	14	1	21	58	23	5
7	37	14	0	22	51	18	5
8	42	17	1	23	32	15	1
9	38	20	10	24	44	23	4
10	33	6	2	25	39	13	2
11	24	6	3	26	26	3	1
12	26	7	5	27	41	17	1
13	36	12	0	28	51	25	5
14	32	10	3	29	46	18	1
15	83	25	2	30	59	23	6

a. What constitutes a "subgroup" in the context of this problem?
b. Is this an attributes data or a variables data scenario? Why?
c. Make a retrospective control chart to assess consistency of the proportion of packages failing both scans. Comment.
d. Make a retrospective control chart to assess consistency of the proportion of packages that are not read on the first scan. Comment.
e. Make a retrospective control chart to assess consistency of the proportion of all packages in a given minute that are not read on the first scan and are read on the second scan. Comment.
f. Calculate the proportions of those packages failing the first scan that also fail the second scan.
g. Make a retrospective control chart to assess consistency of the proportions in (f). Comment.

3.27. Jet Engine Visual Inspection. The data below are representative of counts of nonconformances observed at final assembly of jet engines by GE Aircraft Engines and are taken from a statistical process control training book used by the company. Suppose that one final assembly is inspected per day.

Day	Number of Nonconformances	Day	Number of Nonconformances	Day	Number of Nonconformances
7/5	15	7/15	18	7/29	16
7/6	19	7/16	4	8/1	30
7/7	12	7/19	16	8/2	34
7/8	24	7/20	24	8/3	30
7/9	18	7/21	16	8/4	40
7/10	10	7/22	12	8/5	30
7/11	16	7/25	0	8/6	36
7/12	26	7/26	16	8/8	32
7/13	16	7/27	26	8/9	42
7/14	12	7/28	12	8/10	34

a. State whether the above values come from a variables data or from an attributes data study. Explain.

b. In the context of the problem, what is a "subgroup"?

c. What probability distribution is a likely model for counts of nonconformances on these engines? Briefly, defend your answer.

d. Find an estimated mean number of visually identified nonconformances and the corresponding estimated standard deviation.

e. Find appropriate upper and lower control limits and center line to apply to the counts. Make the corresponding control chart for these data. Does it appear that the process was stable over the period of the study? Why or why not? Identify any out-of-control points.

f. Suppose two inspectors were involved in the data collection. Briefly discuss what must be true (in terms of the data collection protocol) to assure that the chart and analysis in (e) are credible.

3.28. Refer to the **Jet Engine Visual Inspection** case in problem (3.27).

a. When possible causes for out-of-control points on a control chart are addressed and physically eliminated, it is common practice to discard the data associated with those out-of-control points and recalculate control limits. Apply this approach to problem (3.27e) assuming causes of the out-of-control points have been addressed.

b. Suppose the following data are obtained in visual inspection of final engine assemblies over the next three days.

Day	Assemblies Inspected	Number of Nonconformances
1	.5	8
2	2	31
3	1.5	26

(Partial inspection of final engine assemblies could possibly occur because of unforeseen labor problems. More than one engine assembly might be inspected on days 2 and 3 to, in some sense, make up for the partial inspection on day 1.) Using the information from (a) above, find control limits for nonconformance rates on these three days. Also give the center line and three values to plot (nonconformances per engine assembly inspected).

c. Do your values from part (b) suggest process instability? Why or why not?

d. Your center line should be constant across the three days represented in (b). Defend this in mathematical terms.

3.29. When recording nonconformances per unit, the number of standard units inspected may vary from period to period. Let

X_i = the number of nonconformances observed at period i,
k_i = the number of standard units inspected at period i, and
$\hat{u}_i = X_i / k_i$.

The following values were obtained for 9 periods:

i	1	2	3	4	5	6	7	8	9
k_i	1	2	1	3	2	1	1	3	1
$\hat{u}_i$	0	3	0	1.33	4	0	0	.67	1

a. From these values, what conclusions can you make about stability of the process being monitored? Make the appropriate control chart.

b. Suppose that in the future k_i will be held constant at 1 and that 2.4 nonconformances per inspection unit is considered to be "standard quality." Find the probability of an out-of-control signal on a 3-sigma Shewhart control chart, if the true nonconformance rate is at the standard quality level. Find the probability of an out-of-control signal if the true nonconformance rate is 4.8.

c. Continuing, suppose that in the future k_i will be held constant at 2. Find the probability of an out-of-control signal if the true nonconformance rate is at the standard quality level. Find the probability of an out-of-control signal if the true nonconformance rate is 4.8.

d. Compare your solutions in (b) and (c). Which subgroup size ($k = 1$ or $k = 2$) is more appealing? Why?

3.30. Electrical Switches. The following scenario is taken from GE Aircraft Engine training material. One hundred electrical switches are sampled from each of 25 consecutive lots. Each sampled switch is tested and the number failing the test is recorded.

Sample	Number Failing	Sample	Number Failing
1	11	14	18
2	9	15	7
3	15	16	10
4	11	17	8
5	22	18	11
6	14	19	14
7	7	20	21
8	10	21	16
9	6	22	4
10	2	23	11
11	11	24	8
12	6	25	9
13	9		

a. Find the sample fractions of switches failing the test.
b. What is a plausible probability model for describing the count of switches in a particular sample failing the test? Explain.
c. Plot the number failing against the sample period. Plot an appropriate center line and control limits on the same graph.
d. What, in the context of this problem, does your plot in (c) monitor?
e. Interpret your plot in (c). Identify any out-of-control points.
f. What is the usual name of the chart you prepared in part (c)?
g. Suppose causes for out-of-control points identified in (e) were physically removed. It would then make sense to delete the out-of-control points and recalculate limits. Do this recalculation and redo (c).
h. Suppose the number of switches sampled and the number failing for the next three consecutive lots are as follows:

Number Sampled	Number Failing
75	8
144	12
90	11

Using your estimated fraction failing from (g) find control limits and center lines for the three new sets of electrical switches. Are the three sets of control limits and center lines the same? Why is this to be expected?

3.31. A data set in the book *Elementary Statistical Quality Control* by Burr indicates that in the magnaflux inspection for cracks in a type of malleable casting, about $p \approx .11$ of the castings will have detectable cracks. Consider the examination of 12 such castings. Let X be the number of castings from the set of 12 identified as being cracked.

a. Find $P[X = 5]$.
b. Find $P[X > 5]$.
c. Find EX.
d. Find $\text{Var } X$.
e. Ten sets of 12 castings are to be inspected. What is the probability that at least one set of 12 will have one or more cracked castings?

3.32. Refer to the **Plastic Packaging** case in problems (1.9) and (1.10). The plastic bags in question were supposed to hold three bagels each. An ideal bag is 6.75 inches wide, has a 1.5 inch lip, and has a total length of 12.5 inches (including the lip). The ideal hole positions are on the lip. The hole position on selected bags was measured as the distance from the bottom of the bag to the hole. Five bags were obtained at six times on each of three days. Hole position, bag width, bag length, and lip width were measured and recorded for each bag. The data for hole position (in.) are below:

Day	Time	Hole Position
1	10:10 am	1.87500, 1.84375, 1.87500, 1.84375, 1.84375
1	10:25 am	1.90625, 1.90625, 1.90625, 1.87500, 1.90625
1	10:55 am	1.87500, 1.93750, 1.93750, 1.93750, 1.96875
1	11:12 am	2.09375, 2.12500, 2.21875, 2.15625, 2.12500
1	11:35 am	2.00000, 2.00000, 2.00000, 2.00000, 2.03125
1	11:41 am	1.87500, 1.90625, 1.90625, 1.87500, 1.93750
2	8:15 am	1.62500, 1.62500, 1.59375, 1.65625, 1.59375
2	8:54 am	1.62500, 1.62500, 1.59375, 1.68750, 1.65625
2	9:21 am	1.62500, 1.59375, 1.62500, 1.59375, 1.62500
2	9:27 am	1.62500, 1.59375, 1.62500, 1.65625, 1.65625
2	9:51 am	1.56250, 1.59375, 1.56250, 1.56250, 1.56250
2	9:58 am	1.56250, 1.56250, 1.56250, 1.53125, 1.56250
3	10:18 am	1.50000, 1.56250, 1.53125, 1.53125, 1.50000
3	10:33 am	1.53125, 1.53125, 1.53125, 1.53125, 1.50000
3	10:45 am	1.50000, 1.53125, 1.50000, 1.53125, 1.46875
3	11:16 am	1.50000, 1.50000, 1.50000, 1.53125, 1.50000
3	11:24 am	1.53125, 1.53125, 1.50000, 1.50000, 1.50000
3	11:39 am	1.50000, 1.50000, 1.53125, 1.53125, 1.53125

a. What is a natural subgroup for this situation?
b. How many items are in each subgroup described in (a)? How many subgroups are there here in total?
c. Calculate the subgroup means and subgroup ranges.
d. Make a retrospective control chart for mean hole position. Give the center line, control limits, and zone limits.
e. Make a retrospective control chart for variability in position using your values from (c). Give the control limits and zone limits.

f. What is the usual name of the chart in (d)? What is the usual name of the chart in (e)?

g. Is it important which of the charts developed in (d) and (e) is analyzed first? Why or why not?

h. Find the estimated standard deviation of hole position based on the ranges.

3.33. Refer to the **Plastic Packaging** case in problems (1.9) and (3.32).

a. Calculate the 18 subgroup means and 18 subgroup ranges.

b. Make retrospective control charts for mean hole position *one day at a time*. Give center lines, control limits, and zone limits.

c. Make retrospective control charts for variability of hole position *one day at a time.*

d. Based on your answer to (c), is variability of hole location constant within a day? Why or why not?

e. According to your charts in (c), is there a day in which a single standard deviation of hole position is plausible? Why or why not?

f. Suppose your answer in (e) is "yes" for each day. Find estimated σ's for the three different days treated separately. (Base your estimates on sample ranges.)

g. Comment on how your estimates in (f) compare to the estimate in problem (3.32h).

3.34. Refer to the **Plastic Packaging** case in problems (3.32) and (3.33). The ideal lip width is 1.5 inches. The following lip width data (in.) were taken on the same bags described in problem (3.32).

Day	Time	Lip Width
1	10:10 am	1.75000, 1.62500, 1.62500, 1.65625, 1.62500
1	10:25 am	1.62500, 1.62500, 1.62500, 1.65625, 1.65625
1	10:55 am	1.53125, 1.53125, 1.50000, 1.50000, 1.50000
1	11:12 am	1.40625, 1.43750, 1.43750, 1.46875, 1.46875
1	11:35 am	1.46875, 1.46875, 1.46875, 1.46875, 1.40625
1	11:41 am	1.43750, 1.43750, 1.46875, 1.50000, 1.46875
2	8:15 am	1.37500, 1.40625, 1.37500, 1.40625, 1.37500
2	8:54 am	1.37500, 1.43750, 1.43750, 1.40625, 1.40625
2	9:21 am	1.40625, 1.37500, 1.43750, 1.40625, 1.40625
2	9:27 am	1.50000, 1.46875, 1.43750, 1.46875, 1.43750
2	9:51 am	1.43750, 1.43750, 1.43750, 1.43750, 1.43750
2	9:58 am	1.53125, 1.46875, 1.53125, 1.50000, 1.53125
3	10:18 am	1.53125, 1.56250, 1.50000, 1.50000, 1.53125
3	10:33 am	1.50000, 1.53125, 1.53125, 1.50000, 1.50000
3	10:45 am	1.34375, 1.34375, 1.34375, 1.37500, 1.37500
3	11:16 am	1.46875, 1.46875, 1.46875, 1.43750, 1.43750
3	11:24 am	1.37500, 1.40625, 1.40625, 1.40625, 1.40625
3	11:39 am	1.43750, 1.43750, 1.40625, 1.37500, 1.43750

a. Is this a variables data or an attributes data scenario? Why?
b. Find the subgroup means, ranges, and standard deviations.
c. Make retrospective control charts for lip width variability and lip width mean based on the sample ranges.
d. In view of the appearance of your chart for variability of lip width, does it make sense to seriously examine the chart for mean lip width? Why or why not?
e. Instead of making the completely retrospective charts asked for in (c), is it possible to incorporate some "standards" information and construct a different chart for mean lip width? Explain.
f. Instead of treating all 18 samples at once as in part (c), make retrospective R and $\bar{x}$ charts *a day at a time*.
g. Find three daily estimated lip width standard deviations. How do these estimates compare to that calculated when the complete set of data is used? (See (c) above.)
h. Would there be advantages to using subgroup standard deviations instead of subgroup ranges in parts (c) and (f) above? Explain.

3.35. Refer to the **Plastic Packaging** case in problem (3.32).
a. Make a control chart for the standard deviation of hole position.
b. Make a control chart for mean hole position based on subgroup standard deviations.
c. Make charts for the standard deviation of hole position *a day at a time.*
d. Make charts for mean hole position *a day at a time.*
e. For each of (a), (b), (c), and (d), was it helpful to use the subgroup standard deviations instead of the subgroup ranges as in problem (3.32)? Why or why not?

3.36. Refer to the **Hose Skiving** case of problem (1.11). The plant works two shifts/day. Five hoses were sampled every two hours from each of three production lines and skive length measured. Specifications for skive length are *target* $\pm$ .032 inches. The values in the accompanying tables are in units of .001 inch above target.
a. Explain (possibly using the notions of "rational subgrouping" and "stratification") why it would not make good sense to combine data taken at a particular time period from the three different production lines to make a single "sample." (Particularly in cases where it is only possible to select a single item from each line at a given time period, the urge to make such a "sample" is strong, and this kind of error is a common one.)
b. Compute the 48 sample means and ranges for the data above. Then make an $\bar{x}$ chart and an R chart for each of lines 1, 2, and 3. Comment on what they indicate about the stability of the skiving process on the three lines over the two days of this study.
c. One *could* think about plotting all 48 sample means on a single chart, for example plotting means from lines 1, 2, and 3 in that order at a given time period. Discuss why that is not a terribly helpful way of summarizing the

Line 1		
Day	Time	Skive Length
1	8:00 am	3, 2, 4, −5, 2
1	10:00 am	5, −4, −3, 0, −2
1	12:00 pm	−5, 5, 5, −3, 2
1	2:00 pm	−2, 5, 4, −3, 2
1	4:00 pm	−10, 2, 1, 2, 1
1	6:00 pm	−5, −6, −3, −3, −7
1	8:00 pm	−5, 0, −3, −3, −8
1	10:00 pm	−5, −10, 10, −9, −3
2	8:00 am	2, 4, 1, 0, −5
2	10:00 am	−3, 3, −4, 5, 3
2	12:00 pm	−5, −7, 6, 8, −10
2	2:00 pm	3, −4, 4, 6, −3
2	4:00 pm	−10, −7, −3, −1, −3
2	6:00 pm	0, −1, −6, −2, 0
2	8:00 pm	2, 4, −2, −3, 5
2	10:00 pm	1, 0, −1, 7, −5

Line 2		
Day	Time	Skive Length
1	8:00 am	−17, 3, 2, 10, 4
1	10:00 am	13, 3, −2, 12, 15
1	12:00 pm	14, 6, 10, 5, 1
1	2:00 pm	7, 2, 10, 16, 13
1	4:00 pm	−15, −12, −2, −4, 0
1	6:00 pm	−4, −6, −4, −4, 4
1	8:00 pm	2, −5, −5, −3, −4
1	10:00 pm	0, −1, −2, −1, 0
2	8:00 am	15, 2, 16, 10, 14
2	10:00 am	12, 4, −10, 10, −3
2	12:00 pm	1, −7, 4, −5, −9
2	2:00 pm	−6, 8, −5, 18, 20
2	4:00 pm	−2, −4, −5, −1, −3
2	6:00 pm	−2, −2, −2, −4, −2
2	8:00 pm	0, 2, −1, −1, −2
2	10:00 pm	−1, −2, 0, −1, −1

Line 3		
Day	Time	Skive Length
1	8:00 am	−3, −5, 7, 10, 3
1	10:00 am	3, 5, 5, 8, 1
1	12:00 pm	3, 6, 6, 5, 5
1	2:00 pm	5, −2, 5, 4, 6
1	4:00 pm	2, 5, 4, 1, 1
1	6:00 pm	2, 1, 0, 1, 1
1	8:00 pm	1, 3, 5, −6, −10
1	10:00 pm	−7, −5, 4, 2, −9
2	8:00 am	18, 15, 5, 3, 4
2	10:00 am	3, 2, −2, −5, 2
2	12:00 pm	4, 2, 2, 1, 3
2	2:00 pm	6, 5, 4, 2, 5
2	4:00 pm	2, 0, 1, −3, 5
2	6:00 pm	2, −5, −7, −3, −5
2	8:00 pm	−6, −3, −10, −4, −7
2	10:00 pm	0, −4, −7, −10, −2

data. (Will it be easier or harder to see trends for a given production line on this kind of plot or on the separate charts of part (b)?)

3.37. Consider the following hypothetical data from a process where $T(t)$ is the target value, $Y(t)$ is the realized value of the characteristic of interest, $E(t) = T(t) - Y(t)$, $\Delta E(t) = E(t) - E(t-1)$, and $\Delta^2 E(t) = \Delta E(t) - \Delta E(t-1)$. The PID controller $\Delta X(t) = \kappa_1 \Delta E(t) + \kappa_2 E(t) + \kappa_3 \Delta^2 E(t)$ will be used.

Period, t	$T(t)$	$Y(t)$	$E(t)$	$\Delta E(t)$	$\Delta^2 E(t)$	$\Delta X(t)$
1	4	2	2			
2	4	1	3	1		
3	4	0	4	1	0	18
4	4	2	2	−2	−3	1
5	4	2	2	0	2	
6	4	3	1	−1	−1	
7	5	3	2	1	2	
8	5	4	1	−1	−2	
9	5	5	0	−1	0	
10	5	6	−1	−1	0	
11	5	6	−1	0	1	
12	5	6	−1	0	0	

a. What does $\Delta X(t)$ represent in physical terms? Is the adjustment $\Delta X(3)$ made before or after observing $Y(3) = 0$?

b. Suppose the integral coefficient in the control algorithm is 4. What are the proportional and derivative coefficients?

c. Using your answer in (b), find the complete set of $\Delta X(t)$'s.

d. Find the average squared error for the last 9 periods, the last 8 periods, ..., and the last 3 periods.

e. Make a plot of your values from (d) as follows. Label the horizontal axis with t. For $t = 4$ plot the average squared error for periods 4 through 12, for $t = 5$ plot the average squared error for periods 5 through 12, ..., for $t = 10$ plot the average squared error for periods 10 through 12. Does the appearance of this plot give you hope that any transient or "startup" effects have been eliminated before the last few periods and that those periods adequately represent control algorithm performance? Explain.

3.38. Paper Dry Weight. Before progressing to the collection of the data in Table 3.9, several different PID algorithms were tried out. Initially, Miami University Paper Science Lab Research Associate Doug Hart set up the paper-making machine with 1% de-inked pulp stock to produce 20-lb bond paper. No filler was used. Jobe and Hart began an investigation into how to best control the dry weight variable. Twelve periods of data were obtained to benchmark the process without any pump speed adjustments. (It is well known that the pump speed does affect final dry weight.) A standard setting of 4.5 (45% of maximum speed and known to produce paper with a dry weight close to the target of 70 g/m^2) was used. Paper dry weight measurements were made at roughly 5-minute intervals, and these are presented below. Units are g/m^2.

Time	Period, t	$T(t)$	$Y(t)$	$E(t)$
8:45	1	70	75.3	−5.3
8:50	2	70	75.8	−5.8
8:55	3	70	73.1	−3.1
9:00	4	70	72.4	−2.4
9:05	5	70	73.5	−3.5
9:10	6	70	72.8	−2.8
9:15	7	70	72.6	−2.6
9:20	8	70	71.7	−1.7
9:25	9	70	69.8	.2
9:30	10	70	66.9	3.1
9:45*	11	70	70.9	−.9
9:50	12	70	71.7	−1.7

*There was a paper breakage at 9:33.

a. Plot the measured values $Y(t)$ versus t.
b. Plot the errors $E(t)$ versus t.
c. Find the average squared error for periods 1 through 12, for periods 2 through 12 ... for periods 10 through 12.
d. Make a plot of your values from (c) for $t = 1, 2, \ldots, 10$. (At time t plot the average squared error for periods t through 12.)

3.39. Refer to the **Paper Dry Weight** case of problem (3.38). Hart informed Jobe that for every 5-tick increase on the speed pump dial, paper dry weight increases about 1.5 g/m^2. This means that in rough terms, to increase a dry weight by 1 g/m^2, an increase of pump speed setting of about 3.33 ticks is needed.

a. If one were to consider an "integral only" version (a $\kappa_1 = \kappa_3 = 0$ version) of the control equation (3.49) for use with the paper-making machine, why might $\kappa_2 = 3.33$ be a natural first choice? (X is in ticks, while T and Y are in g/m^2.)
b. The "integral only" controller of part (a) was used for 7 time periods and paper dry weight data collected. This is summarized in the following table. Fill in the $\Delta X(t)$ and $E(t)$ columns in that table for $t = 1, 2, \ldots, 8$. (The machine was running without adjustment with X set at 4.5 until 9:55.) (The measurements were taken far enough apart in time that the entire effect of a pump speed change ordered on the basis of data through a given period was felt at the next measuring period.)

Time	Period, t	$T(t)$	$Y(t)$	$E(t)$	$\Delta X(t)$
9:55	1	70	72.1		
10:08	2	70	70.6		
10:14	3	70	71.3		
10:25	4	70	67.1		
10:32	5	70	71.5		
10:38	6	70	70.3		
10:44	7	70	68.4		
10:50	8	70	71.7		

c. Plot $Y(t)$ versus t.
d. Plot $E(t)$ versus t.
e. Find the average squared error for periods 2 through 8, for periods 3 through 8... for periods 6 through 8.
f. Make a plot of your values from (e) for $t = 2, \ldots, 6$. (At time t plot the average squared error for periods t through 8.) Does the appearance of this plot give you hope that any transient or "startup" effects have been eliminated before the last few periods and that those periods adequately represent control algorithm performance?

3.40. Refer to the **Paper Dry Weight** case in problems (3.38) and (3.39). At 10:50 the speed pump dial was set back to 4.5 (45%) and left there for 5 minutes in order to return the system to the benchmark conditions of problem (3.38). A new coefficient κ_2 in an integral control algorithm was adopted and beginning at 10:55 this new adjustment algorithm was employed for 7 periods with results summarized in the following table.

Time	Period, t	$T(t)$	$Y(t)$	$E(t)$	$\Delta X(t)$
10:55	1	70	72.0	−2	−3.32
11:01	2	70	71.7		
11:13	3	70	71.1		
11:19	4	70	68.8		
11:25	5	70	69.6		
11:31	6	70	71.8		
11:37	7	70	68.2		
11:43	8	70	69.7		

a. Find the value of the new coefficient κ_2 used by Jobe and Hart. Then fill in the $E(t)$ and $\Delta X(t)$ values in the table for $t = 2, \ldots, 8$.
b. Plot $Y(t)$ versus t.
c. Plot $E(t)$ versus t.

d. Find the average squared error for periods 2 through 8, for periods 3 through 8 ... for periods 6 through 8.

e. Make a plot of your values from (d) for $t = 2, \ldots, 6$. (At time t plot the average squared error for periods t through 8.) Does the appearance of this plot give you hope that any transient or "startup" effects have been eliminated before the last few periods and that those periods adequately represent control algorithm performance?

3.41. Refer to the **Paper Dry Weight** case of problems (3.38), (3.39), and (3.40). After making the measurement at 11:43 indicated in the problem (3.40), the speed pump dial was again set back to 4.5 and left there for 5 minutes (from 11:44 to 11:49). (This was again done to in some sense return the system to the benchmark conditions.) Hart and Jobe decided to include both integral and proportional terms in a new control equation and $\kappa_2 = 1.66$ and $\kappa_1 = .83$ were selected for use in equation (3.49). (The same integral control coefficient was employed, and a proportional coefficient half as large as the integral coefficient was added.) This new adjustment algorithm was used to produce the values in the following table.

Time	Period, t	$T(t)$	$Y(t)$	$E(t)$	$\Delta E(t)$	$\Delta X(t)$
11:49	1	70	70.9	−.9		
11:54	2	70	70.3	−.3	.6	0
11:59	3	70	68.8			
12:06	4	70	70.0			
12:12	5	70	69.6			
12:18	6	70	69.3			
12:24	7	70	68.4			
12:30	8	70	68.4			
12:36	9	70	69.8			

a. Find the values of $E(t)$, $\Delta E(t)$, and $\Delta X(t)$ for periods 3 through 9.

b. Plot $Y(t)$ versus t.

c. Plot $E(t)$ versus t.

d. Find the average squared error for periods 3 through 9, for periods 4 through 9 ... for periods 7 through 9.

e. Make a plot of your values from (d) for $t = 3, \ldots, 7$. (At time t plot the average squared error for periods t through 9.) Does the appearance of this plot give you hope that any transient or "startup" effects have been eliminated before the last few periods and that those periods adequately represent control algorithm performance?

3.42. Refer to the **Paper Dry Weight** case of problems (3.38), (3.39), (3.40), and (3.41). Yet another control algorithm was considered. κ_1 from problem (3.41) was halved and the coefficient κ_2 was left at 1.66. The pump speed dial was set to 4.5 at 12:37. Thereafter, the new "PI" control algorithm was used to produce the values in the following table.

Time	Period, t	$T(t)$	$Y(t)$	$E(t)$	$\Delta E(t)$	$\Delta X(t)$
12:42	1	70	66.2			
12:45	2	70	66.4			
12:51	3	70	67.2			
12:58	4	70	69.4			
1:04	5	70	69.5			
1:10	6	70	69.2			
1:16	7	70	70.1			
1:22	8	70	66.2			
1:29	9	70	71.7			

a. Find $E(t)$ for all 9 periods and $\Delta E(t)$ and the corresponding $\Delta X(t)$ for periods 2 through 8.
b. Plot $Y(t)$ versus t.
c. Plot $E(t)$ versus t.
d. Find the average squared error for periods 3 through 9, for periods 4 through 9 ... for periods 7 through 9.
e. Make a plot of your values from (d) for $t = 3, \ldots, 7$. (At time t plot the average squared error for periods t through 9.) Does the appearance of this plot give you hope that any transient or "startup" effects have been eliminated before the last few periods and that those periods adequately represent control algorithm performance?

3.43. Refer to Example 3.6.
a. Plot $Y(t)$ versus t.
b. Plot $E(t)$ versus t.
c. Find the average squared error for periods 4 through 11, for periods 5 through 11 ... for periods 9 through 11.
d. Make a plot of your values from (c) for $t = 4, \ldots, 9$. (At time t plot the average squared error for periods t through 11.) Does the appearance of this plot give you hope that any transient or "startup" effects have been eliminated before the last few periods and that those periods adequately represent control algorithm performance?

3.44. Refer to the **Paper Dry Weight** case and specifically the plots in problems (3.38d), (3.39f), (3.40e), (3.41e), (3.42e), and (3.43d). Which control equation seems to be best in terms of producing small average squared error?

3.45. Rewrite the PID control equation (3.49) so that $\Delta X(t)$ is expressed in terms of a linear combination of $E(t)$, $E(t-1)$, and $E(t-2)$, the current and two previous errors.

3.46. Fill levels of jelly jars are of interest. Every half hour, three jars are taken from a production line and net contents measured and recorded. The range and average of these three measurements are calculated and plotted on charts. One of these two charts is intended to monitor location of the fill distribution and the other is meant to be useful in monitoring the spread of the fill distribution.

a. What is the standard name for the chart used to monitor location of the fill level distribution?
b. What is the standard name for the chart used to monitor spread of the fill level distribution?
c. What is the name and value of the tabled constant used to make retrospective control limits for process location?
d. What are the names and values of the two tabled constants used to make retrospective control limits for process spread or variability?
e. In this context, what constitutes a natural subgroup?
f. Let R stand for a single subgroup range. Give an expression for the usual estimate of process short-term variability (σ) based on an average of such ranges.

3.47. Consider again the scenario of problem (3.46). Suppose that instead of ranges and averages, sample standard deviations and averages are computed and plotted.
a. What is the name and value of the tabled constant used to make retrospective control limits for process location?
b. What are the names and values of the two tabled constants used to make retrospective control limits for process spread or variability?
c. Let s stand for a single subgroup standard deviation. Give an expression for the usual estimate of process short-term variability (σ) based on an average of such standard deviations.

3.48. Consider again the scenario of problems (3.46) and (3.47) and suppose that instead of three jars, 10 jars are sampled every half hour. Redo problems (3.46) and (3.47) with this change. For a given set of ranges or standard deviations, will retrospective control limits be wider or will they be narrower with this new sample size?

3.49. Consider again the scenario of problems (3.46) and (3.47) and suppose that instead of plotting averages to monitor location, the decision is made to plot medians. What multiple of σ (or an estimate of this quantity) would be used to set control limits for medians around some central value in the case that $n = 3$? In the case that $n = 11$?

3.50. Sometimes engineers know σ and/or μ (or have credible approximate values for these) when setting control limits. Consider the drained weights of the contents of cans of Brand X green beans. Believable values for the mean and standard deviation of these weights are 21.0 oz and 1.0 oz respectively. Suppose that in a Brand X canning factory, 8 of these cans are sampled every hour and their net contents determined. Sample means and ranges are then computed and used to monitor stability of the filling process.
a. What is the name and value of the multiplier of $\sigma = 1.0$ that would be used to establish a center line for sample ranges?

b. What are the names and values of the multipliers of $\sigma = 1.0$ that would be used to establish upper and lower control limits for sample ranges?

c. What center line and control limits should be established for sample means in this scenario?

3.51. Consider again the situation of problem (3.50), but suppose that instead of ranges and averages, sample standard deviations and averages are computed and plotted. Answer the questions posed in problem (3.50) in this case.

3.52. Consider again the situation of problems (3.50) and (3.51) and suppose that instead of 8 cans, only 5 cans are sampled every hour. Redo problems (3.50) and (3.51) with this change. Will control limits be wider or will they be narrower with this new sample size?

3.53. **Electronic Card Assemblies**. In a 1995 article in *Quality Engineering*, Ermer and Hurtis discussed applications of control charting to the monitoring of soldering defects on electronic card assemblies. One assembly technology they studied was pin-in-hole (PIH) technology, which uses wave soldering to secure components to printed circuit boards after the leads of components have been inserted through holes drilled in the boards. The most common types of soldering defects encountered using this technology are "shorts" (unwanted electrical continuity between points on an assembly) and "opens" (the absence of desired electrical continuity between points on an assembly).

Production of a particular card is done in "jobs" consisting of 24 cards. All cards from a job are tested and a count is made of the total number defects found on the job. What type of probability model might plausibly be used to describe the number of defects found on a given job? What type of control chart might you use to monitor the production of soldering defects? Suppose that records on 132 jobs show a total of 2 defects recorded. What retrospective control limits might then be applied to the 132 different counts of defects? Does a job with any defect at all signal a lack of control?

CHAPTER 4

Process Monitoring Part 2: Additional Methods

The control charting methods introduced in Chapter 3 are relatively straightforward and can be used even by people with minimal quantitative backgrounds. They are robust tools that can be successfully applied in a wide variety of quality assurance situations. There are occasions, however, where somewhat more specialized tools are needed and cases where somewhat more complicated process-monitoring methods can be much more effective than the elementary methods of Shewhart control charting. Some of these tools and methods are the subject of this chapter.

The chapter begins with two sections discussing process-monitoring techniques that improve on Shewhart charts in situations where it is important to quickly detect small process changes, and modern computing power relieves one of the need to compute and plot by hand. That is, EWMA and CUSUM process-monitoring schemes are introduced. The third section in the chapter introduces a method of multivariate process monitoring that allows one to keep track of several quality variables simultaneously, potentially detecting not only changes in individual variables, but also in the nature of relationships between variables. Finally, the chapter ends with a section treating the charting of individuals and "moving ranges," which is sometimes used where the natural sample size for measurements (the rational subgroup size) is $n = 1$.

4.1 EWMA CHARTS

The introduction to Shewhart charting presented in Section 3.1 essentially takes for granted that the most natural thing to do with values of a statistic Q describing process behavior is to plot them in "raw" form, one at a time. While this may be so, there are ways of processing a sequence of values of Q that from some perspectives turn out to have advantages over the plotting of the raw values. This section discusses one such method, the making of control charts based on **Exponentially Weighted Moving Averages** (EWMAs).

For a sequence of values $Q_1, Q_2, Q_3, \ldots$ the corresponding sequence of exponentially weighted moving averages is defined by

Recursion for EWMAs

$$EWMA_i = \lambda Q_i + (1 - \lambda)EWMA_{i-1}, \tag{4.1}$$

where $0 < \lambda \leq 1$ is some constant and $EWMA_0$ is some starting value for the EWMA sequence. The ith EWMA value is a weighted average of the ith Q and the previous EWMA value. Making a EWMA sequence from a sequence of Q's produces a "smoothed" version of the Q sequence.

Example 4.1 Monitoring Pin Head Diameters on Electrical Components. Davies and Sehili worked with the manufacturer of some small electrical devices, monitoring the performance of a machine forming heads on wires inserted through the base of a particular part. A feature of interest was the diameter of the heads formed. Many thousands of these heads were formed each hour. Samples of $n = 5$ consecutive parts taken about once every 10 minutes produced the means and ranges for measured diameters of heads at a particular position that are presented in Table 4.1. (Units are .001 inch.)

Consider the making of EWMAs based on the $\bar{x}$'s in Table 4.1 beginning with a starting value of $EWMA_0 = 25.0$ (which was the target value for these diameters). Then, for example using $\lambda = .2$, one calculates from equation (4.1) to produce

$$\begin{aligned}
EWMA_1 &= .2(30.8) + .8(25.0) = 26.16, \\
EWMA_2 &= .2(31.4) + .8(26.16) = 27.21, \\
EWMA_3 &= .2(30.2) + .8(27.21) = 27.81, \\
EWMA_4 &= .2(30.4) + .8(27.81) = 28.33,
\end{aligned}$$

and so on. Figure 4.1 shows plots of the raw $\bar{x}$ sequence and EWMAs based on the $\bar{x}$ sequence (and a starting value of $EWMA_0 = 25.0$) for $\lambda = .05, .1, .2,$ and .3.

Notice that the plots of exponentially weighted moving averages are smoother than the plot of raw $\bar{x}$'s, the plots with the smaller values of λ being smoother

Table 4.1 Means and Ranges of Measured Pin Head Diameters for 25 Samples of 5 Consecutive Parts (.001 inch)

Sample (i)	$\bar{x}_i$	R_i	Sample (i)	$\bar{x}_i$	R_i
1	30.8	5	14	28.0	4
2	31.4	4	15	26.0	2
3	30.2	4	16	27.0	2
4	30.4	3	17	28.6	5
5	32.0	0	18	28.6	2
6	32.8	2	19	29.6	2
7	33.2	2	20	29.6	3
8	33.2	4	21	30.2	5
9	34.0	2	22	28.8	3
10	25.8	1	23	28.8	2
11	26.4	3	24	29.6	3
12	26.4	3	25	29.2	4
13	24.8	3			

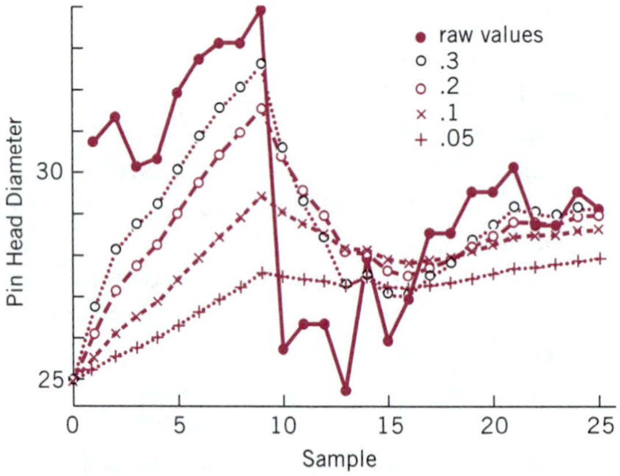

Figure 4.1
Plots of the raw $\bar{x}$ sequence and four different EWMAs for the pin head diameters.

than the ones for the larger values of λ. Notice also that the fairly large change in the level of $\bar{x}$ that occurs between the 9th and 10th samples is tracked by all the EWMAs, but most quickly by those based on the larger values of λ. On the other hand, the fairly gentle general increase in $\bar{x}$'s seen from sample 10 onward is most obvious in the EWMAs for small λ.

It is immediate from equation (4.1) that the limiting form of a EWMA sequence as $\lambda \to 1$ is simply the sequence of raw Q values. The limiting form as $\lambda \to 0$

is a sequence constant at the starting value. Figure 4.1 illustrates that EWMA sequences can be interpreted as smoothed-out versions of the original Q sequence, provided that one remembers that big changes in level for Q are seen only gradually in the EWMA sequence. On the other hand, it seems that small changes and weak trends in a Q sequence may in fact be easier to see in a corresponding EWMA sequence. The motivation for basing a process monitoring scheme on EWMAs is to take advantage of this apparent ability to portray small process changes more clearly than a plot of raw Q's.

The first question to be addressed if one is to plot EWMAs, is exactly how to derive signals of process change from them. A naive answer to this question is to simply look for "3 sigma control limits." It turns out that if stable process conditions hold, $EWMA_i$ defined in equation (4.1) has standard deviation

$$\sigma_{EWMA_i} = \sigma_Q \sqrt{\left(\frac{\lambda}{2-\lambda}\right)\left(1-(1-\lambda)^{2i}\right)}, \tag{4.2}$$

which, as i grows, quickly approaches its limit

$$\sigma_Q \sqrt{\frac{\lambda}{2-\lambda}}. \tag{4.3}$$

Some treatments of EWMA charts simply recommend setting $EWMA_0$ at a standard mean for Q and using standards given limits for $EWMA_i$ set $3\sigma_{EWMA_i}$ away from the starting value. Or, if constant (with i) limits are desired, use of the limiting value (4.3) in place of the exact standard deviation (4.2) is sometimes suggested.

The biggest problem with this "obvious" approach to setting up a EWMA chart is that it doesn't really take into account the likely (ARL) performance of the resulting monitoring scheme. Rational methods for designing and evaluating the ARL performance of EWMA schemes for normally distributed variables Q can be based on work of Crowder that appeared in *Technometrics* in 1987 and the *Journal of Quality Technology* in 1987 and 1989. Crowder's 1989 paper suggests the following method of standards given chart design based on standard values μ_Q and σ_Q.

One first sets $EWMA_0 = \mu_Q$, then turns to the choice of λ. One may, in fact, use any (arbitrary) value one likes. But it is also possible to pick a value that is in some sense "optimal." That is, if one has in mind a desired "all-OK" ARL for the EWMA chart and wants to minimize the ARL for a shift of size δ in the mean of Q (away from μ_Q), Table 4.2 (which contains values read from graphs in Crowder's 1989 paper) can be used to select a λ doing the job, say λ^{opt}. (λ^{opt} from Table 4.2, when used with control limits producing the desired on-target ARL, will give the smallest possible ARL when the mean of Q is $\mu_Q \pm \delta$.) To use Table 4.2 one must pick an appropriate column according to

Standardized "Shift" Needed to Enter Table 4.2

$$shift = \frac{\delta}{\sigma_Q}. \tag{4.4}$$

Then, using this value and the desired all-OK ARL, λ^{opt} is read directly from the body of the table. (In practice, rather than using an "exact" but unpleasant value of λ^{opt} taken from the table, it makes sense to pick some "round" value of λ close to λ^{opt} and use it instead.)

Notice in passing that it is only for large values of *shift* that Table 4.2 suggests values of λ near 1. Remember that the case of $\lambda = 1$ amounts to plotting of individual Q's (regular Shewhart charting). So this is clear indication that when a monitoring scheme is to protect against small shifts in the mean of Q, EWMA charting is preferable to regular Shewhart charting.

Finally, with λ chosen one may choose control limits. Rather than simply using a "3 sigma" chart, it is possible to use Table 4.3 to find a "number of sigmas" necessary to provide the desired all-OK ARL. That is, one uses λ (λ^{opt} if Table 4.2

Table 4.2 Approximately Optimal Values of λ for Detecting Shifts of Various Sizes, λ^{opt}

All-OK	*shift*							
ARL	**.5**	**1.0**	**1.5**	**2.0**	**2.5**	**3.0**	**3.5**	**4.0**
50	.08	.22	.38	.56	.71	.83	.92	.96
100	.07	.18	.32	.49	.66	.78	.88	.94
250	.05	.15	.27	.41	.58	.73	.84	.91
370	.05	.14	.25	.37	.54	.70	.82	.90
500	.05	.13	.24	.36	.52	.68	.80	.89
750	.04	.12	.23	.34	.48	.65	.78	.87
1000	.04	.12	.22	.33	.46	.62	.76	.86

Table 4.3 Factors $\mathcal{K}$ for Setting EWMA Control Limits to Produce a Desired All-OK ARL

All-OK	λ							
ARL	**.05**	**.1**	**.2**	**.3**	**.4**	**.5**	**.75**	**1.0**
50	1.52	1.81	2.06	2.17	2.23	2.27	2.32	2.33
100	1.88	2.15	2.36	2.45	2.50	2.53	2.57	2.58
250	2.32	2.55	2.72	2.79	2.83	2.85	2.87	2.88
370	2.49	2.70	2.86	2.93	2.96	2.98	3.00	3.00
500	2.62	2.81	2.96	3.02	3.05	3.07	3.09	3.09
750	2.78	2.96	3.10	3.15	3.18	3.20	3.21	3.21
1000	2.88	3.06	3.19	3.24	3.26	3.28	3.29	3.29

was used to pick this) and the desired ARL to read a value of $\mathcal{K}$ from Table 4.3, and then sets

$$UCL_{EWMA} = \mu_Q + \mathcal{K}\sigma_Q\sqrt{\frac{\lambda}{2-\lambda}} \quad \text{and} \quad LCL_{EWMA} = \mu_Q - \mathcal{K}\sigma_Q\sqrt{\frac{\lambda}{2-\lambda}}. \tag{4.5}$$

EWMA Chart Control Limits

Example 4.1 continued. The ideal pin head diameter in the situation of Davies and Sehili was 25.0 (.001 inch). A process short-term standard deviation roughly consistent with the R values in Table 4.1 is $\sigma = 1.3$. Engineering specifications for these diameters were in fact 25.0 ± 3.0 so that even with a process mean diameter μ exactly on target at 25.0, it appears that some diameters would be outside specifications. And any real change in the process mean would put a large fraction outside specifications. So this is a situation where detecting small shifts in the process mean (and therefore in the mean of $Q = \bar{x}$) might well be important. So EWMA charting could be an attractive alternative to regular Shewhart charting of the sample averages.

Thus, consider the problem of setting up a EWMA chart for $\bar{x}$'s based on samples of $n = 5$ measured pin head diameters. Suppose that an all-OK ARL of 370 is sensible, but it is desirable to have quickest possible detection of a change in process mean of size $\delta = .5$. Then since $Q = \bar{x}$, $\sigma_Q = \sigma/\sqrt{n}$ and using equation (4.4), the shift parameter to be used in conjunction with Table 4.2 is

$$shift = \frac{.5}{1.3/\sqrt{5}} = .86.$$

So reading from the table, an optimal value of λ for this situation is approximately .1. Then using $\lambda = .1$ and the desired ARL of 370, Table 4.3 shows that $\mathcal{K} = 2.70$ is appropriate. So the control limits for EWMAs prescribed by formulas (4.5) are

$$UCL_{EWMA} = 25.0 + 2.70\left(\frac{1.3}{\sqrt{5}}\right)\sqrt{\frac{.1}{2-.1}} = 25.36$$

and

$$LCL_{EWMA} = 25.0 - 2.70\left(\frac{1.3}{\sqrt{5}}\right)\sqrt{\frac{.1}{2-.1}} = 24.64.$$

Figure 4.2 shows the $\lambda = .1$ EWMA plot from Figure 4.1 with these standards given control limits superimposed. It is clear from this plot, that had these limits

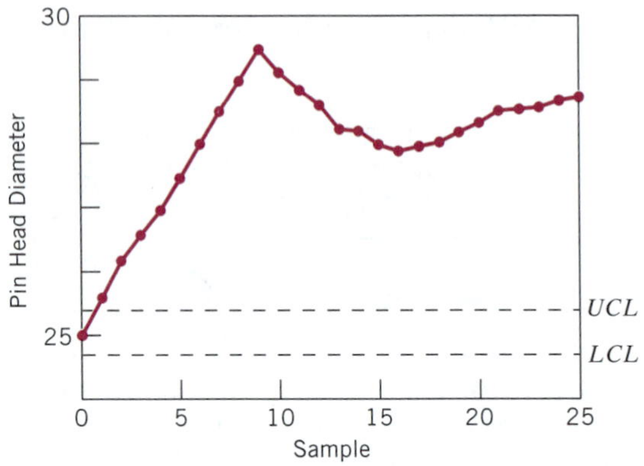

Figure 4.2
Standards given EWMA chart of $\bar{x}$'s ($\lambda = .1$) for the pin head diameter data.

been applied to the samples as they were collected, the fact that the head-forming process was not operating at standard values of process parameters would have been obvious as early as the first sample. (Actually, the first $\bar{x}$ is so far from 25.0 that essentially *any* sensible process-monitoring scheme, including a Shewhart $\bar{x}$ chart, would have detected this.)

Crowder's 1987 *Technometrics* paper and corresponding computer program published in the *Journal of Quality Technology* in 1987 provide bases for evaluating ARLs for EWMA schemes based on normally distributed Q's. (Most often, it is individual measurements x, or sample means $\bar{x}$, that are turned into EWMA sequences. So Crowder's work covers many important situations.) Table A.3 of this book is based on use of a small modification of Crowder's published computer program. To use it to evaluate ARLs for a EWMA scheme, one needs the value of λ and two other variables, $\mathcal{D}^*$ and $\mathcal{H}^*$. If Q's have mean μ_Q and standard deviation σ_Q, the EWMA sequence is subjected to control limits UCL_{EWMA} and LCL_{EWMA} and is started with $EWMA_0 = (UCL_{EWMA} + LCL_{EWMA})/2$, then these variables may be computed as

Standardized "Mean Shift" Needed to Enter Table A.3

$$\mathcal{D}^* = \frac{\left|\mu_Q - \dfrac{UCL_{EWMA} + LCL_{EWMA}}{2}\right|}{\sigma_Q} \tag{4.6}$$

and

"Number of Sigmas" Parameter Needed to Enter Table A.3

$$\mathcal{H}^* = \frac{UCL_{EWMA} - LCL_{EWMA}}{2\sigma_Q}\sqrt{\frac{2-\lambda}{\lambda}}. \tag{4.7}$$

($\mathcal{H}^*$ is the actual "number of sigmas" the control limits are from the center line of the chart.) Armed with λ, $\mathcal{D}^*$, and $\mathcal{H}^*$ one can interpolate in Table A.3 to get EWMA ARLs.

Example 4.1 continued. In order to illustrate the use of formulas (4.6) and (4.7) (and Table A.3) consider the EWMA scheme for $\bar{x}$'s based on $n = 5$ with $\lambda = .1$, $EWMA_0 = 25.0$, $UCL_{EWMA} = 25.36$ and $LCL_{EWMA} = 24.64$. Suppose that in fact the head-forming process is stable at $\mu = 26.0$ and $\sigma = 2.0$. To find the ARL for this situation one computes

$$\mathcal{D}^* = \frac{\left|26.0 - \dfrac{25.36 + 24.64}{2}\right|}{\dfrac{2.0}{\sqrt{5}}} = 1.12$$

and

$$\mathcal{H}^* = \frac{25.36 - 24.64}{2\left(\dfrac{2.0}{\sqrt{5}}\right)}\sqrt{\frac{2 - .1}{.1}} = 1.75.$$

Logarithmic interpolation (over $\mathcal{D}^*$) in Table A.3 for $\lambda = .1$ and $\mathcal{H}^* = 1.75$ produces an ARL of 5.0 for this situation. The shift in mean and increase in process standard deviation (compared to the standard values used to set up the monitoring scheme) will on average lead to a fairly quick signal from the EWMA monitoring scheme.

The treatment of EWMA schemes in this section has purposely been from a standards given perspective. While it is possible, for example, to set $EWMA_0 = \bar{\bar{x}}$ and replace $\mu_{\bar{x}}$ by $\bar{\bar{x}}$ and $\sigma_{\bar{x}}$ by $\hat{\sigma}/\sqrt{n}$ (for $\hat{\sigma}$ some estimate of σ) in limits (4.5), EWMA charts are not commonly used for retrospective analyses. And they are not really suited to such use.

Before closing this section it will be well to review briefly the purposes, pros, and cons of EWMA schemes. Because the arithmetic required to use display (4.1) can be slightly unpleasant, EWMA charts are not really ones for hand calculation on the shop floor. But where calculation and plotting are handled by a computer, they offer ease of interpretation that is close to that for a corresponding Shewhart chart, accompanied by better (on average) sensitivity to small process upsets.

The price to be paid for this increased sensitivity to small process upsets is that (because of the smoothing a EWMA does) detection of very large process changes takes somewhat longer on average than with plotting of raw Q values. Further, EWMA schemes have unfortunate worst-case-scenario behavior. To see this, suppose that at the beginning of monitoring μ_Q and σ_Q are standard, but sometime later it happens that the EWMA wanders in the direction of one control limit. If at that point there is a process change causing μ_Q to shift from its standard value in the direction of the other control limit, it can take a very long time for the EWMA plot to first return to the standard value and then move to that control limit. This worst-case-scenario/inertia problem is not shared by the Shewhart chart (or the CUSUM method discussed in the next section).

4.2 CUSUM CHARTS

The EWMA idea discussed in the previous section is one way to use a sequence of values of Q describing process behavior to produce a monitoring scheme with some advantages over a simple Shewhart chart based on Q's. This section discusses a second method aimed in this direction, the use of **Cumulative Sums** (CUSUMs) in process monitoring. The basics of making and evaluating the ARL properties of both one-sided and "combined" CUSUM schemes are discussed first. Then some additional issues are considered, including sample size choice where sample means are being CUSUMed, Shewhart-CUSUM combinations, and "fast initial response" CUSUM schemes.

4.2.1 Basic CUSUM Charting

For a sequence of values $Q_1, Q_2, Q_3, \ldots$ a corresponding sequence of (raw) cumulative sums is defined by

Recursion for Raw CUSUMs

$$CUSUM_i = (Q_i - k) + CUSUM_{i-1}, \tag{4.8}$$

where k is some constant and $CUSUM_0$ is some starting value for the CUSUM sequence. Often (but not always) $CUSUM_0$ is set to 0. k can be taken to be 0, but more often it is (or is related to) some ideal value for Q.

Example 4.2 CUSUMs and Pin Head Diameter Means. Consider again the sample mean pin head diameters of Davies and Sehili recorded in Table 4.1. With $CUSUM_0 = 0$ and $k = 25.0$ (the ideal pin head diameter), the CUSUM values derived from the $\bar{x}$'s via equation (4.8) are

$$CUSUM_1 = (\bar{x}_1 - 25.0) + CUSUM_0 = 5.8 + 0 = 5.8,$$
$$CUSUM_2 = (\bar{x}_2 - 25.0) + CUSUM_1 = 6.4 + 5.8 = 12.2,$$

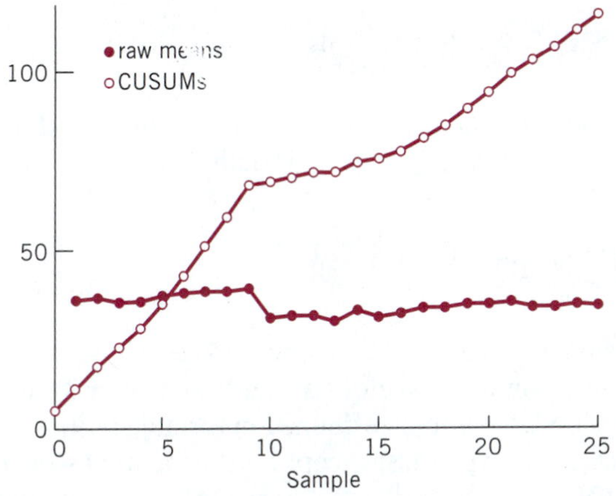

Figure 4.3
Plots of $\bar{x}$'s and a corresponding sequence of CUSUMs for pin head diameters.

$$CUSUM_3 = (\bar{x}_3 - 25.0) + CUSUM_2 = 5.2 + 12.2 = 17.4,$$
$$CUSUM_4 = (\bar{x}_4 - 25.0) + CUSUM_3 = 5.4 + 17.4 = 22.8,$$

and so on. Figure 4.3 shows a plot of the raw $\bar{x}$ sequence and the corresponding CUSUMs.

Until one has had some practice with them, plots of cumulative sums can seem harder to interpret than plots of raw $\bar{x}$'s. On the plot of $\bar{x}$'s in Figure 4.3 the (vertical) level of plotted points indicates the size of the observed $\bar{x}$'s, while on the CUSUM plot it is the **slope** of the plot that indicates the size of pin head diameters. Where the trend on a CUSUM plot is up (the slope is positive), the values being CUSUMed are larger than k. Where the trend on a CUSUM plot is down, the values being CUSUMed are smaller than k. And changes in the size of values being CUSUMed are manifested in terms of changes in slope.

But once one gets used to thinking "slope equals size of CUSUMed values" it is easier to see small changes in size on a CUSUM plot than on a corresponding plot of raw values. (And retrospectively identifying the timing of a level change in a variable is often easier on a CUSUM plot than on a plot of raw values.) It is this ability to show clearly small changes in level that motivates the use of CUSUMs in process monitoring.

There are ways of using CUSUMs defined exactly as in equation (4.8) to set up process-monitoring schemes. A "V-mask" can be placed over a graph of raw CUSUMs like that in Figure 4.3 as each new point is plotted and be used to decide whether a plotted slope has changed from a standard value. But these methods are physically cumbersome and rarely used in practice. What is used instead is a slight modification of the raw CUSUM idea defined in expression (4.8).

For k_1 an appropriate constant and U_0 a starting value, a **high side decision interval CUSUM** sequence based on $Q_1, Q_2, Q_3, \ldots$ is defined by

Recursion for High Side CUSUMs

$$U_i = \max[0, (Q_i - k_1) + U_{i-1}]. \tag{4.9}$$

Similarly, for k_2 an appropriate constant and L_0 a starting value, a **low side decision interval CUSUM** sequence based on $Q_1, Q_2, Q_3, \ldots$ is defined by

Recursion for Low Side CUSUMs

$$L_i = \min[0, (Q_i - k_2) + L_{i-1}]. \tag{4.10}$$

For the time being, this discussion will be limited to the choices $U_0 = 0$ and $L_0 = 0$ for the starting values, and proper choice of constants k_1 and/or k_2 will be discussed soon. The high side CUSUM sequence defined in formula (4.9) is essentially a raw CUSUM sequence of the type (4.8) except for the fact that should a CUSUM ever take a negative value, it is "reset" to 0. In similar fashion, the low side CUSUM sequence defined in formula (4.10) is essentially a raw CUSUM sequence of the type (4.8) except for the fact that should a CUSUM ever take a positive value, it is reset to 0.

Table 4.4 Mean Diameters, High and Low Side CUSUMs Based on 20 Hourly Samples of $n = 4$ Machined Parts (.0001 inch above 1.1800 inch)

Sample (i)	$\bar{x}_i$	U_i	L_i
0	—	0	0
1	9.25	max[0, −.75 + 0] = 0	min[0, 1.25 + 0] = 0
2	8.5	max[0, −1.5 + 0] = 0	min[0, .5 + 0] = 0
3	9.5	max[0, −.5 + 0] = 0	min[0, 1.5 + 0] = 0
4	6.25	max[0, −3.75 + 0] = 0	min[0, −1.75 + 0] = −1.75
5	5.25	max[0, −4.75 + 0] = 0	min[0, −2.75 + (−1.75)] = −4.50
6	5.25	max[0, −4.75 + 0] = 0	min[0, −2.75 + (−4.50)] = −7.25
7	5.75	max[0, −4.25 + 0] = 0	min[0, −2.25 + (−7.25)] = −9.50
8	19.5	max[0, 9.5 + 0] = 9.5	min[0, 11.5 + (−9.50)] = 0
9	10.0	max[0, 0 + 9.5] = 9.5	min[0, 2.0 + 0] = 0
10	9.5	max[0, −.5 + 9.5] = 9.0	min[0, 1.5 + 0] = 0
11	9.5	max[0, −.5 + 9.0] = 8.5	min[0, 1.5 + 0] = 0
12	9.75	max[0, −.25 + 8.5] = 8.25	min[0, 1.75 + 0] = 0
13	12.25	max[0, 2.25 + 8.25] = 10.5	min[0, 4.25 + 0] = 0
14	12.75	max[0, 2.75 + 10.5] = 13.25	min[0, 4.75 + 0] = 0
15	14.5	max[0, 4.5 + 13.25] = 17.75	min[0, 6.5 + 0] = 0
16	8.0	max[0, −2.0 + 17.75] = 15.75	min[0, 0 + 0] = 0
17	10.0	max[0, 0 + 15.75] = 15.75	min[0, 2.0 + 0] = 0
18	10.25	max[0, .25 + 15.75] = 16.0	min[0, 2.25 + 0] = 0
19	8.75	max[0, −1.25 + 16.0] = 14.75	min[0, .75 + 0] = 0
20	10.0	max[0, 0 + 14.75] = 14.75	min[0, 2.0 + 0] = 0

Example 4.3 Decision Interval CUSUMs and Outer Diameters Turned on a CNC Lathe. Allan, Robbins and Wyckoff worked with a machine shop that employed a computer numerically controlled (CNC) lathe to turn an outer diameter on a part it was making for a heavy equipment manufacturer. $\bar{x}$ values for the measured diameters on 20 hourly samples of $n = 4$ parts are given in Table 4.4, along with high side and low side CUSUM values for the $\bar{x}$'s based on $U_0 = 0$, $L_0 = 0$, $k_1 = 10.0$, and $k_2 = 8.0$. The units in Table 4.4 are 10^{-4} inches above 1.1800 inch, and the target for this diameter was 1.1809 inch.

Figure 4.4 shows plots of the "raw" CUSUM sequence defined in display (4.8) made with $CUSUM_0 = 0$ and $k = 9$ and the two decision interval CUSUM sequences defined in displays (4.9) and (4.10) and computed in Table 4.4. Notice that when they are not "zeroed out" the decision interval CUSUMs move in ways similar to the raw CUSUM sequence.

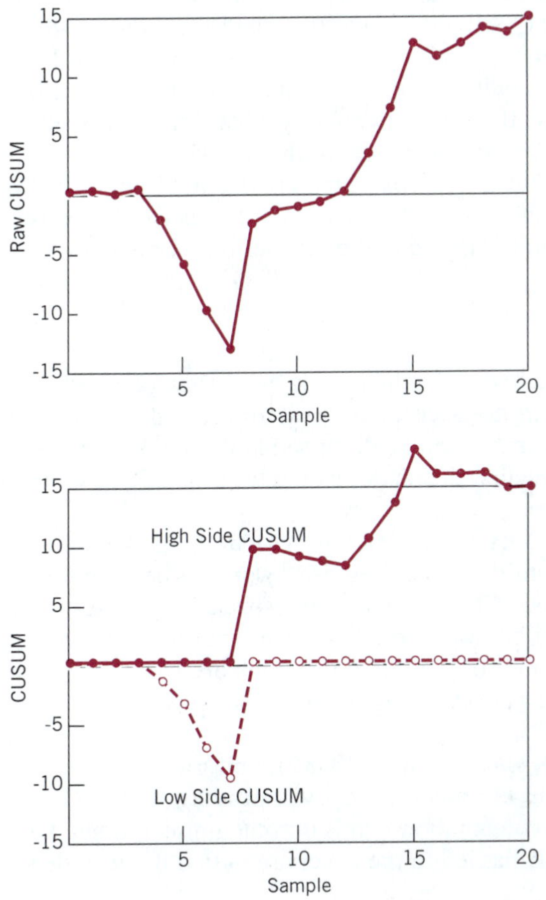

Figure 4.4
Plots of a raw CUSUM sequence and corresponding high and low side CUSUMs.

The choices made for the constants k, k_1, and k_2 in Example 4.3 are typical of CUSUM control charting practice. For a raw CUSUM plot, the value k in display (4.8) is usually chosen to be an "ideal" or standard value of Q, so that slope up indicates an average for Q above the ideal, and slope down indicates a mean below the ideal. For a high side CUSUM, k_1 in display (4.9) is usually chosen to be larger than the ideal value of Q. So if the sequence of values being CUSUMed is typically around or below the standard value, the $(Q_i - k_1)$ terms will tend to be negative and the U_i will tend to stay zeroed out. Similarly, for a low side CUSUM, k_2 in display (4.10) is usually chosen to be smaller than the ideal value of Q. So if the sequence of values being CUSUMed is typically around or above the standard value, the $(Q_i - k_2)$ terms will tend to be positive and the L_i will tend to stay zeroed out. A decision interval CUSUM that then wanders away from 0 is one that suggests that the mean of Q is not at its standard value. A wandering high side CUSUM indicates an increase in average Q and a wandering low side CUSUM indicates a decrease in average Q.

It is worth noting at this point that one could conceivably use either of the two CUSUMs (4.9) and (4.10) by itself in monitoring a process. Or both CUSUMs can be kept on the same sequence of Q's. If one is only concerned about the possibility of an increase in average Q, keeping only the high side CUSUM will suffice. If one is only concerned about the possibility of a decrease in average Q, keeping only the low side CUSUM will suffice. If both the possibility of an increase in mean Q or a decrease in mean Q are of practical concern, both should be kept.

The precise means by which out-of-control signals are derived from CUSUMs (4.9) and/or (4.10) is the following. For a positive constant h, called the **decision interval**, an out-of-control warning is issued at the first period i for which

Decision Interval CUSUM Alarm Rules

$$U_i > h \quad \text{or} \quad L_i < -h. \tag{4.11}$$

(For cases where both high and low side schemes are used simultaneously, it is possible to use different values of the decision interval for the high and low side CUSUMs, but this is not a matter that will be discussed in this book.) That is, an out-of-control signal is issued the first time a decision interval CUSUM wanders too far from 0.

Rational choice of the parameters h and k_1 and/or k_2 for a CUSUM monitoring scheme can be based on ARL considerations. Extensive work has been done computing ARLs for decision interval CUSUM schemes, especially for cases where the variables $Q_1, Q_2, Q_3, \ldots$ can be modeled as independent observations from a normal distribution. The methods presented in this book are based on normal CUSUM ARLs computed using a Fortran program published by Gan in the *Journal of Quality Technology* in 1993.

If one has in mind an all-OK ARL for a CUSUM monitoring scheme, there are many combinations of h and k_1 and/or k_2 that will produce that ARL. It is thus possible to try to choose a combination that is in some sense optimal. Work by a number of different authors has led to the observation that if one wishes to minimize (for a fixed all-OK ARL) the ARL if the mean of Q shifts by an amount δ, then approximately optimal values for k_1 and/or k_2 are

$$k_1^{\text{opt}} = \mu_Q + \frac{\delta}{2} \quad \text{and/or} \quad k_2^{\text{opt}} = \mu_Q - \frac{\delta}{2}, \tag{4.12}$$

Optimal CUSUM Reference Values

where μ_Q is the standard mean for the statistic Q.

Whether or not one uses display (4.12), once one has chosen a value for k_1 or k_2, Table 4.5 can be used to choose h if only one of the decision interval CUSUMs is to be used. Similarly, if both high and low side CUSUMs are to be employed and one has chosen k_1 and k_2 symmetrically about μ_Q, Table 4.6 can be used to choose h. To use these tables, one computes

Table 4.5 Values of $\mathcal{H}$ for Designing One-sided CUSUM Schemes for Normal Variables

All-OK ARL	$\mathcal{H}$ .25	.50	.75	1.00	1.25	1.50
50	3.34	2.23	1.60	1.18	.85	.57
100	4.42	2.85	2.04	1.53	1.16	.86
250	5.99	3.72	2.63	1.98	1.55	1.22
370	6.71	4.10	2.88	2.18	1.71	1.36
500	7.27	4.39	3.08	2.32	1.83	1.47
750	8.03	4.79	3.35	2.52	1.99	1.61
1000	8.59	5.07	3.54	2.67	2.11	1.71

Table 4.6 Values of $\mathcal{H}$ for Designing Combined High and Low Side CUSUM Schemes for Normal Variables

All-OK ARL	$\mathcal{H}$ .25	.50	.75	1.00	1.25	1.50
50	4.42	2.85	2.04	1.53	1.16	.86
100	5.60	3.50	2.48	1.87	1.46	1.13
250	7.27	4.39	3.08	2.32	1.83	1.47
370	8.01	4.77	3.34	2.52	1.99	1.60
500	8.59	5.07	3.54	2.67	2.11	1.71
750	9.37	5.47	3.81	2.87	2.27	1.84
1000	9.93	5.76	4.00	3.01	2.38	1.94

Standardized Reference Value Needed to Enter Table 4.5 or 4.6

$$\mathcal{K} = \frac{k_1 - \mu_Q}{\sigma_Q} \quad \text{or } \mathcal{K} = \frac{\mu_Q - k_2}{\sigma_Q}, \tag{4.13}$$

and armed with $\mathcal{K}$ and a desired all-OK ARL, reads a value of $\mathcal{H}$ out of the table. An appropriate decision interval is then

CUSUM Decision Interval h

$$h = \mathcal{H}\sigma_Q. \tag{4.14}$$

Example 4.3 continued. Consider again the turning example of Allan, Robbins, and Wyckoff. The standard value for the mean part diameter was (in the units of Table 4.4) $\mu_Q = 9.0$. Based on information not shown in Table 4.4, a plausible value for the process short-term standard deviation of individual diameters was $\sigma = 1.6$. Suppose then that one is planning to monitor means of samples of size $n = 4$ as in the students' data. Suppose further that an all-OK ARL of approximately 370 is desired for a monitoring scheme consisting of both high and low side CUSUM schemes run simultaneously. And suppose finally, that subject to the 370 ARL condition, one wishes to optimize the monitoring scheme for quickest possible detection of a change in mean part diameter of size $\delta = 2.0$. How then should CUSUM scheme parameters be set?

To begin with, relationships (4.12) say that in order to provide quickest possible detection of a change in mean part diameter of size 2.0, one should set

$$k_1 = 9.0 + \frac{2.0}{2} = 10.0 \quad \text{and} \quad k_2 = 9.0 - \frac{2.0}{2} = 8.0.$$

These are, of course, the values used earlier in the example calculations in Table 4.4. Then relationships (4.13) say that in using Table 4.6, one will use

$$\mathcal{K} = \frac{10.0 - 9.0}{1.6/\sqrt{4}} = 1.25,$$

so reading from Table 4.6 on the $ARL = 370$ line, $\mathcal{H} = 1.99$ is in order. Then finally, an appropriate decision interval is from display (4.14),

$$h = 1.99\left(\frac{1.6}{\sqrt{4}}\right) = 1.592.$$

Returning to Table 4.4, it is clear that had this decision interval been applied to the CUSUMs of sample means as they were observed, an out-of-control signal would have been issued as early as sample four (by the low side scheme).

Evaluation of likely performance of a particular CUSUM scheme depends upon translating the parameters k_1 and/or k_2 and h and the mean and standard deviation of Q under stable process conditions into ARLs. For cases where Q's may be assumed to be normally distributed, this can be done using Table A.4 for one-sided CUSUM schemes and using Table A.5 for combinations of high and low side schemes.

To enter Table A.4, one needs values of parameters $\mathcal{H}^*$ and $\mathcal{S}^*$ computed from chart parameters h and k_1 or k_2 together with relevant μ_Q and σ_Q. These are

$$\mathcal{H}^* = \frac{h}{\sigma_Q}, \tag{4.15}$$

Standardized Decision Interval Needed to Enter Table A.4 or A.5

and for high side CUSUM schemes

$$\mathcal{S}^* = \frac{\mu_Q - k_1}{\sigma_Q}, \tag{4.16}$$

High Side Standardized Reference Value Needed to Enter Table A.4

while for low side schemes

$$\mathcal{S}^* = \frac{k_2 - \mu_Q}{\sigma_Q}. \tag{4.17}$$

Low Side Standardized Reference Value Needed to Enter Table A.4

To enter Table A.5, one computes $\mathcal{H}^*$ according to formula (4.15), but then uses parameters

$$\mathcal{D}^* = \frac{\left|\mu_Q - \frac{k_1+k_2}{2}\right|}{\sigma_Q} \tag{4.18}$$

Standardized Mean Shift Needed to Enter Table A.5

and

$$\mathcal{K}^* = \frac{k_1 - k_2}{2\sigma_Q}. \tag{4.19}$$

Standardized Reference Value Needed to Enter Table A.5

Example 4.4 Evaluation of a High Side CUSUM ARL. As a synthetic but simple illustration of the use of Table A.4, consider a hypothetical high side CUSUM scheme with $U_0 = 0$, $k_1 = 3.0$, and $h = 5.0$. Suppose then that $Q_1, Q_2, Q_3, \ldots$ can be treated as independent normal variables with mean $\mu_Q = 2.5$ and standard deviation $\sigma_Q = 2.0$. Then using equations (4.15) and (4.16), one enters Table A.4 with

$$\mathcal{H}^* = \frac{5.0}{2.0} = 2.5 \quad \text{and} \quad \mathcal{S}^* = \frac{2.5 - 3.0}{2.0} = -.25\,.$$

Reading directly from Table A.4, the desired ARL is 27.

Example 4.3 continued. To illustrate the finding of CUSUM ARLs using Table A.5, consider the combined high and low side CUSUM scheme set up earlier for the problem of Allan, Robbins, and Wyckoff. That is, consider monitoring mean diameters of samples of size $n = 4$ using *both* decision interval CUSUM schemes (4.9) and (4.10) with $k_1 = 10$, $k_2 = 8$, and $h = 1.592$. Suppose that beginning from startup the turning process is physically stable, but not at standard process parameters. In fact, suppose the process mean is $\mu = 10.0$ and the process standard deviation is $\sigma = 2.5$. Then $Q = \bar{x}$ has mean $\mu_{\bar{x}} = 10.0$ and standard deviation $\sigma_{\bar{x}} = 2.5/\sqrt{4} = 1.25$. So, using displays (4.15), (4.18), and (4.19) one needs the values

$$\mathcal{H}^* = \frac{1.592}{1.25} = 1.27\,,$$

$$\mathcal{D}^* = \frac{\left|10.0 - \dfrac{10.0 + 8.0}{2}\right|}{1.25} = .80\,,$$

and

$$\mathcal{K}^* = \frac{10.0 - 8.0}{2(1.25)} = .80$$

in order to enter Table A.5 to find the desired ARL. Interpolation in that table (or "exact" calculation with the program of Gan used to create it) then yields $ARL \approx 6.0$. The CUSUM scheme designed earlier will take on average about 6 periods to detect this set of nonstandard process parameters.

It is worth noting before going on to other considerations, that actually only Table A.4 is really needed even to find ARLs for the combined high and low side CUSUMs considered here. That is because for the schemes considered here, if

ARL_{high}, ARL_{low}, and ARL_{combined} are respectively the ARLs associated with the high side, the low side, and the combined CUSUM schemes, it turns out that

$$ARL_{\text{combined}} = \frac{ARL_{\text{high}} ARL_{\text{low}}}{ARL_{\text{high}} + ARL_{\text{low}}}. \tag{4.20}$$

Since the terms on the right of display (4.20) can all be evaluated using Table A.4, Table A.5 is convenient, but not absolutely necessary.

This introduction to basic CUSUM charting has intentionally been phrased primarily in standards given terms. Raw CUSUMs (4.8) (without the application of any formal signaling rules) are useful informal data-snooping tools. But CUSUM schemes are not particularly suited to the kind of more formal retrospective analysis for which Shewhart charts were used in Chapter 3.

4.2.2 Some Additional Considerations Regarding CUSUM Charts

There are several additional matters that sometimes arise in the application of CUSUM schemes that, while not essential to an introduction to CUSUM charts, do deserve some mention before ending this section. The first has to do with situations where one has control over the standard deviation of the variables being CUSUMed. Principally, this concerns situations where $Q = \bar{x}$ and one may choose the sample size upon which means are going to be based. In such cases, since under stable process conditions $\sigma_Q = \sigma_{\bar{x}} = \sigma/\sqrt{n}$, one may make σ_Q small by choice of n large. Control over the size of σ_Q allows one not only to choose chart parameters to guarantee a large all-OK ARL, but also to produce a desired (small) ARL at a mean δ units away from the all-OK mean for Q.

Sample Size Choice

Tables 4.7 and 4.8 can in some cases be used to guide the choice of σ_Q, that is the **choice of n** when $Q = \bar{x}$. The tables show for respectively one-sided and

Table 4.7 Approximate Values of $\frac{\delta}{\sigma_Q} = \sqrt{n}\frac{\delta}{\sigma}$ Required to Produce Given All-OK and Off-Target ARLs at Means Separated by δ Units for a One-sided CUSUM Chart for $\bar{x}$'s

Off-Target ARL	All-OK ARL				
	100	250	370	500	750
10	.69	.85	.90	.95	1.00
7.5	.86	1.03	1.09	1.14	1.19
5.0	1.16	1.35	1.42	1.47	1.54
2.5	1.87	2.12	2.21	2.28	2.37

Table 4.8 Approximate Values of $\frac{\delta}{\sigma_Q} = \sqrt{n}\frac{\delta}{\sigma}$ Required to Produce Given All-OK and Off-Target ARLs at Means Separated by δ Units for a Combined High and Low Side CUSUM Scheme for $\bar{x}$'s

	All-OK ARL				
Off-Target ARL	**100**	**250**	**370**	**500**	**750**
10	.81	.95	1.00	1.03	1.08
7.5	.99	1.14	1.19	1.23	1.28
5.0	1.30	1.47	1.54	1.58	1.64
2.5	2.01	2.28	2.37	2.43	2.52

combined CUSUM schemes what value of the ratio

$$\frac{\delta}{\sigma_Q} = \frac{\delta}{\sigma/\sqrt{n}} = \sqrt{n}\frac{\delta}{\sigma}$$

is required to produce the indicated pairs of all-OK and off-target ARLs at process means separated by a distance $\delta > 0$. Armed with values for δ and the process standard deviation σ, one may

1. read a value from the table and solve for the sample size necessary to produce it, and then
2. based on that sample size use the procedures already discussed to pick chart parameters guaranteeing the all-OK ARL and producing the minimum possible ARL for a process mean δ units away from the target mean.

This two-step procedure will produce a CUSUM scheme meeting both ARL goals.

Example 4.3 continued. The two-sided CUSUM scheme set up earlier in this section for the monitoring of the part diameters turned on the CNC lathe was derived under the fixed choice of hourly sample size, $n = 4$. As a modification of the earlier analysis, suppose that what is needed for the practical monitoring of part diameters is a two-sided CUSUM scheme with all-OK ARL of 370 and an ARL of no more than 5.0 when the process mean diameter shifts away from the ideal value of 9.0 by as much as $\delta = 2.0$ units. Then Table 4.8 suggests that

$$\sqrt{n}\frac{\delta}{\sigma} \geq 1.54$$

is needed. Using the process standard deviation $\sigma = 1.6$ employed earlier, this means that one wants

$$\sqrt{n}\frac{2.0}{1.6} \geq 1.54.$$

That is, after a little algebra

$$n \geq 1.52$$

is required.

But this says that the stated ARL goals can be met using a sample size of only $n = 2$. In fact, if before doing this sample size calculation, one had evaluated the $\mu = 11.0$ ARL of the scheme developed earlier ($n = 4$, $k_1 = 10.0$, $k_2 = 8.0$, and $h = 1.592$), it would have been found to be less than 5.0. (It is, in fact, 2.3.) For that matter, since $n = 2$ is larger than the minimum (nonrealizable noninteger) sample size of 1.52 derived from Table 4.8, even an $n = 2$ scheme designed for an on-target ARL of 370 will have ARL below 5.0 if $\mu = 11.0$. (It turns out that such a scheme has $n = 2$, $k_1 = 10.0$, $k_2 = 8.0$, and $h = 3.22$ and produces $ARL = 4.0$ when $\mu = 11.0$.)

Various "enhancements" of the basic CUSUM ideas already discussed have been suggested and are on occasion important in applications. One consideration often raised is related to the picture of CUSUM scheme performance given in Table 4.9. Table 4.9 compares ARLs for one two-sided CUSUM scheme designed (using Table 4.6) to have an all-OK ARL of 370 for normal observations Q, to those of a 3 sigma Shewhart chart. ARLs are given for various values of the difference between the "designed for" mean and the actual mean μ_Q (expressed as a multiple of σ_Q). The comparison in Table 4.9 is quite typical of what one finds when studying CUSUM process-monitoring schemes. For comparable

Table 4.9 ARLs for a Two-sided CUSUM Scheme and a 3 Sigma Shewhart Chart for Normal Q

	$\frac{\lvert\mu_Q - \text{target}\rvert}{\sigma_Q}$							
Monitoring Scheme	**0**	**.5**	**1.0**	**1.5**	**2.0**	**2.5**	**3.0**	**3.5**
Shewhart	370	155	43	15	6.5	3.2	2.0	1.4
CUSUM	370	35	10	5.5	3.9	3.0	2.5	2.2

all-OK ARLs, CUSUM charts have much better (smaller) ARLs than Shewhart charts for small changes in μ_Q, but somewhat worse (larger) ARLs for very large changes in mean.

In applications where there is some chance of an occasional disastrous large process change, the ARL advantage Shewhart charts hold over CUSUM schemes for very large changes in the mean of Q could conceivably be important. So the possibility of simultaneously using both kinds of charts suggests itself. In a 1982 *Journal of Quality Technology* paper, Lucas studied the ARL properties of combined Shewhart-CUSUM monitoring schemes. He found that using both a CUSUM scheme designed for a particular all-OK ARL and a 3 sigma Shewhart scheme usually produces an all-OK ARL that is much (unacceptably) smaller than that of the CUSUM scheme alone. But if one uses a 4 sigma or perhaps even 3.5 sigma Shewhart chart along with a CUSUM scheme, the combination has all-OK ARL behavior close to that of the CUSUM scheme alone. It also has the CUSUM's good ability to detect small shifts in μ_Q and some of the regular Shewhart chart's ability to warn of very large process changes. So adding a loose Shewhart-type alarm rule to a well-designed CUSUM monitoring scheme is one way of attempting to "failsafe" its behavior.

Shewhart/CUSUM Combinations

Another idea for enhancing basic CUSUM charts derives from the fact that in applications, one restarts a process-monitoring scheme after attempts to remedy some process ill. Such an attempt may or may not have been effective. In the event that the "fix" was not effective, it is desirable to have very quick detection of that fact. And this somehow needs to be enabled without changing the monitoring scheme in a way that seriously reduces the all-OK ARL.

Lucas and Crosier, in a 1982 *Technometrics* paper, advocate adjusting the CUSUM starting value(s) U_0 and/or L_0 as a means of providing **fast initial response** for a CUSUM scheme without much changing the all-OK ARL behavior. They advocate instead of 0 starting values, the values

Fast Initial Response (FIR) CUSUM Schemes

$$U_0 = \frac{h}{2} \quad \text{and/or} \quad L_0 = -\frac{h}{2}. \tag{4.21}$$

Starting values or "headstarts" (4.21) start CUSUM schemes half way to an out-of-control signal. What Lucas and Crosier found was that with such choices, the all-OK ARLs of CUSUM schemes are not much reduced from what they would be with 0 headstarts, while the off-target ARLs are reduced substantially. Roughly speaking, when under stable process conditions with μ_Q and σ_Q at their standard values, even with starting values (4.21) CUSUM schemes will tend to initially zero out. And thereafter, the behavior of a scheme is no different than if it had 0 headstarts to begin with.

It is even possible to decide exactly how one should choose h in order to get a desired on-target ARL when the headstarts (4.21) are used. That is, Tables 4.10 and 4.11 can be used exactly like Tables 4.5 and 4.6, except that the starting values (4.21) are employed instead of 0 starting values for the prescribed schemes.

Table 4.10 Values of $\mathcal{H}$ for Designing One-sided CUSUM Schemes for Normal Variables and $h/2$ Headstarts

All-OK ARL	$\mathcal{H}$ .25	.50	.75	1.00	1.25	1.50
50	3.57	2.32	1.65	1.21	.87	.58
100	4.64	2.93	2.08	1.55	1.18	.87
250	6.18	3.78	2.66	2.00	1.56	1.22
370	6.88	4.15	2.91	2.19	1.72	1.36
500	7.42	4.44	3.10	2.34	1.84	1.47
750	8.17	4.83	3.37	2.53	2.00	1.61
1000	8.71	5.11	3.56	2.68	2.11	1.71

Table 4.11 Values of $\mathcal{H}$ for Designing Combined High and Low Side CUSUM Schemes for Normal Variables and $h/2$ Headstarts

All-OK ARL	$\mathcal{H}$ .25	.50	.75	1.00	1.25	1.50
50	4.88	3.01	2.12	1.58	1.19	.88
100	6.00	3.64	2.55	1.91	1.48	1.14
250	7.59	4.49	3.13	2.35	1.85	1.48
370	8.29	4.86	3.38	2.54	2.00	1.61
500	8.84	5.14	3.57	2.69	2.12	1.72
750	9.59	5.53	3.84	2.88	2.28	1.85
1000	10.12	5.81	4.02	3.02	2.39	1.95

Example 4.3 continued. For a final time consider the CNC lathe example of Allan, Robbins, and Wyckoff. Suppose that (as earlier) a combined CUSUM scheme for $\bar{x}$'s based on samples of size $n = 4$ is to be designed with $ARL = 370$ when process parameters are $\mu = 9.0$ and $\sigma = 1.6$ and minimum possible ARL for cases where the process mean changes by as much as $\delta = 2.0$. But with the fast initial response feature added, how should chart parameters be chosen?

As before, relationships (4.12) say that in order to provide quickest possible detection of a change in mean part diameter of size 2.0, one should set

$$k_1 = 9.0 + \frac{2.0}{2} = 10.0 \quad \text{and} \quad k_2 = 9.0 - \frac{2.0}{2} = 8.0 .$$

Then relationships (4.13) say that in using Table 4.11, one will use

$$\mathcal{K} = \frac{10.0 - 9.0}{1.6/\sqrt{4}} = 1.25\,.$$

So reading from Table 4.11 on the $ARL = 370$ line, $\mathcal{H} = 2.00$ is in order. Finally, an appropriate decision interval is from display (4.14),

$$h = 2.00\left(\frac{1.6}{\sqrt{4}}\right) = 1.600\,.$$

Notice that this decision interval to be used with headstarts $U_0 = .80$ and $L_0 = -.80$ is only slightly larger than the no-headstart decision interval of $h = 1.592$ found earlier. Very little increase is required to compensate (in terms of holding the all-OK ARL at 370) for the use of the nonzero headstarts.

Comparison of Tables 4.10 and 4.11 to Tables 4.5 and 4.6 respectively shows the small inflation of the $\mathcal{H}$ values needed to accommodate the $h/2$ headstart(s). This book does not include ARL tables for the fast initial response CUSUMs. (The line has to be drawn somewhere regarding what to document!) If the reader has occasion to need them, ARLs for fast initial response CUSUM schemes can be found in the Lucas and Crosier paper or can be computed using the program of Gan referred to earlier in this section (and, in the case of two-sided schemes, a nonzero headstart generalization of relationship (4.20)).

4.3 MULTIVARIATE CHARTS

This section discusses control charts that can be used to monitor several different variables simultaneously. It begins with a discussion of why this is sometimes helpful and introduces a charting method for standards given contexts. Then there is a treatment of retrospective multivariate charting.

4.3.1 Motivation and Standards Given Multivariate Control Charts

The process-monitoring methods discussed thus far in this book have been univariate methods. That is, as discussed thus far, they are applicable one variable at a time. Although most real process-monitoring applications are highly multivariate, these methods are often still quite adequate when applied separately to the individual variables of importance. For example, in the machining of steel cylinders, it will often be adequate to keep one set of control charts for the diameters of cylinders and another for the lengths of cylinders.

Exceptions to this rule occur in those cases where not only are the variables monitored important in terms of their individual values, but also in terms of **relationship(s) between their values.** Figure 4.5 shows a schematic of a rolled product (like steel or microwave pizza dough) as it comes through a pair of rollers. On that schematic, the variable x_1 is a measured product thickness on the "left" side of the sheet and x_2 is a measured thickness on the "right" side of the product sheet. In rolling the product, it is not only important that x_1 be about right and separately that x_2 be about right, it is also important that x_1 and x_2 not be in a "funny relationship" to each other. If both x_1 and x_2 are slightly above target, the resulting sheet of rolled material is slightly thick. If both are slightly below target, the resulting sheet is slightly thin. But if one is above target and the other is below target, the resulting sheet has the unpleasant wedge-shaped cross section shown in Figure 4.5. (Such a shape would cause part of a microwave pizza to burn while the other side remained uncooked.)

Another (quite different) example where relationships between variables are of high importance is that of multicolor printing. Such printing is done in several impressions laid over one another. These are typically aligned using sets of cross hairs (one for each layer and located at some unimportant spot on the material being printed) that ideally fall exactly on top of one another. One might think of the first set of cross hairs as establishing a coordinate system on which subsequent layers can be located in terms of, say,

$$h_i = \text{the horizontal coordinate of the layer } i \text{ cross hairs}$$

and

$$v_i = \text{the vertical coordinate of the layer } i \text{ cross hairs.}$$

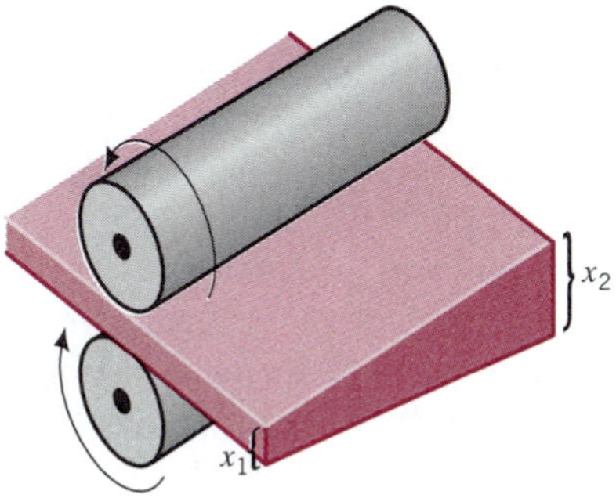

Figure 4.5
Left-side thickness x_1 and right-side thickness x_2 for a sheet of rolled product.

Then if there are, for example, two colors to be printed after the initial one, not only are h_1, v_1, h_2, and v_2 important individually (they need to each be near 0), but relationships between them are also important. A deviation from 0 in h_1 that might be tolerable in and of itself can become intolerable when paired with a deviation from 0 in h_2 of opposite sign. Such circumstances can cause strange-looking "shadows" on the resulting printed images.

Cases such as these, where relationships between several variables are important, bring into focus the need for process-monitoring methods that treat **several variables at once**. The simplest available method uses the basic Shewhart charting idea introduced in Chapter 3, but applies it to a statistic Q calculated from several variables. The statistic Q discussed in this section is one that is sensitive not only to deviations from standard values of variables being monitored, but also to nonstandard relationships between the variables.

If one is to be in a position to detect nonstandard relationships between variables, some numerical measure for expressing degree of relationship is needed. Most elementary introductions to statistics include a discussion of such a measure, the sample correlation between variables x and y measured on n units,

Sample Correlation Between Variables x and y

$$r_{xy} = \frac{\sum xy - \dfrac{\sum x \sum y}{n}}{\sqrt{\left(\sum x^2 - \dfrac{(\sum x)^2}{n}\right)\left(\sum y^2 - \dfrac{(\sum y)^2}{n}\right)}}. \tag{4.22}$$

This measure is always between -1 and 1. $r_{xy} = -1$ indicates a perfect linear observed relationship between x and y with negative slope and $r_{xy} = 1$ indicates a perfect linear observed relationship with positive slope. Where r is close to 1, x and y tend to be big together (relative to their means) and small together (relative to their means). When r_{xy} is close to -1, where one of the variables is small the other tends to be large. Multivariate control charts take into account not only means and standard deviations for the variables involved, but correlations between them as well.

Example 4.5 Monitoring Two Inside Diameters on Machined Steel Sleeves. Loveland, Rahardja, and Rainey worked on the monitoring of a turret lathe used to produce certain steel sleeves. They measured two identifiable diameters of the (nominally circularly cylindrical) sleeves. The first was along what will here be called the "A" axis of the bore and the second was along an axis (the "B" axis) perpendicular to the first. Table 4.12 contains both A and B "measurements" for 28 sleeves. (What is actually represented there are the averages of two measurements of each diameter made by the same technician. The original data set had 30 elements, but two "obviously" aberrant sleeves were dropped from consideration in order to arrive at Table 4.12.)

Table 4.12 Measured "A" and "B" Diameters for a Sample of 28 Steel Sleeves (.0001 inch above nominal)

Sleeve (j)	A Diam (x_{1j})	B Diam (x_{2j})	Sleeve (j)	A Diam (x_{1j})	B Diam (x_{2j})
1	125.0	75.0	15	112.5	67.5
2	135.0	87.5	16	125.0	67.5
3	142.5	95.0	17	127.5	57.5
4	130.0	102.5	18	112.5	67.5
5	125.0	72.5	19	95.0	55.0
6	137.5	55.0	20	122.5	50.0
7	120.0	72.5	21	102.5	45.0
8	115.0	87.5	22	120.0	82.5
9	125.0	65.0	23	115.0	57.5
10	157.5	82.5	24	100.0	65.0
11	100.0	17.5	25	100.0	47.5
12	105.0	40.0	26	100.0	57.5
13	107.5	45.0	27	107.5	72.5
14	110.0	65.0	28	120.0	82.5

Figure 4.6 is a scatterplot of the (x_1, x_2) data pairs in Table 4.12. It is possible to verify (and readers without recent experience using formula (4.22) should probably do so) that the sample correlation between the A and B diameters in Table 4.12 is about $r_{12} = .60$. This positive value says that when the A diameter is larger than its average, the B diameter also tends to be larger than its average. (And when the A diameter is smaller than its average, the B diameter also tends to be smaller than its average.) This fact is at least weakly suggestive of the possibility that while the sizes of the bored holes may vary, their *shapes* tend to remain the same.

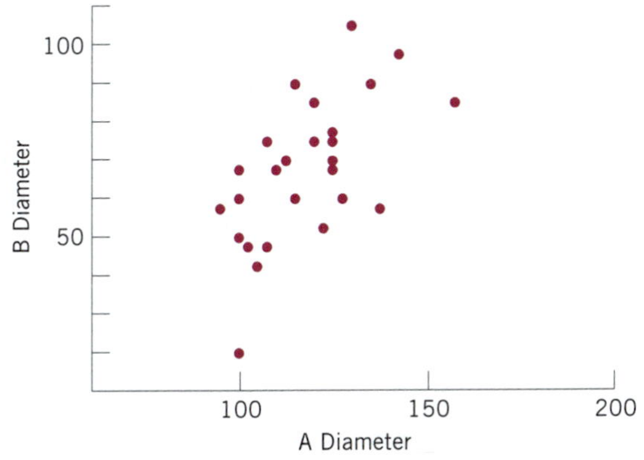

Figure 4.6
Scatterplot of two measured distinguishable diameters on 28 steel sleeves.

As an aside, it should probably be remarked that measuring two diameters is only the crudest of checks of approximate circularity of a figure (or hole). After all, a square has all pairs of perpendicular diameters the same! And there are whole classes of decidedly noncircular figures that have *all* diameters constant. Further, in the present example the facts that $\bar{x}_1 = 177.7$ while $\bar{x}_2 = 65.6$ already indicate that the A diameter of the holes is slightly larger than the B diameter, suggesting that the "average" shape of the holes is somewhat elliptical (or worse) to begin with. The positive correlation between x_1 and x_2 simply suggests that shape is relatively constant sleeve to sleeve. A negative correlation, on the other hand, would suggest that holes in some sleeves tend to look more circular than average, while others look even worse (in terms of circularity) than the averages portray.

The symbol ρ_{xy} is commonly used for the population or long-run value corresponding to the sample correlation r_{xy}, in the same way that μ corresponds to $\bar{x}$ and σ corresponds to s. And it is convenient to summarize all of the variance and correlation values for p variables $x_1, x_2, \ldots, x_p$ in terms of a $p \times p$ matrix called the **variance-covariance matrix** for the variables. This is the matrix with variances down the diagonal and products of correlations and standard deviations off the diagonals. That is, if $\rho_{ij} (= \rho_{ji})$ is the correlation between x_i and x_j and σ_i is the standard deviation of x_i, then the variance-covariance matrix for the p variables x_i is

$$\underset{p\times p}{\boldsymbol{V}} = \begin{bmatrix} \sigma_1^2 & \rho_{12}\sigma_1\sigma_2 & \cdots & \rho_{1p}\sigma_1\sigma_p \\ \rho_{21}\sigma_1\sigma_2 & \sigma_2^2 & \cdots & \rho_{2p}\sigma_2\sigma_p \\ \vdots & \vdots & \ddots & \vdots \\ \rho_{p1}\sigma_1\sigma_p & \rho_{p2}\sigma_2\sigma_p & \cdots & \sigma_p^2 \end{bmatrix} = [\rho_{ij}\,\sigma_i\,\sigma_j]. \tag{4.23}$$

The last equality in display (4.23) says that the value in the ith row and jth column of $\boldsymbol{V}$ is $\rho_{ij}\,\sigma_i\,\sigma_j$. This is correct even for the diagonal elements of $\boldsymbol{V}$ since the correlation of any variable with itself is 1.

So now consider a situation where a sample of size n will produce sample averages $\bar{x}_1, \bar{x}_2, \ldots, \bar{x}_p$. For multivariate process monitoring, one wants a statistic Q that will be large if any $\bar{x}_i$ is far away from its mean μ_i or if the $\bar{x}_i$'s are in a pattern that suggests that standard relationships between the variables are not in effect. One such statistic can be written compactly in matrix terms after a bit more notation is introduced. Let

$$\underset{p\times 1}{\bar{\boldsymbol{x}}} = \begin{bmatrix} \bar{x}_1 \\ \bar{x}_2 \\ \vdots \\ \bar{x}_p \end{bmatrix} \quad \text{and} \quad \underset{p\times 1}{\boldsymbol{\mu}} = \begin{bmatrix} \mu_1 \\ \mu_2 \\ \vdots \\ \mu_p \end{bmatrix}.$$

Then the basic statistic most often used in multivariate process monitoring is

$$X^2 = n(\bar{\boldsymbol{x}} - \boldsymbol{\mu})' \boldsymbol{V}^{-1} (\bar{\boldsymbol{x}} - \boldsymbol{\mu}). \tag{4.24}$$

Basic Measure of Multivariate Process Performance

It is probably not at all obvious to most readers what the quantity (4.24) intends to measure. A way to get some feeling for the meaning of the formula is to consider the special cases of $p = 1$ and $p = 2$ variables. In the special case of $p = 1$,

$$X^2 = \left(\frac{\bar{x}_1 - \mu_1}{\sigma_1 / \sqrt{n}} \right)^2,$$

the square of the usual "z-score" associated with the observed sample mean $\bar{x}_1$. In the special case of $p = 2$,

$$X^2 = \frac{1}{(1 - \rho_{12}^2)} \left(\left(\frac{\bar{x}_1 - \mu_1}{\sigma_1 / \sqrt{n}} \right)^2 - 2\rho_{12} \left(\frac{\bar{x}_1 - \mu_1}{\sigma_1 / \sqrt{n}} \right) \left(\frac{\bar{x}_2 - \mu_2}{\sigma_2 / \sqrt{n}} \right) + \left(\frac{\bar{x}_2 - \mu_2}{\sigma_2 / \sqrt{n}} \right)^2 \right), \tag{4.25}$$

which is a quadratic function of the two sample means (and also of the two z-scores). Figure 4.7 shows (for the case of $p = 2$) plots of the sets of $(\bar{x}_1, \bar{x}_2)$ pairs satisfying the equation

$$X^2 = c$$

for sample size $n = 1$ under the three sets of parameters

1. $\mu_1 = 0$, $\mu_2 = 0$, $\sigma_1 = 1$, $\sigma_2 = 1$, and $\rho_{12} = .5$,
2. $\mu_1 = 0$, $\mu_2 = 0$, $\sigma_1 = 1$, $\sigma_2 = 1$, and $\rho_{12} = 0$, and
3. $\mu_1 = 0$, $\mu_2 = 0$, $\sigma_1 = 1$, $\sigma_2 = 1$, and $\rho_{12} = -.5$.

The message conveyed by Figure 4.7 is one that holds in general. The set of $\bar{\boldsymbol{x}}$ (points in p dimensional space) satisfying the inequality

$$X^2 \leq c$$

is a multivariate ellipsoid (a multidimensional solid "football") centered at the point $\boldsymbol{\mu}$, with shape, size, and orientation determined by n and $\boldsymbol{V}$. The larger n, the smaller the ellipsoid. The larger the variances σ_i^2, the larger the ellipsoid. In the $p = 2$ case with equal variances, $\rho_{12} > 0$ indicates that the ellipsoid is oriented with its major axis "lower-left to upper-right," while $\rho_{12} < 0$ produces a "lower-right to upper-left" orientation.

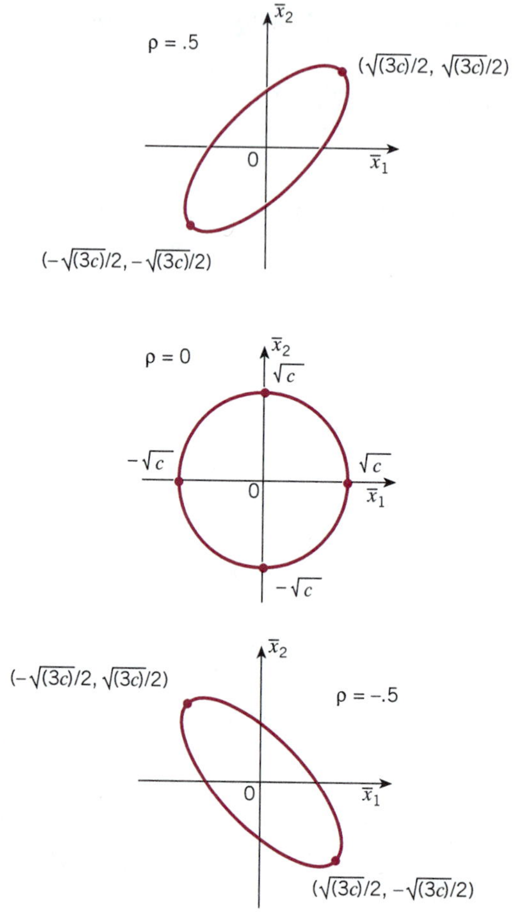

Figure 4.7
Sets of $(\overline{x}_1, \overline{x}_2)$ points satisfying the equation $X^2 = c$ for $\rho_{12} = .5, 0$, and $-.5$.

A Shewhart control chart may be based on the statistic (4.24) by plotting X^2 values and declaring a process to be out-of-control if X^2 is too large.

That is, an out-of-control signal is generated if it requires too large a multidimensional football centered at the vector of standard means $\boldsymbol{\mu}$ to touch the vector of sample means $\overline{\boldsymbol{x}}$. The question is then what "too large" should be. Probability theory provides some guidance here.

It turns out that under (multivariate) normal stable process conditions for the variables $x_1, x_2, \ldots, x_p$, if $\boldsymbol{\mu}$ and $\boldsymbol{V}$ contain standard values for process parameters, the quantity X^2 defined in equation (4.24) has a χ^2 distribution with p associated degrees of freedom. The χ^2 distribution with p degrees of freedom has mean p and standard deviation $\sqrt{2p}$. So applying the generic Shewhart control limits (3.2), a sensible center line and upper control limit for a standards given multivariate control chart are then

$$CL_{X^2} = p \quad \text{and} \quad UCL_{X^2} = p + 3\sqrt{2p}\,. \tag{4.26}$$

Center Line and *UCL* for a Multivariate Chart

Typically no lower control limit is used for X^2. Introducing a lower control limit would mean allowing for out-of-control signals when $\bar{\boldsymbol{x}}$ is "too close" to $\boldsymbol{\mu}$ as measured by the size of an ellipsoid centered at $\boldsymbol{\mu}$ required to touch $\bar{\boldsymbol{x}}$. There seems to be little practical motivation for this. So only an upper control limit is used and the set of possible $\bar{\boldsymbol{x}}$'s leading to an "in-control" judgment consists of an entire solid multidimensional ellipsoid.

Example 4.5 continued. Returning to the situation of monitoring two inside diameters on steel sleeves, suppose that it is plausible to regard the 28 sleeves represented in Table 4.12 as coming from stable process conditions. Suppose further that based on target values of 0 for both x_1 and x_2 and the precision and correlation information given by the data, one wishes to set up a standards given multivariate control chart for "measurements" on individual sleeves of the type represented in Table 4.12.

The $n = 1$ and $p = 2$ version of statistic (4.24) is needed for appropriate $\boldsymbol{\mu}$ and $\boldsymbol{V}$. To begin with, the ideal or standard values of μ_1 and μ_2 are both 0, so that

$$\boldsymbol{\mu} = \begin{bmatrix} 0 \\ 0 \end{bmatrix} \quad \text{and thus} \quad \bar{\boldsymbol{x}} - \boldsymbol{\mu} = \begin{bmatrix} x_1 \\ x_2 \end{bmatrix} - \begin{bmatrix} 0 \\ 0 \end{bmatrix} = \begin{bmatrix} x_1 \\ x_2 \end{bmatrix}$$

are in order. Then, from the data in Table 4.12, one has approximately $r_{12} = .60$, $s_1 = 15$, and $s_2 = 18$. So a plausible standard variance-covariance matrix for this situation is

$$\boldsymbol{V} = \begin{bmatrix} (15)^2 & .6(15)(18) \\ .6(15)(18) & (18)^2 \end{bmatrix}.$$

Then using these expressions with formula (4.24) one has

$$X^2 = 1[x_1, x_2]\frac{1}{(1-(.6)^2)(15)^2(18)^2}\begin{bmatrix} (18)^2 & -.6(15)(18) \\ -.6(15)(18) & (15)^2 \end{bmatrix}\begin{bmatrix} x_1 \\ x_2 \end{bmatrix}.$$

That is, either simplifying this matrix expression or using formula (4.25) for the special case of $p = 2$, one has

$$X^2 = \frac{1}{1-(.6)^2}\left(\frac{x_1^2}{(15)^2} - 2(.6)\frac{x_1x_2}{(15)(18)} + \frac{x_2^2}{(18)^2}\right) = \frac{x_1^2}{92.16} - \frac{x_1x_2}{144} + \frac{x_2^2}{207.36}\,.$$

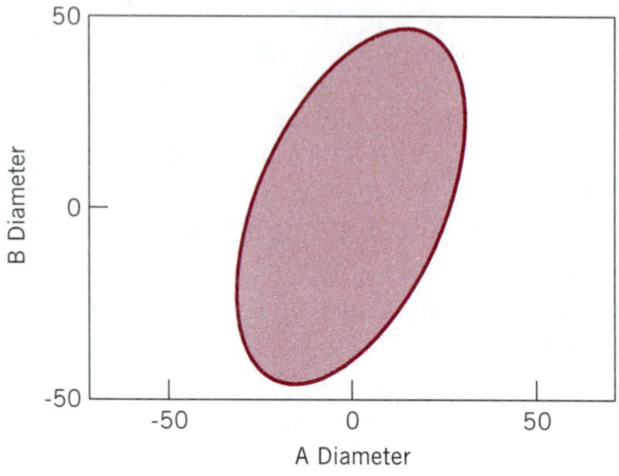

Figure 4.8
The elliptical in-control region for two (coded) measured diameters of a single sleeve.

From display (4.26), it follows that an appropriate center line and upper control limit for a Shewhart control chart based on X^2 are respectively

$$CL_{X^2} = 2 \quad \text{and} \quad UCL_{X^2} = 2 + 3\sqrt{2(2)} = 8\,.$$

That is, for any particular sleeve there will be no out-of-control signal issued provided

$$\frac{x_1^2}{92.16} - \frac{x_1 x_2}{144} + \frac{x_2^2}{207.36} \leq 8\,.$$

Figure 4.8 shows the set of possible (x_1, x_2) pairs that will lead to an in-control judgment based on measurements on one sleeve. Notice the upper-right to lower-left orientation of the ellipse, related to the positive (.6) standard correlation between x_1 and x_2.

For the sake of illustrating the usefulness of multivariate control charts, consider two hypothetical data pairs, (22.5, 27.0) and (22.5, −27.0). For both data pairs, x_1 is $1.5\sigma_1$ above its standard mean. For the first pair, x_2 is $1.5\sigma_2$ above its standard mean, while in the second pair, x_2 is $1.5\sigma_2$ below its standard mean. Notice that *neither* of the sleeves would produce an out-of-control signal on *separate* individuals charts for x_1 or x_2 (i.e. separate $\bar{x}$ charts for $n = 1$). Finding an individual measurement 1.5 standard deviations from its mean is not a particularly surprising or rare event. But consider how a multivariate control chart would treat the two pairs. For the case of $x_1 = 22.5$ and $x_2 = 27.0$,

$$X^2 = \frac{(22.5)^2}{92.16} - \frac{(22.5)(27.0)}{144} + \frac{(27.0)^2}{207.36} = 4.79$$

which does not exceed the upper control limit for X^2 of 8.0. Where both measurements are 1.5 sigma high, the multivariate chart does not signal an alarm. On the other hand, for the case of $x_1 = 22.5$ and $x_2 = -27.0$,

$$X^2 = \frac{(22.5)^2}{92.16} - \frac{(22.5)(-27.0)}{144} + \frac{(-27.0)^2}{207.36} = 13.23$$

which *does* exceed the upper control limit for X^2. Where one measurement is 1.5 sigma high and the other is 1.5 sigma low, the multivariate chart signals an alarm. Why? Not because the individual measurements considered by themselves are so unusual compared to their means and standard deviations, but because the *relationship* between them is not in accord with the one that is expected based on the standard correlation of .6. The first point falls inside the ellipse in Figure 4.8 while the second one falls outside it.

The example illustrates well what a multivariate control chart offers over p separate $\bar{x}$ charts, namely the ability to detect unusual relationships among the p variables. There is much largely misguided and irrelevant material to be found in some textbooks that attempts to motivate the use of multivariate control charts on other grounds. This issue is the genuinely relevant one.

4.3.2 Retrospective Multivariate Control Charts

A final matter for consideration regarding the use of multivariate control charts is how one should go about a **retrospective analysis**. That is, suppose one has r samples of respective sizes $n_1, n_2, \ldots, n_r$ taken from the output of some process, and that each item in a sample consists of measurements of p different characteristics $x_1, x_2, \ldots, x_p$. If one then wishes to ask whether it is plausible that over the period of observation the process was stable at some values of process parameters, how should estimates of $\boldsymbol{\mu}$ and $\boldsymbol{V}$ be generated for use in formula (4.24) with the data in hand?

To begin with, each sample ℓ of size n_ℓ produces a sample mean for the ith variable, $\bar{x}_{i\ell}$, and these can be appropriately averaged to produce an estimator of μ_i. That is,

Pooled Sample Mean for Variable x_i

$$\bar{x}_{i\,\text{pooled}} = \frac{\sum_{\ell=1}^{r} n_\ell \bar{x}_{i\ell}}{\sum_{\ell=1}^{r} n_\ell} = \frac{\text{total of the observed values of } x_i}{\text{total number of data vectors}} \tag{4.27}$$

is a sensible estimator of μ_i. In the case that all sample sizes n_ℓ are the same, this is nothing but $\bar{\bar{x}}_i$, the sample mean of the sample means for variable x_i. The values (4.27) may then be assembled into a $p \times 1$ vector and used in place of $\boldsymbol{\mu}$ in the statistic (4.24).

Producing an appropriate estimator for the matrix $\boldsymbol{V}$ is only slightly more complicated. Each sample ℓ of size $n_\ell > 1$ produces its own sample variance for each variable x_i, say $s_{i\ell}^2$, and its own sample correlation between variables x_i and x_j, say $r_{ij\ell}$. (See problem 4.35 for a possible approach when the n_ℓ are all 1.) These can be assembled into a sample version of a variance-covariance matrix based only on sample ℓ,

$$\hat{\boldsymbol{V}}_\ell = \begin{bmatrix} s_{1\ell}^2 & r_{12\ell}s_{1\ell}s_{2\ell} & \cdots & r_{1p\ell}s_{1\ell}s_{p\ell} \\ r_{21\ell}s_{1\ell}s_{2\ell} & s_{2\ell}^2 & \cdots & r_{2p\ell}s_{2\ell}s_{p\ell} \\ \vdots & \vdots & \ddots & \vdots \\ r_{p1\ell}s_{1\ell}s_{p\ell} & r_{p2\ell}s_{2\ell}s_{p\ell} & \cdots & s_{p\ell}^2 \end{bmatrix} = \left[r_{ij\ell}s_{i\ell}s_{j\ell}\right].$$

These can then be properly averaged to produce a pooled sample variance-covariance matrix. That is, one can define

Pooled Sample Variance-Covariance Matrix

$$\hat{\boldsymbol{V}}_{\text{pooled}} = \frac{1}{\sum_{\ell=1}^{r}(n_\ell - 1)}\left((n_1 - 1)\hat{\boldsymbol{V}}_1 + (n_2 - 1)\hat{\boldsymbol{V}}_2 + \cdots + (n_r - 1)\hat{\boldsymbol{V}}_r\right) \tag{4.28}$$

and use this in place of $\boldsymbol{V}$ in the statistic (4.24). Notice that again if all sample sizes are the same, the pooled estimator (4.28) is simply the usual average of the sample variance-covariance matrices.

The $(n_\ell - 1)$ weighting in the weighted average (4.28) should not surprise readers. It is the same weighting that is employed when one wishes to combine sample variances of a single variable into one pooled sample variance. Equation (4.28) simply says that this weighting is applied to all sample variances and to the products of sample correlations with sample standard deviations.

One caution needs to be offered regarding the use of the estimator (4.28). If it is to replace $\boldsymbol{V}$ in formula (4.24), it needs to be nonsingular. If r and the sample sizes are small and p is large, it can fail to be so. At a minimum, $\sum n_\ell - r$ must be at least p in order to have any hope of inverting $\hat{\boldsymbol{V}}_{\text{pooled}}$. (And ideally, $\sum n_\ell - r$ will be much larger than p when using $\hat{\boldsymbol{V}}_{\text{pooled}}$. An approach to handling the case where all $n_\ell = 1$ and thus $\sum n_\ell - r = 0$ is presented in problem 4.35.)

Example 4.5 continued. For purposes of illustrating the retrospective use of multivariate control charts, consider the data and summary statistics in Table 4.13. The data were artificially generated from a model of stable process behavior consistent with the picture of the steel sleeve inside diameters already used in this section. (So the data are not expected to produce out-of-control signals.)

It is straightforward to check that simply averaging the $10 \times 3 = 30$ values x_1 in Table 4.13 or (since all sample sizes are the same) averaging the 10 values $\bar{x}_{1\ell}$,

Table 4.13 Hypothetical Measured "A" and "B" Diameters and Summary Statistics for 10 Samples of 3 Steel Sleeves

Sample (ℓ)	A Diam (x_1)	B Diam (x_2)	$\bar{x}_{1\ell}$	$\bar{x}_{2\ell}$	$s_{1\ell}$	$s_{2\ell}$	$r_{12\ell}$
1	99	32	108.7	47.0	9.5	13.0	.898
	109	54					
	118	55					
2	142	71	133.7	88.7	9.1	16.0	−.934
	124	102					
	135	93					
3	116	58	116.0	70.7	16.0	29.1	.926
	132	104					
	100	50					
4	111	31	118.7	47.3	16.9	14.8	.654
	138	60					
	107	51					
5	110	67	118.7	78.3	21.4	17.1	.968
	143	98					
	103	70					
6	111	56	125.7	64.0	18.9	25.0	.898
	119	44					
	147	92					
7	129	84	112.7	62.3	15.6	18.8	.872
	111	50					
	98	53					
8	145	96	128.7	80.7	21.0	21.6	.997
	136	90					
	105	56					
9	120	92	116.3	72.3	4.0	17.1	.728
	117	61					
	112	64					
10	103	70	105.3	64.7	14.6	17.6	.920
	92	45					
	121	79					

one has $\bar{x}_{1\text{pooled}} = 118.43$. Similarly, averaging the values x_2 in Table 4.13 or averaging the 10 values $\bar{x}_{2\ell}$, one has $\bar{x}_{2\text{pooled}} = 67.60$. So in a retrospective analysis of the data of Table 4.13, $\boldsymbol{\mu}$ in statistic (4.24) will be replaced by

$$\begin{bmatrix} 118.43 \\ 67.60 \end{bmatrix}.$$

Further, since the sample sizes are all the same, the upper-left entry of $\hat{\boldsymbol{V}}_{\text{pooled}}$ is simply the average of the 10 sample variances $s_{1\ell}^2$, namely 244.00. Similarly, the lower-right entry of $\hat{\boldsymbol{V}}_{\text{pooled}}$ is the average of the 10 sample variances $s_{2\ell}^2$, namely 383.17. And the off-diagonal elements of $\hat{\boldsymbol{V}}_{\text{pooled}}$ are the average of the 10 products $r_{12\ell}s_{1\ell}s_{2\ell}$, namely 234.37. That is,

$$\hat{\boldsymbol{V}}_{\text{pooled}} = \begin{bmatrix} 244.00 & 234.37 \\ 234.37 & 383.17 \end{bmatrix}.$$

So then, the retrospective version of statistic (4.24) for the data of Table 4.13 is

$$X^2 = 3\left[\begin{bmatrix} \bar{x}_1 \\ \bar{x}_2 \end{bmatrix} - \begin{bmatrix} 118.43 \\ 67.60 \end{bmatrix}\right]' \begin{bmatrix} 244.00 & 234.37 \\ 234.37 & 383.17 \end{bmatrix}^{-1} \left[\begin{bmatrix} \bar{x}_1 \\ \bar{x}_2 \end{bmatrix} - \begin{bmatrix} 118.43 \\ 67.60 \end{bmatrix}\right].$$

Table 4.14 contains the 10 values of X^2 for the samples listed in Table 4.13 computed according to this retrospective formula. Figure 4.9 is a plot of the retrospective multivariate control chart (using the limit from display (4.26)). Only the point for sample 4 approaches the control limit, and in all, there is no multivariate evidence of process instability in these data.

There is one final matter that deserves comment in this discussion of retrospective multivariate control charts. That is the choice of an upper control limit. Some authors recommend an upper control limit, based on the so-called T^2 distribution, that is looser than the one in display (4.26). Its use is analogous to the use of critical points based on the t distribution when doing a hypothesis test for a mean with estimated standard deviation in place of corresponding standard normal critical points. Your authors find discussion of T^2 control limits to be rather a "fine point," not really worth the trouble it generates. To introduce this complication when discussing multivariate charts, when it is universally ignored in univariate Shewhart

Table 4.14 Observed Values of X^2 for a Retrospective Analysis of the Data of Table 4.13

ℓ	1	2	3	4	5	6	7	8	9	10
X^2	3.56	3.64	.63	7.97	2.10	2.76	.41	1.49	.92	3.88

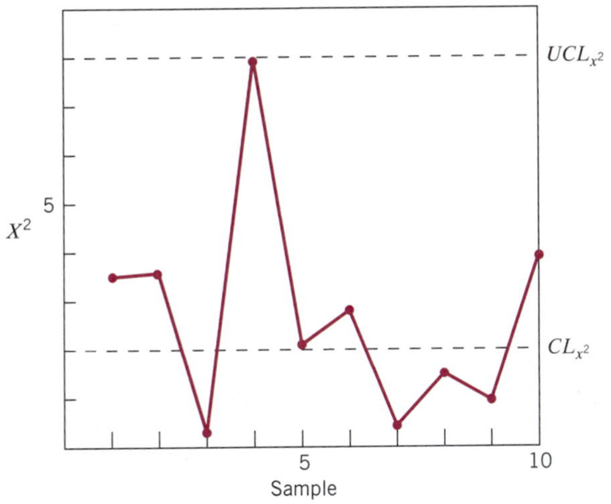

Figure 4.9
Retrospective multivariate control chart for the data of Table 4.13.

control charting, seems especially inconsistent and unjustified. Shewhart control charts are not tools for making fine distinctions, but rather tools for rough-and-ready process analysis, and the limit (4.26) is adequate for such work.

4.4 CHARTS FOR INDIVIDUALS AND MOVING RANGES

There are occasions where the natural sample size in a "variables data" process-monitoring context is $n = 1$. When that is true (there is no way to produce rational subgroups of size two or larger), exactly how to monitor process location and spread is not completely obvious. One sensible possibility is to simply plot individual observations on their own chart. There is no logical problem applying the $\bar{x}$ chart idea with $n = 1$ to get a chart for individuals. Such a chart deals directly with the issue of process location, and indirectly with the matter of process spread. (An increase in σ will eventually create an out-of-control signal on an individuals chart by virtue of the fact that observations are noisier than expected.)

However, practitioners are often not satisfied with the indirect indication of process spread provided by an individuals chart and want to supplement it with one specifically geared to the monitoring of process variability. But samples of size one provide no ranges or standard deviations. So what is often done is to try to monitor process spread using the "moving ranges" of consecutive observations. That is, provided with a sequence of observations $x_1, x_2, x_3, \ldots$ one plots these in an ($n = 1$) $\bar{x}$ chart format, and at times $i = 2, 3, 4, \ldots$ also plots in a range chart format $MR_2, MR_3, MR_4, \ldots$ where

Moving Range at Period i

$$MR_i = |x_i - x_{i-1}|. \tag{4.29}$$

(MR_i is the distance between the individual at period i and the immediately preceding individual.) The resulting monitoring scheme is sometimes called an ***X*/*MR* monitoring scheme.**

Example 4.6 Monitoring the Inside Diameters of Machined Steel Sleeves (Example 4.5 Revisited). Return again to the problem of Loveland, Rahardja, and Rainey and the monitoring of diameters of steel sleeves. For present purposes, monitoring of one (not two) measured diameter(s) will be considered. The amount of machining (and time) required to produce each sleeve was substantial, and there was *a priori* no grounds for assuming that the process was physically stable over the period required to produce even a few sleeves. This is a situation where the natural sample size is $n = 1$. Table 4.15 shows measured inside diameters for 30 consecutive sleeves and the corresponding moving ranges. The units of measurement are .00001 inch above a nominal value.

Table 4.15 Measured Individual Inside Diameters and Moving Ranges for 30 Consecutive Machined Steel Sleeves (.00001 inch above nominal)

Sleeve (i)	x_i	MR_i	Sleeve (i)	x_i	MR_i
1	140	–	16	110	65
2	140	0	17	125	15
3	140	0	18	125	0
4	130	10	19	115	10
5	125	5	20	90	25
6	145	20	21	125	35
7	125	20	22	110	15
8	105	20	23	110	0
9	125	20	24	120	10
10	125	0	25	90	30
11	105	20	26	90	0
12	110	5	27	80	10
13	115	5	28	100	20
14	120	5	29	125	25
15	45	75	30	200	75

A moving range computed as in display (4.29) should not be thought of as a "sample range" equivalent to the range of a rational subgroup (data collected under what one is willing to guarantee are constant process conditions). It will be affected not only by common-cause/random variation, but also potentially by special-cause variation as well. On the other hand, it is as close as one can come to producing an empirical measure of process short-term variation when the natural sample size is one.

Once one agrees to do process monitoring based on individuals and moving ranges, the question of what to use for control limits for x and MR must be addressed. The standard answer to this question is to simply treat the MRs as if they were "regular" ranges of samples of size two. Then the usual formulas for $\bar{x}$ chart limits specialized to the case of $n = 1$ suggest standards given control limits for individuals

$$UCL_x = \mu + 3\sigma \text{ and } LCL_x = \mu - 3\sigma, \tag{4.30}$$

and the usual formulas (3.15) for R charts in the case of $n = 2$ suggest the standards given value for moving ranges

$$UCL_{MR} = 3.686\sigma. \tag{4.31}$$

Many textbooks present exactly formulas (4.30) and (4.31) for use with X/MR chart combinations. But there is a practical problem with these somewhat naive control limits. They tend to produce many more false alarms than most people would like. Crowder, in two 1987 papers in the *Journal of Quality Technology*, computed ARLs for the X/MR chart combination. Among other things, he found that when observations are normal, process parameters are at their standard values, and limits (4.30) and (4.31) are used, the ARL for the X/MR combination is only about 105. This is much smaller than most people would consider appropriate or desirable. (This theoretical result is also borne out in practical experience with the limits (4.30) and (4.31). Using an X/MR scheme with these limits is not at all equivalent in terms of frequency of false alarms to using an $\bar{x}/R$ combination based on samples of size larger than one.)

A sensible modification of the standard treatment of X/MR chart combinations is to replace the number 3 in equations (4.30) and the number 3.686 in equation (4.31) by some other (larger) values, say respectively $\mathcal{M}$ and $\mathcal{R}$. This produces

$$UCL_x = \mu + \mathcal{M}\sigma \text{ and } LCL_x = \mu - \mathcal{M}\sigma, \tag{4.32}$$

Standards Given Control Limits for Individuals

and

$$UCL_{MR} = \mathcal{R}\sigma, \tag{4.33}$$

Standards Given Control Limit for Moving Ranges

Table 4.16 Constants for Designing X/MR Chart Combinations

All-OK ARL	Smallest $\mathcal{M}$ Possible				Smallest $\mathcal{R}$ Possible
50	$\mathcal{M} = 2.33$ ($\mathcal{R} = 4.5^+$)	$\mathcal{M} = 2.40$ $\mathcal{R} = 3.66$	$\mathcal{M} = 2.55$ $\mathcal{R} = 3.41$	$\mathcal{M} = 2.80$ $\mathcal{R} = 3.28$	($\mathcal{M} = 3.3^+$) $\mathcal{R} = 3.24$
100	$\mathcal{M} = 2.58$ ($\mathcal{R} = 5.0^+$)	$\mathcal{M} = 2.65$ $\mathcal{R} = 4.04$	$\mathcal{M} = 2.80$ $\mathcal{R} = 3.77$	$\mathcal{M} = 3.00$ $\mathcal{R} = 3.67$	($\mathcal{M} = 3.5^+$) $\mathcal{R} = 3.60$
250	$\mathcal{M} = 2.88$ ($\mathcal{R} = 5.5^+$)	$\mathcal{M} = 2.95$ $\mathcal{R} = 4.47$	$\mathcal{M} = 3.10$ $\mathcal{R} = 4.22$	$\mathcal{M} = 3.30$ $\mathcal{R} = 4.11$	($\mathcal{M} = 3.8^+$) $\mathcal{R} = 4.05$
370	$\mathcal{M} = 3.00$ ($\mathcal{R} = 6.0^+$)	$\mathcal{M} = 3.10$ $\mathcal{R} = 4.57$	$\mathcal{M} = 3.20$ $\mathcal{R} = 4.40$	$\mathcal{M} = 3.40$ $\mathcal{R} = 4.29$	($\mathcal{M} = 3.8^+$) $\mathcal{R} = 4.23$
500	$\mathcal{M} = 3.09$ ($\mathcal{R} = 6.0^+$)	$\mathcal{M} = 3.20$ $\mathcal{R} = 4.67$	$\mathcal{M} = 3.30$ $\mathcal{R} = 4.53$	$\mathcal{M} = 3.50$ $\mathcal{R} = 4.42$	($\mathcal{M} = 4.0^+$) $\mathcal{R} = 4.36$
750	$\mathcal{M} = 3.21$ ($\mathcal{R} = 6.0^+$)	$\mathcal{M} = 3.30$ $\mathcal{R} = 4.88$	$\mathcal{M} = 3.45$ $\mathcal{R} = 4.66$	$\mathcal{M} = 3.60$ $\mathcal{R} = 4.59$	($\mathcal{M} = 4.0^+$) $\mathcal{R} = 4.55$
1000	$\mathcal{M} = 3.29$ ($\mathcal{R} = 6.5^+$)	$\mathcal{M} = 3.40$ $\mathcal{R} = 4.96$	$\mathcal{M} = 3.50$ $\mathcal{R} = 4.82$	$\mathcal{M} = 3.65$ $\mathcal{R} = 4.72$	($\mathcal{M} = 4.0^+$) $\mathcal{R} = 4.65$

and intuitively should increase the all-OK ARL and make the resulting scheme behave more sensibly.

But exactly how to choose $\mathcal{M}$ and $\mathcal{R}$? Crowder's work on ARLs for X/MR monitoring schemes provides guidance here. For a given all-OK ARL, it is possible to find a whole continuum of possible $(\mathcal{M}, \mathcal{R})$ pairs yielding that ARL value when process parameters are at their standard values. Any of these can be used, and choice between them can be based on trading off tight limits for individuals against tight limits for moving ranges. Table 4.16 gives some combinations of $\mathcal{M}$ and $\mathcal{R}$ for all-OK ARLs of 50, 100, 250, 370, 500, 750, and 1000.

The values in Table 4.16 were derived using a slight modification of Crowder's published ARL program for normal observations. For each ARL value there are five $(\mathcal{M}, \mathcal{R})$, combinations listed. The first has the smallest value of $\mathcal{M}$ possible and the last has the smallest value of $\mathcal{R}$ possible for that all-OK ARL. The plus superscripts in the first and last columns are meant to indicate that the printed value (of $\mathcal{R}$ or $\mathcal{M}$) or any larger value of that variable will produce essentially the same ARL if process parameters are at their standard values. As one moves left to right across a row of the table, one trades off small $\mathcal{M}$ against small $\mathcal{R}$.

Example 4.6 continued. Return to the situation of monitoring the inside diameters of steel sleeves. To illustrate the computation of standards given control limits for X/MR combinations, consider the possibility that process standard values are $\mu = 0$ and $\sigma = 17$. (This latter value might, for example, be considered barely consistent with the actual ± 50 engineering specifications for this measured dimension.) Also suppose that on average, one would like to go 370 sleeves between out-of-control signals if in fact $\mu = 0$ and $\sigma = 17$. Then the "370" row of Table 4.16 should be consulted. Taking, for example, the middle $(\mathcal{M}, \mathcal{R})$ pair from that row, one has from displays (4.32) and (4.33)

$$UCL_x = 0 + 3.20(17) = 54.4 \text{ and } LCL_x = 0 - 3.20(17) = -54.4$$

and

$$UCL_{MR} = 4.40(17) = 74.8.$$

Clearly, according to these limits, there is no way that the machining process was operating at standard values of process parameters over the production of the 30 sleeves represented in Table 4.15. Essentially none of the individuals in Table 4.15 are inside standards given control limits, and moving ranges 15 and 30 are also above their control limit. Whether it is plausible that the process was stable at *some* set of process parameters over the production of the 30 sleeves is, of course, another matter (to be taken up in a retrospective analysis).

The matter of producing a sensible **retrospective analysis** for data like those in Table 4.15 turns on finding an appropriate way to estimate σ, the process short-term standard deviation. In truth, there is no really good way to go about this, as there is no honest way to estimate σ based on samples of size one. But some "dishonest" methods are better than others! Often, quality assurance beginners naively try to estimate σ using the standard deviation of all data in hand. (In the case of the sleeve diameters this would mean simply plugging 30 measurements into a calculator and pressing the "s" button.) The problem with such an approach is that unless the process mean is constant, changes in it will contribute to overall variation and inflate the standard deviation of the data in hand much beyond the size of the real short-term standard deviation. In fact, using the naive standard deviation of the data in hand to produce an s upon which to base "control limits" to judge the individual observations is completely circular. Almost never will one find points outside limits operating in this fashion.

About the best that one can do toward estimating σ based on successive samples of size one is to make use of the mean moving range. If one has r individuals x_i, this is

$$\overline{MR} = \frac{1}{r-1}\sum_{i=2}^{r} MR_i \qquad (4.34)$$

Average Moving Range

and a semiplausible estimator of σ is

Estimator of σ Based on Successive Samples of Size $n = 1$

$$\hat{\sigma} = \frac{\overline{MR}}{1.128}. \tag{4.35}$$

The number 1.128 appearing in formula (4.35) is d_2 for samples of size $n = 2$, and this estimator is thus an analog of the estimator $\bar{R}/d_2$ used in Section 3.2. The rationale for using the quantity (4.35) is that while it will be affected (inflated) by movement of the process mean, it is less affected than anything else one might use. That is, if observations are normal and successive means do not differ too much, the estimator (4.35) has expected value close to σ. This is true even if *overall* (from period 1 to period r) there is a large change in means (that will, for example, tend to dramatically inflate a standard deviation calculated from all data in hand).

Example 4.6 continued. Returning once more to the sleeve inside diameters, note that the 29 moving ranges in Table 4.15 have mean $\overline{MR} = 540/29 = 18.62$. So equation (4.35) then produces

$$\frac{18.62}{1.128} = 16.51$$

as a provisional estimate of the (supposedly constant) process short-term standard deviation. Note also that the 30 individual measurements in Table 4.15 have mean $\bar{x} = 3510/30 = 117.0$. So substituting $\bar{x}$ for μ and $\overline{MR}/1.128$ for σ in equations (4.32) and (4.33), and using $\mathcal{M} = 3.20$ and $\mathcal{R} = 4.40$ as before, one has retrospective control limits

$$UCL_x = 117.0 + 3.20(16.51) = 169.8 \text{ and } LCL_x = 117.0 - 3.20(16.51) = 64.2,$$

and

$$UCL_{MR} = 4.40(16.51) = 72.6.$$

Figure 4.10 shows the retrospective control charts for individuals and moving ranges for the sleeve inside diameter data. The message conveyed by these charts is pretty clear. The machining process was *not* stable over the period of the students' study. Sleeves 15 and 30 were detectably different from the others and provide clear indication of process change.

It is also worth noting that (as might be anticipated based on the discussion before this example) the standard deviation of the 30 inside diameters is substantially larger than 16.51, namely 25.92. $\overline{MR}/1.128$ is typically inflated less by overall change in process mean than is the naive grand standard deviation.

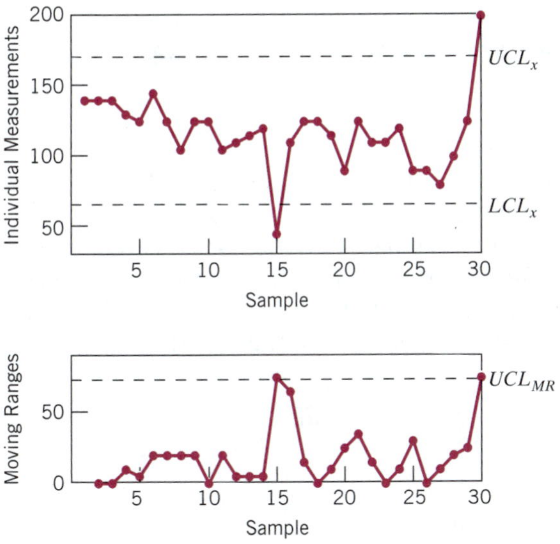

Figure 4.10
Retrospective individual and moving range charts for the diameters in Table 4.15.

There is one final issue that needs discussion in this section. That is the matter of evaluating what an X/MR chart combination can be expected to produce in terms of **ARLs**. Again Crowder's work becomes relevant. Table A.6 gives ARLs computed using a version of his published ARL program for normally distributed data. To enter that table and read out ARLs, one needs values of three parameters, $\mathcal{M}^*$, $\mathcal{R}^*$, and $\mathcal{D}^*$. For process mean μ and standard deviation σ and control limits UCL_x, LCL_x, and UCL_{MR}, these are computed as

$$\mathcal{M}^* = \frac{UCL_x - LCL_x}{2\sigma}, \tag{4.36}$$

Number of Sigmas on the Individuals Chart Needed to Enter Table A.6

$$\mathcal{R}^* = \frac{UCL_{MR}}{\sigma}, \tag{4.37}$$

Standardized Moving Range Control Limit Needed to Enter Table A.6

and

$$\mathcal{D}^* = \frac{\left|\mu - \dfrac{UCL_x + LCL_x}{2}\right|}{\sigma}. \tag{4.38}$$

Standardized "Mean Shift" Parameter Needed to Enter Table A.6

$\mathcal{M}^*$ is the "number of sigmas" for the individuals chart, $\mathcal{R}^*$ is a standardization of the moving range limit, and $\mathcal{D}^*$ is the displacement of the process mean (from the center value for the individuals chart) measured in terms of process standard deviations. (It is straightforward to see that when equations (4.32) and (4.33) are used to set up the charts and parameters are at their standard values, $\mathcal{M}^* = \mathcal{M}$, $\mathcal{R}^* = \mathcal{R}$, and $\mathcal{D}^* = 0$.)

Example 4.6 continued. To illustrate the use of equations (4.36) through (4.38), consider the standards given limits derived earlier for sleeve inside diameters based on a mean of 0 and a process standard deviation of 17, $UCL_x = 54.4$, $LCL_x = -54.4$, and $UCL_{MR} = 74.8$. Suppose that instead of being at standard values, process parameters are actually $\mu = 10$ and $\sigma = 20$. Then, using equations (4.36) through (4.38), one needs to enter the ARL table with

$$\mathcal{M}^* = \frac{54.4 - (-54.4)}{2(20)} = 2.72,$$

$$\mathcal{R}^* = \frac{74.8}{20} = 3.74,$$

and

$$\mathcal{D}^* = \frac{\left|10 - \dfrac{54.5 + (-54.5)}{2}\right|}{20} = .5.$$

These lead (via "exact" calculation with Crowder's ARL program or interpolation in Table A.6) to an ARL of about 56 in this situation. That is, the facts that the mean is off target and the process standard deviation is above its standard value lead to a decrease in mean time to alarm from the all-OK value of 370 to 56.

The X/MR procedure has of late been the subject of several studies by statistical researchers. The growing evidence is that at least for normal observations, it has little to recommend it over the simple charting of individuals (in Shewhart, EWMA, or CUSUM form) *without* the use of moving ranges. However, it *is* a popular tool, especially in the chemical industry. And *if* it is to be used, the kind of modifications of standard "3 sigma charts" practice discussed in this section should definitely be employed.

4.5 CHAPTER SUMMARY

This chapter has introduced some more or less "advanced" methods of statistical process monitoring. The discussions of EWMA and CUSUM schemes in the first two sections show them to be more effective than Shewhart control charts at detecting relatively small process changes. Where process-monitoring computations are handled automatically these methods (and particularly the CUSUM schemes and CUSUM-Shewhart combinations) deserve routine application. The discussion of multivariate process monitoring in the third section provides important means for taking into account relationships between several variables in statistical process control efforts. And the last section of the chapter addresses the issue of process monitoring in situations where the natural subgroup or sample size is $n = 1$. While it seems that the notion of *charting* moving ranges introduced in the section ultimately deserves less enthusiam than it is currently afforded by some practitioners, the discussion of retrospective analysis does bring to light the important idea of estimating σ based on an average moving range where samples are of size $n = 1$.

4.6 CHAPTER 4 EXERCISES

4.1. In what circumstances is a CUSUM or EWMA chart a viable alternative to the Shewhart $\bar{x}$ chart?

4.2. **Milling Operation**. A student group studied a milling operation used in the production of a screen fixture mounting. Of primary importance was a "deviation from flatness" measurement. The units of measurement were .001 in. In the past, deviations from flatness had an average of 2.45 and a standard deviation of 1.40. Suppose that deviation from flatness is approximately normally distributed under stable process conditions.

a. Set up standards given control limits for process location and variability under the following conditions. Monitoring is to be done on the basis of subgroups of size one. An all-OK ARL of about 250 is desirable and the possibility of change in the process mean is of primary concern.

b. Suppose it is very important to be able to detect a shift in the process mean from 2.45 to 3.15. If σ remains at its standard value of 1.40, on average how many subgroups of size one will pass, through the period at which the scheme constructed in part (a) first signals?

c. Ten consecutive mountings produced the deviation from flatness values below (units are .001 inch):

.5, 4.5, 2.0, 2.0, 3.0, 3.0, 2.0, 4.5, 3.0, 0.0.

Find the moving ranges of adjacent observations. Plot the individual observations and moving ranges with the control limits produced in (a).

d. Do your plots in (c) indicate that there has been a process change from standard conditions? Why or why not?

e. Redo part (a) supposing that a possible increase in σ is of more concern than a shift in μ.

f. Answer (b) for the control limits found in (e). Explain why the answer here is consistent with the one you produced in part (b).

4.3. Refer to the **Milling Operation** case in problem (4.2).

a. For purposes of process monitoring only, let a target deviation from flatness be $\mu = 5$, and suppose the process standard deviation is $\sigma = 1.40$, as in problem (4.2). (Clearly, in functional terms a 0 deviation from flatness is ideal.) Compute X/MR control limits for this set of standards, using the same ARL goals as in problem (4.2a).

b. Plot the individuals and moving ranges from problem (4.2c) using your new limits from (a). Does it appear that there has been a process change from standard conditions? Why or why not?

c. Find retrospective X/MR control limits for the data given in problem (4.2c). Use the constants $\mathcal{M}$ and $\mathcal{R}$ called for in problem (4.2a) and estimate σ with $\overline{MR}/1.128$.

d. Plot the individuals and moving ranges from problem (4.2c) using control limits from (c). Is there evidence of process instability? Why or why not?

e. Discuss the practical meaning of the terms "stability," "shift," and "out-of-control" in light of problem (4.2d) and parts (b) and (d) above.

4.4. Refer to the **Lab Carbon Blank** case in problem (2.22). Suppose that repeated measurements of the same blank are normally distributed.

a. Find retrospective control limits and center lines for X/MR plots. Use the "naive" or "3 sigma" approach to setting these, and use $\overline{MR}$ in place of a real average range in the formulas for retrospective limits from Chapter 3.

b. Plot the individuals and moving ranges using the limits from (a). Do you detect any measurement process instability? Why or why not?

c. In your analysis for part (b), which chart should be considered first? Why?

d. Find the estimated mean carbon content measurement and estimated standard deviation of measured carbon content, $\overline{MR}/1.128$.

4.5. Refer to the **Lab Carbon Blank** case in problems (2.22) and (4.4). Suppose the nominal or "real" carbon content is 1.0.

a. Find X/MR control limits and center lines to apply to the data of problem (2.22). Use the naive or 3 sigma approach to setting these and $\overline{MR}$ in place of a real average range in the retrospective formulas, but make use of the nominal value 1.0 in place of $\overline{\overline{x}}$.

b. Does knowledge of a nominal or target value for individuals change the way one should compute MR chart limits? Why or why not?

c. Plot the X/MR values from problem (2.22) using your limits from (a).

d. What dilemma is revealed by your charts in (c) above and problem (4.4b)? Discuss this using phrases such as "consistency of location," "shift in the mean," "off-target process," and "unstable process."

4.6. Refer to the **Lab Carbon Blank** case in problems (2.22), (4.4), and (4.5).

a. Find retrospective X/MR control limits for the data of problem (2.22). Instead of the naive (or 3 sigma) control limits use $\mathcal{M}$ and $\mathcal{R}$ values chosen to correspond to an all-OK ARL of 500 in a standards given context, where the major concern is mean measurement (as opposed to the variability of measurement).

b. In light of your answer to (a), is there evidence of lack of measurement process stability in the *MR* values? In the individuals? Why or why not?

c. Suppose that $\overline{MR}/1.128$ is a good estimate of measurement precision here (i.e., $\sigma \approx \overline{MR}/1.128$) and that your values of $\mathcal{M}$ and $\mathcal{R}$ from (a) are used to set up standards given control limits on individuals and moving ranges for future monitoring. About what ARL is produced if μ changes from its standard value by 2.73 instrument units? Show your work.

d. It is unknown whether the carbon measurements in problem (2.22) were made by the same person or by as many as 14 different people. What configuration of operators would be most effective in isolating instrument changes in the measurement of carbon content? Defend your answer in terms of the concept of "sources of variability."

4.7. Reflect on the possible use of standards given charts in place of retrospective charts in the analysis of historical data. (Clearly, *only* standards given charts can be applied to the real time monitoring of current processes.)

a. Why might an engineer be reluctant to use a standards given chart in a retrospective context?

b. Describe a version of this scenario where use of a standards given chart makes sense.

4.8. Refer to the **Paper Dry Weight** case in problem (3.38). Recall that the target for dry weight of 20-lb bond paper is 70 g/m^2. The pump speed controlling the flow of liquid pulp mixture onto the conveyor-roller mechanism was held fixed at 4.5 (45% of maximum flow) in the production of the data in problem (3.38). Assume that under stable process conditions dry weights are normally distributed.

a. Find X/MR control limits to apply to the data of problem (3.38). Use the nominal dry weight of 70 as a target value, employ $\overline{MR}/1.128$ as an estimate of σ, and use $\mathcal{M}$ and $\mathcal{R}$ values chosen to correspond to an all-OK ARL of 100 in a standards given context, where the major concern is change in mean dry weight (as opposed to increase in the variability of dry weight).

b. 100 measurements correspond to how many subgroups in the context of problem (3.38)?

c. Suppose that the limits of (a) are applied to the future monitoring of individuals and moving ranges. About what ARL is produced if σ is as in part (a), but μ changes from its standard value by 3.5 g/m^2?

d. Make two graphs. First plot the measured dry weight versus period, and second plot the moving ranges versus period. Draw in the control limits found in (a) on the appropriate plots. Evaluate your plots in the context of the original problem.

e. If completely retrospective control limits were used ($\bar{x}$ was used in place of the target value for dry weight) would your conclusion in (d) change?

Why or why not? (Use the same multiples from Table 4.16 as were selected for use in (a).)

4.9. Refer to the **Paper Dry Weight** case of problems (3.38), (3.39), (3.40), (3.41), (3.42), and (4.8). As additional background on the case, note that Hart and Jobe knew the last two runs of paper (discussed in problems (3.41) and (3.42)) were transition runs. The first batch of liquid pulp mixture was being completed and a new batch was being mixed in, until essentially all the dry weight measurements given in problem (3.42) were from a second batch.

a. Consider the two charts constructed in problem (4.8a). If there were out-of-control points, throw those points out and recalculate control limits and center lines. (If no points were outside limits, there is no need to recalculate.)

b. In the context of problems (3.38), (3.39), (3.40), (3.41), and (3.42) why does it make sense to extend the limits you found in (a) into the future?

c. Plot the individual Y's given in problems (3.39), (3.40), (3.41), and (3.42) versus time order of observation. (The Y's from problem (3.39) should be numbered 1 through 8, the ones from problem (3.40) should be numbered 9 through 16, and so on.) Also, plot the moving ranges of the Y's versus time order on another graph. (For each new control algorithm begin making the moving ranges afresh. Don't bridge across data sets.)

d. Draw in the control limits found in (a) above on the two plots made in (c).

e. Do you find any indications of process instability on the plots of (d)? Explain why what you have seen is reasonable in light of the physical facts of this case.

4.10. **Sheet Metal Slips**. Slusher and Mitchell worked with a company that uses an automatic machine to fabricate connectors called "slips" and "drives" for sheet metal duct work. Length is a critical dimension of slips. The two students obtained length measurements for a large number of "acceptable" slips. The sample mean and standard deviation were calculated. It was agreed to treat these two values as standards, $\mu = 9.63$ and $\sigma = .036$ (units were inches). Assume slip length is normally distributed under stable process conditions.

a. Suppose that individual slips are going to be sampled once every five minutes. Set up control limits for individual slip lengths and moving ranges of slip lengths if one desires an all-OK ARL of about 370. Quick detection of a change in mean slip length is more important than detection of a change in standard deviation of slip length.

b. Instead of plotting individual lengths and moving ranges to evaluate stability of the fabrication process, the two analysts decided to plot exponentially weighted moving averages of individual slip lengths (one every 5 minutes). Find control limits, λ, and a center line to give an all-OK ARL of about 370 and quickest possible detection of a shift in mean length of size .018 inch.

c. Next, the two analysts wondered how many five-minute periods would pass until detection of a shift in the average slip length to 9.648 inches or to 9.612 inches. Using the chart you set up in (b) how many periods would pass on average?

d. Finally, the two investigators wished to say how many slips (one taken every five minutes) would pass on average (until detection), should the mean length remain stable at μ, but the process standard deviation increase to .072 inch. Using the chart you set up in (b) how many periods would pass on average?

4.11. Pump End Cap Depths. Depth measurements on pump end caps were of interest to a manufacturer. The target depth was 4.999 inches and manufacturing records indicated that a sensible short-term standard deviation for the metal-cutting process used to machine this dimension was .00023 inch. It was financially feasible for process engineers to obtain four consecutive pump end caps each hour and measure their depths. Suppose that under stable process conditions this part dimension is normally distributed.

a. Set up a EWMA chart (of $\bar{x}$'s) that will give an all-OK ARL of about 250 and quickest possible detection of a shift in mean depth of size .0000575 inch.

b. If, in fact, a shift of .0000575 inch occurs at startup, how many hours of production will pass on average until this change is flagged by an observed $EWMA_i$ falling outside the control limits found in (a)? (Assume that σ remains constant.)

c. Suppose a .000115-inch shift in mean depth occurs at startup. How many hours of production will pass, on average, until an observed $EWMA_i$ falls outside the limits constructed in (a)? (Again assume that σ remains constant.)

d. Consider the chart set up in (a) above. What is the ARL if the process standard deviation is .00046 inch and the process mean is 4.9990575 inches?

4.12. Boiler Nozzle Diameter. A student group investigated the manufacture of fluidized boiler nozzles. The outside diameter of these nozzles is a critical dimension and important to monitor during production. The ideal diameter is 4.600 cm, and there is a maximum standard deviation for diameters of .008 cm. One nozzle per hour will be measured for process monitoring purposes. Assume that under stable process conditions measured nozzle diameters are normally distributed.

a. Find X/MR control limits producing an all-OK ARL of 500, if it is of primary importance to detect a change in the process mean (as opposed to a change in process short-term variability).

b. If the process mean shifts to 4.604 cm (while the process standard deviation remains at .008 cm), what can be said about the mean number of hours that will pass (on average) until detection of the shift? (Give the necessary parameter values for entering the ARL table. Does the table in this book give the desired ARL?)

4.13. Refer to **Boiler Nozzle Diameter** case in problem (4.12).

a. Instead of the X/MR combination considered in problem (4.12) design a EWMA chart for individuals that will produce an all-OK ARL of 500 and quickest possible detection of the eventuality that the mean nozzle diameter shifts off target by .004 cm. (Give values for $EWMA_0$, λ, UCL_{EWMA}, and LCL_{EWMA}.)

b. Redo problem (4.12b) for your EWMA chart from (a). (If the exact ARL is not available from the table in this book, give bounds for it taken from the table.)

c. Compare your answers to problems (4.12b) and (4.13b).

d. Consider again the process-monitoring schemes from problems (4.12b) and (a) above. Compare their abilities to detect a change in mean nozzle diameter to 4.606 (again presuming that σ remains at .008 cm).

e. What general conclusions do the specific calculations here suggest about how EWMA and X/MR schemes compare?

4.14. Consider the monitoring of a process that we will assume under stable conditions produces normally distributed observations with $\sigma = .04$. Let a single observation make up a subgroup, i.e., consider process monitoring on the basis of samples of size $n = 1$.

a. Set up both a two-sided CUSUM scheme and a EWMA scheme for monitoring the process, using a target value of .13 and an all-OK ARL of 370. Quickest possible detection of a .02 change in the mean (from the target) is desired.

b. For both charts in (a), find the ARL if the standard deviation remains at $\sigma = .04$ but the mean changes by .02, .04, .06, and .08. (That is, give four ARLs for both charts.)

c. Find control limits for individuals and moving ranges using Crowder's approach. Use the target value of .13 and an all-OK of 370. Use $\mathcal{R} = 4.57$ as the multiplier for the moving range chart.

d. For your scheme from (c), find the ARL if the standard deviation remains at $\sigma = .04$ but the mean changes by .02, .04, .06, and .08. (That is, give four ARLs for the monitoring scheme in (c).)

e. Plot on a single set of axes the natural logarithms of the ARL values found in parts (b) and (d) above versus shift in mean. Connect the four plotted points for each of the three monitoring schemes with a curve, and compare the three curves. (Note that all three curves should run through the point (0, log(370)).)

4.15. Consider the monitoring of a process that we will assume under stable conditions produces normally distributed observations with $\sigma = .001$ inch. Suppose that the target value for a dimension of interest is .025 inch and that subgroups of size $n = 3$ are going to provide sample means $\bar{x}$.

a. Set up a EWMA (of $\bar{x}$'s) monitoring scheme for this situation, if an on-target ARL of 370 is to be used and quickest possible detection of a change in the process mean of magnitude .0005 inch is desired. (Give values for $EWMA_0$, λ, UCL_{EWMA}, and LCL_{EWMA}.)

b. Suppose the process mean is really .0252 inch (but σ remains at .001 inch). Find the values of the parameters needed to enter the EWMA ARL table. Using these values, interpolate and give an approximate ARL.

c. Find the upper and lower control limits and center line for the standard Shewhart $\bar{x}$ chart here. What is the all-OK ARL for this chart?

d. What is the ARL for the chart from (c) if the process conditions are those of part (b)?

e. Compare the ARLs from (b) and (d) and comment on what they suggest about EWMA schemes in general.

4.16. The diameter of a certain machined cylindrical part has specifications 1.9365 $\pm$ .0005 inch. Suppose diameter means based on $n = 9$ consecutive cylindrical parts are calculated periodically. Past data suggests that a sensible standard for the short-term variability of measured diameters is $\sigma = .00015$ inch. (Note then that under a normal distribution assumption, a perfectly aimed process will place most parts within specifications, but that any significant shift in μ away from the mid-specification will create many nonconforming parts.) Suppose that an all-OK ARL of around 100 is desirable, as is quickest possible detection of the eventuality that the stream of cylindrical parts has a mean diameter off target by .00005 inch.

a. Set up a EWMA chart for this situation. Also, give the values of the parameters you would use to enter the table to find the ARL of your scheme if in fact $\mu = 1.93655$ inch.

b. Set up a two-sided CUSUM scheme for this problem. (Find k_1, k_2, and h.) Also give the values of the parameters you would use to enter this book's table to find the ARL of your scheme if in fact $\mu = 1.93655$ inch.

c. Briefly comment on the relative merits of your monitoring schemes in (a) and (b).

4.17. Refer to the scenario in problem (4.16). Suppose the following are deviations of 10 succesive sample mean diameters from 1.9365 inches (units are 10^{-5} inches above target). (Each of the subgroups producing the means were of size $n = 9$.)

i	1	2	3	4	5	6	7	8	9	10
$\bar{x}_i$	4.0	1.1	−4.9	3.2	4.1	.2	−1	−1.1	2.1	2.0

a. Using your answer to problem (4.16a), find the 10 values $EWMA_i$ corresponding to the sample means in the table. (Be careful to handle the values correctly. For example, the value 4.0 corresponds to a raw measurement of 1.93654.)

b. Does a EWMA chart set up as in problem (4.16a) suggest there has been a change from standard process conditions? Why or why not?

c. Find the 10 high side CUSUMs and 10 low side CUSUMs for the sample means in the table, applying your scheme from problem (4.16b). (Again be careful to handle the values correctly.)

d. Does the CUSUM scheme suggest there has been a change from standard process conditions? Why or why not?

4.18. Refer to the scenario in problem (4.16).

a. Set up an optimal high side CUSUM (of $\bar{x}$'s) scheme if a shift in mean diameter from 1.9365 inches to 1.9366 inches is of concern, short-term variability is believed to be $\sigma = .00015$ inch, subgroups are of size $n = 9$, and an all-OK ARL of 100 is desired. (Find h and k_1.)

b. Suppose the increase in diameter identified in part (a) must (on average) be detected by the time five subgroups are inspected. For the σ and all-OK ARL specified in part (a), how large must n be to meet this requirement?

c. Using your answer to (b) as a new sample size, redo part (a).

d. Set up an optimal low side CUSUM scheme if a shift in mean diameter from 1.9365 inches to 1.9364 inches is of concern, short-term variability is believed to be $\sigma = .00015$ inch, subgroups are of size $n = 9$, and an all-OK ARL of 100 is desired. (Find h and k_2.)

e. Suppose the decrease in diameter identified in part (d) must (on average) be detected by the time five subgroups are inspected. For the σ and all-OK ARL specified in part (a), how large must n be to meet this requirement?

f. Using your answer to (e) as a new sample size, redo part (d).

4.19. Refer to the scenario in problem (4.16).

a. Suppose it is important to detect a change in mean diameter of .00005 inch by the time five subgroups are inspected. Using an appropriate two-sided CUSUM chart that produces an all-OK ARL of 100 and assuming short-term process variability is described by $\sigma = .00015$, how large must the subgroup size be to produce such average performance?

b. Using the sample size found in (a), find the optimal two-sided CUSUM scheme (that has an all-OK ARL of 100 and an ARL of no more than 5.0 when the process mean shifts by .00005 inch). (Find the new combination of k_1, k_2, and h.)

c. How do your ARLs from (b) above and from problem (4.16) compare? Argue that their relationship is sensible in light of the sample sizes in problem (4.16) and above.

d. A standard Shewhart $\bar{x}$ chart has an all-OK ARL of 370. How many "sigmas" (of $\bar{x}$) would be used to make Shewhart control limits for $\bar{x}$ in order to produce an all-OK ARL of 100? Find standards given Shewhart control limits for $\bar{x}$ based on this number of "sigmas," $\sigma = .00015$ and the sample size from (b).

e. Compare the ARL of your chart from (d) to that of the CUSUM scheme from (b) if the process mean is off target by .00005 inch.

4.20. Refer to the scenario in problem (4.16).

a. Suppose engineers have attempted to fix the process following an out-of-control signal from the two-sided CUSUM scheme. They are going to restart the process and continue CUSUM monitoring, but in view of the circumstances wish to use a "fast initial response" scheme, hoping that if their fix was not effective they will get quick detection of this fact. Find parameters for a two-sided FIR CUSUM scheme that they can use and maintain an all-OK ARL of 100. (Find k_1, k_2, h, U_0, and L_0.) Use the sample size of problem (4.16) ($n = 9$) and continue to suppose that under stable process conditions, machined diameters are normally distributed with $\sigma = .00015$ inch, and that changes (from target) in mean diameter of size .00005 inch are important.

b. Using the mean diameters from problem (4.17) and your answer to (a), find high side CUSUMs and low side CUSUMs. (Be careful to handle the values correctly. For example, the value 4.0 corresponds to a raw measurement of 1.93654. Use the restart feature for U_1 and L_1 only.)

4.21. Refer to the scenario of problem (4.18). Increases (from target) of mean diameter of the cylinders continue to be of primary concern, and a change of size .0001 inch is an important one.

a. Suppose engineers have attempted to fix the process following an out-of-control signal from the high side CUSUM scheme. They are going to restart the process and continue CUSUM monitoring, but in view of the circumstances wish to use a "fast initial response" scheme, hoping that if their fix was not effective they will get quick detection of this fact. Find parameters for a high side FIR CUSUM scheme that they can use and maintain an all-OK ARL of 100. (Find k_1, h, and U_0.) Use the sample size of problem (4.18a) ($n = 9$) and continue to suppose that under stable process conditions machined diameters are normally distributed with $\sigma = .00015$ inch.

b. Using the mean diameters from problem (4.17) and your answer to (a), find the high side CUSUMs. (Be careful to handle the values correctly. For example, the value 4.0 corresponds to a raw measurement of 1.93654. Use the restart feature for U_1 only.)

4.22. Tablet Hardness. Cordero, Frueh, and Zeiner studied the production of pharmaceutical tablets with the aim of process improvement. Process engineers knew that both tablet compression and moisture content have an effect on tablet hardness. Two levels ("high" and "low") were identified for both factors. Data from the tablet-manufacturing process was obtained with compression set at its high level and moisture set at its low level. Tablet hardness was measured using a Scheuniger 6d hardness tester and expressed in Standard Cobb Units (SCUs). The tester slowly applies pressure to the sides of a tablet until it fractures. The pressure in SCUs required to fracture the tablet is shown on a digital readout. $n = 10$ consecutive tablets were selected from the process about once every 30 minutes of production. Some summary statistics for 17 periods are collected below. (Units are SCUs.)

Day	Period	Average	Minimum	Maximum
1	1	15.74	13.7	19.0
1	2	17.07	15.9	19.0
1	3	16.03	13.2	18.5
2	4	15.38	13.2	17.8
2	5	16.54	13.8	18.4
2	6	16.66	14.6	18.3
2	7	17.64	15.3	20.0
2	8	16.82	15.3	19.8
2	9	17.20	14.9	20.0
2	10	16.80	14.0	19.6
2	11	17.60	14.7	21.1
2	12	17.33	14.2	20.2
2	13	17.16	16.0	18.7
2	14	17.05	15.5	19.4
2	15	17.70	13.3	21.8
2	16	15.68	14.2	18.0
2	17	16.39	14.9	19.0

The target hardness for these tablets is 17 SCUs with specifications set at 17 ± 5 SCUs. Suppose that under stable process conditions tablet hardness is normally distributed.

a. Is enough information given above to set up a standards given R or s control chart? Why or why not?

b. Make a retrospective R chart for the tablet hardnesses. Is there evidence of nonconstant σ? Explain.

c. Give a reasonable estimate of σ, the short-term variability of tablet hardness. (Base your estimate on the average subgroup range.)

Use your estimate of σ from part (c) in the balance of this problem as if it were σ itself.

d. Suppose it is important to be able to detect a change in mean tablet hardness of 1.2273 SCUs. If an all-OK ARL of 500 is desired, set up a EWMA (for $\bar{x}$'s based on $n = 10$) scheme for tablet hardness. (Give values for $EWMA_0$, λ, UCL_{EWMA}, and LCL_{EWMA}.)

e. Apply the scheme from (d) retrospectively to the sample mean hardnesses. Is there evidence of a shift (away from 17 SCUs) in mean hardness in these averages? Why or why not?

f. Consider again your scheme from (d). If a shift in mean hardness of size 1.2273 SCUs occurs before a first sample is drawn, on average approximately how many 30-minute periods will pass until this condition is flagged? (Give the parameter values you use to enter the ARL table.)

g. Repeat (d) through (f) above using an all-OK ARL of 50.

h. Replace 17 in your scheme from (d) with $\bar{\bar{x}}$ and give new (and fully retrospective) values of $EWMA_0$, UCL_{EWMA}, and LCL_{EWMA}. Apply this scheme to the sample mean hardnesses. Is there evidence of a change in mean hardness in these averages? Why or why not?

i. Comment on how (in the analysis of historical data) an engineer should interpret a standards given EWMA chart as opposed to the kind of retrospective chart produced for part (h) above. Discuss what an out-of-control signal means and doesn't mean in each setup.

4.23. Refer to the **Tablet Hardness** case in problem (4.22).

a. Adopt the goals of problem (4.22d) and design a two-sided CUSUM scheme to meet these criteria. (Find k_1, k_2, and h.)

b. Apply your scheme from (a) retrospectively to the means given in problem (4.22). Is there evidence of a shift (away from 17 SCUs) in mean hardness in these averages? Why or why not?

c. Consider again your scheme from (a). If a shift in mean hardness of size 1.2273 SCUs occurs before a first sample is drawn, on average how many 30-minute periods will pass until this condition is flagged? (Give the parameter values you use to enter the ARL table.)

d. Repeat (a) through (c) above using an all-OK ARL of 50.

e. Replace 17 in your scheme from (a) with $\bar{\bar{x}}$ and give new (and fully retrospective) values of k_1 and k_2. Apply this scheme to the sample mean hardnesses. Is there evidence of a change in mean hardness in these averages? Why or why not?

f. Comment on how (in the analysis of historical data) an engineer should interpret a standards given CUSUM chart as opposed to the kind of retrospective chart produced for part (e) above. Discuss what an out-of-control signal means and doesn't mean in each setup.

4.24. Refer to the **Tablet Hardness** case of problems (4.22) and (4.23), and in particular the CUSUM monitoring of $\bar{x}$'s.

a. Suppose that (while maintaining an all-OK ARL of 500) one wants an ARL of only 5 should a shift of 1.2273 SCUs occur in the process mean. Is the original subgroup size of $n = 10$ adequate to meet these goals? Why or why not? What sample size is required?

b. Repeat (a) if the all-OK ARL is reduced to 100.

c. Using an all-OK ARL of 370, does a CUSUM scheme (for $\bar{x}$'s) designed to produce the quickest possible detection of a shift of 1.2273 SCUs in mean tablet hardness produce a significant reduction in off-target ARL in comparison to a standard Shewhart $\bar{x}$ chart? Why or why not? (Assume a subgroup size of $n = 10$.)

d. Consider, now, designing a CUSUM scheme with an all-OK ARL of 370 and an ARL of 5 if a shift in the mean tablet hardness of 1.2273 SCUs occurs. What subgroup size will enable one to meet these criteria?

e. How does a CUSUM scheme (for $\overline{x}$'s) with sample size from (d) and parameters chosen to meet the goals of part (d) compare to a standard Shewhart $\overline{x}$ chart? Does the CUSUM produce "gains"? Why or why not? What characterizes a "gain"?

4.25. Refer to the **Tablet Hardness** case in problem (4.22) and (4.23).

a. Suppose engineers have attempted to fix the process following an out-of-control signal from a two-sided CUSUM scheme. They are going to restart the process, and continue CUSUM monitoring, but in view of the circumstances wish to use a "fast initial response" scheme, hoping that if their fix was not effective they will get quick detection of this fact. Find parameters for a two-sided FIR CUSUM scheme that they can use and meet the ARL goals set for the scheme designed in problem (4.23a). (Find k_1, k_2, h, U_0, and L_0.)

b. Using the mean hardnesses from problem (4.22) and your answer to (a) above find high and low side CUSUMs. (Use the restart feature for U_1 and L_1 only.)

4.26. Refer to the **Window Frame** case in problem (1.16). Process engineers periodically sampled one window frame from production and measured offsets for the four identifiable corner joints. "Left" and "right" offsets can occur at each location, and recorded offsets do not reflect direction. The upper specification limit for corner offset was .025 inch. (If an offset outside specification is found, the product is rectified.) Offset data (in inches) are given below for 12 windows.

Window	Corner 1	Corner 2	Corner 3	Corner 4
1	.0035	.0085	.0190	.0130
2	.0235	.0025	.0070	.0110
3	.0040	.0080	.0135	.0120
4	.0030	.0075	.0095	.0180
5	.0175	.0085	.0090	.0080
6	.0070	.0085	.0100	.0210
7	.0135	.0045	.0025	.0010
8	.0065	.0215	.0010	.0070
9	.0175	.0120	.0240	.0065
10	.0020	.0150	.0165	.0160
11	.0080	.0095	.0040	.0150
12	.0020	.0045	.0016	.0130

a. Make retrospective moving range charts for the offsets above, one corner at a time. Use $\mathcal{R}$ producing an all-OK ARL of 370 for the MR charts used alone in a standards given context. (No X charts need be considered at this point.) For each corner, turn your average moving range into an estimate of σ for that corner.

b. For a given corner, does it make sense to make any kind of a two-sided chart for process location? Why or why not?

c. For purposes of process monitoring only, use .01 inch as a "target" offset. (Clearly, in functional terms 0 offset is the ideal.) Suppose that engineers wish to have quick detection of an increase in mean offset of .01377 inch (above the target of .01 inch). For each corner, design an appropriate one-sided CUSUM scheme (for individual responses) using an all-OK ARL of 370 and treating your estimates of process short-term variability from (a) as if they were exact values of σ's. (Find k_1 and h for each corner.) Then apply these schemes retrospectively to the four data streams and say what they suggest about process stability.

d. For each of your schemes in (c) find the parameters needed in order to enter the ARL table and evaluate the scheme's ARL if indeed σ is as estimated in part (a) and the mean offset is in fact .02377 inch. Find the resulting ARL for each corner. (If necessary, use appropriate interpolation and show your work.)

e. All of the calculations in this problem are based on a normal distribution assumption for measured offset. Discuss why this is potentially troublesome. (Hint: What is the smallest possible offset? Do you expect offset to have a symmetric distribution under stable process conditions?)

4.27. Refer to the **Window Frame** case in problems (1.16) and (4.26). Suppose the question arises as to how many window frames in a subgroup would be needed to produce a CUSUM scheme (for $\bar{x}$'s) with both an all-OK ARL of 370 and also an ARL of 2.5 should the mean offset increase to .02377 inch. (Use .01 inch as a target offset.)

a. Consider the corners one at a time and find the necessary sample sizes.

b. What subgroup size will guarantee that separate optimally designed CUSUM (for $\bar{x}$'s) schemes for the four corners will each have an all-OK ARL of 370 and an ARL of no more than 2.5 if in fact the mean offset is .02377 inch?

4.28. Refer to the **Window Frame** case in problems (1.16), (4.26), and (4.27).
Suppose engineers have attempted to fix the process following an out-of-control signal from a high side CUSUM scheme. (In the present case, they might, for example, have employed what they know about how tongue-and-groove widths of frame pieces impact offsets.) They are going to restart the process, and continue CUSUM monitoring, but in view of the circumstances wish to use a fast initial response scheme, hoping that if their fix was not effective they will get quick detection of this fact. Find parameters for high side FIR CUSUM schemes that will meet the ARL goals of problem (4.26c). (Find k_1 and h values for each corner separately.)

4.29. Refer to the **Rolled Paper** case introduced in problem (3.18). Use $\bar{s}/c_4$ from problem (3.18) as an approximate value for the process short-term standard deviation, σ, and take $\bar{\bar{x}} = 11.2$ as a process mean under typical operating conditions. Suppose that under stable process conditions sheet lengths are normally distributed.

a. Design a EWMA monitoring scheme (for $\bar{x}$'s based on $n = 5$) that has an all-OK ARL of 750 and affords quickest possible detection of a shift of 1.26 away from the typical process mean. (Give values for $EWMA_0$, λ, UCL_{EWMA}, and LCL_{EWMA}.)

b. Apply your scheme from (a) retrospectively to the sample means given in problem (3.18). What is indicated about the stability of the sheet-cutting process?

c. Instead of using 11.2 as a target, consider adopting the nominal sheet length as a target value. (Note that in the units of the table in problem (3.18) this is a target of 0.) Redo (a) and (b) with this change.

d. Are the conclusions for (c) different than those obtained in (a) and (b)? Discuss.

e. Design a two-sided CUSUM scheme to meet the criteria specified in (a) above. (Find h, k_1, and k_2).

f. Apply your scheme from (e) retrospectively to the sample means given in problem (3.18). What is indicated about the stability of the sheet-cutting process?

g. Let the nominal length (0 in the units of the table) be the target value. Redo (e) and (f) with this change.

h. Are the conclusions for (g) different than those obtained in (e) and (f)? Discuss.

i. Which analysis do you prefer, the EWMA analysis or the CUSUM analysis? Explain in the context of this problem.

4.30. Refer to the **Plastic Packaging** case of problem (3.34). Use $\bar{R}/d_2$ for the data given in problem (3.34) as an approximation for the lip width standard deviation, σ, and take $\bar{\bar{x}}$ from that problem as a process mean under typical operating conditions. Suppose that under stable process conditions lip widths are normally distributed.

a. Design a EWMA monitoring scheme (for $\bar{x}$'s based on $n = 5$) that has an all-OK ARL of 500 and affords quickest possible detection of a shift of $.4472\sigma$ away from the typical process mean. (Give values for $EWMA_0$, λ, UCL_{EWMA}, and LCL_{EWMA}.)

b. Apply your scheme from (a) retrospectively to the sample means given in problem (3.34). What is indicated about the stability of the process?

c. Instead of using the value of $\bar{\bar{x}}$ from problem (3.34) as a target, consider adopting the ideal lip width of 1.5 inches as a target value. Redo (a) and (b) with this change.

d. Are the conclusions for (c) different than those obtained in (a) and (b)? Discuss.

e. Design a two-sided CUSUM scheme to meet the criteria identified in (a) above. (Find h, k_1, and k_2.)

f. Apply your scheme from (e) retrospectively to the sample means given in problem (3.34). What is indicated about the stability of the process?

g. Let the ideal lip width be the target value. Redo (e) and (f) with this change.

h. Are the conclusions for (g) different than those obtained in (e) and (f)? Discuss.

i. Which analysis do you prefer, the EWMA analysis or the CUSUM analysis? Explain in the context of this problem.

4.31. Based on past experience, an engineer develops

$$\boldsymbol{V} = \begin{pmatrix} 5 & 4 \\ 4 & 5 \end{pmatrix}$$

as a standard variance-covariance matrix for bivariate measurements, x_1 and x_2. Desired means for x_1 and x_2 are both 20. Suppose samples of $n = 4$ items are taken hourly and x_1 and x_2 are measured on each of these items. Consider the following 8 pairs of subgroup means.

Subgroup	1	2	3	4	5	6	7	8
$\bar{x}_1$	20.0	22.0	22.0	16.5	20.0	19.0	18.0	18.0
$\bar{x}_2$	20.0	22.0	20.0	16.5	20.0	21.0	22.0	18.0

a. Make standards given $\bar{x}$ charts for variables x_1 and x_2 separately. Are there any out-of-control signals?

b. Make a standards given multivariate control chart. (Find the upper control limit and center line and use either equation (4.24) or equation (4.25) to compute values to plot.) Are there any out-of-control signals?

c. If in (b) there are points that plot out-of-control on the multivariate chart but not on either $\bar{x}$ chart, explain that phenomenon in terms of the nature of the pair of means and the standard correlation between x_1 and x_2.

4.32. A manufacturer of heavy equipment monitors the torques required to loosen 4 distinguishable bolts holding a face plate on a particular machinery component being mass produced in one of its facilities.

a. Why (given enough automation of the torque monitoring process) might there be physical reasons to simultaneously monitor torques from the 4 distinguishable bolts using a multivariate chart in this setting?

b. In fact, all 4 bolts on a given face plate are tightened simultaneously by a pneumatic wrench having 4 heads that are fed off a single compressed air line. Explain why this suggests that a standard correlation between torques at different positions ought to be positive.

c. Find standards given control limits and a matrix expression that you could use to compute a value for judging whether four torques $x_1 = 17.0, x_2 = 18.5, x_3 = 20.0$, and $x_4 = 20.0$ from the bolts on a single face plate constitute evidence of process movement away from standard values of $\mu_1 = \mu_2 = \mu_3 = \mu_4 = 18.5$ ft lbs, $\sigma_1 = \sigma_2 = \sigma_3 = \sigma_4 = 1.0$ ft lb, and $\rho = .4$ (for all pairs of bolts). (Plug the numbers here into a correct formula, but you need not do the matrix inversion or multiplication needed to produce a numerical value for X^2.)

4.33. Refer to the **Plastic Packaging** case in problems (1.9), (1.10), (3.32), (3.33), and (3.34). Data for hole position are given in problem (3.32), and for those same bags, data for lip width are given in problem (3.34). (Values at corresponding positions in the tables for these problems came from the same bag.)

a. Find an estimated variance-covariance matrix for hole position and lip width for each of the 18 periods represented in problems (3.32) and (3.34). Then pool these (as in display (4.28)) to produce a single estimated variance-covariance matrix.

b. Find a two-dimensional vector of sample means of hole position and lip width for each of the 18 periods represented in problems (3.32) and (3.34). Then average these (as in display (4.27)) to find a single vector of $\bar{\bar{x}}$'s.

c. Make a retrospective multivariate control chart based on your computations from (a) and (b). (List the values you plot and say what you use for the upper control limit and the center line.)

d. Does your chart from (c) produce any indications of process change over the three days of the study? Why or why not?

e. Suppose the target value for hole position is 1.69 inches, and (as stated before) the target for lip width is 1.5 inches. Using these standards instead of the vector of $\bar{\bar{x}}$'s in your computation of values to plot, redo (c) and (d).

Day	Time	Bag Length				
1	10:10 am	12.34375,	12.31250,	12.37500,	12.37500,	12.40625
1	10:25 am	12.28125,	12.37500,	12.25000,	12.28125,	12.28125
1	10:55 am	12.31250,	12.34375,	12.40625,	12.34375,	12.40625
1	11:12 am	12.25000,	12.25000,	12.21875,	12.21875,	12.25000
1	11:35 am	12.25000,	12.28125,	12.21875,	12.31250,	12.31250
1	11:41 am	12.21875,	12.21875,	12.28125,	12.31250,	12.25000
2	8:15 am	12.46875,	12.43750,	12.37500,	12.37500,	12.37500
2	8:54 am	12.40625,	12.40625,	12.31250,	12.40625,	12.37500
2	9:21 am	12.37500,	12.37500,	12.43750,	12.34375,	12.40625
2	9:27 am	12.37500,	12.31250,	12.34375,	12.43750,	12.43750
2	9:51 am	12.43750,	12.37500,	12.34375,	12.37500,	12.34375
2	9:58 am	12.40625,	12.34375,	12.37500,	12.34375,	12.37500
3	10:18 am	12.40625,	12.43750,	12.50000,	12.46875,	12.46875
3	10:33 am	12.37500,	12.40625,	12.37500,	12.37500,	12.37500
3	10:45 am	12.53125,	12.56250,	12.56250,	12.56250,	12.53125
3	11:16 am	12.40875,	12.50000,	12.50000,	12.46875,	12.46875
3	11:24 am	12.50000,	12.50000,	12.50000,	12.53125,	12.53125
3	11:39 am	12.50000,	12.50000,	12.50000,	12.50000,	12.50000

4.34. Refer to the **Plastic Packaging** case in problems (1.9), (1.10), (3.32), (3.33), (3.34), and (4.33). In addition to hole position and lip width, bag length was also measured. The bag lengths corresponding to the data given in problems (3.32) and (3.34) are given on page 192. The ideal bag length is 12.5 inches. Redo problem (4.33e) now considering not just hole position and lip width, but length as well. (This is now a three-dimensional problem, not just a two-dimensional one.)

4.35. Grit Composition. In a 1993 *Quality Engineering* article, Holmes and Mergen discussed a multivariate process monitoring problem involving grit (like that used to make sandpaper). Their article gives the percent by weight of particles of three sizes in 56 samples of such material. These values are reproduced in the table below, x_1 being the percent of large particles, x_2 being the percent of medium particles, and x_3 being the percent of small particles in the mixture. Notice that in a problem that like this one involves "compositional data," since the sum of percentages must be the same for every sample ($x_1 + x_2 + x_3 = 100$) there is no more information in the vector (x_1, x_2, x_3) than there is in any vector of only two of the percentages, for example (x_1, x_2). So for purposes of grit process analysis, it suffices to consider only the variables x_1 and x_2. Henceforth confine attention to these two variables, the percentages of large and medium particles in the samples.

a. Suppose that it makes sense to think of the 56 multivariate observations above as coming from the same process conditions. It is then sensible to use them to produce a sample mean vector $(\overline{x}_1, \overline{x}_2)'$ and sample variance-covariance matrix

$$\hat{\boldsymbol{V}} = \begin{pmatrix} s_1^2 & rs_1s_2 \\ rs_1s_2 & s_2^2 \end{pmatrix}$$

to summarize process performance. Find these.

b. If one thinks of the data here as (bivariate) samples of size $n = 1$, taken over 56 time periods, the possibility of investigating process stability using a multivariate control chart arises. Note, however, that no standards are given here, so any analysis one might do must be entirely retrospective. A naive possibility is then to use the $n = 1$ version of formula (4.24) with the estimates from part (a) in place of standards for $\boldsymbol{\mu}$ and $\boldsymbol{V}$ to compute 56 points to plot. Why is this method of retrospective charting likely to prove ineffective? (Hint: Consider the discussion of *univariate* retrospective Shewhart charting based on samples of size $n = 1$ found in Section 4.4. What would movement of the bivariate process mean across the 56 periods tend to do to the estimated variance-covariance matrix? The suggested procedure is just the multivariate generalization of using $\overline{x}$ in place of μ and s in place of σ in a retrospective Shewhart individuals chart.)

Sample	x_1	x_2	x_3	Sample	x_1	x_2	x_3
1	5.4	93.6	1.0	29	7.4	83.6	9.0
2	3.2	92.6	4.2	30	6.8	84.8	8.4
3	5.2	91.7	3.1	31	6.3	87.1	6.6
4	3.5	86.9	9.6	32	6.1	87.2	6.7
5	2.9	90.4	6.7	33	6.6	87.3	6.1
6	4.6	92.1	3.3	34	6.2	84.8	9.0
7	4.4	91.5	4.1	35	6.5	87.4	6.1
8	5.0	90.3	4.7	36	6.0	86.8	7.2
9	8.4	85.1	6.5	37	4.8	88.8	6.4
10	4.2	89.7	6.1	38	4.9	89.8	5.3
11	3.8	92.5	3.7	39	5.8	86.9	7.3
12	4.3	91.8	3.9	40	7.2	83.8	9.0
13	3.7	91.7	4.6	41	5.6	89.2	5.2
14	3.8	90.3	5.9	42	6.9	84.5	8.6
15	2.6	94.5	2.9	43	7.4	84.4	8.2
16	2.7	94.5	2.8	44	8.9	84.3	6.8
17	7.9	88.7	3.4	45	10.9	82.2	6.9
18	6.6	84.6	8.8	46	8.2	89.8	2.0
19	4.0	90.7	5.3	47	6.7	90.4	2.9
20	2.5	90.2	7.3	48	5.9	90.1	4.0
21	3.8	92.7	3.5	49	8.7	83.6	7.7
22	2.8	91.5	5.7	50	6.4	88.0	5.6
23	2.9	91.8	5.3	51	8.4	84.7	6.9
24	3.3	90.6	6.1	52	9.6	80.6	9.8
25	7.2	87.3	5.5	53	5.1	93.0	1.9
26	7.3	79.0	13.7	54	5.0	91.4	3.6
27	7.0	82.6	10.4	55	5.0	86.2	8.8
28	6.0	83.5	10.5	56	5.9	87.2	6.9

Recognizing the difficulty raised in part (b) concerning retrospective control charting of individual multivariate observations, Holmes and Mergen suggested the following way of producing an estimated variance-covariance matrix from a sequence of samples of size $n = 1$. (This method is an analog of the use of $\overline{MR}/1.128$ as an estimate of σ that was discussed in Section 4.4.)

As an estimate of the variance of any single variable, x_i, Holmes and Mergen suggested using the average of the sample variances of all $r-1 = 55$ pairs of adjacent observations, which turns out to be

$$\frac{1}{2(r-1)}\sum_{\ell=2}^{r}(x_{i(\ell-1)} - x_{i\ell})^2.$$

Then as an estimate of the covariance between variable x_i and variable x_j (that is, the quantity $\rho_{ij}\,\sigma_i\,\sigma_j$) Holmes and Mergen suggested the corresponding quantity

$$\frac{1}{2(r-1)}\sum_{\ell=2}^{r}(x_{i(\ell-1)}-x_{i\ell})(x_{j(\ell-1)}-x_{j\ell}).$$

Such quantities may then be collected into the necessary $p \times p$ matrix.

c. Use the Holmes and Mergen suggestion and compute an estimated variance-covariance matrix for the percentages of large and medium particles. Then using your sample mean vector from part (a), this estimated variance-covariance matrix and the $p = 2$ and $n = 1$ versions of formulas (4.24) and (4.26), make a retrospective multivariate individuals chart for the grit data. What conclusions about the stability of the physical process are suggested by your chart?

4.36. An example in a GE Aircraft Engines statistical process control workbook concerns the diameters of three different milled holes in a heat sink cover produced by an electronics manufacturer. Ideal values for these three diameters (A, B, and C) are respectively 1.500 inches, .3235 inch, and .0800 inch. Measured diameters from 8 covers are given in the following table.

Part	A Diameter	B Diameter	C Diameter
1	1.499	.3250	.0800
2	1.498	.3235	.0780
3	1.500	.3240	.0790
4	1.502	.3240	.0800
5	1.502	.3240	.0795
6	1.499	.3245	.0795
7	1.499	.3240	.0800
8	1.499	.3240	.0795

a. Letting x_1, x_2, and x_3 stand respectively for the A, B, and C hole diameters of such heat sinks and supposing the manufacturing process to have been physically stable over the period represented by the data above, it is sensible to use the sample mean vector $(\bar{x}_1, \bar{x}_2, \bar{x}_3)'$ and sample variance-covariance matrix

$$\hat{V} = \begin{pmatrix} s_1^2 & r_{12}s_1s_2 & r_{13}s_1s_3 \\ r_{12}s_1s_2 & s_2^2 & r_{23}s_2s_3 \\ r_{13}s_1s_3 & r_{23}s_2s_3 & s_3^2 \end{pmatrix}$$

to represent process performance. Compute these.

b. Suppose now that the 8 heat sinks of this problem were selected from production at 30-minute intervals and it is not necessarily the case that the production process was physically stable over the period represented by the parts. Problem (4.35) introduced a method of Holmes and Mergen for estimating a variance-covariance matrix based on a series of samples of size $n = 1$. Apply that method here and compare your estimate to the "sample" variance-covariance matrix obtained in part (a).

c. Use your "sample" mean vector from (a), your estimated variance-covariance matrix from part (b), and the $p = 3$ and $n = 1$ versions of formulas (4.24) and (4.26) to make a retrospective multivariate individuals chart for the hole diameter data. What conclusions about the stability of the physical process are suggested by your chart?

4.37. Main Lab versus Unit Lab. Tracy, Young, and Mason, in a 1995 *Journal of Quality Technology* article discussed a measurement problem where a physical sample is obtained and immediately measured by the production unit lab, and then sent to a main plant lab for confirmation of the measurement. (For proprietary reasons, the identity of the measured variable was not completely disclosed.) Let x_1 be the unit lab measurement on a sample and x_2 be the corresponding main lab measurement. Historical evidence was used to develop standards $\mu_1 = 7.09$, $\mu_2 = 7.11$, $\sigma_1 = .387$, $\sigma_2 = .155$, and $\rho_{12} = .556$. Describe a circumstance (both in numerical terms and with an appropriate graph) in which a sample has both x_1 and x_2 inside 3 sigma control limits for individual values, but taken together they produce a strong out-of-control signal on a multivariate control chart. (Picture this circumstance on a plot in the (x_1, x_2)-plane.)

CHAPTER 5

Process Characterization and Capability Analysis

The previous two chapters have dealt with tools for monitoring processes and detecting physical instabilities. The ultimate goal of using these methods is finding the source(s) of any process upsets and removing them, creating a process that is consistent/repeatable/predictable in its pattern of variation. When that has been accomplished, it then makes sense to summarize that pattern of variation graphically and/or in terms of numerical summary measures. These descriptions of consistent process behavior can then become the bases for engineering and business decisions about if, when, and how the process should be used.

This chapter discusses methods for characterizing the pattern of variation exhibited by a reasonably predictable system. It begins with a section presenting some more methods of statistical graphics (beyond those in Section 2.4) useful in process characterization. Then there is a section discussing some "process capability indices" and the making of confidence intervals for them. Next, the making of prediction and tolerance intervals for measurements generated by a stable process is presented. A fourth section considers the problem of predicting the output variation for a system simple enough to be described by an explicit equation, in terms of the variability of system inputs. Finally, the topic of estimating variance components is considered for some processes that because of the existence of identifiable sources of variation cannot be called completely "stable."

5.1 MORE STATISTICAL GRAPHICS FOR PROCESS CHARACTERIZATION

The elementary graphical methods of Section 2.4 provide a starting point for picturing the pattern of variability produced by a process. But other, slightly more sophisticated methods are possible and often prove useful. Some of these are the topic of this section. After briefly reviewing the simple ideas of dot plots and stem-and-leaf diagrams, graphical tools based on the concept of distribution quantiles are discussed. The tools of quantile plots, box plots, and both empirical and theoretical Q-Q plots (probability plots) are introduced.

5.1.1 Dot Plots and Stem-and-Leaf Diagrams

Two very simple but quite revealing ways of presenting small to moderate-sized data sets are through the use of dot plots and stem-and-leaf diagrams. A **dot plot** is made by ruling off an appropriate scale and then placing a large dot above the scale for each data point, stacking dots corresponding to points that are identical (to the indicated precision of the data). A **stem-and-leaf** diagram is made by using a vertical line (a stem) to separate the leading digits for data values from the final few (usually one or two) digits. These (sets of) final digits are stacked horizontally to form "leaves" that function like the bars of a histogram, portraying the shape of the data set. The virtue of a stem-and-leaf diagram is that it provides its picture of data set shape without the loss of the exact individual data values.

Example 5.1 **Tongue Thickness on Machined Steel Levers.** Unke, Wayland, and Weppler worked with a machine shop on the manufacture of some steel levers. The ends of these levers were machined to form tongues. Table 5.1 contains measured thicknesses for 20 tongues cut during the period the students worked on the process. The units are inches and engineering specifications for the tongue-thickness were .1775 to .1875. Figure 5.1 is a dot plot of the tongue thickness data

Table 5.1 Measured Tongue Thicknesses for Twenty Machined Steel Levers (inches)

.1825	.1817	.1841	.1813	.1811
.1807	.1830	.1827	.1835	.1829
.1827	.1816	.1819	.1812	.1834
.1825	.1828	.1868	.1812	.1814

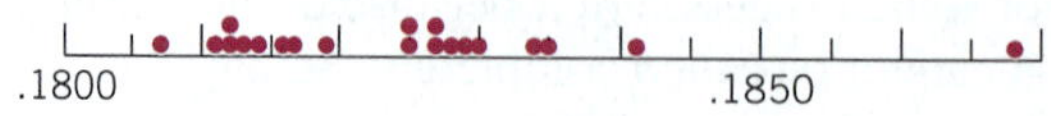

Figure 5.1
Dot plot of the tongue-thickness data.

.186	8
.185	
.184	1
.183	0 4 5
.182	5 5 7 7 8 9
.181	1 2 2 3 4 6 7 9
.180	7

Figure 5.2
Stem-and-leaf diagram of tongue thicknesses.

and Figure 5.2 is a corresponding stem-and-leaf diagram. On the stem-and-leaf diagram, the digits ".18X" have been placed to the left of the stem and only the final digits recorded in Table 5.1 have been used to make the leaves. Notice also that the digits in the leaves have been sorted smallest to largest to further improve the organization of the data.

A useful variation on the basic stem-and-leaf diagram is the notion of placing two of them back-to-back, using the same stem for both, with leaves for the first projecting to the left and leaves for the second projecting to the right. This device is helpful when one needs to make comparisons between two different processes or sets of process conditions.

Example 5.2 **Heat Treating Gears.** The article "Statistical Analysis: Mack Truck Gear Heat Treating Experiments" by P. Brezler (*Heat Treating,* November 1986) describes a company's efforts to find a way to reduce distortion in the heat treating of some gears. Two different methods were considered for loading gears into a continuous carburizing furnace, namely laying the gears flat in stacks and hanging them from rods passing through the gear bores. Table 5.2 contains measurements of "thrust face runout" made after heat treating 38 gears laid and 39 gears hung. Figure 5.3 shows back-to-back stem-and-leaf diagrams for the runouts. (Notice that in this display there are two sets of stems for each leading digit, a "0-4" stem and a "5-9" stem.) The stem-and-leaf diagrams clearly show the laying method to produce smaller distortions than the hanging method.

Table 5.2 Thrust Face Runouts for Gears Laid and Gears Hung (.0001 inch)

Laid	Hung
5, 8, 8, 9, 9, 9, 9, 10, 10, 10	7, 8, 8, 10, 10, 10, 10, 11, 11, 11
11, 11, 11, 11, 11, 11, 11, 12, 12, 12	12, 13, 13, 13, 15, 17, 17, 17, 17, 18
12, 13, 13, 13, 13, 14, 14, 14, 15, 15	19, 19, 20, 21, 21, 21, 22, 22, 22, 23
15, 15, 16, 17, 17, 18, 19, 27	23, 23, 23, 24, 27, 27, 28, 31, 36

Laid Runouts		Hung Runouts
9 9 9 9 8 8 5	0	7 8 8
4 4 4 3 3 3 3 2 2 2 2 1 1 1 1 1 1 1 0 0 0	1	0 0 0 0 1 1 1 2 3 3 3
9 8 7 7 6 5 5 5 5	1	5 7 7 7 7 8 9 9
	2	0 1 1 1 2 2 2 3 3 3 3 4
7	2	7 7 8
	3	1
	3	6

Figure 5.3
Back-to-back stem-and-leaf plots of thrust face runouts (10^{-4} inch).

Dot plots and stem-and-leaf diagrams are not so much graphical methods for formal presentations as they are tools for working data analysis. Histograms are probably more easily understood by the uninitiated than are stem-and-leaf diagrams. As such, they are more appropriate for final reports and presentations. But because of the grouping that is usually required to make a histogram, they do not provide the kind of complete picture of a data set provided by the simple dot plots and stem-and-leaf diagrams discussed here.

5.1.2 Quantiles and Box Plots

The concept of **quantiles** is an especially helpful one for building descriptions of distributions. Roughly speaking, the p quantile (or $100 \times p$th percentile) of a distribution is a number such that a fraction p of the distribution lies to the left and a fraction $1 - p$ lies to the right of the number. If one scores at the .8 quantile (80th percentile) on a national college entrance exam, about 80% of those taking the exam had lower marks and 20% had higher marks. Or, since 95% of the standard normal distribution is to the left of 1.645, one can say that 1.645 is the .95 quantile of the standard normal distribution. The notation $\boldsymbol{Q(p)}$ will be used in this book to stand for the p quantile of a distribution.

Notation for the p Quantile

While the concept of a distribution quantile should be more or less understandable to readers in approximate terms, there remains the job of providing a precise definition of $Q(p)$ for small data sets. For example, faced with a sample consisting of the five values 4, 5, 5, 6, and 8, exactly what should one mean by $Q(.83)$? What number should one announce as placing 83% of the five numbers to the left and 17% to the right? Definition 5.1 gives the two-part convention that will be used in this book for quantiles of data sets.

Definition 5.1 For a data set consisting of n values $x_1 \leq x_2 \leq \cdots \leq x_n$ (x_i is the ith smallest data value)

1. for any positive integer i, if $p = (i - .5)/n$, the ***p* quantile** of the data set is

$$Q\left(\frac{i-.5}{n}\right) = x_i, \qquad \text{and}$$

2. for values of p not of the form $(i - .5)/n$ for any integer i but with $.5/n < p < (n - .5)/n$, the ***p* quantile** of the data set is found by linear interpolation between the two values $Q((i - .5)/n)$ with corresponding $(i - .5)/n$ closest to p.

Definition 5.1 defines $Q(p)$ for all p between $.5/n$ and $(n - .5)/n$. If one wants to find a particular single quantile for a data set, say $Q(p)$, one may solve the equation

$$p = \frac{i - .5}{n}$$

for i, yielding

$$i = np + .5 . \tag{5.1}$$

Index i of the Ordered Data Point that is $Q(p)$

If for the p in question $np + .5$ is an integer, then $Q(p)$ is simply the $(np + .5)$th smallest data point. If relationship (5.1) leads to a noninteger value for i, one interpolates between the ordered data points with indices just smaller than and just larger than $np + .5$.

It is sometimes helpful to plot $Q(p)$ as a function of p. Such a plot is called a **quantile plot** and can be made by plotting the n points $((i - .5)/n, x_i)$ and then drawing in the interpolating line segments. It gives essentially the same information about a distribution as the possibly more familiar "cumulative frequency ogive" of elementary statistics.

Example 5.1 continued. Returning to the tongue-thickness example, Table 5.3 shows the $n = 20$ ordered data values and corresponding values of $(i - .5)/n$.

From the values in Table 5.3 it is clear, for example, that $Q(.425) = .1819$. But should one desire the .83 quantile of the tongue-thickness data set, one must interpolate appropriately between $Q(.825) = .1834$ and $Q(.875) = .1835$. Doing so, one finds that

$$\begin{aligned} Q(.83) &= \frac{.830 - .825}{.875 - .825} Q(.875) + \left(1 - \frac{.830 - .825}{.875 - .825}\right) Q(.825), \\ &= .1(.1835) + .9(.1834), \\ &= .18341 . \end{aligned}$$

Table 5.3 Ordered Tongue Thicknesses and Values of $\frac{i-.5}{20}$

i	$p = \frac{i-.5}{20}$	$x_i = Q\left(\frac{i-.5}{20}\right)$	i	$p = \frac{i-.5}{20}$	$x_i = Q\left(\frac{i-.5}{20}\right)$
1	.025	.1807	11	.525	.1825
2	.075	.1811	12	.575	.1827
3	.125	.1812	13	.625	.1827
4	.175	.1812	14	.675	.1828
5	.225	.1813	15	.725	.1829
6	.275	.1814	16	.775	.1830
7	.325	.1816	17	.825	.1834
8	.375	.1817	18	.875	.1835
9	.425	.1819	19	.925	.1841
10	.475	.1825	20	.975	.1868

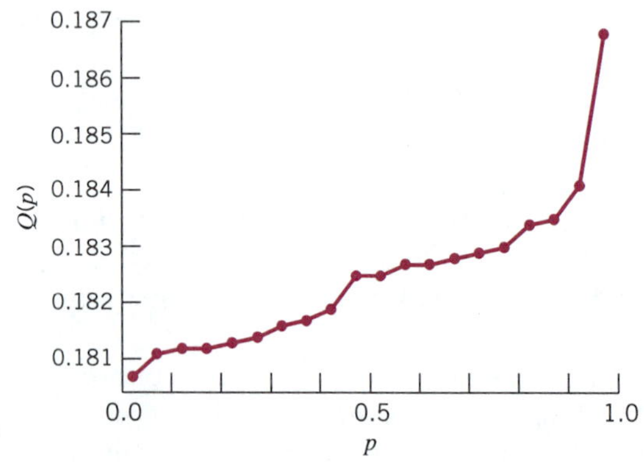

Figure 5.4
Quantile plot for the tongue-thickness data.

And giving a more complete summary of the entire quantile function for the tongue-thickness data set, Figure 5.4 is a quantile plot based on Table 5.3.

Special values of p have corresponding specially named p quantiles. The .5 quantile of a distribution is nothing more than the usual **median,** symbolized as $\tilde{x}$ in Section 3.2. $Q(.25)$ is often called the **first quartile** of a distribution and $Q(.75)$ is called the **third quartile**. And the values $Q(.1), Q(.2), \ldots, Q(.9)$ are called the **deciles** of a distribution.

Example 5.1 continued. Looking carefully at Table 5.3, it is easy to see that the median of the tongue-thickness data set is the simple average of $Q(.475) = .1825$ and $Q(.525) = .1825$. That is, $\tilde{x} = Q(.5) = .1825$. Similarly, the first quartile of the data set is half way between $Q(.225) = .1813$ and $Q(.275) = .1814$. That is, $Q(.25) = .18135$. And the third quartile of the tongue-thickness data is the mean of $Q(.725) = .1829$ and $Q(.775) = .1830$, namely $Q(.750) = .18295$.

Quantiles are basic building blocks for many useful descriptors of a distribution. The median is a well-known measure of location. The difference between the quartiles is a simple measure of spread called the **interquartile range.** In symbols

$$IQR = Q(.75) - Q(.25)\,. \tag{5.2}$$

Interquartile Range

And there are a number of helpful graphical techniques that make use of quantiles. One of these is the **box plot** invented by John Tukey.

Figure 5.5 shows a generic box plot. The box locates the middle 50% of the distribution, with a dividing line drawn at the median. The placement of this dividing line gives some indication of symmetry (or lack thereof) for the center part of the distribution. Lines (or "whiskers") extend out from the box to the most extreme data points that are within $1.5IQR$ (1.5 times the box length) of the box. Any data values that fall more than $1.5IQR$ away from the box are plotted individually and in the process identified as unusual or "outlying."

Example 5.1 continued. As an illustration of the calculations necessary to implement the schematic shown in Figure 5.5, consider the making of a box plot for the tongue thicknesses. Previous calculation has shown the median thickness to be .1825 and the quartiles of the thickness distribution to be $Q(.25) = .18135$ and $Q(.75) = .18295$. Thus from display (5.2), the interquartile range for the thickness data set is

$$IQR = Q(.75) - Q(.25) = .18295 - .18135 = .0016\,.$$

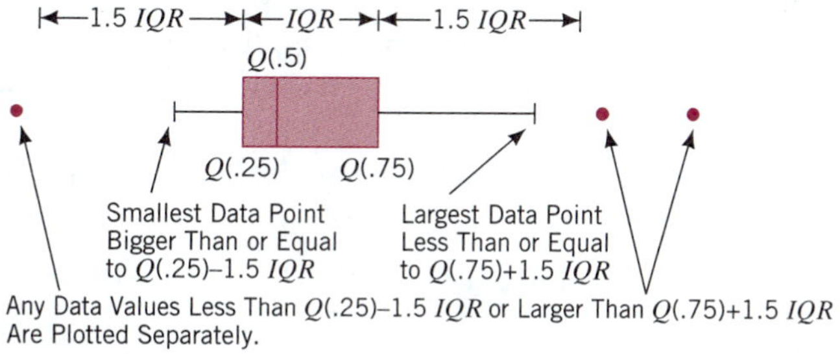

Figure 5.5
Generic box plot.

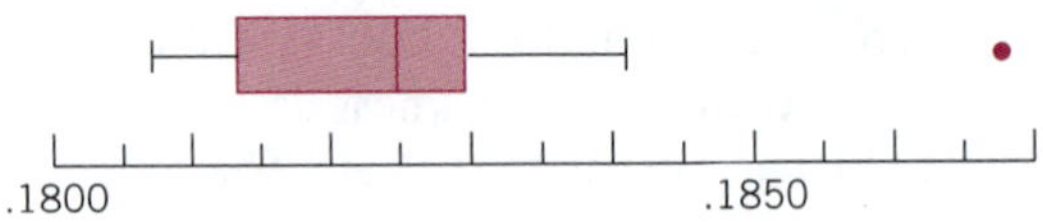

Figure 5.6 Box plot for the tongue-thickness data.

Then, since

$$Q(.25) - 1.5IQR = .18135 - .0024 = .17895$$

and there are no data points less than .17895, the lower whisker extends to the smallest data value, namely .1807. Further, since

$$Q(.75) + 1.5IQR = .18295 + .0024 = .18535$$

and there is one data point larger than this sum, the value .1868 will be plotted individually and the upper whisker will extend to the second largest data value, namely .1841. Figure 5.6 is a box plot for the tongue-thickness data. It reveals some asymmetry in the central part of the data set, its relative short-tailedness to the low side, and the one very large outlying data value.

Box plots carry a fair amount of information about distribution location, spread, and shape. They do so in a very compact way. In fact, one of their chief virtues is that many of them can be placed on a single page to facilitate comparisons among a large number of distributions. The next example illustrates the comparison of three distributions using side-by-side box plots.

Example 5.3 **Comparing Hardness Measurement Methods.** Blad, Sobatka, and Zaug did some hardness testing on a single metal specimen. They tested it on three different machines, 10 times per machine. A dial Rockwell tester, a digital Rockwell tester, and a Brinell tester were used. The Brinell hardnesses they recorded (after conversion in the case of the Rockwell readings) are given in Table 5.4.

Figure 5.7 shows box plots for the measurements produced by the three hardness testers. It is very helpful for comparing them. It shows, among other things, the comparatively large variability and decided skewness of the Brinell machine

Table 5.4 Hardness Values for a Single Specimen Obtained from Three Different Testers (Brinell Hardness)

Dial Rockwell	Digital Rockwell	Brinell
536.6, 539.2, 524.4, 536.6	501.2, 522.0, 531.6, 522.0	542.6, 526.0, 520.5, 514.0
526.8, 531.6, 540.5, 534.0	519.4, 523.2, 522.0, 514.2	546.6, 512.6, 516.0, 580.4
526.8, 531.6	506.4, 518.1	600.0, 601.0

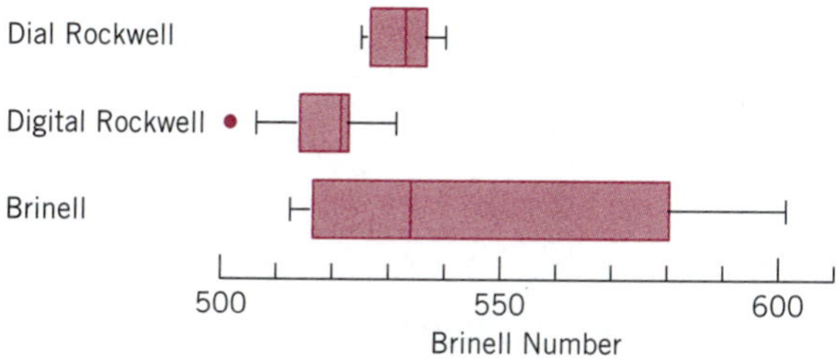

Figure 5.7
Box plots for hardness measurements made on three different testers.

measurements, and the fact that the Dial Rockwell machine seems to read consistently higher than the Digital Rockwell machine.

5.1.3 *Q-Q* Plots and Normal Probability Plots

An extremely important application of the quantile notion is to the careful comparison of the shapes of two distributions through the making and interpretation of ***Q-Q* plots.** The easiest version of the Q-Q plot to understand is that where both distributions involved are empirical, representing data sets. The *most important* version is that where one distribution is empirical and the other is a theoretical distribution, and one is essentially investigating how the shape of a data set matches that of some probability distribution. This discussion will begin with the easy-to-understand (but practically less important) case of comparing shapes for two data sets, and then proceed to the more important case.

Consider the two small data sets given in Table 5.5. Figure 5.8 shows dot plots for them and reveals that by most standards they have the same shape. There are several ways one might try to quantify this fact. For one thing, the relative sizes of the gaps or differences between successive ordered data values are the same for the two data sets. That is, the gaps for the first data set are in the ratios 1:1:0:2 and for the second data set the ratios are 2:2:0:4.

A second (and for present purposes more germane) observation is that the ordered values in the second data set are linearly related to those in the first. In fact, the values in the second data set were derived by doubling those in the first data set and adding 1. This means that the second quantile function is linearly related to the first by the relation

$$Q_2(p) = 2Q_1(p) + 1 .$$

Table 5.5 Two Small Artificial Data Sets

Data Set 1	Data Set 2
2, 3, 4, 4, 6	5, 7, 9, 9, 13

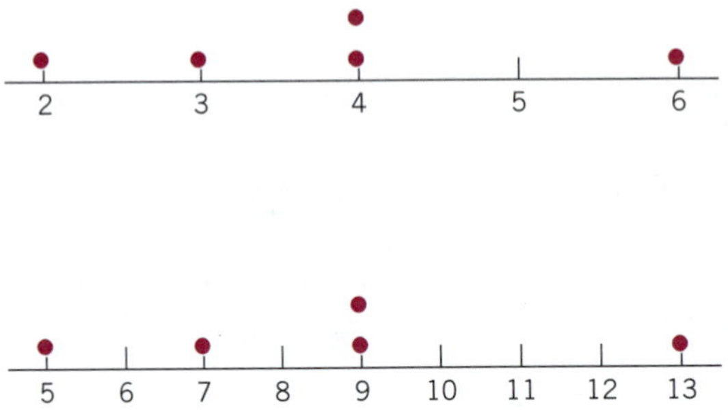

Figure 5.8 Dot diagrams for two small data sets.

Notice then that if one makes up ordered pairs of the form $(Q_1(p), Q_2(p))$ and plots them, all of the plotted points will fall on a single line. Using the values $p = .1, .3, .5, .7$, and $.9$, one has in the present case the five ordered pairs

$$(Q_1(.1),\ Q_2(.1)) = (2, 5),$$
$$(Q_1(.3),\ Q_2(.3)) = (3, 7),$$
$$(Q_1(.5),\ Q_2(.5)) = (4, 9),$$
$$(Q_1(.7),\ Q_2(.7)) = (4, 9), \quad \text{and}$$
$$(Q_1(.9),\ Q_2(.9)) = (6, 13),$$

and the scatterplot in Figure 5.9.

What is true in this highly simplified and artificial example is true in general. Equality of "shape" for two distributions is equivalent to the two corresponding quantile functions being linearly related. A way of investigating the extent to which two distributions have the same shape is to plot for suitable p, ordered pairs of the form

Points for a Q-Q Plot of Distributions 1 and 2

$$(Q_1(p), Q_2(p)), \tag{5.3}$$

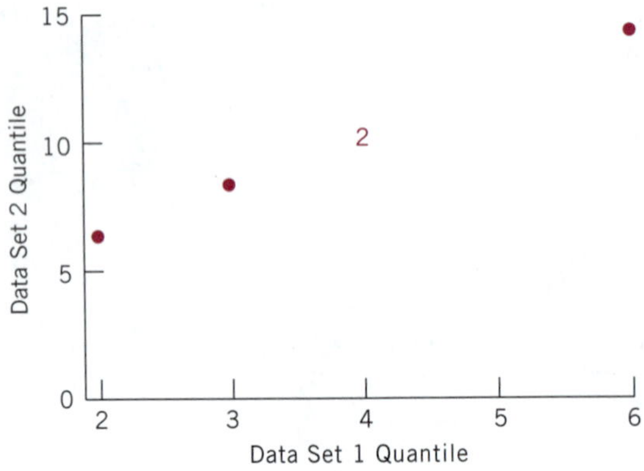

Figure 5.9 Q-Q plot for the data sets of Table 5.5.

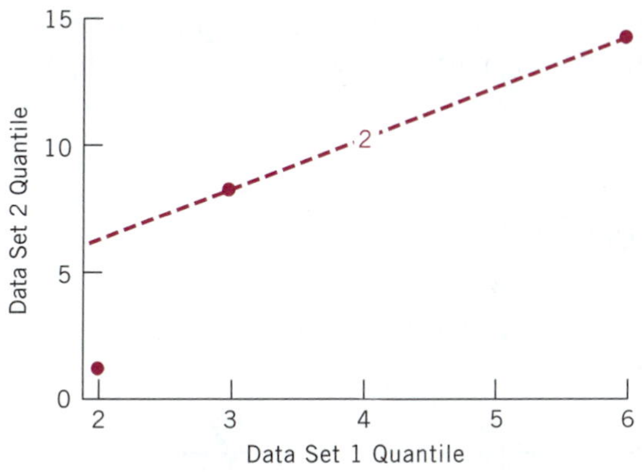

Figure 5.10
Q-*Q* plot after modifying one value in Table 5.5.

looking for linearity. Where there is perfect linearity on such a plot, equality of shape is suggested. Where there are departures from linearity, those departures can often be interpreted in terms of the relative shapes of the two distributions. Consider for example, a modified version of the present example where the value 5 in the second data set is replaced by 1. Figure 5.10 is the Q-Q plot for this modification.

Notice that while four of the plotted points in Figure 5.10 fall on a single straight line, the fifth, (2,1), does not fall on that line. It is below/to the right of the line established by the rest of the points. To bring it "back on line" with the rest, it would have to be moved either up or to the left on the plot. This says that relative to the shape of data set 1, the second data set is long tailed to the low side. Equivalently, relative to the shape of data set 2, the first is short tailed to the low side.

The important version of Q-Q plotting where the first distribution is that of a data set and the second is a theoretical or probability distribution is usually called **probability plotting.** And the most common kind of probability plotting is **normal plotting,** where one is investigating the degree of similarity between the shape of a data set and the archetypal bell shape of the normal distribution.

The values of p typically used in making the points (5.3) for a probability plot are those corresponding exactly to the data points in hand, namely those of the form $(i - .5)/n$ for integer i. Using such values of p, if $Q_z(p)$ is the standard normal quantile function, it follows that a normal plot can be made on regular graph paper by plotting the n points

$$\left(x_i,\ Q_z\left(\frac{i - .5}{n}\right)\right), \qquad (5.4)$$

Points for a Normal Plot of an x Data Set

where as in Definition 5.1, x_i is the ith smallest data value.

Standard normal quantiles for use in display (5.4) can, of course, be found by locating values of p in the body of a cumulative normal probability table like Table A.2, and then reading corresponding quantiles from the table's margin. Statistical packages (like Minitab) provide "inverse cumulative probability" functions that can be used to automate this procedure. And there are approximations for $Q_z(p)$ that are quite adequate for plotting purposes. One particularly simple approximation (borrowed from *Probability and Statistics for Engineers and Scientists* by Walpole and Myers) is

Approximate Standard Normal p Quantile

$$Q_z(p) \approx 4.91[p^{.14} - (1-p)^{.14}],$$

which returns values that are accurate to within .01 for $.005 \leq p \leq .995$ (and to within .05 for $.001 \leq p \leq .999$).

Example 5.1 continued. Consider the problem of assessing how normal/bell-shaped the tongue-thickness data of Table 5.1 are. A normal probability plot (theoretical Q-Q plot) can be used to address this problem. Table 5.6 shows the formation of the necessary ordered pairs and Figure 5.11 is the resulting normal

Table 5.6 Coordinates for Points of a Normal Plot of the Tongue-Thickness Data

i	$x_i = Q\left(\frac{i-.5}{20}\right)$	$Q_z\left(\frac{i-.5}{20}\right)$	i	$x_i = Q\left(\frac{i-.5}{20}\right)$	$Q_z\left(\frac{i-.5}{20}\right)$
1	.1807	−1.96	11	.1825	.06
2	.1811	−1.44	12	.1827	.19
3	.1812	−1.15	13	.1827	.32
4	.1812	−.93	14	.1828	.45
5	.1813	−.76	15	.1829	.60
6	.1814	−.60	16	.1830	.76
7	.1816	−.45	17	.1834	.93
8	.1817	−.32	18	.1835	1.15
9	.1819	−.19	19	.1841	1.44
10	.1825	−.06	20	.1868	1.96

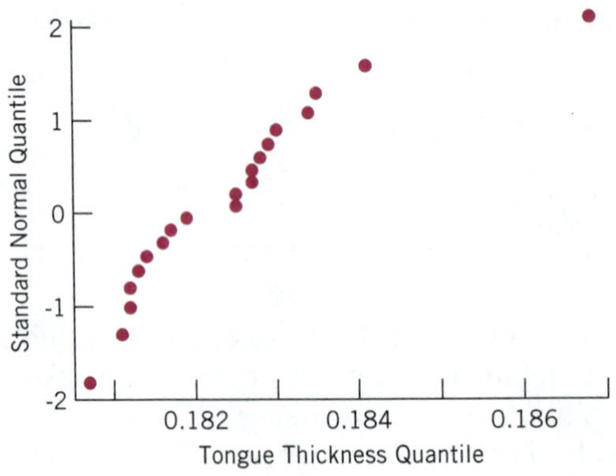

Figure 5.11 Normal plot of the tongue-thickness data.

plot. The plot might be called roughly linear, except for the point corresponding to the largest data value. In order to get that point back in line with the others, one would need to move it either to the left or up. That is, relative to the normal distribution shape, the data set is long tailed to the high side. The tongue thickness of .1868 simply doesn't fit into the somewhat normal-looking pattern established by the rest of the data.

Theoretical Q-Q plotting (probability plotting) is important for several reasons. To begin with, it helps one judge how much faith to place in calculations based on a probability distribution, and suggests in what ways the calculations might tend to be wrong. For example, Figure 5.11 suggests that if one uses a normal distribution to describe tongue thickness, the frequency of very large data values might well be underpredicted.

A second way in which probability plotting is often helpful is in providing graphical estimates of distribution parameters. For example, it turns out that if one makes a normal plot of an exactly normal distribution, the slope of the plot is the reciprocal of σ and the horizontal intercept is μ. That suggests that for a real data set whose normal plot is fairly linear, one might infer that

Estimates of a Mean and Standard Deviation from a Normal Plot

1. the horizontal intercept of an approximating line is a sensible estimate of the mean of the process generating the data, and
2. the reciprocal of the slope is a sensible estimate of the standard deviation of the process generating the data.

Example 5.4 Angles of Holes Drilled by Electrical Discharge Machining (EDM). Duren, Ling, and Patterson worked on the production of some small, high-precision metal parts. Holes in these parts were being drilled using an electrical discharge machining technique. The holes were to be at an angle to one flat surface of the parts and engineering specifications on that angle were 45° ± 2°. The actual angles produced were measured on 50 consecutive parts and are given in Table 5.7. (The units there are degrees and the values are in decimal form. The data were originally in degrees and minutes.)

Figure 5.12 is a normal plot of the hole angle data. Notice that the plot is fairly linear and that the horizontal intercept of an approximating line is near the sample mean $\bar{x} = 44.117$, while the slope of an approximating line is approximately the reciprocal of $s = .983$.

Table 5.7 Angles of Fifty Holes Drilled by Electrical Discharge Machining

46.050	45.250	45.267	44.700	44.150	44.617	43.433	44.550	44.633	45.517
44.350	43.950	43.233	45.933	43.067	42.833	43.233	45.250	42.083	44.067
43.133	44.200	43.883	44.467	44.600	43.717	44.167	45.067	44.000	42.500
45.333	43.467	43.667	44.000	44.000	45.367	44.950	45.100	43.867	43.000
42.017	44.600	43.267	44.233	45.367	44.267	43.833	42.450	44.650	42.500

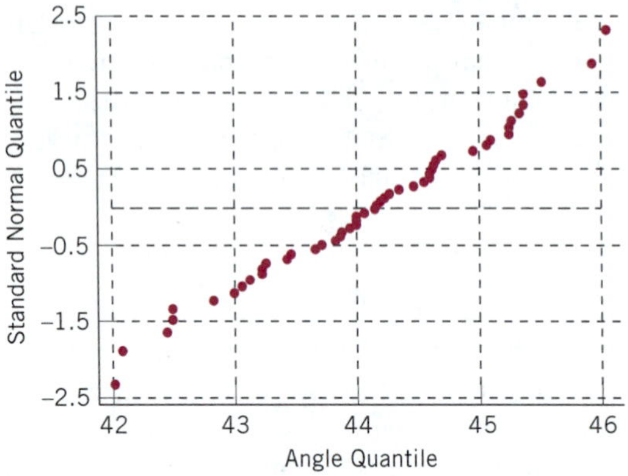

Figure 5.12
Normal plot of the hole angle data.

The facts that (for bell-shaped data sets) normal plotting provides a simple way of approximating a standard deviation and that 6σ is often used as a measure of the intrinsic spread of measurements generated by a process, together lead to the common practice of basing **process capability analyses** on normal plotting. Figure 5.13 shows a very common type of industrial form that essentially facilitates the making of a normal plot by removing the necessity of evaluating the standard normal quantiles $Q_z(p)$. (On the special vertical scale one may simply use the plotting position p rather than $Q_z(p)$, as would be required when using regular graph paper.) After plotting a data set and drawing in an approximating straight line, 6σ can be read off the plot as the difference in horizontal coordinates for points on the line at the "$+3\sigma$" and "-3σ" vertical levels (i.e., with $p = .0013$ and $p = .9987$).

The virtues of using a form like the one in Figure 5.13 are that it encourages the *plotting* of process data (always a plus) and also allows even fairly nonquantitative people to easily estimate and develop some intuition about "the process spread." Normal plotting is certainly not the last word in process characterization, but it *is* a very important tool that can and should be used alongside some of the other (numerical) methods presented in the following sections.

5.2 PROCESS CAPABILITY MEASURES AND THEIR ESTIMATION

The methods of Section 5.1 are ones that can be used to give a visual picture of the pattern of variation associated with a process. Often, in addition to these it is convenient to have some numerical summary measures to quote as more or less representing/condensing the graphics. Of course, the usual mean and standard deviation from elementary statistics are helpful in this regard. But there are also slightly more specialized measures that have come into common use in quality assurance circles. Some of these are the subject of this section.

Capability Analysis Sheet

Supplier	Date
Part	Specifications
Operation	3 sigma =
Technician	4 sigma =
Characteristic	Units

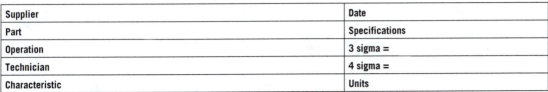

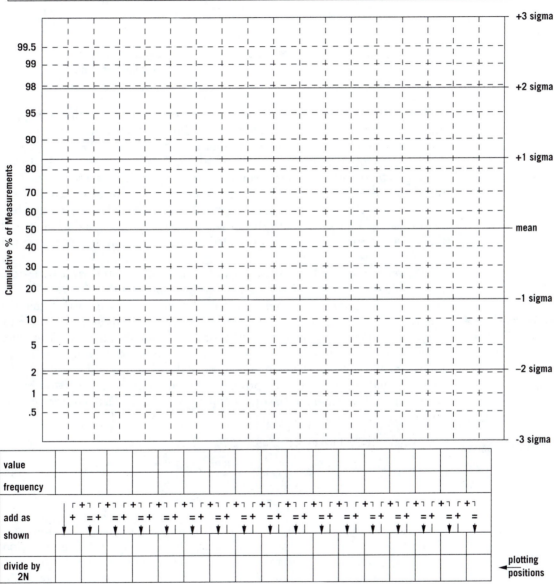

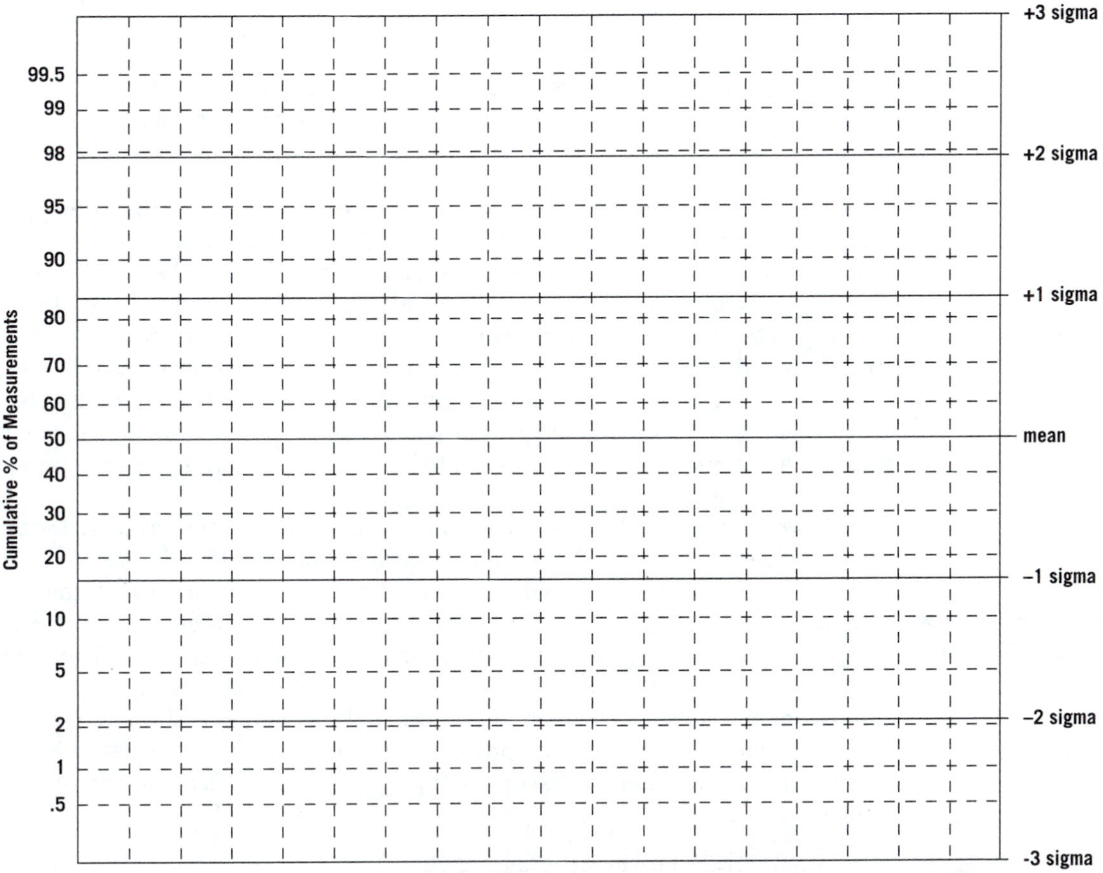

value																	
frequency																	
add as shown	+	= +	= +	= +	= +	= +	= +	= +	= +	= +	= +	= +	= +	= +	= +	= +	=
divide by 2N																	

plotting positions

Figure 5.13
Capability analysis form.

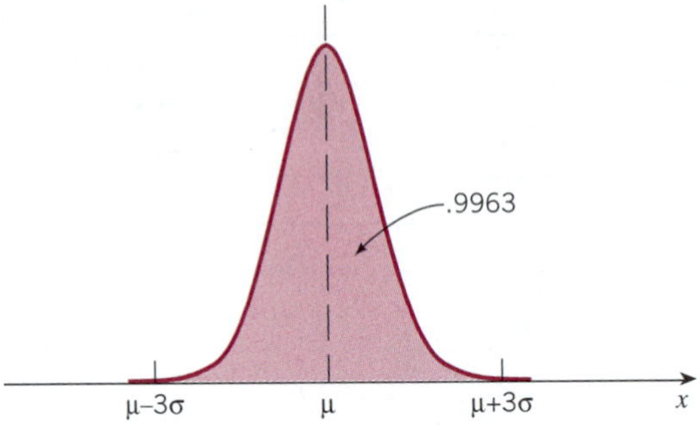

Figure 5.14
Normal distribution.

This section presents some measures of process performance that are appropriate for processes that generate at least roughly normally distributed data. The "process capability" and the two "capability ratios", C_p and C_{pk}, are discussed and methods for making confidence intervals for them are presented. But before turning to these tasks, it is important to offer a disclaimer: Unless a normal distribution makes sense as a description of process output, these measures are of very dubious relevance. Further, the confidence interval methods presented here for estimating them are completely unreliable unless a normal model is appropriate. So the normal plotting idea presented in the last section is a very important prerequisite for these methods.

It is well known that the vast majority of a normal distribution is located within three standard deviations of its mean. Figure 5.14 illustrates this elementary point. In light of the picture given by Figure 5.14, it makes some sense to say that (for a normal distribution) 6σ is a measure of process spread, and to call 6σ the **process capability** for a stable process generating normally distributed measurements. Notice that the fact that there are elementary statistical methods for estimating the standard deviation of a normal distribution implies that it is easy to give confidence limits for the process capability. That is, if one has in hand a sample of n observations with corresponding sample standard deviation s, then confidence limits for 6σ are simply

Process Capability

Confidence Limits for 6σ

$$6s\sqrt{\frac{n-1}{\chi^2_{\text{upper}}}} \quad \text{and/or} \quad 6s\sqrt{\frac{n-1}{\chi^2_{\text{lower}}}}, \tag{5.5}$$

where χ^2_{upper} and χ^2_{lower} are upper and lower percentage points of the χ^2 distribution with $n-1$ degrees of freedom. If the first limit in display (5.5) is used alone as a lower confidence bound for 6σ, the associated confidence level is the probability that a χ^2_{n-1} random variable takes a value less than χ^2_{upper}. If the second limit in display (5.5) is used alone as an upper confidence bound for 6σ, the associated confidence is the probability that a χ^2_{n-1} random variable exceeds χ^2_{lower}. If both limits in display (5.5) are used to make a two-sided interval for the process capability, the associated confidence level is the probability that a χ^2_{n-1} random variable takes a value between χ^2_{lower} and χ^2_{upper}.

Example 5.5 Process Capability for the Angles of EDM Drilled Holes (Example 5.4 Revisited). Figure 5.12 shows the angle data of Table 5.7 to be reasonably described by a normal distribution. As such, it makes sense to consider estimating the process capability for the angles at which holes are drilled. Recall that the students' data had $n = 50$ and $s = .983$. From the approximation provided with Table A.7 or a statistical package, the .05 and .95 quantiles of the χ^2 distribution for $\nu = n - 1 = 49$ degrees of freedom are respectively 33.93 and 66.34. Thus, from display (5.5), the interval with end points

$$6(.983)\sqrt{\frac{50-1}{66.34}} \quad \text{and} \quad 6(.983)\sqrt{\frac{50-1}{33.93}},$$

that is,

$$5.07^\circ \quad \text{and} \quad 7.09^\circ$$

is a 90% confidence interval for the process capability. (One is "90% sure" that the process spread is at least 5.07° and no more than 7.09°.)

Where there are both an upper specification U and a lower specification L for measurements generated by a process, it is common to compare process variability to the spread in those specifications. One way of doing this is through **process capability ratios.** And a popular process capability ratio is

$$C_p = \frac{U - L}{6\sigma}. \tag{5.6}$$

Capability Ratio C_p

When this measure is 1, the process output will fit more or less exactly inside specifications *provided the process mean is exactly on target at* $(U + L)/2$. When C_p is larger than 1, there is some "breathing room" in the sense that a process would not need to be perfectly aimed in order to produce essentially all measurements inside specifications. On the other hand, where C_p is less than 1, no matter how well a process producing normally distributed observations is aimed, a significant fraction of the output will fall outside specifications.

The very simple form of equation (5.6) makes it clear that once one knows how to estimate 6σ, one may simply divide the known difference in specifications by confidence limits for 6σ in order to derive confidence limits for C_p. That is, lower and upper confidence limits for C_p are respectively

Confidence Limits for C_p

$$\frac{(U-L)}{6s}\sqrt{\frac{\chi^2_{\text{lower}}}{n-1}} \quad \text{and/or} \quad \frac{(U-L)}{6s}\sqrt{\frac{\chi^2_{\text{upper}}}{n-1}}, \tag{5.7}$$

where again χ^2_{upper} and χ^2_{lower} are upper and lower percentage points of the χ^2 distribution with $n-1$ degrees of freedom. If the first limit in display (5.7) is used alone as a lower confidence bound for C_p, the associated confidence level is the probability that a χ^2_{n-1} random variable exceeds χ^2_{lower}. If the second limit in display (5.7) is used alone as an upper confidence bound for C_p, the associated confidence is the probability that a χ^2_{n-1} random variable is less than χ^2_{upper}. If both limits in display (5.7) are used to make a two-sided interval for C_p, the associated confidence level is the probability that a χ^2_{n-1} random variable takes a value between χ^2_{lower} and χ^2_{upper}.

Example 5.5 continued. Recall from Example 5.4 in Section 5.1 that the engineering specifications on the angles for the EDM drilled holes were $45° \pm 2°$. That means that for this situation $U - L = 4°$. So, since from before one is 90% confident that 6σ is between 5.07° and 7.09°, one can be 90% confident that C_p is between

$$\frac{4}{7.09} \quad \text{and} \quad \frac{4}{5.07},$$

that is, between

$$.56 \quad \text{and} \quad .79.$$

Of course, this same result could have been obtained beginning directly with expressions (5.7) rather than starting from the limits (5.5) for 6σ.

It is important to note that C_p is more a measure of process *potential* than it is a measure of current performance. Since the process aim is not considered in the computation of C_p, it is possible for a misaimed process with very small intrinsic variation to have a huge value of C_p and yet currently be turning out essentially no product in specifications. C_p attempts only to measure "what could be" were the process perfectly aimed. This is not necessarily an undesirable feature of C_p, but it is one that users need to understand.

Another process capability index that does take account of the process mean (and is more a measure of current process performance than of potential performance) is one commonly known as C_{pk}. This measure can be described in words as "the number of 3σ's that the process mean is to the good side of the closest specification." For example, if $U - L$ is 10σ, and μ is 4σ below the upper specification, then C_{pk} is $4\sigma/3\sigma = 1.33$. On the other hand, if $U - L$ is 10σ and μ is 4σ above the upper specification, then C_{pk} is -1.33.

In symbols,

Capability Index C_{pk}

$$C_{pk} = \min\left\{\frac{U-\mu}{3\sigma}, \frac{\mu-L}{3\sigma}\right\} = \frac{U - L - 2\left|\mu - \frac{U+L}{2}\right|}{6\sigma}. \tag{5.8}$$

This quantity will be positive as long as μ is between L and U. It will be large if μ is between L and U (preferably centered between them) and $U - L$ is large compared to σ.

Making a confidence interval for C_{pk} is more difficult than making one for C_p. In fact, currently the best available method is only appropriate for large samples and provides a real confidence level that only approximates the nominal one. The method is based on the natural single number estimate of C_{pk},

Estimate of C_{pk}

$$\hat{C}_{pk} = \min\left\{\frac{U-\bar{x}}{3s}, \frac{\bar{x}-L}{3s}\right\} = \frac{U - L - 2\left|\bar{x} - \frac{U+L}{2}\right|}{6s}. \tag{5.9}$$

If z is the p quantile of the standard normal distribution ($z = Q_z(p)$), an approximate $p \times 100\%$ lower confidence bound for C_{pk} is then

Lower Confidence Bound for C_{pk}

$$\hat{C}_{pk} - z\sqrt{\frac{1}{9n} + \frac{\hat{C}_{pk}^2}{2n-2}}. \tag{5.10}$$

Example 5.5 continued. Recall from Example 5.4 of Section 5.1 that the 50 EDM hole angles had corresponding sample mean $\bar{x} = 44.117$. Then from relationship (5.9),

$$\hat{C}_{pk} = \min\left\{\frac{47 - 44.117}{3(.983)}, \frac{44.117 - 43}{3(.983)}\right\} = \min\{.98, .38\} = .38.$$

So, for example, since the .95 quantile of the standard normal distribution is 1.645, an approximate 95% lower confidence bound for C_{pk} is from expression (5.10)

$$.38 - 1.645\sqrt{\frac{1}{9(50)} + \frac{(.38)^2}{2(50)-2}} = .28.$$

One can be in some sense approximately 95% sure that C_{pk} for the angles in the EDM drilling process is at least .28.

Overreliance upon process capability measures like the ones discussed here has come under a fair amount of criticism in the past few years. Critics have correctly noted that

1. 6σ, C_p, and C_{pk} have only dubious relevance when a process distribution is not normal,
2. "one-number summaries" like those discussed here can leave much unsaid about what a process is doing or even the shape of a distribution of measurements it is generating, and
3. the whole business of really going to work tuning a process, monitoring for and removing upsets, and determining what it is really "capable" of doing involves much more than the simple estimation of 6σ or one of the measures C_p or C_{pk}.

To these objections, your authors would add the observation that the capability ratios C_p and C_{pk} depend upon specifications that are sometimes subject to unannounced change (even arbitrary change). This makes it difficult to know from one reporting period to the next what has happened to process variability if estimates of C_p or C_{pk} are all that are provided. It thus seems that for purposes of comparisons across time, if any of the measures of this section are to be used, the simple process capability 6σ is most attractive.

Despite the criticism their use has received, the measures of this section are very popular. Provided one understands their limitations and simply views them as one of many tools for summarizing process behavior, they have their place. But the wise engineer will not assume that computing and reporting one of these figures is in any way the last word in assessing process performance.

5.3 PREDICTION AND TOLERANCE INTERVALS

The methods of the previous section and those of elementary statistics represent one way of characterizing the pattern of variation produced by a stable process, namely through the estimation of process parameters and summary measures. Another approach is to provide intervals likely to contain either the next measurement from the process or a large portion of all additional values it might generate. The making of such intervals is the subject of this section. So-called prediction interval and tolerance interval methods for normal processes are considered first. Then some simple "nonparametric" methods are considered. These can be used to produce prediction and tolerance intervals for any stable process, regardless of whether or not the process distribution is normal.

5.3.1 Intervals for a Normal Process

The usual confidence interval methods presented in elementary statistics courses concern the estimation of process parameters like the mean μ and standard deviation σ. The estimation methods for 6σ, C_p, and C_{pk} presented in Section 5.2 essentially concern the empirical approximation of interesting functions of these process parameters. A completely different approach to process characterization is to use data, not to approximate process parameters or functions of them, but rather to provide intervals in some sense representing where additional observations are likely to fall. Two different formulations of this approach are the making of prediction intervals and the making of tolerance intervals.

A **prediction interval** based on a sample of size n from a stable process is an interval thought likely to contain a single additional observation drawn from the process. Suppose one uses a formula intended to produce 90% prediction intervals. The associated confidence guarantee is that if the whole business of "selecting a sample of n, making the corresponding interval, observing an additional value and checking to see if the additional value is in the interval" is repeated many times, about 90% of the repetitions will be successful.

Where it is sensible to assume that one is sampling from a normal process, a very simple formula can be given for prediction limits. That is, prediction limits for a single additional observation from a normal distribution are

Normal Distribution Prediction Limits

$$\bar{x} - ts\sqrt{1 + \frac{1}{n}} \quad \text{and/or} \quad \bar{x} + ts\sqrt{1 + \frac{1}{n}}, \tag{5.11}$$

where t is a quantile of the t distribution with $\nu = n - 1$ associated degrees of freedom. If t is the p quantile and the first limit in display (5.11) is used alone as a lower prediction bound, the associated confidence level is $p \times 100\%$. Similarly, if t is the p quantile and the second limit in display (5.11) is used alone as an upper prediction bound, the associated confidence level is $p \times 100\%$. And if t is the p quantile and both limits in display (5.11) are used as the end points of a two-sided prediction interval, the associated prediction confidence level is $(2p - 1) \times 100\%$.

Example 5.6 Predicting the Angle of an Additional EDM Drilled Hole (Examples 5.4 and 5.5 Revisited). The normal plot of Figure 5.12 shows the angle data of Table 5.7 to be reasonably normal looking. Capability figures for the EDM drilling process were estimated in Example 5.5. Here consider the prediction of a single additional angle measurement. Either from the t distribution quantiles given in Table A.8, or from a statistical package, the $p = .95$ quantile of the t distribution with $\nu = 50 - 1 = 49$ degrees of freedom is 1.6766. Then recalling that for the angle data $\bar{x} = 44.117$ and $s = .983$, formulas (5.11) show that a 90% two-sided prediction interval for a single additional angle has end points

$$44.117 - 1.6766(.983)\sqrt{1 + \frac{1}{50}} \quad \text{and} \quad 44.117 + 1.6766(.983)\sqrt{1 + \frac{1}{50}},$$

that is,

$$42.45° \quad \text{and} \quad 45.78°.$$

One can be in some sense 90% sure that a single additional angle generated by the stable EDM drilling process will be between 42.45° and 45.78°.

A second formulation of the problem of producing an interval locating where additional observations from a process are likely to fall is that of making a **tolerance interval** for some (typically large) fraction of the process distribution. A tolerance interval for a fraction p of a distribution is a data-based interval thought likely to bracket at least that much of the distribution. Suppose one uses a formula intended to produce 95% tolerance intervals for a fraction $p = .90$ of a distribution. Then the associated confidence guarantee is that if the whole business of "selecting a sample of n, computing the associated interval and checking to see what fraction of the distribution is in fact bracketed by the interval" is repeated many times, about 95% of the intervals will bracket at least 90% of the distribution.

Where it is sensible to assume that one is sampling from a normal process distribution, very simple formulas can be given for tolerance limits. That is, using constants τ_1 or τ_2 given in Tables A.9, a one-sided tolerance limit for a fraction p of an entire normal process distribution is

One-sided Normal Distribution Tolerance Limits

$$\bar{x} - \tau_1 s \quad \text{or} \quad \bar{x} + \tau_1 s, \tag{5.12}$$

while a two-sided tolerance interval for a fraction p of such a distribution can be made using end points

Two-sided Normal Distribution Tolerance Limits

$$\bar{x} - \tau_2 s \quad \text{and} \quad \bar{x} + \tau_2 s. \tag{5.13}$$

The constant τ_1 or τ_2 may be chosen (from Tables A.9b or A.9a respectively) to provide a 95% or a 99% confidence level.

Example 5.6 continued. Rather than as before predicting a single additional EDM drilled angle, consider the problem of announcing an interval likely to contain 95% of the angle distribution. In fact, for purposes of illustration, consider the making of both a 99% one-sided lower tolerance bound for 95% of all additional angles and a 99% two-sided tolerance interval for 95% of all additional angles.

Beginning with the one-sided problem, the first formula in display (5.12) and the $n = 50, p = .95$, 99% confidence level entry of Table A.9b produce the lower tolerance bound

$$44.117 - 2.269(.983) = 41.89^\circ.$$

One can be "99% sure" that at least 95% of all angles are 41.89° or larger.

In a similar fashion, using the formulas (5.13) and the $n = 50, p = .95$, 99% confidence level entry of Table A.9a, one has the end points

$$44.117 - 2.580(.983) = 41.58^\circ \quad \text{and} \quad 44.117 + 2.580(.983) = 46.65^\circ.$$

That is, one can be in some sense 99% sure that at least 95% of all angles are in the interval (41.58,46.65).

It is instructive to compare this second interval to the 95% prediction interval obtained earlier, namely (42.45,45.78). The tolerance interval is clearly larger than the prediction interval, and this is typical of what happens using common (large) confidence levels for tolerance intervals. A tolerance interval is simply designed to do a more ambitious task than a corresponding prediction interval. That is, a prediction interval aims to locate a single additional measurement while a tolerance interval intends to locate most of *all* additional observations. It is therefore not surprising that the tolerance interval would need to be larger.

5.3.2 Intervals Based on Maximum and/or Minimum Sample Values

The prediction and tolerance intervals prescribed by displays (5.11), (5.12), and (5.13) are very definitely normal distribution intervals. If a normal distribution is not a good description of the stable process data-generating behavior of a system under consideration, the confidence guarantees associated with these formulas are null and void. If measurements x are not normal, on occasion it is possible to find a "transformation" $g(\cdot)$ such that transformed measurements $g(x)$ are normal. When this can be done, one can then simply find prediction or tolerance intervals for $g(x)$ and then "untransform" the end points of such intervals (using the inverse function $g^{-1}(\cdot)$) to provide prediction or tolerance intervals for raw values x. This approach to making intervals really amounts to finding a convenient scale upon which to express the variable of interest when the original one turns out to be inconvenient.

A second approach to making prediction and tolerance intervals when a process distribution does not seem to be normal, is to use limits that carry the same confidence level guarantee for *any* (continuous) stable process distribution. Such limits can be based on minimum and/or maximum values in a sample. Because their applicability is not limited to the normal "parametric family" of distributions, these limits are sometimes called **nonparametric limits**.

If one has a sample of n measurement from a stable process, the most obvious of all statistical intervals based on those measurements are

General One-sided Prediction or Tolerance Interval (Lower Bound)

$$(\min x_i, \infty), \tag{5.14}$$

General One-sided Prediction or Tolerance Interval (Upper Bound)

$$(-\infty, \max x_i), \tag{5.15}$$

and

General Two-sided Prediction or Tolerance Interval

$$(\min x_i, \max x_i). \tag{5.16}$$

It turns out that any of these intervals can be used as either a prediction interval for a single additional observation from a process, or as a tolerance interval for a fraction p of the process distribution.

Where either of the one-sided intervals (5.14) or (5.15) is used as a prediction interval, the associated prediction confidence level is

Prediction Confidence of One-sided Nonparametric Intervals

$$\frac{n}{n+1}. \tag{5.17}$$

Where the two-sided interval (5.16) is used, the associated prediction confidence level is

Prediction Confidence of Two-sided Nonparametric Intervals

$$\frac{n-1}{n+1}. \tag{5.18}$$

Where either of the one-sided intervals (5.14) or (5.15) is used as a tolerance interval for a fraction p of the output from a stable process, the associated confidence level is

Confidence Level of One-sided Nonparametric Tolerance Intervals

$$1 - p^n. \tag{5.19}$$

And where the two-sided interval (5.16) is used as a tolerance interval, the associated confidence level is

$$1 - p^n - n(1-p)p^{n-1}. \tag{5.20}$$

Confidence Level of Two-sided Nonparametric Tolerance Intervals

Example 5.7 Prediction and Tolerance Intervals for Tongue Thicknesses of Machined Levers (Example 5.1 Revisited). The normal plot in Figure 5.11 shows the tongue-thickness data of Table 5.1 to be long tailed to the high side, and clearly not adequately described as approximately normal. As such, the normal distribution formulas (5.11) through (5.13) are not appropriate for making prediction or tolerance intervals for tongue thicknesses. But if one assumes that the machining process represented by those data is stable, the methods represented in formulas (5.14) through (5.20) can be used.

Consider first the problem of announcing a two-sided prediction interval for a single additional tongue thickness. Reviewing the data of Table 5.1, it is easy to see that for the sample of size $n = 20$ represented there,

$$\min x_i = .1807 \quad \text{and} \quad \max x_i = .1868.$$

So, in view of displays (5.16) and (5.18), the interval with end points

$$.1807 \text{ inch} \quad \text{and} \quad .1868 \text{ inch}$$

can be used as a prediction interval for a single additional tongue thickness. And the associated prediction confidence is

$$\frac{20-1}{20+1} = .905 = 90.5\% .$$

One can in some sense be 90% sure that an additional tongue thickness generated by this machining process would be between .1807 inch and .1868 inch.

As a second way of expressing what the data say about other tongue thicknesses, consider the making of a two-sided tolerance interval for 90% of all tongue thicknesses. The method represented in displays (5.16) and (5.20) implies that the interval with end points

$$.1807 \text{ inch} \quad \text{and} \quad .1868 \text{ inch}$$

has associated confidence level

$$1 - (.9)^{20} - 20(1-.9)(.9)^{19} = .608 = 60.8\%.$$

One can be only about 61% sure that 90% of all tongue thicknesses generated by the machining processes are between .1807 inch and .1868 inch.

The prediction and confidence interval methods presented here are a very small fraction of those available. In particular, there are methods specifically crafted for other families of process distributions besides the normal family. The reader is referred to the book *Statistical Intervals: A Guide for Practitioners* by Hahn and Meeker for a more comprehensive treatment of the many available methods, should the ones presented in this section not prove adequate for his or her purposes.

5.4 PROBABILISTIC TOLERANCING AND PROPAGATION OF ERROR

The methods of the previous three sections have had to do with characterizing the pattern of variation associated with a stable process on the basis of a sample from that process. There are, however, occasions where one needs to predict the pattern of variation associated with a stable system before such data are available. (This is quite often the case in engineering design contexts, where one is in the business of choosing between a number of different possible designs for a process or product, without having many systems of each type available for testing.)

Where the product or process of interest can be described in terms of a relatively simple equation involving the properties of some components or system inputs, and information is available on variabilities of the components or inputs, it is often possible to do the necessary prediction. This section presents methods for accomplishing this task. There is first a brief discussion of the use of simulations. Then a simple result from probability theory concerning the behavior of linear combinations of random variables is applied to the problem. And finally, a very useful method of approximation is provided for situations where the exact probability result cannot be invoked.

In abstract terms, the problem addressed in this section can be phrased as follows: Given k random system inputs $X, Y, \ldots, Z$, an output of interest U, and the form of a function g giving the exact value of the output in terms of the inputs,

$$U = g(X, Y, \ldots, Z), \tag{5.21}$$

how does one infer properties of the random variable U from properties of $X, Y, \ldots, Z$? For particular joint distributions for the inputs and fairly simple functions g, the methods of multivariate calculus can sometimes be invoked to find formulas for the distribution of U. But problems that yield easily to such an approach are rare, and much more widely applicable methods are needed for engineering practice.

One fairly crude but quite general tool for this problem is that of **probabilistic simulations**. These are easily accomplished these days using widely available statistical software. What one does is to use (pseudo-)random number generators to produce many (say n) realizations of the vector $(X, Y, \ldots, Z)$. Upon plug-

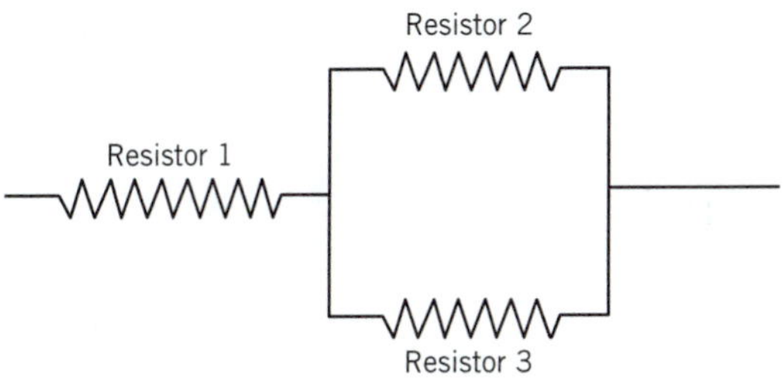

Figure 5.15
Schematic of a simple assembly of three resistors.

ging each of these realizations into the function g, one obtains realizations of the random variable U. Properties of the distribution of U can then be inferred in approximate fashion from the empirical distribution of these realizations. This whole program is especially easy to carry out when it is sensible to model the inputs $X, Y, \ldots, Z$ as independent. Then the realizations of the inputs can be generated separately from the k marginal distributions. (The generation of realizations of *dependent* variables $X, Y, \ldots, Z$ is possible, but beyond the scope of this discussion.)

Example 5.8 **Approximating the Distribution of the Resistance of an Assembly of Three Resistors.** The "laws" of physics often provide predictions of the behavior of simple systems. Consider the schematic of the assembly of three resistors given in Figure 5.15. Elementary laws of physics lead to the prediction that if R_1, R_2, and R_3 are the resistances of the three resistors in the schematic, then the resistance of the assembly will be

$$R = R_1 + \frac{R_2 R_3}{R_2 + R_3}.$$

Suppose that one is contemplating the use of such assemblies in a mass-produced product. And suppose further that resistors can be purchased so that R_1 has mean 100 Ω and standard deviation 2 Ω, and that both R_2 and R_3 have mean 200 Ω and standard deviation 4 Ω. What can one then predict about the variability of the assembly resistance?

Printout 5.1 shows a simple simulation done using the Minitab statistical package assuming that the three resistances are independent and normally distributed.

Printout 5.1

```
MTB > random 200 c1;
SUBC> normal 100 2.
MTB > random 200 c2 c3;
SUBC> normal 200 4.
MTB > let c4=c1+c2*c3/(c2+c3)
MTB > describe c4
```

```
                  N      MEAN    MEDIAN    TRMEAN     STDEV    SEMEAN
C4              200    199.84    199.87    199.80      2.41      0.17

                MIN       MAX        Q1        Q3
C4           194.08    207.10    198.23    201.27

MTB > histogram c4

Histogram of C4   N = 200

Midpoint   Count
     194       3  ***
     195       3  ***
     196      13  *************
     197      15  ***************
     198      24  ************************
     199      26  **************************
     200      38  **************************************
     201      38  **************************************
     202      14  **************
     203      12  ************
     204       8  ********
     205       3  ***
     206       1  *
     207       2  **

MTB > stem-and-leaf c4

Stem-and-leaf of C4        N  = 200
Leaf Unit = 0.10

    4  194 0149
   12  195 44567889
   25  196 0013344566888
   45  197 01111222367778888999
   70  198 0001222233444577888888999
  (36) 199 000012233334445555556666677778899999
   94  200 0000001111333333555666777777888888
   60  201 00011222222233344444677889
   34  202 111122346778899
   19  203 1222355566
    9  204 2349
    5  205 135
    2  206 7
    1  207 1
```

It is evident from the printout that assembly resistances are predicted to average on the order of 199.8 Ω and to have a standard deviation on the order of 2.4 Ω. And the histogram and stem-and-leaf diagram hint that while the distribution of assembly resistances may be roughly bell shaped, there may be some slight skewness (long-tailedness) to the high side.

Of course, rerunning the simulation would produce slightly different results. But for purposes of obtaining a quick, rough-and-ready picture of predicted assembly variability, this technique is very convenient and powerful.

For the very simplest of functions g relating system inputs to the output U, it is possible to provide expressions for the mean and variance of U in terms of the means and variances of the inputs. Consider the case where g is linear in the inputs, that is, for constants $a_0, a_1, \ldots, a_k$ suppose that

$$U = a_0 + a_1 X + a_2 Y + \cdots + a_k Z. \tag{5.22}$$

Then if the variables $X, Y, \ldots, Z$ are independent with respective means μ_X, $\mu_Y, \ldots, \mu_Z$ and variances $\sigma_X^2, \sigma_Y^2, \ldots, \sigma_Z^2$, U has mean

Mean of a Linear Function of k Random Variables

$$\mu_U = a_0 + a_1 \mu_X + a_2 \mu_Y + \cdots + a_k \mu_Z \tag{5.23}$$

and variance

Variance of a Linear Function of k Independent Random Variables

$$\sigma_U^2 = a_1^2 \sigma_X^2 + a_2^2 \sigma_Y^2 + \cdots + a_k^2 \sigma_Z^2. \tag{5.24}$$

These facts represented in displays (5.22) through (5.24) are not directly applicable to problems like that in Example 5.6, since there the assembly resistance is not linear in either R_2 or R_3. But they *are* directly relevant to many engineering problems involving geometrical dimensions. That is, often geometrical variables of interest on discrete parts or assemblies of such parts are sums and differences of more fundamental variables. Clearances between shafts and ring bearings are differences between inside diameters of the bearings and the shaft diameters. Thicknesses of five-ply sheets of plywood are sums of thicknesses of the individual layers. Tongue thicknesses of steel levers machined on both sides are the original steel bar thicknesses minus the depths of cut on both sides of the bar. And so on.

Example 5.9 **Choosing a Box Size in a Packaging Problem.** Miles, Baumhover, and Miller worked with a company that was having a packaging problem. The company bought cardboard boxes nominally 9.5 inches in length intended to hold exactly 4 units of a product they produced, stacked side by side in the boxes. They were finding that many boxes were unable to accommodate the full 4 units and a sensible figure was needed for a new target dimension on the boxes.

The students measured the thicknesses of 25 units of product. They found that these had $\bar{x} = 2.577$ inches and $s = .061$ inch. They also measured several of the nominally 9.5-inch boxes and found their (inside) lengths to have mean $\bar{x} = 9.556$ inches and $s = .053$ inch. Consider applying the results (5.23) and (5.24) to the problem of finding a workable new target dimension for the inside length of boxes ordered by this company.

Let X_1, X_2, X_3, and X_4 be the thicknesses of 4 units to be placed in a box, and Y be the inside length of the box. Then the clearance or "head space" in the box is

$$U = Y - X_1 - X_2 - X_3 - X_4.$$

Based on the students' measurements, a plausible model for the variables Y, X_1, X_2, X_3, X_4 is one of independence where Y has mean to be chosen and standard deviation .053 and each of the X variables has mean 2.577 and standard deviation .061. Then, since U is of form (5.22), display (5.24) implies that U has

$$\sigma_U^2 = 1^2\sigma_Y^2 + (-1)^2\sigma_{X_1}^2 + (-1)^2\sigma_{X_2}^2 + (-1)^2\sigma_{X_3}^2 + (-1)^2\sigma_{X_4}^2,$$

that is,

$$\sigma_U^2 = (.053)^2 + 4(.061)^2 = .0177,$$

so that

$$\sigma_U = .133 \text{ inch.}$$

One might then hope for at least approximate normality of U, and reason that if the mean of U were set at $3\sigma_U$, essentially all of the boxes would be able to hold the required 4 units of product (few values of U would be negative). Again remembering that U is of form (5.22), display (5.24) implies that

$$\mu_U = \mu_Y - \mu_{X_1} - \mu_{X_2} - \mu_{X_3} - \mu_{X_4} = \mu_Y - 4(2.577).$$

So setting $\mu_U = 3\sigma_U = 3(.133)$, one has

$$3(.133) = \mu_Y - 4(2.577),$$

that is, one wants

$$\mu_Y = 10.707 \text{ inch.}$$

(Given that nominally 9.5-inch boxes were running with mean inside lengths of 9.556 inch, it might be possible to order boxes with nominal lengths .056 inch below the value for μ_Y found above without creating packing problems.)

Example 5.9 is a very nice example of a real "probabilistic tolerancing" problem. Such problems can be effectively attacked using the relationships (5.23) and (5.24). It is evident from the example that if the standard deviation of the head space (namely $\sigma_U = .133$ inch) is unacceptably large, then either the uniformity of the thicknesses of the units of product or the uniformity of the inside lengths of the boxes will need substantial improvement. (In fact, setting $\sigma_Y = 0$ and recalculating σ_U will show the reader that the potential for variance reduction associated with the box length is small. It is product uniformity that will require attention.)

The simple exact probability result represented in displays (5.22) through (5.24) is not of direct help where U is nonlinear in one or more of $X, Y, \ldots, Z$. But it suggests how one might proceed to develop approximate formulas for the mean and variance of U for even nonlinear g. That is, provided g is smooth near the point $(\mu_X, \mu_Y, \ldots, \mu_Z)$ in k-dimensional space, if the point $(x, y, \ldots, z)$ is not too far from $(\mu_X, \mu_Y, \ldots, \mu_Z)$ a first-order multivariate Taylor expansion of g implies that

$$\left.\begin{aligned} g(x, y, \ldots, z) \approx g(\mu_X, \mu_Y, \ldots, \mu_Z) + g'_x(\mu_X, \mu_Y, \ldots, \mu_Z)(x - \mu_X) \\ + g'_y(\mu_X, \mu_Y, \ldots, \mu_Z)(y - \mu_Y) + \cdots + g'_z(\mu_X, \mu_Y, \ldots, \mu_Z)(z - \mu_Z) \end{aligned}\right\} \tag{5.25}$$

where the subscripted g' functions are the partial derivatives of g. Now the function on the right of approximation (5.25) is linear in the variables $x, y, \ldots, z$. So plugging the random variables $X, Y, \ldots, Z$ into approximation (5.25) and then applying the probability result indicated in displays (5.22) through (5.24) one can arrive at approximations for μ_U and σ_U^2. These turn out to be

$$\boxed{\mu_U \approx g(\mu_X, \mu_Y, \ldots, \mu_Z)} \tag{5.26}$$

Approximate Mean of a Function of k Random Variables

and

$$\boxed{\sigma_U^2 \approx \left(\frac{\partial g}{\partial x}\right)^2 \sigma_X^2 + \left(\frac{\partial g}{\partial y}\right)^2 \sigma_Y^2 + \cdots + \left(\frac{\partial g}{\partial z}\right)^2 \sigma_Z^2,} \tag{5.27}$$

Approximate Variance of a Function of k Independent Random Variables

where the partial derivatives indicated in display (5.27) are evaluated at the point $(\mu_X, \mu_Y, \ldots, \mu_Z)$. (The notation for the partial derivatives used in display (5.27) is more compact but less complete than the g' notation used on the right of approximation (5.25). The same partials are involved.) The formulas (5.26) and (5.27) are often called **the propagation of error formulas** in that they provide a simple approximate view of how "error" or variation "propagates" through a function g. And it is also worth noting that the exact result (5.24) for linear g is (upon realizing that $a_1, a_2, \ldots, a_k$ are the partial derivatives of a linear g) essentially a special case of relationship (5.27).

Example 5.10 Uncertainty in the Measurement of Viscosity of S.A.E. no. 10 Oil. One technique for measuring the viscosity of a liquid is to place it in a cylindrical container and determine the force needed to turn a cylindrical rotor of nearly the same diameter as the container at a given velocity. If F is the force, D_1 is the diameter of the rotor, L is the length of the rotor, D_2 is the inside diameter of the container, and ν is the velocity at which the rotor surface moves, then the implied viscosity is

$$\eta = \frac{F(D_2 - D_1)}{\pi \nu D_1 L} = \frac{F}{\pi \nu L}\left(\frac{D_2}{D_1} - 1\right).$$

Suppose that one wishes to measure the viscosity of S.A.E. no. 10 oil and the basic measurement equipment available has precision adequate to provide standard deviations for the variables F, D_1, D_2, L, and ν,

$$\sigma_F = .05\text{N}, \sigma_{D_1} = \sigma_{D_2} = \sigma_L = .05 \text{ cm} \quad \text{and} \quad \sigma_\nu = 1 \text{ cm/sec.}$$

Further, suppose that approximate values for the quantities F, D_1, D_2, L, and ν are

$$F \approx 151 \text{ N}, D_1 \approx 20.00 \text{ cm}, D_2 \approx 20.50 \text{ cm}, L \approx 20.00 \text{ cm} \quad \text{and} \quad \nu \approx 30 \text{ cm/sec.}$$

These approximate values will be used as means for the variables and formula (5.27) employed to find an approximate standard deviation to use in describing the precision with which the viscosity can be determined.

To begin with, the partial derivatives of η with respect to the various measured quantities are

$$\frac{\partial \eta}{\partial F} = \frac{(D_2 - D_1)}{\pi \nu D_1 L},$$

$$\frac{\partial \eta}{\partial D_1} = \frac{F}{\pi \nu L}\left(-\frac{D_2}{D_1^2}\right),$$

$$\frac{\partial \eta}{\partial D_2} = \frac{F}{\pi \nu D_1 L},$$

$$\frac{\partial \eta}{\partial L} = -\frac{F(D_2 - D_1)}{\pi \nu D_1 L^2},$$

and

$$\frac{\partial \eta}{\partial \nu} = -\frac{F(D_2 - D_1)}{\pi \nu^2 D_1 L}.$$

And it is straightforward to check that if the approximate values of F, D_1, D_2, L, and ν are plugged into these formulas, then (in the appropriate units)

$$\frac{\partial \eta}{\partial F} = 1.326 \times 10^{-5}, \quad \frac{\partial \eta}{\partial D_1} = -4.106 \times 10^{-3}, \quad \frac{\partial \eta}{\partial D_2} = 4.005 \times 10^{-3},$$

$$\frac{\partial \eta}{\partial L} = -1.001 \times 10^{-4}, \quad \text{and} \quad \frac{\partial \eta}{\partial \nu} = -6.676 \times 10^{-5}.$$

Then from expression (5.27) it is apparent that an approximate variance for η is

$$\sigma_\eta^2 \approx (1.326 \times 10^{-5})^2(.05)^2 + (-4.106 \times 10^{-3})^2(.05)^2 + (4.005 \times 10^{-3})^2(.05)^2$$
$$+ (-1.001 \times 10^{-4})^2(.05)^2 + (-6.676 \times 10^{-5})^2(1)^2,$$

and doing the arithmetic and taking the square root, one finds

$$\sigma_\eta \approx 2.9 \times 10^{-4} \text{ N sec/cm}^2.$$

Making use of relationship (5.26), this standard deviation of measurement accompanies a "true" or mean measured viscosity of about

$$\eta = \frac{151(20.50 - 20.00)}{\pi 30(20)(20)} = 20.0 \times 10^{-4} \text{ N sec/cm}^2.$$

There are several points that need to be made about the practical use of the methods presented here before closing this section. For one thing, it needs to be emphasized that formulas (5.26) and (5.27) are only approximations (based on the linearization of g at the point $(\mu_X, \mu_Y, \ldots, \mu_Z)$). It is a good idea to cross check results one gets using these propagation of error formulas with results of modest simulations (a different kind of approximation). For example, Printout 5.2 is a small Minitab simulation for the problem of Example 5.10 and shows substantial agreement with the calculations in the example.

Printout 5.2

```
MTB > random 200 c1;
SUBC> norm 151 .05.
MTB > random 200 c2;
SUBC> norm 20 .05.
MTB > random 200 c3;
SUBC> norm 20.5 .05.
MTB > random 200 c4;
```

```
SUBC> norm 20 .05.
MTB > random 200 c5;
SUBC> norm 30 1.
MTB > let c6=c1*(c3-c2)/(c5*c2*c4)
MTB > let c7=c6/3.14159265
MTB > describe c7

                N       MEAN     MEDIAN     TRMEAN      STDEV     SEMEAN
C7            200    0.00201    0.00201    0.00201    0.00028    0.00002

              MIN        MAX         Q1         Q3
C7        0.00129    0.00269    0.00184    0.00220

MTB > stem-and-leaf c7

Stem-and-leaf of C7          N  = 200
Leaf Unit = 0.000010

     1    12 9
     2    13 5
     9    14 0267789
    16    15 0224679
    26    16 0112346677
    42    17 0111222345555679
    67    18 1123334444455556667888999
    97    19 001111223344445666666667778899
  (23)    20 00112234455556778888899
    80    21 001111223333444555555556789999
    50    22 000000001112233445556799
    26    23 1244456779
    16    24 00111246
     8    25 04799
     3    26 559
```

Propagation of Error and Variance Partitioning

A reasonable question to ask is "If one is going to do simulations anyway, why bother to do the hard work to use the propagation of error formulas?" One answer to this question lies in the important extra insight into the issue of variance transmission provided by the use of formula (5.27). Formula (5.27) can be thought of as providing a partition of the variance of the output U into separate parts attributable to the various inputs individually. That is, each of the terms on the right side of formula (5.27) is related to a single one of the inputs $X, Y, \ldots, Z$ and can be thought of as that variable's impact on the variation in U. Comparing these can, for example, lead to the identification of the biggest source(s) of (unwanted) variability in a system and allow the consideration of where engineering resources might best be invested in order to try and reduce the size of σ_U. To get these kinds of insights from simulations would require the comparison of many different simulations using various hypothetical values of the standard deviations of the inputs.

Example 5.10 continued. Returning to the approximation for σ_η^2 and displaying some of the intermediate arithmetic, one has

$$\left(\frac{\partial \eta}{\partial F}\right)^2 \sigma_F^2 = (1.326 \times 10^{-5})^2(.05)^2 = 4.4 \times 10^{-13},$$

$$\left(\frac{\partial \eta}{\partial D_1}\right)^2 \sigma_{D_1}^2 = (-4.106 \times 10^{-3})^2(.05)^2 = 4.21 \times 10^{-8},$$

$$\left(\frac{\partial \eta}{\partial D_2}\right)^2 \sigma_{D_2}^2 = (4.005 \times 10^{-3})^2(.05)^2 = 4.01 \times 10^{-8},$$

$$\left(\frac{\partial \eta}{\partial L}\right)^2 \sigma_L^2 = (-1.001 \times 10^{-4})^2(.05)^2 = 2.51 \times 10^{-11},$$

and

$$\left(\frac{\partial \eta}{\partial \nu}\right)^2 \sigma_\nu^2 = (-6.676 \times 10^{-5})^2(1)^2 = 4.46 \times 10^{-9}.$$

It is then evident from these values that the biggest contributors to the variance of η are the two diameter measurements. (Next in order of importance is the velocity measurement, whose contribution is an order of magnitude smaller.) The single most effective method of improving the precision of the viscosity measurement would apparently be to find a more precise method of measuring the diameters of the rotor and cylinder.

It is instructive to note that although the standard deviation of the length measurement is exactly the same as those of the two diameter measurements (namely .05 cm), the length measurement contribution to σ_η^2 is much smaller than those of the two diameter measurements. This is because the partial derivative of η with respect to L is much smaller than those with respect to D_1 and D_2. That is, the contributions to the overall variance involve not only the variances of the inputs but the "gains" or rates of change of the output with respect to the inputs. This is only sensible. After all, if g is constant with respect to an input variable, whether or not it varies should have no impact on the output variation. On the other hand, if g changes rapidly with respect to an input, any variation in that variable will produce substantial output variation.

A final caution regarding the methods of this section is that one should not expect the impossible from them. They are tools for predicting the pattern of variation *within a particular model.* But even the best of equations we use to describe physical phenomena are only approximations to reality. They typically ignore variables whose effects on a response of interest are "small." They are often good only over limited ranges of the inputs (and of other variables that don't even appear in the equations). And so on. The end result of this is that the kinds of predictions that have been illustrated in this section should be thought of as typically producing *underpredictions* of the kinds of variability that will be seen, should one observe a number of realizations of the output variable over a period of time.

5.5 VARIANCE COMPONENT ESTIMATION IN BALANCED HIERARCHICAL STUDIES

The goal of process monitoring is to identify and remove assignable causes of variation, producing a physically stable process that yields data that can be thought of as random draws from a single fixed universe. Like it or not, there are situations where this goal is not completely achievable, either because one is early in a process improvement program, or because of legitimate physical or resource constraints. But it can happen that while a process is not completely stable, its pattern of variation can nevertheless be broken down into identifiable consistent components. This section concerns quantifying the impact of such sources of variation.

5.5.1 Balanced Hierarchical Data Structures

Figure 5.16 is a schematic of a type of symmetrical, tree-structured data set often potentially available in quality assurance contexts. In the figure, a series of factors A, B, C ... are such that each represents a subdivision or refinement of the immediately preceding one. The I different levels of Factor A each give rise to their own sets of J levels of Factor B, which in turn give rise to their own sets of K levels of Factor C, and so on. The total number of data values represented by Figure 5.16 is $I \cdot J$ for the case of Factors A and B alone, $I \cdot J \cdot K$ for the case of A, B, and C, and so on. The data structure is called a **balanced hierarchical** (or balanced nested) **structure** because each level of a given factor gives rise to

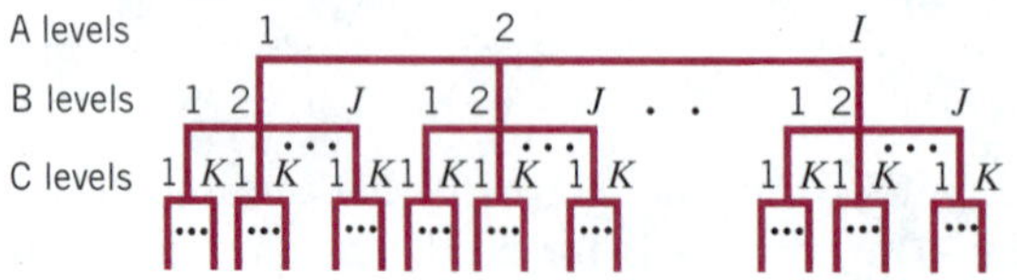

Figure 5.16
Tree diagram of a balanced hierarchical data structure.

the same number of levels of the next. An equivalent way of describing this is to say that (equal-sized) samples of data points are grouped naturally into (equal-sized) groups of samples, which are themselves grouped into (equal-sized) groups of groups, and so on.

A data structure like that in Figure 5.16 might arise where heats (Factor A) of a metal are poured into ingots (Factor B), from which are cut samples (Factor C), each of which is tested for content of a certain trace element. Or in the mass production of a particular type of widget, production might be segmented into work weeks (Factor A), which are segmented into days of the week (Factor B), which are segmented into hours (Factor C), from each of whose output some fixed number of widgets (Factor D) are selected for testing.

When there are important physical differences between levels of the various hierarchical factors, it will not be appropriate to think of the ultimate observations represented by Figure 5.16 as random draws from the same universe. And the thrust of this section is that it is often useful to quantify the various contributions to the overall variability seen in the ultimate responses. For example, in the case of heats of a metal poured into ingots and cut into pieces for analysis, it may be that unavoidable differences in ore mined in different locations lead to legitimate heat-to-heat variability in trace elements. At the same time, poor mixing of molten metal may produce differences between ingots within a heat and pieces within an ingot. An important process improvement activity might then be determining the relative magnitudes of heat-to-heat and mixing-induced differences in trace elements. If, for example, the ingot-to-ingot and piece-to-piece variations turn out to be of smaller order of magnitude than the heat-to-heat variation, there would be little incentive to find a means to improve mixing of heats.

Example 5.11 Lengths of Sawed Steel Rods. Baik, Johnson, and Umthun worked with a small metal fabrication company on the cutting of some steel rods. Bundles of 80 rods were welded together at one end and the whole bundle cut to length at once. Engineering specifications for rod lengths were 33.69 ± .03 inches. Table 5.8 gives the measured lengths for 4 rods from each of 10 consecutive bundles.

The data in Table 5.8 can be thought of as 10 samples of size 4, and as a simple example of the hierarchical structure of Figure 5.16, where there is only a single level of nesting. Figure 5.17 is a plot of the rod lengths versus bundle number. It is evident from the plot (as it would be from a slightly more formal retrospective Shewhart $\bar{x}$ and s chart analysis of the $\bar{y}_i$ and s_i) that it does not make sense to think of the 40 numbers in Table 5.8 as 40 random draws from a single universe. The 4 lengths from a given bundle tend to be much more alike than are lengths from different bundles. This is no doubt due to slightly different bundle-to-bundle setups made for the cuts. To the extent that these differences in setup are unavoidable, it may be of interest to quantify the sizes of the "between-bundle" and "between-rods-within-a-bundle" variabilities.

Table 5.8 Measured Lengths of Four Rods Taken From Each of Ten Consecutive Bundles (inches above 33.69 inches)

Bundle (i)	Length (y)	$\bar{y}_i$	s_i
1	−.0115, −.0110, −.0085, −.0105	−.010375	.00132
2	−.0080, −.0070, −.0060, −.0045	−.006375	.00149
3	−.0095,−.0100, −.0130, −.0165	−.012250	.00323
4	.0090, .0125, .0125, .0080	.010500	.00235
5	−.0105, −.0100,−.0150, −.0075	−.010750	.00312
6	.0115, .0150, .0175, .0180	.015500	.00297
7	.0020, .0005, .0010, .0010	.001125	.00063
8	−.0010, −.0025, −.0020, −.0030	−.002125	.00085
9	−.0020, .0015, .0025, .0025	.001125	.00214
10	−.0010, −.0015, −.0020, −.0045	−.002250	.00156

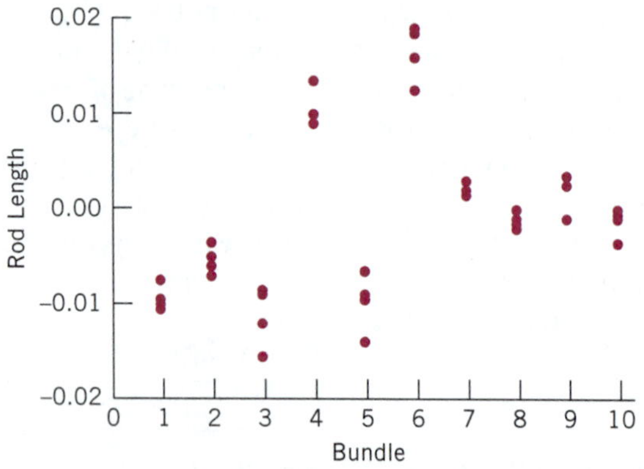

Figure 5.17 Measured lengths of sawed steel rods.

Example 5.12 Copper Contents of Bronze Castings. The article "Statistical Quality Control in the Chemical Laboratory" by G. Wernimont that appeared in *Quality Engineering* in 1989 discusses the copper contents of some bronze castings. $I = 11$ castings were selected for study. From each of these, $J = 2$ physical specimens were cut. And for each of the $I \cdot J = 22$ different specimens, $K = 2$ laboratory determinations of copper content, y, were made. The data and some summary statistics are given in Table 5.9. Note that the $I \cdot J \cdot K = 44$ data values comprise $I \cdot J = 22$ samples of size $K = 2$, which are naturally aggregated into $I = 11$ groups of size $J = 2$.

Figure 5.18 is a plot of the copper contents y_{ijk} versus ij combination (the ij pairs listed in lexicographical order). It shows some substantial variation between the specimens (the two j's) for the various given castings (fixed i's). This suggests that much of the variation seen in measured copper content is traceable to internal heterogeneity within the various castings.

Table 5.9 Measured Copper Contents for Eleven Bronze Castings (%)

Casting (i)	Specimen (j)	Analysis (k)	y_{ijk} Copper Content	Specimen Mean	Casting Mean
1	1	1	85.54	85.550	85.5375
1	1	2	85.56		
1	2	1	85.51	85.525	
1	2	2	85.54		
2	1	1	85.54	85.570	85.4100
2	1	2	85.60		
2	2	1	85.25	85.250	
2	2	2	85.25		
3	1	1	85.72	85.745	85.3450
3	1	2	85.77		
3	2	1	84.94	84.945	
3	2	2	84.95		
4	1	1	85.48	85.490	85.2450
4	1	2	85.50		
4	2	1	84.98	85.000	
4	2	2	85.02		
5	1	1	85.54	85.555	85.6975
5	1	2	85.57		
5	2	1	85.84	85.840	
5	2	2	85.84		
6	1	1	85.72	85.790	85.8250
6	1	2	85.86		
6	2	1	85.81	85.860	
6	2	2	85.91		
7	1	1	85.72	85.740	85.7825
7	1	2	85.76		
7	2	1	85.81	85.825	
7	2	2	85.84		
8	1	1	86.12	86.120	86.1400
8	1	2	86.12		
8	2	1	86.12	86.160	
8	2	2	86.20		
9	1	1	85.47	85.480	85.6200
9	1	2	85.49		
9	2	1	85.75	85.760	
9	2	2	85.77		
10	1	1	84.98	85.040	85.4700
10	1	2	85.10		
10	2	1	85.90	85.900	
10	2	2	85.90		
11	1	1	85.12	85.145	85.1775
11	1	2	85.17		
11	2	1	85.18	85.210	
11	2	2	85.24		

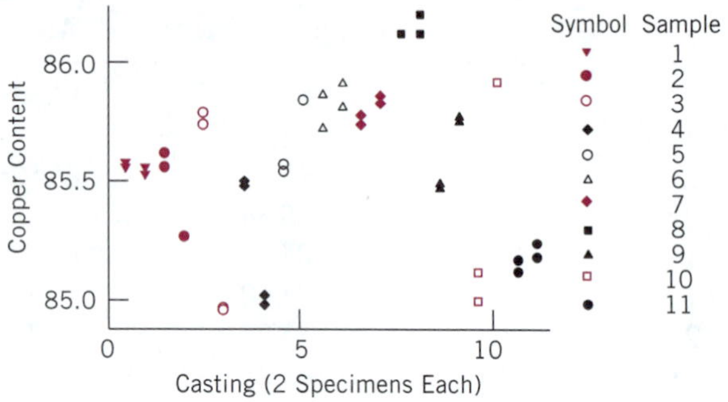

Figure 5.18
Measured copper contents of 2 specimens from each of 11 bronze castings.

5.5.2 A Random Effects Model and Estimators of Variance Components

The usual goal of data analysis for balanced hierarchical studies is to make quantitative the kinds of qualitative insights already gained in the two examples. The basis for doing this is a **random effects model** for the observations. One type of random effects model has already been used in this book (in Section 2.2) as the basis for gage R&R calculations. The model that can be used here is similar in spirit but appropriate to the present hierarchical data structure. Data analysis can be done for nested studies with any number of levels of hierarchy, but for sake of concreteness, this section will treat primarily the case of three hierarchical factors. Once the reader sees how an analysis goes for factors A, B, and C, it will be easy enough to reason by analogy how to treat other numbers of factors.

So, suppose that

$$y_{ijk} = \text{the observation at the } k\text{th level of C within the } j\text{th level of B within the } i\text{th level of A}.$$

A random effects model for the $I \cdot J \cdot K$ observations is then

Hierarchical Random Effects Model

$$y_{ijk} = \mu + \alpha_i + \beta_{ij} + \epsilon_{ijk} \tag{5.28}$$

where μ is an (unknown) constant, the α_i are I normal random variables with mean 0 and variance σ_α^2, the β_{ij} are $I \cdot J$ normal random variables with mean 0 and variance σ_β^2, the ϵ_{ijk} are $I \cdot J \cdot K$ normal random variables with mean 0 and variance σ^2, and all of the α's, β's, and ϵ's are independent. In this model μ is an overall average measurement, the α's are (random) effects of different levels of Factor A, the β's are (random) effects of different levels of Factor B within levels

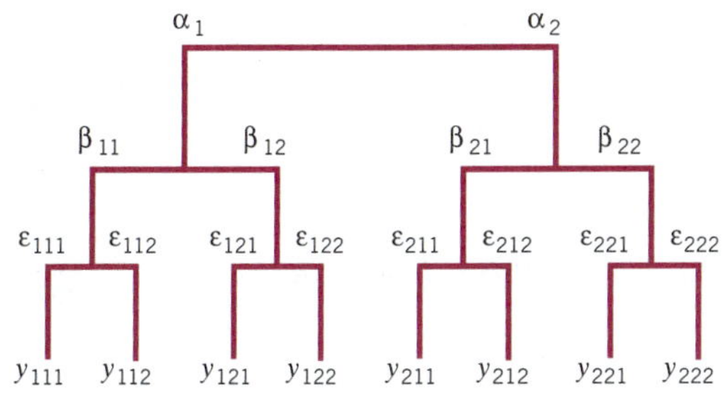

Figure 5.19
Tree diagram corresponding to model equation (5.28) for $I = J = K = 2$.

of A, and the ϵ's are (random) effects of different levels of Factor C within levels of B within A. The variances σ_α^2, σ_β^2, and σ^2 are the **variance components** and govern how much variation there is in the random effects and is eventually seen in the y_{ijk}.

To perhaps make model equation (5.28) somewhat more understandable, consider a hypothetical situation with $I = 2$, $J = 2$, and $K = 2$. The model equation says that there is a normal distribution with mean 0 and variance σ_α^2 from which α_1 and α_2 are drawn. And there is another normal distribution with mean 0 and variance σ_β^2 from which β_{11}, β_{12}, β_{21}, and β_{22} are drawn. And finally, there is a third normal distribution with mean 0 and variance σ^2, from which eight ϵ's are drawn. Eight observations y_{ijk} are then assembled by adding terms down the branches of the tree shown in Figure 5.19.

As a first step toward the estimation of the variance components, note that at the first level of data aggregation (at fixed levels of both Factor A, and B within A) the data within a sample differ only by virtue of having different corresponding ϵ's. In Example 5.12 for instance, measurements on a fixed specimen (j) from a fixed casting (i) differ only in terms of the analysis effects ϵ_{ijk} for different k, the size of which are governed by the variance component σ^2. This suggests that any sample variance s_{ij}^2 could be used to estimate σ^2. Or better yet, all $I \cdot J$ of these could be averaged to produce a single estimator of σ^2. That is, the most commonly used estimator of the variance component σ^2 is

$$\hat{\sigma}^2 = \frac{1}{IJ} \sum_{i,j} s_{ij}^2 . \tag{5.29}$$

Estimator of σ^2 (the Error Variance) for Two Levels of Nesting

Next, consider measuring variation at the second level of aggregation of data values in a balanced hierarchical structure. It is reasonable to take a sample variance *of means* for a given level of Factor A (a fixed i) as a measure of variation between levels of Factor B within that level of Factor A. In Example 5.12 for instance, sample variances of specimen means measure variation between specimens within a casting and are influenced by σ_β^2, the specimen-to-specimen variance component. With this in mind let

$$\begin{aligned}\bar{y}_{ij} &= \text{the sample mean of data values at level } i \\ &\quad \text{of Factor A and level } j \text{ of B within A,} \\ &= \frac{1}{K}\sum_{k} y_{ijk}\,,\end{aligned}$$

$$\begin{aligned}\bar{y}_{i.} &= \text{the sample mean of data values at level } i \text{ of Factor A,} \\ &= \frac{1}{JK}\sum_{j,k} y_{ijk}\,, \\ &= \frac{1}{J}\sum_{j} \bar{y}_{ij}\,,\end{aligned}$$

and

$$\begin{aligned}s^2_{\mathrm{B}i} &= \text{the sample variance of the } J \text{ "B level" sample means } \bar{y}_{ij} \text{ at level } i \text{ of A,} \\ &= \frac{1}{J-1}\sum_{j}(\bar{y}_{ij} - \bar{y}_{i.})^2\,.\end{aligned}$$

The sample variances $s^2_{\mathrm{B}i}$ are the basic building blocks of an estimator of the variance component σ^2_β, but some care must be exercised in arriving at a final form. The fact is that under the model (5.28), not only does σ^2_β influence the size of the $s^2_{\mathrm{B}i}$, but so also does σ^2. (In the context of Example 5.12, analysis-to-analysis variation can inflate a sample variance of specimen means.) In fact, the expected value of each of the random variables $s^2_{\mathrm{B}i}$ is

$$\mathrm{E}s^2_{\mathrm{B}i} = \sigma^2_\beta + \frac{\sigma^2}{K}\,. \tag{5.30}$$

So simple averaging of the $s^2_{\mathrm{B}i}$ would tend to overestimate σ^2_β.

In light of relationship (5.30), a first correction to an average of the $s^2_{\mathrm{B}i}$ as an estimator of σ^2_β is made by subtracting from it an estimator of σ^2/K, namely $\hat{\sigma}^2/K$. But then, such a difference can on occasion be negative, while the variance component σ^2_β is nonnegative. When this happens, it is common to replace the negative estimate with 0. So ultimately, one arrives at the estimator

Estimator of σ^2_β (the Factor B Variance Component) for Two Levels of Nesting

$$\hat{\sigma}^2_\beta = \max\left(0, \frac{1}{I}\sum_{i} s^2_{\mathrm{B}i} - \frac{\hat{\sigma}^2}{K}\right) = \max\left(0, \frac{1}{I}\sum_{i} s^2_{\mathrm{B}i} - \frac{1}{K}\left(\frac{1}{IJ}\sum_{i,j} s^2_{ij}\right)\right). \tag{5.31}$$

Finally, consider variation at the highest level of aggregation in a three-factor hierarchical data set, and corresponding estimation of the variance component σ^2_α. A sample variance of the I means $\bar{y}_{i.}$ measures variation between levels of Factor A. In Example 5.12, for instance, a sample variance of casting means measures casting-to-casting variation and is surely influenced by σ_α^2, the casting to casting variance component. With this in mind let

$$\begin{aligned}\bar{y}_{..} &= \text{the grand sample mean of the data}, \\ &= \frac{1}{IJK}\sum_{i,j,k} y_{ijk}, \\ &= \frac{1}{I}\sum_i \bar{y}_{i.},\end{aligned}$$

and

$$\begin{aligned}s_A^2 &= \text{the sample variance of the } I \text{ "A level" sample means } \bar{y}_{i.}, \\ &= \frac{1}{I-1}\sum_i (\bar{y}_{i.} - \bar{y}_{..})^2 .\end{aligned}$$

The sample variance s_A^2 is the basic building block of an estimator of the variance component σ_α^2. But again some care must be exercised in arriving at a final form. The fact is that under the model (5.28), not only does σ_α^2 influence the size of s_A^2, but so also do σ_β^2 and σ^2. (In the context of Example 5.12, analysis-to-analysis variation and specimen-to-specimen variation can inflate a sample variance of casting means.) In fact, the expected value of the random variable s_A^2 is

$$\mathrm{E}s_A^2 = \sigma_\alpha^2 + \frac{1}{J}\left(\sigma_\beta^2 + \frac{\sigma^2}{K}\right). \tag{5.32}$$

So using s_A^2 to estimate σ_α^2 would tend to overstate the size of this variance.

A first correction to s_A^2 as an estimator of σ_α^2 is motivated by comparison of equations (5.30) and (5.32). One might subtract from s_A^2 an average of sample variances s_{Bi}^2 divided by J. But such a difference can on occasion be negative, while the variance component σ_α^2 is nonnegative. When this happens, it is common to replace the negative estimate with 0. So ultimately, one arrives at

$$\hat{\sigma}_\alpha^2 = \max\left(0,\ s_A^2 - \frac{1}{J}\left(\frac{1}{I}\sum_i s_{Bi}^2\right)\right). \tag{5.33}$$

Estimator of σ_α^2 (the Factor A Variance Component) for Two Levels of Nesting

---

Example 5.12 continued. Consider again the casting example of Wernimont. Table 5.9 contains the raw data and specimen and casting sample means. These can be used to produce the sample variances given in Table 5.10 and to verify that $s_A^2 = .080076$.

Then, using formula (5.29) one has the estimated variance for repeated analyses of any given specimen,

Table 5.10 Sample Variances of Individual Measurements and Specimen Sample Means for the Casting Data

Casting (i)	Specimen (j)	s_{ij}^2	s_{Bi}^2
1	1	.00020	.00031
1	2	.00045	
2	1	.00180	.05120
2	2	0	
3	1	.00125	.32000
3	2	.00005	
4	1	.00020	.12005
4	2	.00080	
5	1	.00045	.04061
5	2	0	
6	1	.00980	.00245
6	2	.00500	
7	1	.00080	.00361
7	2	.00045	
8	1	0	.00080
8	2	.00320	
9	1	.00020	.03920
9	2	.00020	
10	1	.00720	.36980
10	2	0	
11	1	.00125	.00211
11	2	.00180	

$$\begin{aligned}\hat{\sigma}^2 &= \frac{1}{11(2)}(.00020 + .00045 + .00180 + \cdots + 0 + .00125 + .00180)\,,\\ &= .001595\,.\end{aligned}$$

So on a standard deviation scale, one estimates σ by $\sqrt{.001595} = .0399\%$.

Next, equation (5.31) indicates that the variance component σ_β^2 measuring specimen-to-specimen variation within a casting can be empirically approximated by

$$\begin{aligned}\hat{\sigma}_\beta^2 &= \max\left(0,\ \frac{1}{11}(.00031 + .05120 + \cdots + .36980 + .00211) - \frac{1}{2}(.001595)\right),\\ &= .085579\,.\end{aligned}$$

The specimen-to-specimen standard deviation is thus estimated as $\sqrt{.085579} = .2925\%$.

Finally, equation (5.33) shows that the casting-to-casting variance component, σ_α^2, can be estimated as

$$\hat{\sigma}_{\alpha}^{2} = \max\left(0,\ .080076 - \frac{1}{2}\left(\frac{1}{11}(.00031 + .05120 + \cdots + .36980 + .00211)\right)\right),$$
$$= .036888\,,$$

and the casting-to-casting standard deviation is estimated as $\sqrt{.036888} = .1921\%$.

Comparison of the estimated variance components here identifies the specimen-to-specimen variation as the major contributor to variability in measured copper contents. It is even larger than the casting-to-casting contribution. And both the specimen-to-specimen variability and the casting-to-casting variability dwarf the analysis-to-analysis variation. This last observation has the important practical implication that the present lab analysis method is probably adequate to allow one to clearly see casting-to-casting and specimen-to-specimen changes in copper content.

5.5.3 Issues of Interpretation and Calculation

There are a number of comments that need to be made about the data analysis method introduced here before closing this section. One set of these has to do with the interpretation of the variance components and their estimates. The most straightforward interpretation of σ_{α}^{2}, σ_{β}^{2}, and σ^{2} individually, is as variances that would be observed in observations y_{ijk} *in the absence of other kinds of variation*. If, in Example 5.12, all castings were alike and internally completely homogeneous, one should expect to still see variability in copper analysis results describable in terms of σ^{2}. If all castings were alike but not internally homogeneous, and laboratory analyses were perfectly repeatable, then (due to specimen-to-specimen variation) one could still expect to see variability in copper analysis results describable in terms of σ_{β}^{2}. And if castings varied but were each completely internally homogeneous and laboratory analyses were perfectly repeatable, one could expect to see variation in copper analysis results describable in terms of σ_{α}^{2}.

It is also possible to think of the variance components as parts of an "overall" variance. For example, model (5.28) says that if one were to make up a data set using $J = K = 1$ (that is, only one level of each factor except the first) then the observations would have variance

$$\sigma_{\alpha}^{2} + \sigma_{\beta}^{2} + \sigma^{2}\,,$$

the sum of the variance components. (In the context of Example 5.12, one would be thinking of cutting one specimen from each casting and making only one analysis on it.) The ratios

$$\frac{\sigma_{\alpha}^{2}}{\sigma_{\alpha}^{2} + \sigma_{\beta}^{2} + \sigma^{2}},\quad \frac{\sigma_{\beta}^{2}}{\sigma_{\alpha}^{2} + \sigma_{\beta}^{2} + \sigma^{2}},\quad \text{and} \quad \frac{\sigma^{2}}{\sigma_{\alpha}^{2} + \sigma_{\beta}^{2} + \sigma^{2}} \qquad (5.34)$$

Fractions of Variance Attributable to 3 Hierarchical Factors

are then naturally thought of as fractions of an overall variance attributable to, respectively, Factors A, B, and C.

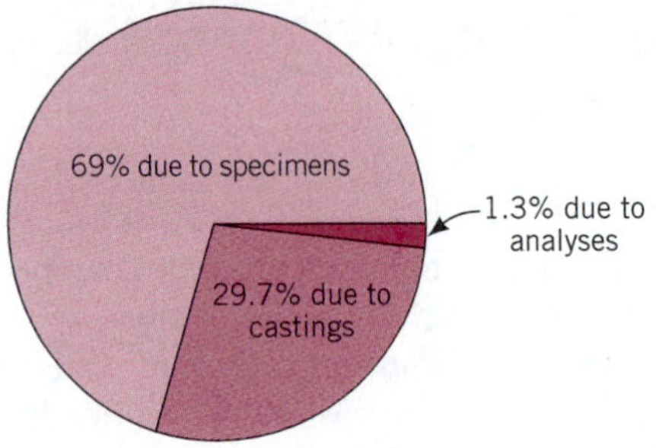

Figure 5.20
Pie chart for fractions of overall variance contributed by castings, specimens, and analyses.

Example 5.12 continued. In the copper content study, estimates of the fractions of overall variance contributed by castings, specimens, and analyses can be computed in accord with display (5.34). One estimates $\sigma_\alpha^2 + \sigma_\beta^2 + \sigma^2$ as $.036888 + .085579 + .001595 = .124062$. So the fraction of variance contributed by castings is approximately

$$\frac{.036888}{.124062} = .297\,,$$

the fraction contributed by specimens is approximately

$$\frac{.085579}{.124062} = .690\,,$$

and the fraction contributed by analyses is approximately

$$\frac{.001595}{.124062} = .013\,.$$

Figure 5.20 is a pie chart representation of these fractions and is a particularly effective means of communicating the fact that specimen-to-specimen variation dominates in the data of Table 5.9.

General Prescription for Estimating Variance Components

Another set of comments about the analysis of this section concerns how the ideas here generalize to more levels of hierarchy. The story is this. One first computes sample means at each level of aggregation and then corresponding sample variances. The estimator of the lowest-level variance, σ^2, is just the average of sample variances of individual observations. At any higher level, the estimator of a variance component is (except for correction to zero in case of a negative result) the difference between "the average of sample variances of means at that level of aggregation" and "the average of sample variances of means at the next lower-level aggregation divided by the number of levels of that next factor (nested within levels of the one under discussion)." Following this prescription, it is possible to do calculations for any number of levels of hierarchy using nothing more than a simple hand-held calculator.

Calculation of Estimates from Entries of an ANOVA Table

However, while the kind of calculations illustrated in this section can be done by hand, in practice statistical packages will typically be used. Printout 5.3 shows the results of using two different Minitab procedures on the data of Table 5.9. The first procedure, "NESTED," provides estimated variance components directly. The second, "ANOVA," gives only an "ANOVA Table," from which one must separately deduce the estimates of the variance components. In such a table, the "error" line "MS" value is $\hat{\sigma}^2$. Any other estimated variance component is found (except for correction to 0 in the case of a negative result) by subtracting from the MS value on the line corresponding to the component of interest the MS value immediately below it, and then dividing the difference by the product of the numbers of levels of all factors listed below the one of interest. Many intermediate-level-applied statistics texts (including *Statistics for Engineering Problem Solving* by Vardeman) provide more details on the making and interpretation of ANOVA tables for balanced hierarchical data sets.

Printout 5.3

```
MTB > nested c4=c1 c2 c3


Analysis of Variance for copper

Source      DF          SS          MS          F        P
casting     10      3.2031      0.3203      1.854    0.163
specimen    11      1.9003      0.1728    108.285    0.000
analysis    22      0.0351      0.0016
Total       43      5.1385


Variance Components

Source      Var Comp.    % of Total     Std Dev
casting        0.037          29.73       0.192
specimen       0.086          68.98       0.293
analysis       0.002           1.29       0.040
Total          0.124                      0.352


Expected Mean Squares

 1 casting    1.00(3) +  2.00(2) +  4.00(1)
 2 specimen   1.00(3) +  2.00(2)
 3 analysis   1.00(3)
```

```
MTB > ANOVA 'copper' = c1 c2(c1);
SUBC>   Random c1 c2.


Analysis of Variance (Balanced Designs)

Factor              Type Levels Values
casting           random     11    1    2    3    4    5    6    7    8    9   10
11
specimen(casting) random      2    1    2


Analysis of Variance for copper

Source               DF          SS          MS          F        P
casting              10     3.20306     0.32031       1.85    0.163
specimen(casting)    11     1.90030     0.17275     108.29    0.000
Error                22     0.03510     0.00160
Total                43     5.13845
```

Example 5.12 continued. To illustrate the computation of estimated variance components from the entries of an ANOVA table, note first that the tables in the printout provide $\hat{\sigma}^2$ directly as the MS values on the "Analysis" and "Error" lines. Then observe that

$$\frac{1}{2}(.17275 - .00160) = .08558 = \hat{\sigma}_\beta^2 ,$$

and

$$\frac{1}{2 \cdot 2}(.32031 - .17275) = .03689 = \hat{\sigma}_\alpha^2$$

as advertised.

The last point to be made in this discussion is that it is possible not only to produce single-number estimates of the variance components, but also to give *confidence intervals* for them, and thereby quantify the precision with which they are estimated. The book *Confidence Intervals on Variance Components* by Burdick and Graybill gives a number of useful formulas for intervals on individual variance components and on important functions of them (like those in display (5.34)). Although the formulas provided by Burdick and Graybill are not impossibly complicated, they do go beyond what is reasonable to present in a limited introduction like the present one. However, for purposes of illustrating what comes out of the

Table 5.11 Estimates and Confidence Limits for Variance Components Based on the Casting Data

Variance Component	Estimate	90% Two-Sided Confidence Limits
σ_α^2	.0369	0 to .1602
σ_β^2	.0856	.0475 to .2069
σ^2	.0016	.0010 to .0028

formulas, Table 5.11 gives both this section's estimates and 90% two-sided confidence intervals for the variance components in the copper content study. The confidence limits in Table 5.11 could, of course, be converted to confidence limits for standard deviations instead of variances by taking square roots.

5.6 CHAPTER SUMMARY

When a process is behaving consistently, it makes sense to try to characterize or describe its behavior in quantitative terms. This chapter has discussed methods for this enterprise. Graphical methods including the important tool of normal plotting were discussed first. Then measures of process capability and their estimation were considered in the second section. Prediction and tolerance intervals were presented next, as means of projecting output of a stable process based on a sample. Then propagation of error and simulations were introduced as tools for engineering design that can be used to predict variation in system performance from variation in component characteristics. Finally, the last section of the chapter discussed the problem of assessing the importance of several identifiable sources of variation in some cases where a process is not completely stable.

5.7 CHAPTER 5 EXERCISES

5.1. Weld Pulls. The following scenario and data are used by GE Aircraft Engines as part of a data analysis demonstration. A shop uses spot welding to join two pieces of an assembly. Welds have been failing in the field and some have been observed to pop apart in a 280°F electrostatic paint oven. Process-monitoring efforts are going to be applied to these welds, and as a part of a preliminary "snapshot" of current process performance, 25 weld strengths are measured for each of two machines. (Specifications on weld strength are that an individual weld button hold up under a 1100 psi pull without tearing. No upper specification limit is used, but manufacturing personnel believe that "blue welds" with strengths larger than 1800 psi are brittle and difficult to finish because of excessive dimpling.) The strength data are below in psi.

Machine 1

1368	1129	1020	1157	1531
1022	1195	1288	1220	1792
1313	1764	989	1666	1643
1703	1764	1952	1706	2004
1135	1946	1105	1502	1629

Machine 2

1187	1862	1821	1713	1887
1110	1376	1871	1315	1498
1206	1736	1904	1873	1208
1696	1307	1965	1305	1744
1358	1215	1551	1369	1375

a. For the machine 1 weld strengths, make a dot diagram, a stem-and-leaf plot, and a frequency table, where the first category begins with 900 psi and the last category ends with 2100 psi. (Use six categories of equal length.) Make the relative frequency histogram corresponding to your frequency table.
b. Redo (a) for machine 2.
c. Make back-to-back stem-and-leaf plots for the two machines.
d. Make back-to-back relative frequency histograms for the two machines.
e. Suppose engineering management decides that since current production goals do not require full use of both welders, all welding will be done with the better of the two machines. (The other machine will not be used.) Which machine should be used? Why?

5.2. Refer to the **Weld Pull** case in problem (5.1).
a. Find the 25 values of $(i - .5)/25$ for $i = 1, 2, \ldots, 25$ and (by ordering the observations from smallest to largest) find the corresponding quantiles of the machine 1 data.
b. Find the .10, .25, .50, and .90 quantiles of the machine 1 data.
c. Do you expect your answers in part (b) to be exactly the corresponding quantiles for all welds made by machine 1? Why or why not?
d. Find the 25 standard normal quantiles $Q_z(\frac{i-.5}{25})$ for $i = 1, 2, \ldots, 25$.
e. Use your answers to (a) and (d) to make a normal plot for the machine 1 pull strengths. Does it appear the pull strengths are coming from a normal distribution? Why or why not?
f. Apply the natural logarithm (ln) transformation to each of the raw strength quantiles from (a). (The results are quantiles of the distribution of log strengths.)

g. Plot standard normal quantiles in (d) versus the log strength quantiles from (f). Does it appear that the log strength distribution is normal? Why or why not?

5.3. Refer to the **Weld Pull** case of problems (5.1) and (5.2).

a. Redo problem (5.2) for machine 2.

b. Make side-by-side box plots for the original data from problem (5.1). (Make one for data from machine 1 and one for data from machine 2.)

c. Making use of the facts of the case given in problem (5.1) and your graph from (b), compare machine 1 and machine 2 weld quality.

5.4. Oil Field Production. Geologists and engineers from a large oil company considered drilling new wells in a field where 64 wells had previously been drilled. The oil production figures for each of the 64 wells were available for analysis and are given below (units of the data are 1000 barrels).

217.1	43.4	79.5	82.2	56.4	36.6	12.0	12.1*
53.2	69.5	26.9	35.1	49.4	64.9	28.3	20.1*
46.4	156.5	13.2	47.6	44.9	14.8	104.9	30.5*
42.7	34.6	14.7	54.2	34.6	17.6	44.5	7.1*
50.4	37.9	32.9	63.1	92.2	29.1	10.3*	10.1*
97.7	12.9	196.0	69.8	37.0	61.4	37.7*	18.0*
103.1	2.5	24.9	57.4	58.8	38.6	33.7*	3.0*
51.9	31.4	118.2	65.6	21.3	32.5	81.1*	2.0*

Twelve of the wells were "completed" using a different technique than the other 52 wells. (The process of completing a well involves stimulation of the rock formation to draw the last "hard to get" oil.) The data values with an "*" correspond to the 12 wells that were completed using the second or alternative method.

Knowledge of the .10, .50, and .90 quantiles of a field's production distribution is very useful to geologists and engineers as they decide whether it is economically feasible to drill again.

a. Make a stem-and-leaf plot of oil production for the 64 wells. Then make back-to-back stem-and-leaf plots for the two groups of wells with different completion methods. Does it appear there is a difference in the distributions of total production for the two groups? Explain.

b. Make side-by-side box plots for the two groups of wells with different completion methods. Compare the two distributions based on these plots.

c. Find and plot the 52 points that make up a normal probability plot for the oil production of the wells completed by the first method. Does it appear that there is any serious departure from the normal distribution shape in these data? Explain.

d. Find and graph the 12 points that make up a normal probability plot for the oil production of the wells completed by the alternative method. Does it appear there is any serious departure from the normal distribution shape in these data? Explain.

e. Find and graph the 64 points that make up a normal probability plot for the oil production of the whole set of wells. Does it appear there is any serious departure from the normal distribution shape in these data? Explain.

5.5. Refer to the **Oil Field Production** case in problem (5.4).

a. Find the .10, .50, and .90 quantiles of the standard normal distribution, $Q_z(.1)$, $Q_z(.5)$, and $Q_z(.9)$.

Note that the p quantile of a normal distribution with mean μ and standard deviation σ is $\mu + \sigma Q_z(p)$.

b. Find the sample mean and sample standard deviation for the 52 wells completed by the first method. Use these, a normal distribution assumption, the formula above, and your answer to (a) to estimate the .1, .5, and .9 quantiles of the production distribution for this field (under the first completion method).

c. Find directly the .10, .50, and .90 quantiles of the production data for the first well completion method (represented by the 52 wells).

d. Find the sample mean and sample standard deviation for the 12 wells completed by the alternative method. Use these, a normal distribution assumption, the formula above, and your answer to (a) to estimate the .1, .5, and .9 quantiles of the production distribution for this field (under the alternative completion method).

e. Find directly the .10, .50, and .90 quantiles of the production data for the alternative well completion method (represented by the 12 wells).

f. For the first completion method, which set of estimates do you recommend for the .10, .50, and .90 quantiles, the set from (b) or the one from (c)? Why?

g. For the alternative completion method, which set of estimates do you recommend for the .10, .50, and .90 quantiles, the set from (d) or the one from (e)? Why?

5.6. Refer to the **Oil Field Production** case in problems (5.4) and (5.5). Apply the natural log (ln) transformation to every data value in problem (5.4). Redo (b)–(g) in problem (5.5) for the transformed data (the log productions).

5.7. Consider the following small ordered data set for the variable x:

$$1,\ 4,\ 5,\ 10,\ 11$$

a. Apply the natural logarithm (ln) transformation to each of the data values. Does the order of the values change as the x's are transformed to $\ln(x)$'s?

b. Find the .30, .50, and .70 quantiles of both the x distribution and the $\ln(x)$ distribution.

c. Let y be the .50 quantile of the $\ln(x)$ distribution. Find $\exp(y)$.

d. Let w be the .30 quantile of the $\ln(x)$ distribution. Find $\exp(w)$.

e. What relationship do you see between quantiles of the x distribution and quantiles of the $\ln(x)$ distribution?

5.8. Part Hardness. Measured hardness values for eight heat-treated steel parts produced on a single day were (in units of mm):

3.175, 3.200, 3.100, 3.200,
3.150, 3.100, 3.100, 3.175

a. What must have been true about the heat-treating process on the day of data collection before an estimate of C_{pk} derived from these data could possibly have any practical relevance or use?

b. What other quantities (besides raw data values like those above) are needed in order to estimate C_{pk}? Where should these quantities come from?

The analysts that obtained the data above had previously done a gage R&R study on the hardness measurement process. Their conclusion was that for this method of hardness measurement, the repeatability standard deviation is $\sigma \approx .044$ mm.

c. Suppose that hardness specifications were $\pm.150$ mm around an ideal value. Is the gaging method used here adequate to check conformance to such specifications? Explain.

d. Compare the sample standard deviation of the eight measurements above to the estimate of σ developed in the gage R&R study. Based on this comparison, do you see clear evidence of any real hardness differences in the eight parts?

e. Formula (5.8) involves μ, σ, U, and L. If these parameters refer to "true" (as opposed to values measured with error) hardnesses for parts, how optimistic are you about getting a sensible estimate of C_{pk} using the data in this problem? Explain.

f. If one uses the hardness values from this problem to estimate "σ" and set up a monitoring scheme for part hardness (say a EWMA or CUSUM of individuals), will your scheme allow for any "natural manufacturing variability" as part of all-OK conditions? Explain in light of your answer to (d). Do you see this as necessarily good or bad if the eight parts used in this study fairly represent an entire day's production of such parts?

5.9. Refer to the **Oil Field Production** case in problems (5.4), (5.5), and (5.6).

a. In problems (5.6b) and (5.6c) you were asked to find estimates of population .10, .50, and .90 quantiles of the log production distribution (for the first completion method) using two different methods. Exponentiate those estimates (i.e., plug them into the $\exp(\cdot)$ function). These exponentiated values can be treated as estimates of quantiles of the raw production distribution. Compare them to the values obtained in problems (5.5b) and (5.5c).

b. In problems (5.6d) and (5.6e) you were asked to find estimates of population .10, .50, and .90 quantiles of the log production distribution (for the alternative completion method) using two different methods. Exponentiate those estimates (i.e., plug them into the $\exp(\cdot)$ function). These exponentiated values can be treated as estimates of quantiles of the raw production distribution. Compare them to the values obtained in problems (5.5d) and (5.5e).

c. In all, you should have three different sets of estimates in part (a). Which set do you find to be most credible? Why?

d. In all, you should have three different sets of estimates in part (b). Which set do you find to be most credible? Why?

5.10. Refer to the **Oil Field Production** case in problems (5.4) and (5.6). Consider the 52 wells completed using the first method and assume that oil production from wells like those on the field in question is approximately normally distributed.

a. Find a two-sided interval that has a 95% chance of containing oil production values from 95% of all wells like the 52.

b. Find a two-sided interval that has a 95% chance of containing the total oil production of the next well drilled in the same manner as the 52.

c. Find a lower bound, B, such that one can be 99% sure that 90% of all wells drilled in the same manner as the 52 will produce B or more barrels of oil.

5.11. Refer to the **Oil Field Production** case in problems (5.4), (5.6), and (5.10). Redo (a)–(c) in problem (5.10) assuming that the logarithm of total oil production for a well like the 52 is normally distributed. (Do the computations called for in (a) through (c) on the log production values, then exponentiate your results to get limits for raw production values.)

5.12. Refer to problems (5.10) and (5.11). Which of the two sets of intervals produced in problems (5.10) and (5.11) do you find to be more credible? Why?

5.13. Refer to the **Oil Field Production** case in problems (5.10) and (5.11). Redo (a)–(c) in problem (5.10) and then problem (5.11) for wells like the 12 completed using the alternative method. Which set of intervals (those derived from the original data or from the transformed data) do you find to be more credible? Why?

5.14. Refer to the **Oil Field Production** case in problems (5.10), (5.11), and (5.13). One important feature of the real problem has been ignored in making projections of the sort in problems (5.10), (5.11), and (5.13). What is this? (Hint: Is the "independent random draws from a fixed population" model necessarily sensible here? When using a straw to drink liquid from a glass, how do the first few "draws" compare to the last few in terms of the amount of liquid obtained per draw?)

5.15. Sheet Metal Saddle Diameters. Eversden, Krouse, and Compton investigated the fabrication of some sheet metal saddles. These are rectangular pieces of sheet metal that have been rolled into a "half tube" for placement under an insulated pipe to support the pipe without crushing the insulation. Three saddle sizes were investigated. Nominal diameters were 3.0 inch, 4.0 inch, and 6.5 inch. Twenty saddles of each size were sampled from production. Measured diameters (in inches) are below.

3-Inch Saddles				4-Inch Saddles				6.5-Inch Saddles			
3.000	3.031	2.969	3.063	4.625	4.250	4.250	4.313	6.625	6.688	6.406	6.438
3.125	3.000	3.000	3.125	4.313	4.313	4.313	4.250	6.469	6.469	6.438	6.375
3.000	3.063	3.000	3.063	4.313	4.313	4.125	4.063	6.500	6.469	6.375	6.375
3.031	2.969	2.969	3.094	4.094	4.125	4.094	4.125	6.469	6.406	6.375	6.469
3.063	3.188	3.031	2.969	4.156	4.156	4.156	4.125	6.469	6.438	6.500	6.563

a. Make side-by-side box plots for the data from the three saddle sizes.

b. Make a frequency table for each of the three saddle sizes. (Use categories of equal length. For the 3-inch saddle data, let 2.95 inches be the lower limit of the first category and 3.20 inches be the upper limit of the last category and employ five categories. For the 4-inch saddle data, let 4.05 inches be the lower limit of the first category and 4.65 inches be the upper limit of the last category and use five categories. For the 6.5-inch saddle data, let 6.35 inches be the lower limit of the first category and 6.70 inches be the upper limit of the last category and use five categories.)

c. Make the relative frequency histograms corresponding to the three frequency tables requested in (b).

d. Supposing specification limits are $\pm.20$ inches around the respective nominal diameters, draw in the corresponding specifications on the three relative frequency histograms made in (c). What do you conclude about the saddle-making process with respect to meeting these specifications?

e. Make a quantile plot for each saddle size.

f. Find the .25, .50, and .75 quantiles for data from each saddle size. Give the corresponding IQR values.

5.16. Refer to the **Saddle Diameters** case in problem (5.15).

a. Find the 20 values of $(i - .5)/20$ for $i = 1, 2, \ldots, 20$ and (by ordering values in each of the data sets from smallest to largest) find the corresponding quantiles of the three data sets in problem (5.15).

b. Find the standard normal quantiles $Q_z(\frac{i-.5}{20})$ for $i = 1, 2, \ldots, 20$. Then for each of the three data sets from problem (5.15) plot the standard normal quantiles versus corresponding diameter quantiles.

c. Draw in straight lines summarizing your graphs from (b). Note that $Q_z(.25) \approx -.67$, $Q_z(.50) = 0$, and $Q_z(.75) \approx .67$, and from your lines drawn on the plots, read off diameters corresponding to these standard normal quantiles. These values can function as additional estimates of quantiles of the diameter populations. Give the corresponding estimated IQRs.

d. Suppose the object is to make inferences about quantiles of all saddles of a given size. For each saddle size, which set of estimates do you find to be more credible, those from problem (5.15f) or those from (c) above? Why?

5.17. Refer to the **Saddle Diameters** case in problem (5.15). Assume diameters for each saddle size are approximately normally distributed.

a. For each saddle size, find a two-sided interval that you are 95% sure will include 90% of all saddle diameters.

b. For each saddle size, find a two-sided interval that you are 95% sure will contain the diameter of the next saddle of that type fabricated.

c. For each saddle size, find a numerical value so that you are 99% sure that 90% of all saddles of that type have diameters at least as big as your value.

d. For each saddle size, find a numerical value so that you are 95% sure the next saddle of that type will have a diameter at least as big as your value.

5.18. Refer to the **Saddle Diameters** case in problems (5.15) and (5.17). Consider first (for the nominally 3-inch saddles) the use of the two-sided statistical interval

$$(\min x_i, \max x_i).$$

a. Thought of as an interval hopefully containing 90% of all diameters (of nominally 3-inch saddles), what confidence should be attached to this interval?

b. Thought of as an interval hopefully containing the next diameter (of a nominally 3-inch saddle), what confidence should be attached to this interval?

Consider next (for the nominally 3-inch saddles) the use of the one-sided statistical interval

$$(\min x_i, \infty).$$

c. How sure are you that 90% of all diameters (of nominally 3-inch saddles) lie in this interval?

d. How sure are you that the next diameter (of a nominally 3-inch saddle) will lie in this interval?

5.19. Refer to problems (5.17) and (5.18).

a. The intervals requested in (5.17) and (5.18) are based on a mathematical model of "independent random draws from a fixed universe." Such a model makes sense if the saddle-forming process is physically stable. Suppose the data in problem (5.17) are in fact listed in the order of fabrication (read left to right, then top to bottom for each size) of the saddles. Investigate (using a retrospective Shewhart X chart with σ estimated by $\overline{MR}/1.128$ as in Section 4.4) the stability of the fabrication process (one saddle size at a time). Does your analysis indicate any problems with the relevance of the basic "stable process" model assumptions? Explain.

b. The intervals requested in problem (5.17) are based on a normal distribution model, while the ones in problem (5.18) are not. Which set of intervals seems most appropriate for the nominally 3-inch saddles? Why?

5.20. Casehardening in Red Oak. Kongable, McCubbin, and Ray worked with a sawmill on evaluating and improving the quality of kiln-dried wood for use in furniture, cabinetry, flooring, and trim. Casehardening (one of the problems that arises in drying wood) was the focus of their work. Free water evaporates first during the wood-drying process until about 30% (by weight) of the wood is water. Bound water then begins to leave cells and the wood begins to shrink as it dries. Stresses produced by shrinkage result in casehardening. When casehardened wood is cut, some stresses are eliminated and the wood becomes distorted. To test for casehardening of a board, prongs are cut in the board and a comparison of the distances between the prongs before and after cutting reveals the degree of casehardening. A decrease indicates casehardening, no change is ideal, and an increase indicates reverse casehardening. The engineers sampled 15 dried 1-inch red oak boards and cut out prongs from each board. Distances between the prongs were measured and are recorded below. (Units are inches.)

Board	Distance Before	Distance After
1	.80	.58
2	.79	.29
3	.77	.50
4	.79	.77
5	.90	.55
6	.77	.36
7	.90	.57
8	.80	.55
9	.80	.64
10	.79	.27
11	.79	.79
12	.80	.80
13	.79	.41
14	.80	.72
15	.79	.37

a. The data in the table above are most appropriately thought of as which of the following: (i) two samples of $n = 15$ univariate data points or (ii) a single sample of $n = 15$ bivariate data points? Explain.

b. One might wish to look at either the "before" or "after" width distribution by itself. Make dot plots useful for doing this.

c. As an alternative to the dot plots of part (b), make two box plots.

d. Make normal probability plots for assessing how bell shaped the "before" and "after" distributions appear to be. Comment on the appearance of these plots.

A natural means of reducing the before-and-after measurements to a single univariate data set is to take differences (say *after* − *before*).

e. Suppose that the intention was to mark all sets of prongs to be cut at a distance of .80 inch apart. Argue that despite this good intention, the *after* − *before* differences are probably a more effective measure of casehardening than the "after" measurements alone (or the "after" measurements minus .80). Compute these 15 values.

f. Make a dot plot and a box plot for the differences. Mark on these the ideal difference. Do these plots indicate that casehardening has taken place?

g. Make a normal probability plot for the differences. Does this measure of casehardening appear to follow a normal distribution for this batch of red oak boards? Explain.

h. If there were no casehardening, what horizontal intercept would you expect to see for a line summarizing your plot in (g)? Explain.

i. When one treats the answers in parts (a) through (h) above as characterizations of "casehardening" one is really making an implicit assumption that there are no unrecognized assignable/nonrandom causes at work in the data. (For example, one is tacitly assuming that the boards represented in the data came from a single lot that was processed in a consistent fashion, etc.) In particular, there is an implicit assumption that there were no important time-order-related effects. Investigate the reasonableness of this assumption (using a retrospective Shewhart X chart with σ estimated by $\overline{MR}/1.128$ as in Section 4.4) supposing that the board numbers given in the table indicate order of sawing and other processing. Comment on the practical implications of your analysis.

5.21. Refer to the **Casehardening** case in problem (5.20). Assume both the "after" values and the *after* − *before* differences are approximately normal.

a. Find a two-sided interval that you are 99% sure will contain the next "after" measurement for such a board.

b. Find a two-sided interval that you are 95% sure contains 95% of all "after" measurements.

c. Find a two-sided interval that you are 95% sure will contain the next difference for such a board.

d. Find a two-sided interval that you are 99% sure contains 99% of all differences for such boards.

5.22. Refer to the **Casehardening** case in problem (5.20), and in particular to the *after* − *before* differences, x. Consider the use of the two-sided statistical interval

$$(\min x_i, \max x_i).$$

a. Thought of as an interval hopefully containing the next difference, what confidence should be attached to this interval?

b. Thought of as an interval hopefully containing 95% of all differences, what confidence should be attached to this interval?

c. Thought of as an interval hopefully containing 99% of all differences, what confidence should be attached to this interval?

5.23. Refer to the **Casehardening** case in problems (5.20), (5.21), and (5.22), and in particular to the *after* – *before* differences.

The prediction interval in problem (5.21c) relies on a normal distribution assumption for the differences, while the corresponding interval in problem (5.22a) does not. Similarly, the tolerance intervals in problems (5.21b) and (5.21d) rely on a normal distribution assumption for the differences, while the corresponding intervals in problems (5.22b) and (5.22c) do not.

a. Which set of intervals (the one in problem (5.21) or the one in problem (5.22)) is most appropriate in this problem? Why?

b. Compare the intervals (based on the differences) found in problem (5.21) with those from problem (5.22) in terms of lengths and associated confidence levels. When a normal distribution model is appropriate, what does its use provide in terms of the practical effectiveness of statistical intervals?

c. If one were to conclude that differences cannot be modeled as normal and were to find the results of problem (5.22) to be of little practical use, what other possibility remains open for finding prediction and tolerance intervals based on the differences?

5.24. Bridgeport Numerically Controlled Milling Machine. Field, Lorei, Micklavzina, and Stewart studied the performance of a Bridgeport numerically controlled milling machine. Positioning accuracy was of special concern. Published specifications for both x and y components of positioning accuracy were "$\pm$ .001 inch." One of the main problems affecting positioning accuracy is "backlash." (Backlash is the inherent play that a machine has when it stops movement or reverses direction of travel in a given plane.) The group conducted an experiment aimed at studying the effects of backlash.

A series of holes were reamed, moving the machine's head in the x direction only. Then a series of holes were reamed, moving the machine's head in the y direction only. (The material used was a 1/4-inch-thick acrylic plate. It was first spot faced with a 3/8-inch drill bit to start and position the hole. The hole was then drilled with a regular 15/64-inch twist drill. A 1/4-inch reamer was finally used to improve hole size and surface finish. This machining was all done at 1500 RPM and a feed rate of 4 inches/min.) The target x distance between successive holes in the first set was 1.25 inches, while the target y distance between successive holes in the second set was .75 inch. The two tables below contain the 30 measured distances between holes (in inches).

It is not completely obvious how to interpret positioning accuracy specifications like the "*nominal* $\pm$ .001 inch" ones referred to above. But it is perhaps sensible to apply them to distances between successive holes like the ones made by the students.

x Movement					*y* Movement				
1.2495	1.2485	1.2505	1.2495	1.2505	.7485	.7490	.7505	.7500	.7480
1.2490	1.2495	1.2505	1.2500	1.2505	.7515	.7495	.7500	.7490	.7505
1.2495	1.2525	1.2500	1.2520	1.2505	.7490	.7525	.7500	.7490	.7490
1.2485	1.2510	1.2480	1.2480	1.2495	.7510	.7485	.7490	.7485	.7495
1.2505	1.2490	1.2505	1.2500	1.2505	.7480	.7500	.7500	.7485	.7490
1.2500	1.2515	1.2500	1.2495	1.2495	.7500	.7490	.7500	.7510	.7505

a. Make a box plot for the x distances. Indicate on the plot the ideal value of 1.250 inches and the specifications $L = 1.249$ inches and $U = 1.251$ inches.

b. Make a box plot for the y distances. Indicate on the plot the ideal value of .750 inch and the specifications $L = .749$ inch and $U = .751$ inch.

c. Is there any clear indication in your box plots from (a) and (b) that x positioning is better than y positioning (or vice versa)? Explain.

d. Make a normal plot for the x distances. Is it sensible to treat x distances as normally distributed? Why?

e. From the plot in (d), estimate the mean and the standard deviation of x distances. Explain how you got your answer.

f. Make a normal plot for the y distances. Is it sensible to treat y distances as normally distributed? Why?

g. From the plot in (f), estimate the mean and the standard deviation of y distances. Explain how you got your answer.

5.25. Refer to the **Bridgeport Numerically Controlled Milling Machine** case in problem (5.24). Assume both the x and y distances are approximately normally distributed.

a. Make a two-sided interval that you are 95% sure contains 99% of x distances between such reamed holes (nominally spaced 1.250 inches apart on a horizontal line).

b. Does your interval in (a) "sit within" the specification limits for such x distances? Is this circumstance appealing? Why or why not?

c. Make a two-sided interval that you are 99% sure contains 95% of y distances between such reamed holes (nominally spaced .750 inches apart on a vertical line).

d. Does your interval in (c) "sit within" the specification limits for such y distances? Is this circumstance appealing? Why or why not?

e. In light of the results of problem (5.24d) and problem (5.24f), are the intervals made in (a) and (c) above credible? Explain.

5.26. Hose Cleaning. Delucca, Rahmani, Swanson, and Weiskircher studied a process used to ensure that some industrial hoses are free of debris. Specifications were that the inside surfaces of these were to carry no more than 44 mg of contaminant per square meter of hose surface. (The hoses are cleaned by blowing air through them at high pressure.)

Periodically, five of these hoses are tested by rinsing them with trichloroethylene, filtering the liquid, and recovering solids washed out in the cleaning fluid. The data below are mg of solids per m^2 of inside hose surface from 13 such tests.

Sample	Contamination Levels
1	45.00, 47.77, 145.43, 31.84, 45.01
2	24.00, 33.37, 22.87, 27.89, 21.46
3	13.50, 17.75, 13.34, 9.87, 15.42
4	3.0, 5.21, 1.82, 6.97, 11.13
5	8.62, 4.91, 18.42, 6.16, 6.58
6	231.02, 440.32, 136.24, 379.77, 171.78
7	257.00, 207.18, 240.09, 213.93, 389.62
8	107.57, 101.40, 133.49, 141.50, 92.56
9	51.00, 47.72, 59.45, 53.75, 46.51
10	85.00, 58.40, 52.30, 60.50, 46.84
11	44.00, 45.66, 83.30, 47.31, 66.13
12	88.37, 44.35, 35.65, 146.78, 37.50
13	59.30, 55.67, 62.52, 33.66, 34.96

This scenario and data set have a number of interesting features. For one, specifications here are inherently one sided, so that inherently two-sided capability measures like C_p and C_{pk} do not make much sense in this context. For another, it is obvious from a plot of the observed contamination levels against sample number, or from a plot of sample standard deviations against sample means, that the variability in measured contamination level increases with mean contamination level.

Consider first the matter of clear dependence of standard deviation on mean. In a circumstance like this (and particularly where data range over several orders of magnitude), it is often helpful to conduct an analysis not on the raw data scale, but on a logarithmic scale instead. Notice that on a logarithmic scale, the upper specification for solids washed out in a test is $\ln(44) = 3.78$ ($\ln$ mg/m^2).

a. Replace the raw data above by their natural logarithms. Then compute 13 subgroup means and standard deviations.

b. Make retrospective $\bar{x}$ and s charts for these samples. Working on the log scale, is there evidence of process instability in either of these charts? Would an analysis on the original scale of measurement look any more favorable (in terms of process stability)?

c. In light of your answer to (b), explain why it doesn't make sense to try to state process capabilities based on the data presented here.

Suppose that after some process analysis and improvements in the way hoses are cleaned, control charts for logarithms of measured contaminations show no signs of process instability. Suppose further, that then combining 10 samples of 5 measured log contamination rates to produce a single sample of size $n = 50$, one finds $\bar{x} = 3.05$ and $s = 2.10$ for (logged) contamination rates and a normal model to be a good description of the data.

d. Find an interval that (assuming continued process stability) you are 95% sure will include log contamination levels for 95% of all future tests of this type. Transform this interval into one for contamination levels measured on the original scale.

e. Find an interval that (assuming continued process stability) you are 99% sure will contain the next measured log contamination level. Transform this interval into one for the next contamination level measured on the original scale.

f. Give a 95% two-sided confidence interval for the process capability, 6σ, measured on the log scale.

Although C_p and C_{pk} are not relevant in problems involving one-sided specifications, there are related capability indices that can be applied. In cases where there is only an upper engineering specification U, the measure

$$CPU = \frac{U - \mu}{3\sigma}$$

can be used, and in cases where there is only a lower engineering specification L, there is the corresponding measure

$$CPL = \frac{\mu - L}{3\sigma}.$$

As it turns out, the formula (5.10) used to make lower confidence bounds for C_{pk} can be applied to the estimation of CPU or CPL (after replacing $\hat{C}_{pk}$ by $\widehat{CPU} = \frac{U-\bar{x}}{3s}$ or by $\widehat{CPL} = \frac{\bar{x}-L}{3s}$) as well.

g. Find and interpret the estimate of CPU corresponding to the description above of log contamination levels obtained after process improvement.

h. Give a 95% lower confidence bound for CPU.

i. Suppose management raises the upper specification limit to 54 mg/m^2. How does this change your answers to (g) and (h)?

j. When comparing estimated capability ratios from different time periods, what does part (i) show must be true in order to allow a sensible comparison?

5.27. Drilling Depths. Deford, Downey, Hahn, and Larsen measured drill depths in a particular type of pump housing. Specifications on the depth for each hole measured were 1.29 ± .01 inches. Depth measurements for 24 holes were taken and recorded in the order of drilling.

Hole	Measured Depth	Hole	Measured Depth
1	1.292	13	1.292
2	1.291	14	1.290
3	1.291	15	1.292
4	1.291	16	1.291
5	1.291	17	1.290
6	1.291	18	1.291
7	1.290	19	1.290
8	1.291	20	1.291
9	1.292	21	1.290
10	1.291	22	1.290
11	1.291	23	1.291
12	1.290	24	1.291

a. Find the average moving range for these data (as in Section 4.4) and estimate σ by $\overline{MR}/1.128$. Use this estimate of the process short-term variability to make a retrospective "3 sigma" Shewhart X chart for these data. Is there evidence of drilling process instability on this chart?

b. In light of part (a), it perhaps makes sense to treat the measured depths as a single sample of size $n = 24$ from a stable process. Do so and give 90% lower confidence bounds for both C_p and C_{pk}.

c. Give a 95% two-sided confidence interval for the process capability 6σ.

d. Find an interval that you are 95% sure contains 90% of all depth measurements for holes drilled like these.

e. The validity of all the estimates in this problem depends upon the appropriateness of a normal distribution model for measured hole depth. The fact that the gaging seems to have been fairly crude relative to the amount of variation seen in the hole depths (only three different depths were ever observed) makes a completely satisfactory investigation of the reasonableness of this assumption impossible. However, do the best you can in assessing whether normality seems plausible. Make a normal probability plot for the data and discuss how linear the plot looks (making allowance for the fact that the gaging is relatively crude).

5.28. Refer to the **Journal Diameters** case of problem (3.16). Suppose that after some attention to control charts and process behavior, engineers are able to bring the grinding process to physical stability. Further suppose that a sample of $n = 30$ diameters then has sample mean $\bar{x} = 44.97938$, sample standard deviation $s = .00240$, and a fairly linear normal probability plot. In the following, use the mid-specification of 44.9825 as a target diameter.

a. Give 95% lower confidence bounds for both C_p and C_{pk}.

b. Which estimated index from (a) reflects potential process performance? Which one summarizes current performance? Explain.

c. If (say because of heavy external pressure to show "quality improvement," real or illusory) the lower specification for these diameters was arbitrarily lowered and the upper specification was arbitrarily raised, what would happen to C_p and C_{pk}?

d. In light of your answer to part (c), what must be clarified when one is presenting (or interpreting someone else's presentation of) a series of estimated C_p's or C_{pk}'s?

e. Find a two-sided interval that you are 99% sure will include the next measured journal diameter from the process described in problem (3.16). What mathematical assumptions support the making of this interval?

f. Find a two-sided interval that you are 95% sure contains 99% of all measured journal diameters.

5.29. Refer to the **U-bolt Threads** case of problem (3.20).

a. Make retrospective $\bar{x}$ and R control charts for the thread-length data. Is there evidence of process instability in these means and ranges?

b. In light of part (a), it makes sense to treat the $5 \times 18 = 90$ data points in problem (3.20) as a single sample from a stable process. You may check that the grand sample mean for these data is $\bar{x} = 11.13$ and the grand sample standard deviation is $s = 1.97$. (The units are .001 inch above nominal length.) Compare this latter figure to $\overline{R}/d_2$ and note that as expected there is substantial agreement between these figures.

c. Specifications for thread lengths are given in problem (3.20h). Use these and $\bar{x}$ and s from part (b) and give estimates of C_p and C_{pk}.

d. Find a 95% lower confidence bound for C_{pk}.

e. Find a 90% two-sided confidence interval for C_p. What mathematical model assumptions support the making of this interval?

f. Which of the two quantities, C_p or C_{pk}, is a better indicator of the actual performance of the U-bolt thread-making process? Why?

g. Find a 99% two-sided confidence interval for the process short-term standard deviation, σ.

h. Find a two-sided interval that you are 99% sure contains 95% of all thread lengths for bolts from this process.

i. What mathematical model assumptions support the making of the inferences in parts (d), (e), (g), and (h)? How would you check to see if those are sensible in the present situation?

j. Use a statistical package to help you make the check you suggested in part (i) and comment on your result.

5.30. An engineer plans to perform two different tests on a disk drive. Let X be the time needed to perform test 1 and Y be the time required to perform test 2. Further, let μ_X and μ_Y be the respective means of these random variables and σ_X and σ_Y be the standard deviations. Of special interest is the total time required to make both tests, $X + Y$.

a. What is the mean of $X + Y$, $\mu_{X+Y} = \mathrm{E}(X + Y)$?

b. What is required in terms of model assumptions in order to go from σ_X and σ_Y to a standard deviation for $X + Y$, $\sigma_{X+Y} = \sqrt{\mathrm{Var}(X + Y)}$?

c. Make the assumption alluded to in (b) and express σ_{X+Y} in terms of σ_X and σ_Y.

Suppose that in order to set work standards for testing disk drives of a certain model, the engineer runs both test 1 and test 2 on n such drives. Test 1 time requirements have sample mean $\bar{x}$ and sample standard deviation s_x, while test 2 time requirements have sample mean $\bar{y}$ and sample standard deviation s_y.

d. How would you use the information from the n tests to estimate μ_{X+Y}? (Give a formula.)

e. How would you use the information from the n tests to estimate the function of σ_X and σ_Y that you gave as an answer in part (c)? (Give a formula.)

f. Suppose that in the engineer's data, there is a fairly large (positive) sample correlation between the observed values of x and y. (For example, drives requiring an unusually long time to run test 1 also tended to require an unusually long time to run test 2.) Explain why you should then not expect your answer to (e) to serve as a sensible estimate of σ_{X+Y}. In this circumstance, if one has access to the raw data, what is a safer way to estimate σ_{X+Y}? (Hint: Suppose you can reconstruct the n sums $x + y$.)

g. Would the possibility alluded to in part (f) invalidate your answer to part (d)? Explain.

5.31. Let X be the inside length, Y be the inside width, and Z be the inside height of a box. Suppose $\mu_X = 20$, $\mu_Y = 15$, and $\mu_Z = 12$, while $\sigma_X = .5$, $\sigma_Y = .25$, and $\sigma_Z = .3$. (All units are inches.) Assume inside height, inside width, and inside length are "unrelated" to one another.

a. Find the mean area of the inside bottom of the box.

b. Approximate the standard deviation of the area of the inside bottom of the box.

c. Find the mean inside volume.

d. Approximate the standard deviation of the inside volume.

5.32. Impedance in a Micro-Circuit. In their article "Robust Design Through Optimization Techniques," which appeared in *Quality Engineering* in 1994, Lawson and Madrigal modeled impedance (Z) in a thin film redistribution layer as

$$Z = \left(\frac{87.0}{\sqrt{\epsilon + 1.41}}\right) \ln\left(\frac{5.98A}{.80B + C}\right)$$

where A is an insulator thickness, B is a line width, C is a line height, and ϵ is the insulator dielectric constant. ϵ is taken to be known as 3.10, while A, B, and C are treated as random variables (as they would be in the manufacture of such circuits) with means $\mu_A = 25$, $\mu_B = 15$, and $\mu_C = 5$, and standard deviations $\sigma_A = .3333$, $\sigma_B = .2222$, and $\sigma_C = .1111$. (Units for A, B, and C are 10^{-6} inches. The article doesn't give the units of Z.)

a. Find an approximate mean and standard deviation of impedance for this type of device. Assume A, B, and C are independent.

b. Variation in which of the variables is likely to be the largest contributor to variation in impedance of manufactured devices of this type? Explain.

5.33. Refer to Example 5.11. Recall that the case discussed there involves the sawing of bundles of steel rods.

a. The discussion of hierarchical data structures in Section 5.5 used the symbols I and J for respectively the numbers of levels (observed) of A and of B within each level of A. What do these symbols represent in the context of Example 5.11? What are their numerical values?

b. Give a random effects model for the $I \times J$ measured rod lengths in Example 5.11.

c. Find the estimated variance of rod lengths within a bundle.

d. Find an estimated variance reflecting bundle-to-bundle variability.

e. Find an estimate of the variance that would be experienced if a single rod from many different bundles were selected and measured.

f. Find an estimate of the fraction of overall variance (from (e)) attributable to differences between bundles.

5.34. TV Electron Guns. In their article, "Design Evaluation for Reduction in Performance Variation of TV Electron Guns," which appeared in *Quality Engineering* in 1992, Ranganathan, Chowdhury, and Seksaria reported on their efforts to improve consistency of cutoff voltages for TV electron guns. Cutoff voltage (Y) can be modeled in terms of several geometric properties of an electron gun as

$$Y = \frac{KA^3}{(B + D)(C - D)}$$

where K is a known constant of proportionality, A is the diameter of an aperture in the gun's first grid, B is the distance between that grid and the cathode, C is the distance between that grid and a second grid, and D is a measure of lack of flatness of the first grid. The quality team wished to learn how variation in A, B, C, and D was ultimately reflected in variation in Y. (In particular, it was of interest to know the relative importance of the lack of flatness variable. At the beginning of the study, an expensive selective assembly process was being used to try to compensate for problems with consistency in this variable.)

a. Find an approximation for the standard deviation of cutoff voltage assuming A, B, C, and D are independent.

b. In a manufacturing context such as the present one, which of the terms in an expression like you produced for part (a) represent target values set in product design, and which represent item-to-item manufacturing variation? How might realistic values for these latter terms be obtained?

The following means and standard deviations are consistent with the description of the case given in the article. (Presumably for reasons of corporate security, complete details are not given in the paper, but these values are consistent with everything that is said there. In particular, no units for the dimensions A through D are given.)

Dimension	Mean	Standard Deviation
A	.70871	.00443
B	.26056	.00832
C	.20845	.02826
D	0	.00865

c. Using the means and standard deviations given above and your answer to (a), approximate the fractions of variance in cutoff voltage attributable to variability in the individual dimensions A through D.

d. In light of your answer to (c) does it seem that the company's expensive countermeasures to deal with variation in lack of flatness in the first grid are justified? Where should their variance-reduction efforts be focused next? Explain.

e. What advantages does the method of this problem have over first measuring cutoff voltage and then dissecting a number of complete electron guns, in an effort to see which dimensions most seriously affect the voltage?

5.35. **Cast Iron Carbon Content.** In their article "Wet Method of Chemical Analysis of Cast Iron: Upgrading Accuracy and Precision through Experimental Design," which appeared in a 1994–1995 issue of *Quality Engineering,* Anand, Bhadkamkar, and Moghe discussed a study made on the amount of carbon in cast iron. Hardness and machining properties depend on the amount of carbon in the metal. On each of 2 days, 2 heats of cast iron were studied and 3 determinations of carbon content were made for each of these heats, producing a balanced hierarchical data set. (Carbon contents were measured and recorded as percentages.)

a. The discussion in Section 5.5 uses the notations I, J, and K. What do these symbols represent in the present context?

b. What are the numerical values of I, J, and K here?

c. Give a (random effects) model equation for describing measured carbon contents.

d. In terms of the variance components for your model from (c), what is the variance that would be experienced, if on many different days a single heat was selected and a single carbon content measured?

5.36. Refer to the **Cast Iron Carbon Content** case in problem (5.35). An ANOVA table (made using a computer package and in the style of the Minitab output in Printout 5.3) for the cast iron contents measured by Anand, Bhadkamkar, and Moghe is below.

Source	*SS*	*df*	*MS*
day	.0184	1	.0184
heat(day)	.0027	2	.0014
error	.2044	8	.0256
total	.2255	11	

a. What does the parameter σ measure in the present context? Give an estimate of this quantity.

b. Suppose that one would like any carbon content measurement made on a particular heat to be within, say, .05 of the mean for that heat. In light of your answer to (a) do you think this goal is being met? Explain. What would you report as an estimate of a "measurement capability ratio" appropriate here, $\frac{6\sigma}{.10}$?

c. Compute and interpret estimates of σ_α and σ_β for this scenario. Are there important differences between days? Between heats within a day?

d. Make and interpret a pie chart like Figure 5.20 for the present situation.

5.37. The heat conductivity of a circular cylindrical bar of diameter D and length L, connected between two constant temperature devices of temperatures T_1 and T_2, that conducts C calories in τ seconds is

$$\lambda = \frac{4CL}{\pi(T_1 - T_2)\tau D^2}.$$

In a particular laboratory determination of λ for brass, the quantities C, L, T_1, T_2, τ, and D can be measured with means and standard deviations approximately as follows:

Variable	Mean	Standard Deviation	$\frac{\partial\lambda}{\partial \text{variable}}$
C	240 cal	10 cal	.000825
L	100 cm	.1 cm	.199
T_1	100°C	1°C	−.00199
T_2	0°C	1°C	.00199
τ	600 sec	1 sec	.000332
D	1.6 cm	.1 cm	−.249

(The units of the partial derivatives are the units of λ (cal/(cm)(sec)(°C)) divided by the units of the variable in question.)

a. Find an approximate standard deviation for a realized heat conductivity value from a single determination.

b. In this experimental setup, which of the variables do you expect to contribute most to the variation in experimentally determined heat conductivity values? Explain.

5.38. Fluidized Boiler Nozzles. Mietzel, Oliver, and Stout worked to improve the process used in a university machine shop to make the boiler nozzles needed in the cogeneration process that supplies Iowa State University with its heat and electricity. A portion of their efforts involved identifying and estimating sources of variability in nozzle thickness. Three batches of nozzles machined from 304 stainless steel were studied and 9 nozzles were selected from each batch and measured. The resulting thicknesses (in cm) are given in the following table.

Batch	Thickness
1	1.640, 1.655, 1.640, 1.645, 1.645, 1.625, 1.585, 1.635, 1.575
2	1.580, 1.540, 1.580, 1.570, 1.605, 1.575, 1.575, 1.590, 1.560
3	1.555, 1.570, 1.570, 1.540, 1.580, 1.585, 1.570, 1.590, 1.640

Consider an analysis of these data that treats this as a hierarchical data set where Factor A is "batches" and Factor B within A is "nozzles."

a. Give the standard model equation for describing how measured nozzle thickness varies.

b. What do the symbols I and J used in Section 5.5 represent in this context?

c. What are the numerical values of I and J here?

d. Find estimates of σ and σ_α and say what they measure.

e. What does $\sigma_\alpha^2 + \sigma^2$ measure here? (What kind of a data set would have this variance?)

f. Estimate the fractions of "overall variance" from (e) contributed by batches and by nozzles.

5.39. Refer to the **Fluidized Boiler Nozzles** case in problem (5.38). The shop also makes nozzles from 316 stainless steel. The table below gives thickness measurements from three batches of this second material.

Batch	Thickness
1	1.645, 1.670, 1.630, 1.640, 1.645, 1.640, 1.655, 1.645, 1.650
2	1.655, 1.655, 1.645, 1.640, 1.640, 1.650, 1.650, 1.645, 1.645
3	1.655, 1.645, 1.640, 1.640, 1.650, 1.635, 1.645, 1.630, 1.645

Redo parts (d) and (f) of problem (5.38) for this second material. Then compare the shop's ability to consistently machine the two different materials.

5.40. Refer to the **Plastic Packaging** case in problem (3.32) and the hole-position data set given there. One can think of that data set as having hierarchical structure with Factor A being "days," Factor B within A being "sampling time," and Factor C within B within A being "bags."

a. Do an analysis of this data set parallel to that in Example 5.12. Estimate the three variance components σ_α^2, σ_β^2, and σ^2 and say what they indicate about the consistency of hole position in these bags.

b. Compare the conclusions of your "variance component analysis" in (a) to those of the $\bar{x}$ and R chart analysis employed in problem (3.32). How does your estimate of σ here compare to $\bar{R}/d_2$ computed from 18 subgroups? How are your conclusions about the sizes of σ_α^2 and σ_β^2 consistent with the appearance of your control charts?

5.41. An output voltage V_{out} in an audio circuit is a function of an input voltage V_{in} and gains N and K supplied respectively by a transformer and an amplifier via the relationship

$$V_{\text{out}} = V_{\text{in}}NK.$$

Many such circuits are to be built and used in circumstances where V_{in} may be somewhat variable.

a. Considering manufacturing variability in N and K and variation in V_{out} attributable to environmental factors, how would you predict the mean of V_{out} without testing any of these complete circuits? How would you predict the amount of variability to be seen in V_{out} without testing any of these complete circuits?

b. What assumption might you add to your answers to (a) in order to predict the fraction of output voltages from these circuits that will fall in the range 99 V to 100 V?

c. Why might it be "not so smart" to rely entirely on calculations like those alluded to in (a) and (b) to the complete exclusion of product testing?

CHAPTER 6

Experimental Design and Analysis for Process Improvement Part 1: Basics

The first five chapters of this book provide tools for bringing a process to physical stability and then characterizing its behavior. The question of what to do if the resulting picture of the process is not to one's liking, however, remains. This chapter and the next present some tools for addressing this issue. That is, Chapters 6 and 7 concern statistical methods that support intelligent process experimentation, and can provide guidance in improvement efforts.

This chapter begins with a section presenting "one-way" methods for the analysis of experimental data. These treat r samples in ways that do not refer to any special structure in the process conditions leading to the observations. They are thus among the most widely applicable of statistical methods. The second section discusses the analysis of two-way complete factorial data sets, and introduces some of the basic concepts needed in the study of systems where several different factors potentially impact a response variable. Finally, the chapter culminates with a long section discussing p-way factorial analysis, with particular emphasis on those complete factorial studies where each factor appears at 2 levels, known as the 2^p *factorial studies*.

6.1 ONE-WAY METHODS

Figure 6.1 is useful for helping one think about experimental design and data analysis. Pictured is the proverbial "black box," representing a process to be studied. Into that process go many noisy/variable inputs, and out of the process comes at least one noisy output of interest, y. At the time of experimentation, there are "knobs" on the process that the experimenter can manipulate, variables $x_1, x_2, x_3, \ldots$ whose settings or "levels" are under the investigator's control. The engineer can choose values for the variables and observe one or more values of y, choose another set of values for the variables and observe one or more additional values of y, and so on. The object of such data collection and subsequent data analysis is to figure out how the black box/process responds to changes in the knob settings (i.e., how the process output depends upon the variables $x_1, x_2, x_3, \ldots$). Armed with such knowledge, one can then try to

1. optimize the choice of values for the variables $x_1, x_2, x_3, \ldots$ in terms of maximizing (or minimizing) y or,
2. identify those variables that have the greatest effect on y, with the goal of prescribing very careful future control of those, to the end of reducing variation in y.

This section presents methods of data analysis that do not depend upon any specifics of the nature of the variables $x_1, x_2, x_3, \ldots$ nor on the pattern of changes one makes in them when collecting data. (In particular, there is no assumption that the variables $x_1, x_2, x_3, \ldots$ are quantitative. Nor does one need to suppose that any specific number of the variables are changed in any specific way in order to use the basic methods of this section.) All that one assumes is that there are r different sets of experimental conditions under consideration.

The presentation begins with a discussion of a "one-way" normal model for data from r different sets of conditions, and an estimator of variance in that model. Then a method of making confidence intervals for linear combinations of the r means involved is presented. And finally, there is a discussion of some methods of making "simultaneous" confidence intervals.

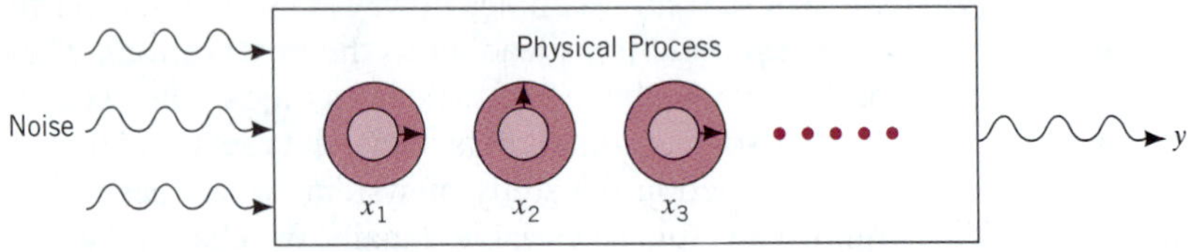

Figure 6.1
Black box with noisy inputs, "knobs" $x_1, x_2, x_3, \ldots$, and noisy output y.

6.1.1 The One-Way Normal Model and a Pooled Estimator of Variance

Example 6.1 is typical of many engineering experiments. Fairly small samples are taken to represent r different sets of process conditions, and one must use the limited information provided by the r samples to make comparisons.

Example 6.1 **Strengths of Solder Joints.** The paper "Fracture Mechanism of Brass/Sn-Pb-Sb Solder Joints and the Effect of Production Variables on the Joint Strength," by Tomlinson and Cooper, appeared in *Journal of Materials Science* in 1986 and contains some data on shear strengths of solder joints (units are megaPascals). Table 6.1 gives part of the Tomlinson/Cooper data for six sets of process conditions (defined in terms of the cooling method employed and the amount of antimony in the solder).

The data in Table 6.1 compose $r = 6$ samples of common size $m = 3$, representing six different ways of making solder joints. In terms of the conceptualization of Figure 6.1, there are $p = 2$ process "knobs" that have been turned in the collection of these data, the cooling method knob and the % antimony knob. Section 6.2 introduces methods of analysis aimed at detailing separate effects of the two factors. But to begin with, it will be useful to simply think of the data (and summary statistics) in Table 6.1 as generated by six different unstructured sets of process conditions.

In order to make statistical inferences from r small samples like those represented in Table 6.1, it is necessary to adopt some model for the data generation process. By far the most widely used, most convenient, and most easily understood such mathematical description is the **one-way normal** (or Gaussian) **model**. In words, this model says that the r samples in hand come from r normal distributions *with possibly different means* $\mu_1, \mu_2, \ldots, \mu_r$, *but a common standard deviation* σ. Figure 6.2 is a graphical representation of these assumptions.

Table 6.1 Shear Strengths and Some Summary Statistics for $r = 6$ Different Soldering Methods (MPa)

Method (i)	Cooling	Sb (% weight)	Strength, y	$\bar{y}_i$	s_i
1	H_2O quench	3	18.6, 19.5, 19.0	19.033	.451
2	H_2O quench	5	22.3, 19.5, 20.5	20.767	1.419
3	H_2O quench	10	15.2, 17.1, 16.6	16.300	.985
4	oil quench	3	20.0, 20.9, 20.4	20.433	.451
5	oil quench	5	20.9, 22.9, 20.6	21.467	1.250
6	oil quench	10	16.4, 19.0, 18.1	17.833	1.320

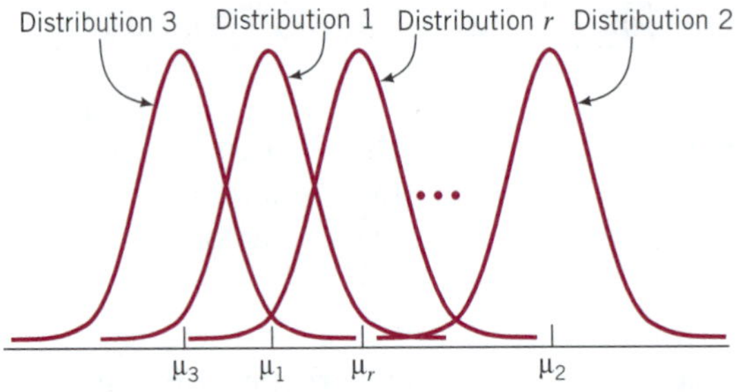

Figure 6.2
r normal distributions with a common standard deviation.

One-Way Normal Model

It is helpful to also have a statement of the basic one-way Gaussian model assumptions in terms of symbols. The one that will be used in this text is that with y_{ij} the jth observation in sample i (from the ith set of process conditions),

One-Way Model Equation

$$y_{ij} = \mu_i + \epsilon_{ij} , \tag{6.1}$$

for $\mu_1, \mu_2, \ldots, \mu_r$ (unknown) means and $\epsilon_{11}, \ldots, \epsilon_{1n_1}, \epsilon_{21}, \ldots, \epsilon_{2n_2}, \ldots, \epsilon_{r1}, \ldots, \epsilon_{rn_r}$ independent normal random variables with mean 0 and (unknown) standard deviation σ. $n_1, n_2, \ldots, n_r$ are the sample sizes, and the means $\mu_1, \mu_2, \ldots, \mu_r$, and the standard deviation σ are the parameters of the model that must be estimated from data.

It is always wise to do one's best to examine the plausibility of a mathematical model before using it to guide engineering decisions. So the question arises as to how one might investigate the appropriateness of the "constant variance, normal distributions" model represented by equation (6.1). Where the sample sizes $n_1, n_2, \ldots, n_r$ are moderate to large (say on the order of at least 6 or 7) one option is to make r normal plots (possibly on the same set of axes) for the different samples. One uses the normal plotting method of Section 5.1 and hopes to see reasonably linear plots with roughly the same slope, linearity indicating normality, and equal slopes indicating constant variance.

While making r different normal plots is a viable option where sample sizes are moderate to large, it is not really very helpful in contexts like Example 6.1 where sample sizes are extremely small. About all that can be done for very small sample sizes is to examine "residuals" in much the same way one studies residuals in regression analysis. That is, for the one-way model (6.1), the ith sample mean $\bar{y}_i$ is a kind of "predicted" or "fitted" response for the ith set of process conditions. One might then (at least for those samples with $n_i > 1$) compute and examine **residuals**

Residual for Data Point y_{ij}

$$e_{ij} = y_{ij} - \bar{y}_i . \tag{6.2}$$

The motivation for doing so is that in light of the model (6.1), e_{ij} is an approximation for ϵ_{ij}, and the e_{ij} might thus be expected to look approximately like

Table 6.2 Residuals for the Soldering Data (MPa)

Method (i)	Residuals, $e_{ij} = y_{ij} - \bar{y}_i$
1	$18.6 - 19.033 = -.433,\ 19.5 - 19.033 = .467,\ 19.0 - 19.033 = -.033$
2	$22.3 - 20.767 = 1.533,\ 19.5 - 20.767 = -1.267,\ 20.5 - 20.767 = -.267$
3	$15.2 - 16.300 = -1.100,\ 17.1 - 16.300 = .800,\ 16.6 - 16.3 = .300$
4	$20.0 - 20.433 = -.433,\ 20.9 - 20.433 = .467,\ 20.4 - 20.433 = -.033$
5	$20.9 - 21.467 = -.567,\ 22.9 - 21.467 = 1.433,\ 20.6 - 21.467 = -.867$
6	$16.4 - 17.833 = -1.433,\ 19.0 - 17.833 = 1.167,\ 18.1 - 17.833 = .267$

Gaussian random variation. (Actually, when sample sizes vary, a slightly more sophisticated analysis would consider **standardized residuals,** which in this context are essentially the e_{ij} divided by $\sqrt{(n_i - 1)/n_i}$. But for sake of brevity this refinement will not be pursued here.) So, for example, if model (6.1) is to be considered completely appropriate, a normal plot of the e_{ij} ought to look reasonably linear.

Example 6.1 continued. Considering again the soldering study, Table 6.2 contains $18 = 6 \times 3$ residuals computed according to formula (6.2).

Figure 6.3 is a normal plot of the 18 residuals of Table 6.2. It is quite linear and raises no great concerns about the appropriateness of an analysis of the soldering data based on the one-way normal model.

After considering the appropriateness of the basic model (6.1), the next order of business is to estimate the parameter σ. This is the (supposedly constant)

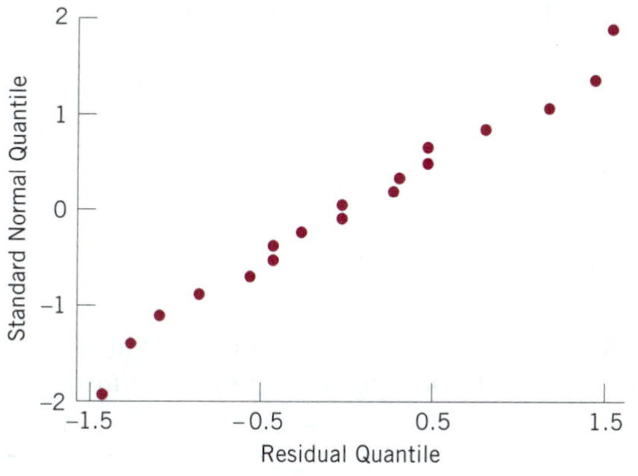

Figure 6.3 Normal plot of the residuals in the soldering study.

standard deviation of observations y from any fixed set of process conditions. Any one of the sample standard deviations $s_1, s_2, \ldots, s_r$ could alone serve as an estimator of σ. But it is reasonable that a better option would be to somehow combine all r of these into a single "pooled" estimator. The path usually followed is to compute a kind of weighted average of the r sample variances, and then take a square root in order to get back to the original units of observation. That is, a pooled estimator of σ^2 is

Pooled Estimator of σ^2

$$s^2_{\text{pooled}} = \frac{(n_1 - 1)s_1^2 + (n_2 - 1)s_2^2 + \cdots + (n_r - 1)s_r^2}{(n_1 - 1) + (n_2 - 1) + \cdots + (n_r - 1)}, \tag{6.3}$$

or setting $n = n_1 + n_2 + \cdots + n_r$ and abbreviating the word "pooled" to the letter "P," one can rewrite display (6.3) as

Pooled Estimator of σ^2

$$s_{\text{P}}^2 = \frac{(n_1 - 1)s_1^2 + (n_2 - 1)s_2^2 + \cdots + (n_r - 1)s_r^2}{n - r}. \tag{6.4}$$

Of course, a corresponding pooled estimator of σ is thus

Pooled Estimator of σ

$$s_{\text{P}} = \sqrt{s_{\text{P}}^2}. \tag{6.5}$$

Example 6.1 continued. Table 6.1 gives the $r = 6$ sample variances for the soldering methods. These may be combined according to formula (6.4) to produce the pooled sample variance

$$s_{\text{P}}^2 = \frac{(3-1)(.451)^2 + (3-1)(1.419)^2 + \cdots + (3-1)(1.320)^2}{18 - 6},$$
$$= 1.116,$$

so that

$$s_{\text{P}} = \sqrt{1.116} = 1.056 \text{ MPa}.$$

This pooled estimate of σ is meant to represent the amount of variability seen in shear strengths for any one of the six conditions included in the Tomlinson and Cooper study.

It is worth noting that the pooled estimator (6.4) is sometimes called the "error mean square" (MSE) where people emphasize ANOVA in the analysis of experimental data. This text does not emphasize ANOVA or the significance tests that it facilitates, preferring to concentrate on confidence intervals and graphical displays in the analysis of experimental data. And from this present point of view, the terminology "pooled estimator of variance" is probably more natural.

6.1.2 Confidence Intervals for Linear Combinations of Means

The estimator s_P is a measure of background noise or observed experimental variability against which differences in the sample means $\bar{y}_1, \bar{y}_2, \ldots, \bar{y}_r$ must be judged. The particular form of the estimator specified in displays (6.4) and (6.5) is convenient because it allows the development of a number of simple formulas of statistical inference. For one thing, it allows for the making of simple confidence intervals for linear combinations of the means $\mu_1, \mu_2, \ldots, \mu_r$.

That is, for constants $c_1, c_2, \ldots, c_r$, consider the problem of estimating the **linear combination of means**

$$L = c_1\mu_1 + c_2\mu_2 + \cdots + c_r\mu_r , \qquad (6.6)$$

based on samples from process conditions $1, 2, \ldots, r$. A number of important quantities are of this form. For example, where all c_i's are 0 except one which is 1, L is just a particular mean of interest. And where all c_i's are 0 except one that is 1 and one that is -1, L is a difference in two particular means of interest. And as this chapter proceeds, it will become evident that there are many other more complicated L's that are useful in the analysis of data from process improvement experiments.

A natural estimator for L specified in display (6.6) is

$$\hat{L} = c_1\bar{y}_1 + c_2\bar{y}_2 + \cdots + c_r\bar{y}_r , \qquad (6.7)$$

Estimator of L

obtained by replacing each population mean by its corresponding sample version. And as it turns out, one can use $\hat{L}$ as the basis of a confidence interval for L. That is, confidence limits for L are

$$\hat{L} \pm t s_P \sqrt{\frac{c_1^2}{n_1} + \frac{c_2^2}{n_2} + \cdots + \frac{c_r^2}{n_r}} , \qquad (6.8)$$

Confidence Limits for L

where t is a quantile of the t distribution with $n - r$ associated degrees of freedom. (n continues to stand for the total of the r sample sizes.) If t is the p quantile and only one of the limits indicated in display (6.8) is used, the associated confidence level is $p \times 100\%$. If both limits in formula (6.8) are employed to make a two-sided confidence interval, the associated confidence is $(2p - 1) \times 100\%$.

Two important special cases of formula (6.8) are those where L is a single mean and where L is a difference of means. That is, confidence limits for the ith mean response μ_i are

Confidence Limits for a Single Mean

$$\bar{y}_i \pm ts_P\sqrt{\frac{1}{n_i}}. \tag{6.9}$$

And confidence limits for the difference in the ith and i'th means, $\mu_i - \mu_{i'}$, are

Confidence Limits for a Difference in Means

$$\bar{y}_i - \bar{y}_{i'} \pm ts_P\sqrt{\frac{1}{n_i} + \frac{1}{n_{i'}}}. \tag{6.10}$$

Example 6.1 continued. Return again to the soldering problem, and first consider the issue of estimating an individual method mean strength, μ_i. There are associated with the pooled standard deviation s_P a total of $n - r = 18 - 6 = 12$ degrees of freedom. So, for example, finding from Table A.8 that the .975 quantile of the t_{12} distribution is 2.179, two-sided 95% confidence limits for any single mean strength in the soldering study are from display (6.9)

$$\bar{y}_i \pm 2.179(1.056)\sqrt{\frac{1}{3}}, \quad \text{that is,} \quad \bar{y}_i \pm 1.328 \text{ MPa}.$$

In some sense, the $\bar{y}_i$ values in Table 6.1 are each good to within 1.328 MPa as representing their respective long-run mean solder joint strengths.

Next, consider the comparison of any two mean strengths. One revealing way to make such a comparison is through a confidence interval for their difference, as indicated in display (6.10). Again using the fact that the .975 quantile of the t_{12} distribution is 2.179, two-sided 95% confidence limits for any particular difference in mean strengths are

$$\bar{y}_i - \bar{y}_{i'} \pm 2.179(1.056)\sqrt{\frac{1}{3} + \frac{1}{3}}, \quad \text{that is,} \quad \bar{y}_i - \bar{y}_{i'} \pm 1.879 \text{ MPa}.$$

For example, $\mu_1 - \mu_4$ (which represents the difference in mean strengths for water-quenched and oil-quenched joints when 3% antimony is used in the solder) can be estimated with 95% confidence as

$$(19.033 - 20.433) \pm 1.879, \quad \text{that is,} \quad -1.400 \text{ MPa} \pm 1.879 \text{ MPa}.$$

The fact that the uncertainty reflected by the ± 1.879 MPa figure is larger in magnitude than the difference between $\bar{y}_1$ and $\bar{y}_4$ says that evidence in the data that methods 1 and 4 produce different joint strengths is not overwhelming. (The confidence interval includes 0.)

Finally, as an example of using the general form of the confidence interval formula presented in formula (6.8), one might contemplate estimation of

$$\begin{aligned} L &= \frac{1}{3}(\mu_4 + \mu_5 + \mu_6) - \frac{1}{3}(\mu_1 + \mu_2 + \mu_3) \\ &= -\frac{1}{3}\mu_1 - \frac{1}{3}\mu_2 - \frac{1}{3}\mu_3 + \frac{1}{3}\mu_4 + \frac{1}{3}\mu_5 + \frac{1}{3}\mu_6, \end{aligned}$$

the difference between the average of the oil-quench mean strengths and the average of the water-quench mean strengths. First, from formula (6.7)

$$\hat{L} = -\frac{1}{3}(19.033 + 20.767 + 16.300) + \frac{1}{3}(20.433 + 21.467 + 17.833) = 1.211.$$

Then, based on the fact that each c_i is $\pm\frac{1}{3}$ and thus has square $\frac{1}{9}$, formula (6.8) gives 95% two-sided confidence limits for L of the form

$$1.211 \pm 2.179(1.056)\sqrt{6\left(\frac{1}{9}\right)\left(\frac{1}{3}\right)}, \quad \text{that is,} \quad 1.211 \text{ MPa} \pm 1.085 \text{ MPa}.$$

This shows that (at least on average across different amounts of antimony) the oil-quenched joints are detectably stronger than water-quenched joints. (The interval includes only positive values.)

6.1.3 Some Simultaneous Confidence Interval Methods

It is important to realize that the confidence level associated with an interval with end points (6.8) (or one of the specializations (6.9) and (6.10)) is an *individual* confidence level, pertaining to a single interval at a time. For example, if one uses formula (6.9) r times, to estimate each of $\mu_1, \mu_2, \ldots, \mu_r$ with 90% confidence, one is 90% confident of the first interval, *separately* 90% confident of the second interval, *separately* 90% confident of the third interval, and so on. One is *not* 90% confident that all r of the intervals are correct. But there are times when it is desirable to be able to say that one is simultaneously "90% sure" of a whole collection of inferences. The purpose of this subsection is to present some relatively simple methods that provide overall or simultaneous confidence levels in the estimation of a number of linear combinations of means, L.

First, if one wishes to make several confidence intervals and announce an overall or simultaneous confidence level, one simple approach is to use something called **Bonferroni's Inequality**. This inequality says that if l intervals have associated individual confidence levels $\gamma_1, \gamma_2, \ldots, \gamma_l$, then the confidence that should be associated with them simultaneously or as a group, say γ, satisfies

Bonferroni's Lower Bound on Overall Confidence

$$\gamma \geq 1 - ((1 - \gamma_1) + (1 - \gamma_2) + \cdots + (1 - \gamma_l)) . \tag{6.11}$$

(This says that the "unconfidence" associated with a group of inferences is no worse than the sum of the individual "unconfidences," $1 - \gamma \leq (1 - \gamma_1) + (1 - \gamma_2) + \cdots + (1 - \gamma_l)$.)

Example 6.1 continued. Consider a simple use of the Bonferroni Inequality (6.11) in the solder joint strength study. The $r = 6$ individual 95% confidence intervals for the means μ_i of the form $\bar{y}_i \pm 1.328$ taken as a group have simultaneous confidence level at least 70%, since

$$1 - 6(1 - .95) = .7.$$

If one wanted to be more sure of intervals for the $r = 6$ means, one could instead make (much wider) 99% individual intervals and be at least 94% confident that all six intervals cover their respective means.

The Bonferroni idea is a simple all-purpose tool that covers many different situations. It is particularly useful when there are a relatively few quantities to be estimated and individual confidence levels are large (so that the lower bound for the joint or simultaneous confidence is not so small as to be practically useless). But it is also somewhat crude, and a number of more specialized methods have been crafted to produce exact simultaneous confidence levels for estimating particular sets of L's. When these are applicable, they will produce narrower intervals (or equivalently, higher stated simultaneous confidence levels) than those provided by the Bonferroni method.

One such method for making simultaneous confidence intervals concerns intervals for $r' \leq r$ of the means μ_i. That is, if one wishes to make intervals for some or all of the means in an r sample study and announce a joint or simultaneous confidence level, it is possible to do so by replacing t in formula (6.9) with another constant. The method is due to Pillai and Ramachandran, and in this text will be referred to as the **P-R method**. To use the P-R method to make r' two-sided intervals for means μ_i with simultaneous 95% confidence, one computes end points

$$\bar{y}_i \pm \kappa_2^* s_P \sqrt{\frac{1}{n_i}}, \tag{6.12}$$

P-R Two-Sided Limits for r' Means

where the constant κ_2^* is taken from Table A.10a and is read from the column corresponding to r' means and row corresponding to $n - r$ degrees of freedom (for s_P). And to use the P-R method to make r' one-sided intervals for means μ_i with simultaneous 95% confidence, one computes end points

$$\bar{y}_i - \kappa_1^* s_P \sqrt{\frac{1}{n_i}} \quad \text{or} \quad \bar{y}_i + \kappa_1^* s_P \sqrt{\frac{1}{n_i}}, \tag{6.13}$$

P-R One-Sided Limits for r' Means

where the constant κ_1^* is taken from Table A.10b and is read from the column corresponding to r' means and row corresponding to $n - r$ degrees of freedom. (The constants κ_2^* are in fact .95 quantiles of the "studentized maximum modulus distributions" and the constants κ_1^* are .95 quantiles of the "studentized extreme deviate distributions.")

Example 6.1 continued. In contrast to the *individual* two-sided 95% confidence intervals made earlier for the method mean joint strengths, consider now the possibility of giving $r = 6$ intervals for the method mean strengths with associated *simultaneous* 95% confidence. Using the $\nu = 12$ line of Table A.10a, formula (6.12) says that an uncertainty of plus or minus

$$3.095(1.056)\sqrt{\frac{1}{3}} = 1.887 \text{ MPa}$$

should be associated with each $\bar{y}_i$ in Table 6.1, if one wishes to have 95% confidence that the resulting intervals simultaneously all cover their corresponding method mean joint strengths. Notice that (not unexpectedly) this ± 1.887 figure is larger than the ± 1.328 figure that is appropriate if one only requires *individual* 95% confidence in the $r = 6$ intervals. Bigger (and thus less informative) intervals are required when one wishes to be reasonably sure that all of a number of inferences are simultaneously correct.

One other simple simultaneous confidence interval method concerns comparison of all pairs of $r' \leq r$ of the means μ_i. That is, if one selects some or all of the means in an r sample study and determines to make intervals for all possible differences of these, a joint confidence level for the set of intervals can be announced if one replaces t in formula (6.10) by another constant. The method is due to **John Tukey** and uses quantiles of the "studentized range distributions" in place of t distribution quantiles. To make simultaneous two-sided confidence intervals for all differences among r' means, one computes end points

Tukey Limits for Comparing r' Means

$$\bar{y}_i - \bar{y}_{i'} \pm \frac{q^*}{\sqrt{2}} s_P \sqrt{\frac{1}{n_i} + \frac{1}{n_{i'}}}, \tag{6.14}$$

where the constant q^* is taken from Tables A.11a&b. These tables are arranged in columns corresponding to the number of means being compared, r', and rows corresponding to the degrees of freedom associated with s_P, $n - r$. If q^* is read from Table A.11a, a minimum 95% simultaneous confidence is guaranteed, while if Table A.11b is consulted the minimum guaranteed simultaneous confidence is 99%. (When all n_i are the same, these minimum confidence levels are exact. When the n_i vary, the stated levels are conservative.)

Example 6.1 continued. In order to illustrate the use of Tukey's method, consider making simultaneous 95% intervals for all differences in the method mean joint strengths for the soldering study. Formula (6.14) indicates that an uncertainty of plus or minus

$$\frac{4.75}{\sqrt{2}}(1.056)\sqrt{\frac{1}{3} + \frac{1}{3}} = 2.896 \text{ MPa}$$

should be associated with any difference in the $\bar{y}_i$ values listed in Table 6.1, if one wishes to have 95% confidence that the resulting intervals simultaneously all cover their corresponding differences in method mean joint strengths. Notice that as expected, the ± 2.896 figure is larger than the ± 1.879 figure found earlier to be appropriate when 95% confidence in individual intervals was of concern. The more broadly one requires that a particular confidence level apply, the more conservative must be the associated inferences.

This book does not emphasize hypothesis testing. But it is worth noting that the Tukey method applied to the differences in all r means can be thought of as providing (in addition to its important interval inferences) some testing insights as well. Where no difference in sample means exceeds its corresponding "uncertainty"

$$\frac{q^*}{\sqrt{2}} s_P \sqrt{\frac{1}{n_i} + \frac{1}{n_{i'}}}$$

in magnitude, one has little evidence against the null hypothesis $H_0 : \mu_1 = \mu_2 = \cdots = \mu_r$. On the other hand, when there are those that do, some people like to identify the two corresponding means as detectably different.

This mention of hypothesis testing again brings up the fact that many textbook treatments of inference for the one-way model center around an ANOVA-based hypothesis test of $H_0 : \mu_1 = \mu_2 = \cdots = \mu_r$. In contrast, a confidence interval-based approach has been taken here in the belief that it is simultaneously simpler and more informative, and has the added virtue of providing an effective means for studying the effects of multiple factors. (Unlike the methods presented in this book, ANOVA-based approaches to the analysis of multifactor experiments encounter severe complications when data sets are unbalanced, i.e., not all n_i are the same.) Those readers curious about the details of ANOVA-based methods are directed to intermediate-level texts on statistical methods (e.g., Vardeman's *Statistics for Engineering Problem Solving*).

6.2 TWO-WAY FACTORIALS

It is perhaps not yet apparent how the analysis tools of the previous section address the facts that most often process improvement experiments involve several factors and that separating their influences is a primary issue. This section begins to show how these vital matters can be handled, taking the case of $p = 2$ factors as a starting point.

The section begins with a discussion of two-way factorial data structures and graphical and qualitative analyses of these. Then the concepts of main effects and interactions are defined and the estimation of these quantities considered. Finally, there is a brief discussion of the fitting of simplified or reduced models to balanced two-way factorial data.

6.2.1 Graphical and Qualitative Analysis for Complete Two-Way Factorial Data

This section concerns situations where on the proverbial black box of Figure 6.1 there are two "knobs" under the control of an experimenter. That is, two-factor experimentation is treated. So for ease of communication, let Factor A and Factor B be two generic names for factors that potentially impact some process output of interest, y. In cases where levels of Factors A and B are defined in terms of values of quantitative process variables x_1 and x_2 respectively, the tool of multiple regression analysis provides a powerful method of data analysis. That tool is applied to multifactor process improvement problems in Sections 7.2 and 7.3 of this book. But here, methods of analysis that can be used even when one or both of Factors A and B is qualitative will first be considered.

The most straightforward analyses of two-factor studies are possible in cases where Factor A has levels $1, 2, \ldots, I$, Factor B has levels $1, 2, \ldots, J$, and every possible combination of a level of A and a level of B is represented in the data set. Such a data set will be called an **$I \times J$ complete factorial** data set. Figure 6.4 illustrates the fact that the $I \times J$ different sets of process conditions in a complete factorial study can be laid out in a rectangular two-way table. Rows correspond to levels of Factor A, columns correspond to levels of Factor B, and the factorial is "complete" in the sense that there are data available corresponding to all $I \times J$ "cells" in the table.

Corresponding to Figure 6.4, let

$$y_{ijk} = \text{the } k\text{th observation from level } i \text{ of Factor A and level } j \text{ of Factor B},$$

$$\bar{y}_{ij} = \frac{1}{n_{ij}} \sum_k y_{ijk}$$

$$= \text{the sample mean from level } i \text{ of Factor A and level } j \text{ of Factor B},$$

and

$$s_{ij}^2 = \frac{1}{n_{ij} - 1} \sum_k (y_{ijk} - \bar{y}_{ij})^2$$

$$= \text{the sample variance from level } i \text{ of Factor A and level } j \text{ of Factor B},$$

where n_{ij} is the sample size corresponding to level i of A and level j of B. It is important to realize that except for the introduction of (i, j) double subscripting to recognize the two-factor structure, there is nothing new in these formulas. One is simply naming the various observations from a data structure like that indicated in Figure 6.4 and the corresponding $r = I \times J$ sample means and variances.

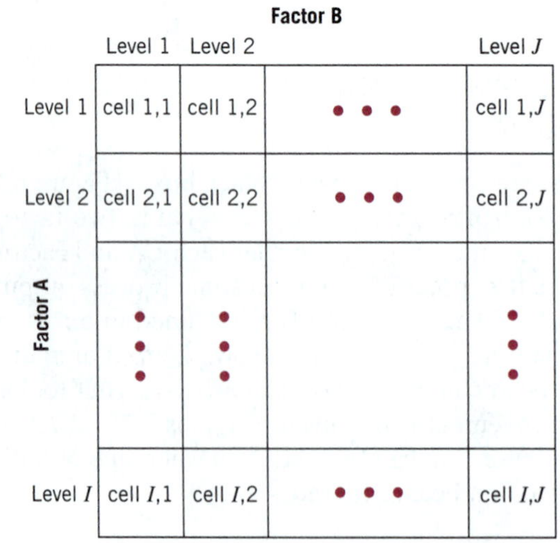

Figure 6.4
$r = I \times J$ combinations in a two-way factorial study.

Table 6.3 Shear Strengths and Summary Statistics for $I \times J = 2 \times 3$ Combinations of Cooling Method and Amount of Antimony (MPa)

Factor A Cooling	i	Factor B Sb (% weight)	j	Strength, y	$\bar{y}_{ij}$	s_{ij}
H_2O quench	1	3	1	18.6, 19.5, 19.0	19.033	.451
H_2O quench	1	5	2	22.3, 19.5, 20.5	20.767	1.419
H_2O quench	1	10	3	15.2, 17.1, 16.6	16.300	.985
oil quench	2	3	1	20.0, 20.9, 20.4	20.433	.451
oil quench	2	5	2	20.9, 22.9, 20.6	21.467	1.250
oil quench	2	10	3	16.4, 19.0, 18.1	17.833	1.320

Example 6.2 **Two-Way Analysis of Solder Joint Strengths (Example 6.1 Revisited).** Table 6.3 is essentially a repeat of Table 6.1, giving the joint strength data of Tomlinson and Cooper and corresponding sample means and standard deviations. What is new here is only that in place of naming $r = 6$ sets of process conditions with indices $i = 1, 2, 3, 4, 5,$ and 6, double subscripts (i, j) corresponding to the 2×3 different combinations of $I = 2$ different cooling methods and $J = 3$ different amounts of antimony are used.

Figure 6.5 shows the six sample means and standard deviations of Table 6.3 laid out in a 2×3 table, with rows corresponding to levels of Factor A (cooling method) and columns corresponding to levels of Factor B (% Sb).

Most basically, data from a complete two-way factorial study are simply observations from $r = I \times J$ different sets of process conditions, and all of the material from the previous section can be brought to bear (as it was in Example 6.1) on their analysis. But in order to explicitly acknowledge the two-way structure, it is common to not only double subscript the samples (for level of A and level of

Factor A Cooling Method	Factor B Sb % Weight: 3	5	10
H_2O	19.033 .451	20.767 1.419	16.300 .985
Oil	20.433 .451	21.467 1.250	17.833 1.320

Figure 6.5
Sample means and standard deviations from Table 6.3 (MPa).

B), but to also double subscript the theoretical mean responses as well, writing μ_{ij} instead of simply the μ_i used in Section 6.1. And so, using the obvious subscript notation for ϵ's the one-way model assumptions (6.1) are rewritten for the two-way factorial context as

Two-Way Model Equation

$$y_{ijk} = \mu_{ij} + \epsilon_{ijk} ,$$

for $\mu_{11}, \mu_{12}, \ldots, \mu_{1J}, \mu_{21}, \ldots, \mu_{2J}, \ldots, \mu_{I1}, \ldots, \mu_{IJ}$ (unknown) means and $\epsilon_{111}, \ldots, \epsilon_{11n_{11}}, \epsilon_{121}, \ldots, \epsilon_{12n_{12}}, \ldots, \epsilon_{IJ1}, \ldots, \epsilon_{IJn_{IJ}}$ independent normal random variables with mean 0 and (unknown) standard deviation σ.

Interaction Plot

Finding interpretable patterns in how the means μ_{ij} change with i and j (with level of Factor A and level of Factor B) is usually a primary goal in a two-way factorial analysis. A very effective first step in achieving that goal is to make a plot of the $I \times J$ sample means $\bar{y}_{ij}$ versus (say) level of Factor B, connecting points having a common level of (say) Factor A with line segments. Such a plot is usually called an **interaction plot** (although the terminology is not terribly descriptive or helpful). It is useful to indicate on such a plot the precision with which the mean responses are known. This can be done by using confidence limits for the μ_{ij} to make **error bars** around the sample means. The basic individual confidence limits (6.9) can be used, or alternatively (and probably more appropriately since one will view all $I \times J$ error bars at once) the P-R simultaneous confidence limits prescribed in display (6.12) can be used.

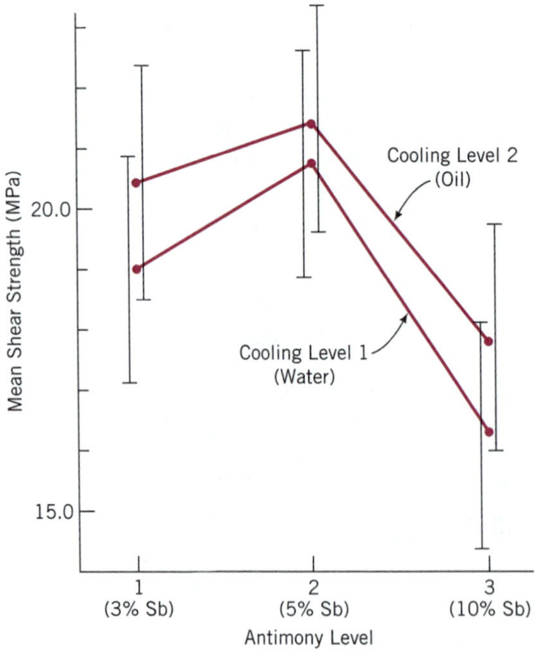

Figure 6.6
Interaction plot for the solder joint strength study.

Example 6.2 continued. Figure 6.6 is an interaction plot for the solder joint strength data. In Example 6.1, simultaneous 95% confidence limits for the six method means μ_{ij} were found to be of the form $\bar{y}_{ij} \pm 1.887$ MPa, and this 1.887 MPa figure has been used to make the error bars in the figure.

Figure 6.6 gives a helpful summary of what the data say about how cooling method and amount of antimony impact solder joint strength. That is, there are strong hints that (1) a large amount of antimony in the solder is not good (in terms of producing large joint strength), (2) oil-quenched joints are stronger than water-quenched joints, and (3) patterns of response to changes in a given factor (A or B) are consistent across levels of the other factor (B or A). But these conclusions are somewhat clouded by the relatively large error bars on the plot (indicating imprecision in one's knowledge about long-run mean joint strengths). The indicated uncertainty doesn't seem large enough to really draw into question the importance of amount of antimony in determining joint strength. But exactly how the two different cooling methods compare is perhaps somewhat murkier. And the extent to which a change in antimony levels possibly produces different changes in mean joint strength for the two different cooling methods is nearly completely clouded by the "experimental noise level" pictured by the error bars.

In two-way factorial experiments where it turns out that one has good precision (small error bars) for estimating all $I \times J$ means, an interaction plot in the style of Figure 6.6 can be all that is really required to describe how the two factors impact y. But where one has learned less about the individual means, finer/more quantitative analyses are helpful. And the next subsection discusses such analyses, both because they can be important in two-factor studies and because they set the pattern for what is done in the analysis of p-way factorial data for $p > 2$.

6.2.2 Defining and Estimating Effects

Figure 6.6 for the solder joint strength data hints at "an effect" of cooling method, Factor A, on y. If one can make a sensible quantitative definition of exactly what the effect of a level of Factor A might mean, the possibility exists of doing inference (giving a confidence interval) for it. To those ends, notations for row and column averages of both μ_{ij}'s and corresponding $\bar{y}_{ij}$'s are needed. These are indicated in Figure 6.7 for the specific case of the solder study and detailed in general in formulas (6.15) through (6.20).

Let

$$\bar{y}_{i.} = \frac{1}{J}\sum_{j} \bar{y}_{ij} = \text{the simple average of the row } i \text{ sample means} \quad (6.15)$$

Average of Row i Sample Means

Factor A Cooling Method	Factor B Sb % Weight: 3	5	10	
H_2O	$\bar{y}_{11}$ μ_{11}	$\bar{y}_{12}$ μ_{12}	$\bar{y}_{13}$ μ_{13}	$\bar{y}_{1.}$ $\mu_{1.}$
Oil	$\bar{y}_{21}$ μ_{21}	$\bar{y}_{22}$ μ_{22}	$\bar{y}_{23}$ μ_{23}	$\bar{y}_{2.}$ $\mu_{2.}$
	$\bar{y}_{.1}$ $\mu_{.1}$	$\bar{y}_{.2}$ $\mu_{.2}$	$\bar{y}_{.3}$ $\mu_{.3}$	$\bar{y}_{..}$ $\mu_{..}$

Figure 6.7
Two-way layout of $\bar{y}_{ij}, \bar{y}_{i.}, \bar{y}_{.j}, \bar{y}_{..}, \mu_{ij}, \mu_{i.}, \mu_{.j}, \mu_{..}$ for the 2×3 solder joint study.

and correspondingly

$$\mu_{i.} = \frac{1}{J}\sum_{j} \mu_{ij} = \text{the simple average of the row } i \text{ theoretical means.} \quad (6.16)$$

Similarly, take

Average of Column j Sample Means

$$\bar{y}_{.j} = \frac{1}{I}\sum_{i} \bar{y}_{ij} = \text{the simple average of the column } j \text{ sample means} \quad (6.17)$$

and correspondingly

$$\mu_{.j} = \frac{1}{I}\sum_{i} \mu_{ij} = \text{the simple average of the column } j \text{ theoretical means.} \quad (6.18)$$

Finally, use the notation

Average of All $I \times J$ Sample Means

$$\bar{y}_{..} = \frac{1}{IJ}\sum_{i,j} \bar{y}_{ij} = \text{the average of all } I \times J \text{ cell sample means} \quad (6.19)$$

and correspondingly

$$\mu_{..} = \frac{1}{IJ}\sum_{i,j} \mu_{ij} = \text{the average of all } I \times J \text{ theoretical means.} \quad (6.20)$$

		Factor B Sb % Weight 3	5	10	
Factor A Cooling Method	H_2O	$\bar{y}_{11} = 19.033$	$\bar{y}_{12} = 20.767$	$\bar{y}_{13} = 16.300$	$\bar{y}_{1.} = 18.700$
	Oil	$\bar{y}_{21} = 20.433$	$\bar{y}_{22} = 21.467$	$\bar{y}_{23} = 17.833$	$\bar{y}_{2.} = 19.911$
		$\bar{y}_{.1} = 19.733$	$\bar{y}_{.2} = 21.117$	$\bar{y}_{.3} = 17.067$	$\bar{y}_{..} = 19.306$

Figure 6.8
Cell, marginal, and overall means from the data in Table 6.3 (MPa).

Example 6.2 continued. Averaging $\bar{y}_{ij}$ values across rows, down columns, and over the whole table summarized in Figure 6.5 produces the $\bar{y}_{i.}, \bar{y}_{.j}$ values and the value of $\bar{y}_{..}$ displayed in Figure 6.8 along with the six cell means. Of course, since the long run mean strengths μ_{ij} are not known, it is not possible to present an analog of Figure 6.8 giving numerical values for the $\mu_{i.}$, $\mu_{.j}$, and $\mu_{..}$.

The row and column means in Figure 6.8 suggest a way of measuring the direct or main effects of Factors A and B on y. One might base comparisons of levels of Factor A on row averages of mean responses and comparisons of levels of Factor B on column averages of mean responses. This thinking leads to definitions of main effects and their estimated or fitted counterparts.

Definition 6.1. The (theoretical) **main effect of Factor A at level i** in a complete $I \times J$ two-way factorial is

$$\alpha_i = \mu_{i.} - \mu_{..}.$$

Definition 6.2. The (estimated or) **fitted main effect of Factor A at level i** in a complete $I \times J$ two-way factorial is

$$a_i = \bar{y}_{i.} - \bar{y}_{..}.$$

Definition 6.3. The (theoretical) **main effect of Factor B at level j** in a complete $I \times J$ two-way factorial is

$$\beta_j = \mu_{.j} - \mu_{..}.$$

Definition 6.4. The (estimated or) **fitted main effect of Factor B at level j** in a complete $I \times J$ two-way factorial is

$$b_j = \bar{y}_{.j} - \bar{y}_{..}.$$

A main effects are row averages of cell means minus a grand average, while B main effects are column averages of cell means minus a grand average. And a very small amount of algebra makes it obvious that differences in main effects of a factor are corresponding differences in row or column averages. That is, from the definitions

$$\alpha_i - \alpha_{i'} = \mu_{i.} - \mu_{i'.} \quad \text{and} \quad a_i - a_{i'} = \bar{y}_{i.} - \bar{y}_{i'.}, \tag{6.21}$$

while

$$\beta_j - \beta_{j'} = \mu_{.j} - \mu_{.j'} \quad \text{and} \quad b_j - b_{j'} = \bar{y}_{.j} - \bar{y}_{.j'}. \tag{6.22}$$

Example 6.2 continued. Some arithmetic applied to the row and column average means in Figure 6.8 shows that the fitted main effects of cooling method and antimony content for the data of Table 6.3 are

$$a_1 = 18.700 - 19.30\bar{5} = -.60\bar{5} \quad \text{and} \quad a_2 = 19.91\bar{1} - 19.30\bar{5} = .60\bar{5}$$

and

$$b_1 = 19.73\bar{3} - 19.30\bar{5} = .42\bar{7},\ b_2 = 21.11\bar{6} - 19.30\bar{5} = 1.81\bar{1}, \quad \text{and}$$
$$b_3 = 17.06\bar{6} - 19.30\bar{5} = -2.23\bar{8}.$$

More decimals have been displayed in Example 6.2 than are really justified on the basis of the precision of the original data. This has been done for the purpose of pointing out (without clouding the issue with round-off error) that $a_1 + a_2 =$

(a)

Level of A	Level of B: 1	2	3	
1	2	6	4	4
2	1	5	3	3
	1.5	5.5	3.5	

(b)

Level of A	Level of B: 1	2	3	
1	3	2	7	4
2	0	9	0	3
	1.5	5.5	3.5	

Figure 6.9
Two hypothetical sets of means μ_{ij} for 2×3 factorials that share the same row and column averages.

0 and $b_1 + b_2 + b_3 = 0$. These relationships are no accident. It is an algebraic consequence of the form of Definitions 6.1 through 6.4 that

$$\sum_i a_i = 0 \quad \text{and} \quad \sum_j b_j = 0, \quad \text{and similarly,} \quad \sum_i \alpha_i = 0 \quad \text{and} \quad \sum_j \beta_j = 0. \tag{6.23}$$

Both fitted and theoretical main effects of any factor sum to 0 over all possible levels of that factor. Notice that, in particular, relationships (6.23) imply that when a factor has only two levels, the two (theoretical or fitted) main effects must have the same magnitude but opposite signs.

The main effects of Factors A and B do not in general "tell the whole story" about how means μ_{ij} depend upon i and j. Figure 6.9 specifies two hypothetical sets of means μ_{ij} for 2×3 factorials that share the same row and column averages and therefore the same main effects. Interaction plots for the two sets of means are given in Figure 6.10.

The qualitative characters of the two plots in Figure 6.10 are substantially different. The first graph represents a situation that is fundamentally simpler than the second. On the first graph, as one changes level of Factor B, the same change in mean response is produced regardless of whether one is looking at the first level of Factor A or at the second level. In some sense it is legitimate to think about how Factor B impacts the mean response as independent of the level of A under discussion. The same is not true for the second plot. On the second graph, what happens to mean response when one changes level of B depends strongly on what level of A is being considered. That implies, for example, that if one were interested in maximizing mean response, the preferred level of B would depend on the level of A being used. No simple blanket recommendation like "level 2 of Factor B is best" can be made in situations like that pictured on the second plot.

A way of describing the feature of the first plot in Figure 6.10 that makes it simple, is to say that the plot exhibits **parallelism** between the profiles (across levels of B) of mean response for various levels of A. It is important to be able to measure the degree to which a set of means departs from the kind of simple parallelism seen on the first plot, and this can be done easily in terms of quantities

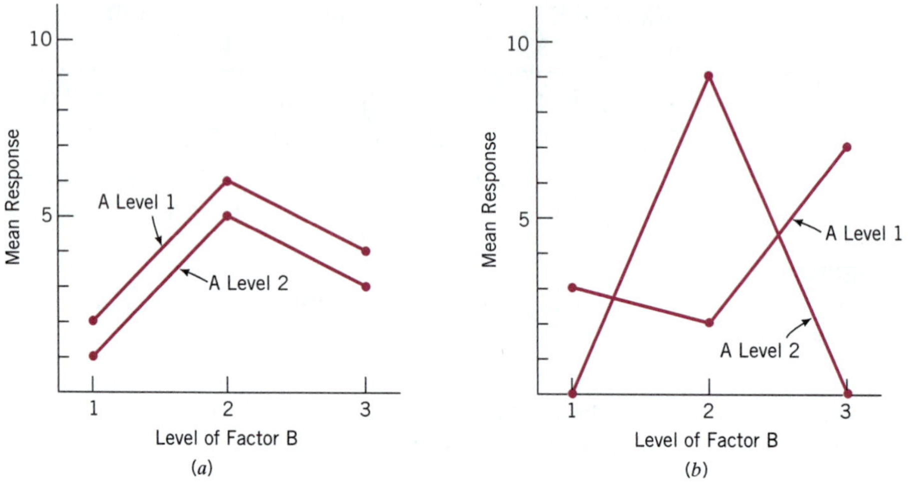

Figure 6.10
Interaction plots for the hypothetical sets of means in Figure 6.9.

already introduced in this section. That is, it turns out that parallelism on an interaction plot of μ_{ij}'s is equivalent to the possibility that for all (i, j), combinations

$$\mu_{ij} = \mu_{..} + \alpha_i + \beta_j .$$

This is exactly the eventuality that the main effects (and the grand mean) completely summarize the μ_{ij}. Departures from this simple state of affairs can then be measured by taking the difference between the left- and right-hand sides of this simple ideal relationship. That is, one is led to two more definitions.

Definition 6.5. The (theoretical) **interaction of Factor A at level *i* and Factor B at level *j*** in a complete $I \times J$ two-way factorial is

$$\alpha\beta_{ij} = \mu_{ij} - (\mu_{..} + \alpha_i + \beta_j).$$

Definition 6.6. The (estimated or) **fitted interaction of Factor A at level *i* and Factor B at level *j*** in a complete $I \times J$ two-way factorial is

$$ab_{ij} = \bar{y}_{ij} - (\bar{y}_{..} + a_i + b_j).$$

To the extent that parallelism or lack thereof can be clearly identified on an interaction plot like Figure 6.6, data-based examination of the possibility of important AB interactions in a two-factor study can proceed graphically. But a more quantitative look at the issue must begin with computation of the fitted interactions defined in Definition 6.6. In this regard, it is worth noting that there is an alternative "computational" formula for fitted interactions that is sometimes useful. That is,

$$ab_{ij} = \bar{y}_{ij} - \bar{y}_{i.} - \bar{y}_{.j} + \bar{y}_{..}$$

Computational Formula for Fitted Interactions

Example 6.2 continued. Returning again to the solder joint strength example, Table 6.4 organizes calculations of the six fitted interactions derived from the data of Table 6.3.

It is worth noting that the largest of the fitted interactions in Table 6.4 is smaller than even the smallest fitted main effect calculated earlier. That is numerical evidence that the lack of parallelism in Figure 6.6 is in some sense smaller than the gap between the oil- and water-quench profiles and the differences between the observed strengths for the different amounts of antimony.

Once again, more decimal places are displayed in Table 6.4 than are really justified on the basis of the precision of the original data. This has been done so that it can be clearly seen that $ab_{11} + ab_{21} = 0$, $ab_{12} + ab_{22} = 0$, $ab_{13} + ab_{23} = 0$, $ab_{11} + ab_{12} + ab_{13} = 0$, and $ab_{21} + ab_{22} + ab_{23} = 0$. These relationships are not special to the particular data set. Fitted interactions sum to 0 down any column or across any row in any two-way factorial. It is an algebraic consequence of Definitions 6.5 and 6.6 that

$$\sum_i ab_{ij} = 0 \quad \text{and} \quad \sum_j ab_{ij} = 0, \quad \text{and similarly that}$$

$$\sum_i \alpha\beta_{ij} = 0 \quad \text{and} \quad \sum_j \alpha\beta_{ij} = 0. \tag{6.24}$$

Table 6.4 Calculation of the Fitted Interactions for the Solder Joint Strength Study

i	j	$\bar{y}_{ij}$	$\bar{y}_{..} + a_i + b_j$	$ab_{ij} = \bar{y}_{ij} - (\bar{y}_{..} + a_i + b_j)$
1	1	$19.03\bar{3}$	$19.30\bar{5} + (-.60\bar{5}) + .42\bar{7} = 19.12\bar{7}$	$-.09\bar{4}$
1	2	$20.76\bar{6}$	$19.30\bar{5} + (-.60\bar{5}) + 1.81\bar{1} = 20.51\bar{1}$	$.25\bar{5}$
1	3	16.300	$19.30\bar{5} + (-.60\bar{5}) + (-2.23\bar{8}) = 16.46\bar{1}$	$-.16\bar{1}$
2	1	$20.43\bar{3}$	$19.30\bar{5} + .60\bar{5} + .42\bar{7} = 20.33\bar{8}$	$.09\bar{4}$
2	2	$21.46\bar{6}$	$19.30\bar{5} + .60\bar{5} + 1.81\bar{1} = 21.72\bar{2}$	$-.25\bar{5}$
2	3	$17.83\bar{3}$	$19.30\bar{5} + .60\bar{5} + (-2.23\bar{8}) = 17.67\bar{2}$	$.16\bar{1}$

To go beyond simply computing single-number estimates of (main and interaction) effects, to making confidence intervals for these, it is only necessary to realize that effects are linear combinations of the means μ_{ij} (L's). And not surprisingly, the fitted effects are the corresponding linear combinations of the sample means $\bar{y}_{ij}$ (the corresponding $\hat{L}$'s). That implies (repeating formula (6.8) in two-way factorial/double subscript notation) that

Confidence Limits for L in Two-Way Notation

$$\hat{L} \pm t s_{\mathrm{P}} \sqrt{\frac{c_{11}^2}{n_{11}} + \cdots + \frac{c_{1J}^2}{n_{1J}} + \frac{c_{21}^2}{n_{21}} + \cdots + \frac{c_{IJ}^2}{n_{IJ}}}, \tag{6.25}$$

can be used to make confidence intervals for the main effects and interactions. The only real question in applying formula (6.25) to the estimation of an effect is what form the sum of squared coefficients over sample sizes takes. That is, one needs to know how to compute the sum

$$\sum_{i,j} \frac{c_{ij}^2}{n_{ij}} \tag{6.26}$$

that goes under the root in formula (6.25). It is possible to derive formulas for these sums where the L's involved are two-way factorial effects or differences in such effects. Table 6.5 collects the very simple formulas that are appropriate when data are **balanced** (all n_{ij} are equal to some number m). Table 6.6 gives the more complicated-looking formulas for the quantities (6.26) needed when the n_{ij} vary.

Table 6.5 Balanced Data Formulas for the Quantities $\sum_{i,j} \frac{c_{ij}^2}{n_{ij}}$ Needed to Make Confidence Intervals for Effects in Two-Way Factorials (All $n_{ij} = m$)

L	$\hat{L}$	$\sum_{i,j} \frac{c_{ij}^2}{n_{ij}}$
$\alpha\beta_{ij}$	ab_{ij}	$\frac{(I-1)(J-1)}{mIJ}$
α_i	a_i	$\frac{I-1}{mIJ}$
$\alpha_i - \alpha_{i'}$	$a_i - a_{i'}$	$\frac{2}{mJ}$
β_j	b_j	$\frac{J-1}{mIJ}$
$\beta_j - \beta_{j'}$	$b_j - b_{j'}$	$\frac{2}{mI}$

Table 6.6 General Formulas for the Quantities $\sum_{i,j} \frac{c_{ij}^2}{n_{ij}}$ Needed to Make Confidence Intervals for Effects in Two-Way Factorials

L	$\hat{L}$	$\sum_{i,j} \frac{c_{ij}^2}{n_{ij}}$
$\alpha\beta_{ij}$	ab_{ij}	$\left(\frac{1}{IJ}\right)^2\left(\frac{(I-1)^2(J-1)^2}{n_{ij}} + (I-1)^2\sum_{j'\neq j}\frac{1}{n_{ij'}} + (J-1)^2\sum_{i'\neq i}\frac{1}{n_{i'j}} + \sum_{i'\neq i, j'\neq j}\frac{1}{n_{i'j'}}\right)$
α_i	a_i	$\left(\frac{1}{IJ}\right)^2\left((I-1)^2\sum_{j}\frac{1}{n_{ij}} + \sum_{i'\neq i, j}\frac{1}{n_{i'j}}\right)$
$\alpha_i - \alpha_{i'}$	$a_i - a_{i'}$	$\frac{1}{J^2}\left(\sum_{j}\frac{1}{n_{ij}} + \sum_{j}\frac{1}{n_{i'j}}\right)$
β_j	b_j	$\left(\frac{1}{IJ}\right)^2\left((J-1)^2\sum_{i}\frac{1}{n_{ij}} + \sum_{i, j'\neq j}\frac{1}{n_{ij'}}\right)$
$\beta_j - \beta_{j'}$	$b_j - b_{j'}$	$\frac{1}{I^2}\left(\sum_{i}\frac{1}{n_{ij}} + \sum_{i}\frac{1}{n_{ij'}}\right)$

Example 6.2 continued. Once again consider the solder joint strength example and now the problem of making confidence intervals for the various factorial effects, beginning with the interactions $\alpha\beta_{ij}$. The fitted interactions ab_{ij} are collected in Table 6.4. Use of the formula (6.25) allows one to associate "plus or minus values" with these estimates. In Example 6.1, the pooled estimate of σ from the joint strength data was found to be $s_P = 1.056$ MPa with $\nu = 12$ associated degrees of freedom. Since the data in Table 6.3 are *balanced* factorial data with $I = 2, J = 3$, and $m = 3$, using the first line of Table 6.5 it follows that 95% two-sided confidence limits for the interaction $\alpha\beta_{ij}$ are

$$ab_{ij} \pm 2.179(1.056)\sqrt{\frac{(2-1)(3-1)}{3(2)(3)}}, \quad \text{that is} \quad ab_{ij} \pm .767 \text{ MPa.}$$

Notice then that all six of the intervals for interactions (centered at the values in Table 6.4) contain positive numbers, negative numbers *and* 0. It is possible that the $\alpha\beta_{ij}$ are all essentially 0 and correspondingly the lack of parallelism on Figure 6.6 is no more than a manifestation of experimental error. By the standard of these 95% individual confidence limits, the magnitude of the uncertainty associated with any fitted interaction exceeds that of the interaction itself and the apparent lack of parallelism is "in the noise range."

Next, using the second row of Table 6.5, it follows that 95% two-sided confidence limits for the cooling method main effects are

$$a_i \pm 2.179(1.056)\sqrt{\frac{2-1}{3(2)(3)}}, \quad \text{that is} \quad a_i \pm .542 \text{ MPa}.$$

This is in accord with the earlier more qualitative analysis of the joint strength data made on the basis of Figure 6.6 alone. The calculation here, together with the facts that $a_1 = -.606$ and $a_2 = .606$, shows that one can be reasonably sure the main effect of water quench is negative and the main effect of oil quench is positive. The oil-quench joint strengths are on average larger than the water-quench strengths. But the call is still a relatively "close" one. The $\pm.542$ value is nearly as large as the fitted effects themselves.

Finally, as an illustration of the use of formula (6.25) in the comparison of main effects, consider the estimation of differences in antimony amount main effects, $\beta_j - \beta_{j'}$. Using the last row of Table 6.5, 95% two-sided confidence limits for differences in antimony main effects are

$$b_j - b_{j'} \pm 2.179(1.056)\sqrt{\frac{2}{3(2)}}, \quad \text{that is} \quad b_j - b_{j'} \pm 1.328 \text{ MPa}.$$

Recall that $b_1 = .428$, $b_2 = 1.811$, and $b_3 = -2.239$ and note that while b_1 and b_2 differ by less than 1.328 MPa, b_1 and b_3 differ by substantially more than 1.328, as do b_2 and b_3. This implies that while the evidence of a real difference between average strengths for levels 1 and 2 of antimony is not sufficient to allow one to make definitive statements, both antimony level 1 and antimony level 2 average joint strengths are clearly above that for antimony level 3. This conclusion is in accord with the earlier analysis based entirely on Figure 6.6. The differences between antimony levels are clearly more marked (and evident above the background/experimental variation) than the cooling method differences.

Example 6.3 Computing Factors from Table 6.6. As a way of illustrating the intended meaning of the components of the formulas in Table 6.6, consider a hypothetical 3×3 factorial where $n_{12} = 1$, $n_{33} = 1$, and all other n_{ij} are 2. For the (i, j)-pair (1,1), sums appearing in the table would be

$$\frac{1}{n_{11}} = .5,$$

$$\sum_{j' \neq 1} \frac{1}{n_{1j'}} = \frac{1}{n_{12}} + \frac{1}{n_{13}} = 1.0 + .5 = 1.5,$$

$$\sum_{i'\neq 1} \frac{1}{n_{i'1}} = \frac{1}{n_{21}} + \frac{1}{n_{31}} = .5 + .5 = 1.0,$$

$$\sum_{i'\neq 1, j'\neq 1} \frac{1}{n_{i'j'}} = \frac{1}{n_{22}} + \frac{1}{n_{23}} + \frac{1}{n_{32}} + \frac{1}{n_{33}} = .5 + .5 + .5 + 1.0 = 2.5,$$

$$\sum_{j} \frac{1}{n_{1j}} = \frac{1}{n_{11}} + \frac{1}{n_{12}} + \frac{1}{n_{13}} = .5 + 1.0 + .5 = 2.0,$$

$$\sum_{i} \frac{1}{n_{i1}} = \frac{1}{n_{11}} + \frac{1}{n_{21}} + \frac{1}{n_{31}} = .5 + .5 + .5 = 1.5,$$

$$\sum_{i'\neq 1, j} \frac{1}{n_{i'j}} = \frac{1}{n_{21}} + \frac{1}{n_{22}} + \frac{1}{n_{23}} + \frac{1}{n_{31}} + \frac{1}{n_{32}} + \frac{1}{n_{33}}$$
$$= .5 + .5 + .5 + .5 + .5 + 1.0 = 3.5,$$

and

$$\sum_{i, j'\neq 1} \frac{1}{n_{ij'}} = \frac{1}{n_{12}} + \frac{1}{n_{13}} + \frac{1}{n_{22}} + \frac{1}{n_{23}} + \frac{1}{n_{32}} + \frac{1}{n_{33}}$$
$$= 1.0 + .5 + .5 + .5 + .5 + 1.0 = 4.0.$$

Using sums of these types, intervals for factorial effects can be computed even in this sort of unbalanced data situation.

Intervals for two-way factorial effects (and their differences) made using display (6.25) and sums computed as indicated in Tables 6.5 and 6.6 carry individual confidence coefficients. It is thus of interest to point out that there are also **simultaneous** interval methods appropriate for two-way factorial situations. For example, a Tukey method for simultaneous estimation of all differences $\alpha_i - \alpha_{i'}$ (or all differences $\beta_j - \beta_{j'}$) basically involves replacing t in formula (6.25) with $q^*/\sqrt{2}$, where q^* is selected from Tables A.11a&b for comparing I (or J) means and is based on $n - IJ$ degrees of freedom. (The reader is referred to Section 8.1 of Vardeman's *Statistics for Engineering Problem Solving* for a more complete discussion of this method.)

6.2.3 Fitting and Checking Simplified Models for Balanced Two-Way Factorial Data

The possibility that interactions in a two-way factorial situation are negligible is one that brings important simplification to interpreting how Factors A and B affect the response y. In the absence of important interactions, one may think of A and B acting on y more or less "independently" or "separately." And if, in addition, the A main effects are negligible, one can think about the "two-knob black box system" as having only one knob (the B knob) that really does anything in terms of changing y.

The confidence intervals for two-way factorial effects introduced in this section are important tools for investigating whether some effects can indeed be ignored. A further step in this direction is to derive *fitted values* for y under assumptions that some of the factorial effects are negligible and to use these to compute *residuals*. The idea here is very much like what is done in regression analysis. Faced with a large list of possible predictor variables, one goal of standard regression analysis is to find an equation involving only a few of those predictors that does an adequate job of describing the response variable. In the search for such an equation, y values predicted by a candidate equation are subtracted from observed y values to produce residuals, and these are plotted in various ways looking for possible problems with the candidate.

The business of finding predicted values for y under an assumption that some of the two-way factorial effects are 0 is in general a problem that must be addressed using a regression program and what are known as "dummy variables." And the matter is subtle enough that treating it here is not feasible. The reader is instead referred for the general story to books on regression analysis like *Applied Linear Statistical Models* by Neter, Kutner, Nachtsheim, and Wasserman and intermediate-level books on statistical methods. What *can* be done here is to point out that in the special case that factorial data are balanced, appropriate fitted values can be obtained by simply adding to the grand sample mean fitted effects corresponding to those effects that one does not wish to assume are negligible.

That is, for balanced data, under the assumption that all $\alpha\beta_{ij}$ are 0, an appropriate estimator of the mean response when Factor A is at level i and Factor B is at level j (a fitted value for any y_{ijk}) is

Balanced Data "No-Interaction" Fitted Values

$$\hat{y}_{ijk} = \bar{y}_{..} + a_i + b_j. \tag{6.27}$$

Further, for balanced data, under the assumption that all $\alpha\beta_{ij}$ are 0 and all α_i are also 0, an appropriate estimator of the mean response when Factor B is at level j (a fitted value for any y_{ijk}) is

Balanced Data "B Effects Only" Fitted Values

$$\hat{y}_{ijk} = \bar{y}_{..} + b_j. \tag{6.28}$$

And again for balanced data, under the assumption that all $\alpha\beta_{ij}$ are 0 and all β_j are also 0, an appropriate estimator of the mean response when Factor A is at level i (a fitted value for any y_{ijk}) is

Balanced Data "A Effects Only" Fitted Values

$$\hat{y}_{ijk} = \bar{y}_{..} + a_i. \tag{6.29}$$

Using one of the relationships (6.27) through (6.29), residuals are then defined as differences between observed and fitted values

Residuals

$$e_{ijk} = y_{ijk} - \hat{y}_{ijk}. \tag{6.30}$$

It is hopefully clear that the residuals defined in equation (6.30) are not the same as those defined in Section 6.1 and used there to check on the reasonableness of the basic one-way normal model assumptions. The residuals (6.30) are for a more specialized model, one of "no interactions," "B effects only", or "A effects only," depending upon which of equations (6.27) through (6.29) is used to find fitted values.

Once residuals (6.30) have been computed, they can be plotted in the same ways that one plots residuals in regression contexts. If the corresponding simplified model for y is a good one, residuals should "look like noise" and carry no obvious patterns or trends (that indicate that something important has been missed in modeling the response). One hopes to see a fairly linear normal plot of residuals, and hopes for "trendless/random scatter with constant spread" plots of residuals against levels of Factors A and B. Departures from these expectations draw into doubt the appropriateness of the reduced or simplified description of y.

Example 6.2 continued. The earlier analysis of the solder joint strength data suggests that a no-interaction description of joint strength might be tenable. To further investigate the plausibility of this, consider the computation and plotting of residuals based on fitted values (6.27). Table 6.7 shows the calculation of the six fitted values (6.27) and lists the 18 corresponding residuals for the no-interaction model (computed from the raw data listed in Table 6.3).

Figure 6.11 is a normal plot of the 18 residuals listed in Table 6.7, and Figures 6.12 and 6.13 are respectively plots of residuals versus level of A and then against

Table 6.7 Fitted Values and Residuals for the No-Interaction Model of Solder Joint Strength

i	j	$\hat{y}_{ijk} = \bar{y}_{..} + a_i + b_j$	$e_{ijk} = y_{ijk} - \hat{y}_{ijk}$
1	1	$\hat{y}_{11k} = 19.30\bar{5} + (-.60\bar{5}) + .42\bar{7} = 19.128$	$-.53, .37, -.13$
1	2	$\hat{y}_{12k} = 19.30\bar{5} + (-.60\bar{5}) + 1.81\bar{1} = 20.511$	$1.79, -1.01, -0.01$
1	3	$\hat{y}_{13k} = 19.30\bar{5} + (-.60\bar{5}) + (-2.23\bar{8}) = 16.461$	$-1.26, .64, .14$
2	1	$\hat{y}_{21k} = 19.30\bar{5} + .60\bar{5} + .42\bar{7} = 20.339$	$-.34, .56, .06$
2	2	$\hat{y}_{22k} = 19.30\bar{5} + .60\bar{5} + 1.81\bar{1} = 21.722$	$-.82, 1.18, -1.12$
2	3	$\hat{y}_{23k} = 19.30\bar{5} + .60\bar{5} + (-2.23\bar{8}) = 17.672$	$-1.27, 1.33, .428$

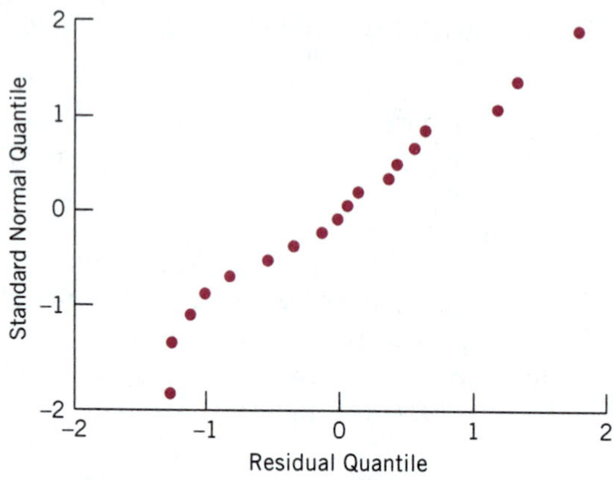

Figure 6.11
Normal plot of the residuals for a no-interaction model of solder joint strength.

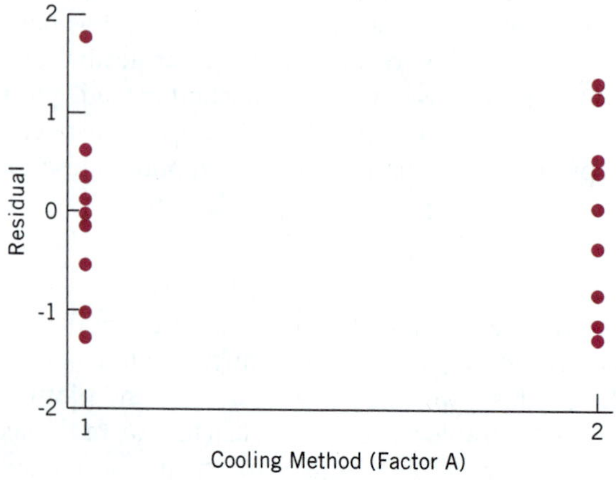

Figure 6.12
Plot of no-interaction residuals versus level of Factor A in the soldering study.

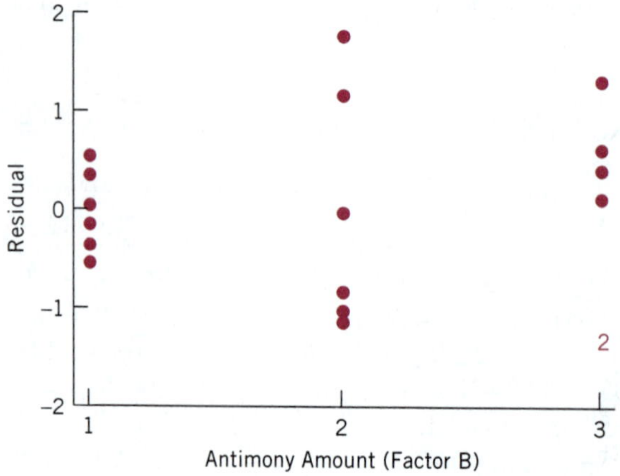

Figure 6.13
Plot of no-interaction residuals versus level of Factor B in the soldering study.

level of B. The normal plot is reasonably linear (except possibly for its extreme lower end) and the plot of residuals against level of Factor A is completely unremarkable. The plot against level of Factor B draws attention to the fact that the first level of antimony has residuals that seem somewhat smaller than those from the other two levels of antimony. But on the whole, the three plots offer no strong reason to dismiss a "normal distributions with constant variance *and no interactions between A and B*" model of joint strength.

As a final comment in this discussion of fitting and checking simplified models for two-way factorials, it should be noted that a two-way model with no AB interactions is sometimes called "the randomized block model." This is in honor of the fact that people often hope that it is appropriate in cases where one of the factors is not really of primary interest, but its levels instead constitute "blocks" of homogeneous conditions in which one might compare the effects of levels of the other factor. In such cases, one hopes that the response to changes in level of the factor of interest is similar across blocks. That is, one hopes that there are no interactions between A and B.

6.3 2^p FACTORIALS

The previous section began discussion of how to profitably conduct and analyze the results of process improvement experiments involving several factors. The subject there was the case of two factors. With the background provided by Section 6.2, it is time to consider the general case of $p \geq 2$ factors. This will be done primarily from the perspective of cases where each factor has only two levels, the $2 \times 2 \times \cdots \times 2$ or 2^p factorial studies. This may at first seem like an unacceptable restriction, but in fact it is not. As p grows, experimenting at more than two levels of many factors in a full p-way factorial arrangement quickly becomes practically impossible because of the large number of combinations involved. And there are some huge advantages associated with the 2^p factorials in terms of ease of data analysis.

The section begins with a general discussion of notation and how effects and fitted effects are defined in a p-way factorial. Then, methods for judging the statistical detectability of effects in the 2^p situation are presented, first for cases where there is some replication and then for cases where there is not. Next, the Yates algorithm for computing fitted effects for 2^p factorials is discussed. Finally, there is a brief discussion of fitting and checking models for balanced 2^p factorials that involve only some of the possible effects.

6.3.1 Notation and Defining Effects in p-Way Factorials

Consider now an instance of the generic process experimentation scenario represented in Figure 6.1 where there are p knobs under the experimenter's control. Naming the factors involved A, B, C, ... and supposing that they have respectively $I, J, K, \ldots$ possible levels, a **full factorial in the p factors** is a study where one has data from all $I \times J \times K \times \cdots$ different possible combinations of levels of these p factors. Figure 6.14 provides a visual representation of these possible combinations laid out in a three-dimensional rectangular array for the case of $p = 3$ factors. Then, Examples 6.4 and 6.5 introduce respectively $p = 3$ and $p = 4$-way factorial data sets that will be used in this section to illustrate methods of 2^p factorial analysis.

Example 6.4 **Packing Properties of Crushed T-61 Tabular Alumina Powder.** Ceramic Engineering researchers Leigh and Taylor, in their 1990 *Ceramic Bulletin* paper "Computer Generated Experimental Designs," present the results of a 2^3 factorial study on the packing properties of crushed T-61 tabular alumina powder. Densities, y, of the material were determined under several different measurement protocols. Two different "mesh sizes" of particles were employed, full flasks of the material of two different volumes were used and the flasks were subjected to one of two vibration conditions before calculating densities. This can be thought of as a three-way factorial situation where the factors and their levels are:

Factor A—Mesh Size	6 mesh vs. 60 mesh
Factor B—Flask	100 cc vs. 500 cc
Factor C—Vibration	none vs. yes

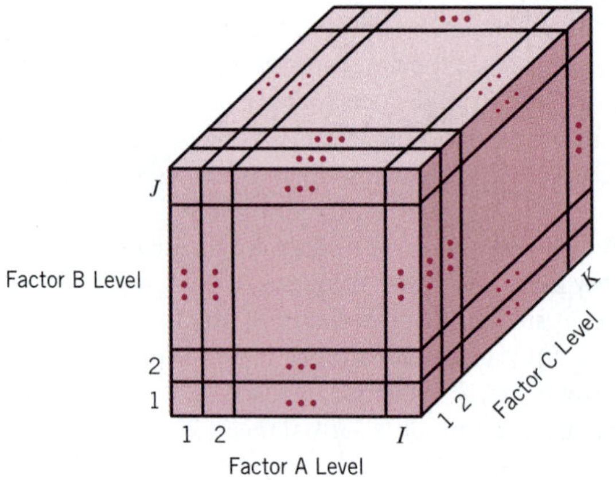

Figure 6.14
IJK cells for a three-way factorial.

Table 6.8 Crushed T-61 Tabular Alumina Powder Densities, Sample Means, and Sample Standard Deviations for 2^3 Different Measurement Protocols (g/cc)

Mesh	Flask	Vibration	Measured Density	$\bar{y}$	s
6	100	none	2.13, 2.15, 2.15, 2.19, 2.20	2.164	.030
60	100	none	1.96, 2.01, 1.91, 1.95, 2.00	1.966	.040
6	500	none	2.23, 2.19, 2.18, 2.21, 2.22	2.206	.021
60	500	none	1.88, 1.90, 1.87, 1.89, 1.89	1.886	.011
6	100	yes	2.16, 2.31, 2.32, 2.22, 2.35	2.272	.079
60	100	yes	2.29, 2.29, 2.23, 2.39, 2.18	2.276	.079
6	500	yes	2.16, 2.39, 2.30, 2.33, 2.43	2.322	.104
60	500	yes	2.35, 2.38, 2.26, 2.34, 2.34	2.334	.044

Table 6.8 gives the $m = 5$ densities reported by the researchers for each of the $r = 2 \times 2 \times 2$ measurement protocols and corresponding sample means and standard deviations.

The goal of a 2^3 factorial analysis of the data in Table 6.8 will be to identify structure that can be interpreted in terms of the individual and joint effects of the three factors Mesh, Flask, and Vibration.

Example 6.5 **Bond Pull-Outs on Dual In-Line Packages.** The article "An Analysis of Means for Attribute Data Applied to a 2^4 Factorial Design" by R. Zwickl that appeared in the Fall 1985 *ASQC Electronics Division Technical Supplement* describes a four-way factorial study done to help improve the manufacture of an electronic device called a dual in-line package. Counts were made of the numbers of bonds (out of 96) showing evidence of ceramic pull-out (small numbers are desirable) on devices made under all possible combinations of levels of four factors. The factors and their levels used in the study were:

Factor A—Ceramic Surface	unglazed vs. glazed
Factor B—Metal Film Thickness	normal vs. 1.5 times normal
Factor C—Annealing Time	normal vs. 4 times normal
Factor D—Prebond Clean	normal clean vs. no clean

Table 6.9 gives Zwickl's data. (We will suppose that the counts recorded in Table 6.9 occurred on one device made under each set of experimental conditions.)

Zwickl's data are unreplicated 2^4 factorial data. (In fact, they are attribute or count data. But for present purposes it will suffice to ignore this fact and treat them as if they were measurements obtained from 16 different process conditions.)

Table 6.9 Counts of Pull-Outs on Dual In-Line Packages Under 2^4 Sets of Experimental Conditions

A	B	C	D	Pull-Outs
unglazed	normal	normal	normal clean	9
glazed	normal	normal	normal clean	70
unglazed	1.5 ×	normal	normal clean	8
glazed	1.5 ×	normal	normal clean	42
unglazed	normal	4 ×	normal clean	13
glazed	normal	4 ×	normal clean	55
unglazed	1.5 ×	4 ×	normal clean	7
glazed	1.5 ×	4 ×	normal clean	19
unglazed	normal	normal	no clean	3
glazed	normal	normal	no clean	6
unglazed	1.5 ×	normal	no clean	1
glazed	1.5 ×	normal	no clean	7
unglazed	normal	4 ×	no clean	5
glazed	normal	4 ×	no clean	28
unglazed	1.5 ×	4 ×	no clean	3
glazed	1.5 ×	4 ×	no clean	6

The object of a 2^4 factorial analysis will be to find simple structure in the data that can be discussed in terms of the separate and joint effects of the four factors.

It should be obvious from analogy with what was done in the previous section that general notation for p-way factorial analyses will involve at least p subscripts, one for each of the factors. For example, for the case of $p = 3$ factors, one will write

y_{ijkl} = the lth observation at the ith level of A, the jth level of B, and the kth level of C,

μ_{ijk} = the long-run mean system response when A is at level i, B is at level j, and C is at level k,

n_{ijk} = the number of observations at level i of A, level j of B, and level k of C,

$\bar{y}_{ijk}$ = the sample mean system response when A is at level i, B is at level j, and C is at level k,

s_{ijk} = the sample standard deviation when A is at level i, B is at level j, C is at level k, and write the one-way model equation (6.1) in three-way factorial notation as

$$y_{ijkl} = \mu_{ijk} + \epsilon_{ijkl}.$$

Three-Way Model Equation

Further, the obvious "dot subscript" notation can be used to indicate averages of sample or long-run mean responses over the levels of factors "dotted out" of the notation. For example, for the case of $p = 3$ factors, one can write

$$\bar{y}_{.jk} = \frac{1}{I}\sum_{i}\bar{y}_{ijk}, \quad \bar{y}_{.j.} = \frac{1}{IK}\sum_{i,k}\bar{y}_{ijk}, \quad \bar{y}_{...} = \frac{1}{IJK}\sum_{i,j,k}\bar{y}_{ijk}$$

and so on.

The multiple subscript notation is in general necessary to write down technically correct formulas for p-way factorials. However, it is extremely cumbersome and unpleasant to use. One of the benefits of dealing primarily with 2^p problems is that something much more compact and workable can be done when all factors have only two levels. In 2^p contexts it is common to designate (arbitrarily if there is no reason to think of levels of a given factor as ordered) a "first" level of each factor as the "low" level and the "second" as the "high" level. (Often the shorthand "−" is used to designate a low level and the shorthand "+" is used to stand for a high level.) Combinations of levels of the factors can then be named by listing those factors which appear at their second or high levels. Table 6.10 illustrates this naming convention for the 2^3 case.

Armed with appropriate notation, one can begin to define effects and their fitted counterparts. The place to start is with the natural analogs of the two-way factorial main effects introduced in Definitions 6.1 through 6.4. These were row or column averages of cell means minus a grand average. That is, they were averages of cell means for a level of the factor under discussion minus a grand average. That same thinking can be applied in p-way factorials, provided one realizes that averaging must be done over levels of $(p - 1)$ other factors. The corresponding definitions will be given here for $p = 3$ factors with the understanding that the

Table 6.10 Naming Convention for 2^3 Factorials

Level of A	i	Level of B	j	Level of C	k	Combination Name
−	1	−	1	−	1	(1)
+	2	−	1	−	1	a
−	1	+	2	−	1	b
+	2	+	2	−	1	ab
−	1	−	1	+	2	c
+	2	−	1	+	2	ac
−	1	+	2	+	2	bc
+	2	+	2	+	2	abc

reader should be able to reason by analogy (simply adding some dot subscripts) to make definitions for cases with $p > 3$.

Definition 6.7. The (theoretical) **main effects of factors A, B, and C** in a complete three-way factorial are

$$\alpha_i = \mu_{i..} - \mu_{...}, \quad \beta_j = \mu_{.j.} - \mu_{...}, \quad \text{and} \quad \gamma_k = \mu_{..k} - \mu_{...}.$$

Definition 6.8. The (estimated or) **fitted main effects of factors A, B, and C** in a complete three-way factorial are

$$a_i = \bar{y}_{i..} - \bar{y}_{...}, \quad b_j = \bar{y}_{.j.} - \bar{y}_{...}, \quad \text{and} \quad c_k = \bar{y}_{..k} - \bar{y}_{...}.$$

It is an algebraic consequence of the form of Definitions 6.7 and 6.8 that main effects and fitted main effects sum to zero over the levels of the factor under consideration. That is, for the case of three factors one has the extension of display (6.23)

$$\sum_i \alpha_i = 0, \sum_j \beta_j = 0, \sum_k \gamma_k = 0, \sum_i a_i = 0, \sum_j b_j = 0, \text{ and } \sum_k c_k = 0.$$

One immediate implication of these relationships is that where factors have only two levels, one need only calculate one of the fitted main effects for a factor. The other is then obtained by a simple sign change.

Example 6.4 continued. Considering again the density measurements of Leigh and Taylor, one might make the "low" versus "high" level designations as:

Factor A—Mesh Size	6 mesh (−) and 60 mesh (+)
Factor B—Flask	100 cc (−) and 500 cc (+)
Factor C—Vibration	none (−) and yes (+)

With these conventions, Table 6.11 gives two sets of notation for the sample means listed originally in Table 6.8. Both the triple subscript and the special 2^3 conventions are illustrated.

Table 6.11 Alternative Notations for the Sample Mean Measured Alumina Powder Densities

Mesh (A)	Flask (B)	Vibration (C)	Sample Mean
−	−	−	$\bar{y}_{111} = \bar{y}_{(1)} = 2.164$
+	−	−	$\bar{y}_{211} = \bar{y}_{a} = 1.966$
−	+	−	$\bar{y}_{121} = \bar{y}_{b} = 2.206$
+	+	−	$\bar{y}_{221} = \bar{y}_{ab} = 1.886$
−	−	+	$\bar{y}_{112} = \bar{y}_{c} = 2.272$
+	−	+	$\bar{y}_{212} = \bar{y}_{ac} = 2.276$
−	+	+	$\bar{y}_{122} = \bar{y}_{bc} = 2.322$
+	+	+	$\bar{y}_{222} = \bar{y}_{abc} = 2.334$

It is then the case that

$$\bar{y}_{...} = \frac{1}{8}(2.164 + 1.966 + \cdots + 2.334) = 2.1783,$$

and, for example,

$$\bar{y}_{2..} = \frac{1}{4}(\bar{y}_{a} + \bar{y}_{ab} + \bar{y}_{ac} + \bar{y}_{abc}) = \frac{1}{4}(1.966 + 1.886 + 2.276 + 2.334) = 2.1155.$$

So using Definition 6.4,

$$a_2 = \bar{y}_{2..} - \bar{y}_{...} = 2.1155 - 2.1783 = -.063.$$

The average of the four "60 mesh" mean densities is .063 g/cc *below* the overall average of the eight sample means. A simple sign change then says that $a_1 = .063$ and the main effect of mesh size at its low level is *positive* .063 g/cc.

Similar calculations then show that for the means of Table 6.11,

$$b_2 = \bar{y}_{.2.} - \bar{y}_{...} = 2.187 - 2.178 = .009 \quad \text{and}$$
$$c_2 = \bar{y}_{..2} - \bar{y}_{...} = 2.301 - 2.178 = .123.$$

Then switching signs for these two-level factors, one also has $b_1 = -.009$ and $c_1 = -.123$. Figure 6.15 is a very common and helpful kind of graphic for displaying the 2^3 factorial means sometimes called a **cube plot**. On the plot for this example, the fact that $a_2 = -.063$ g/cc says that the average of the means on the right face of the cube is .063 g/cc below the overall average of the eight sample means pictured.

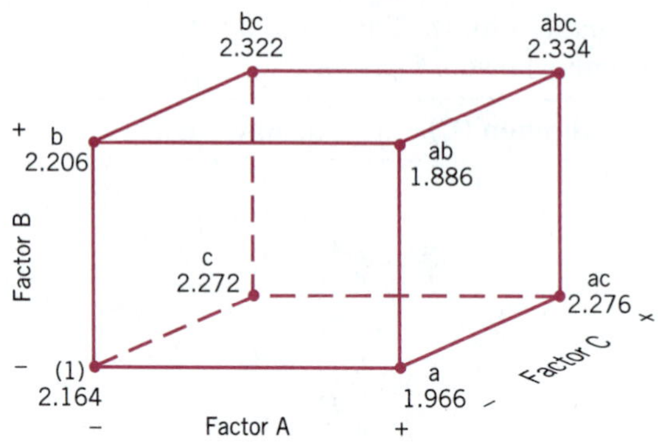

Figure 6.15
Cube plot of sample mean measured alumina powder densities (g/cc).

The fact that $b_2 = .009$ says that the average of the means on the top face of the cube is .009 g/cc above the overall average. And the fact that $c_2 = .123$ says that the average of the means on the back face of the cube is .123 g/cc above the overall average.

Main effects do not completely describe a p-way factorial set of means any more than they completely describe a two-way factorial. There are interactions to consider as well. In a p-way factorial, two-factor interactions are what one would compute as interactions via the methods of the previous section *after averaging out over all levels of all other factors*. For example, in a three-way factorial, two-factor interactions between A and B are what one has for interactions from Section 6.2 *after averaging over levels of Factor C*. The precise definitions for the three-factor situation follow (and the reader can reason by analogy and the addition of subscript dots to corresponding definitions for more than three factors).

Definition 6.9. The (theoretical) **two-factor interactions of pairs of factors A, B, and C** in a complete three-way factorial are

$$\begin{aligned} \alpha\beta_{ij} &= \mu_{ij.} - (\mu_{...} + \alpha_i + \beta_j), \\ \alpha\gamma_{ik} &= \mu_{i.k} - (\mu_{...} + \alpha_i + \gamma_k), \quad \text{and} \\ \beta\gamma_{jk} &= \mu_{.jk} - (\mu_{...} + \beta_j + \gamma_k). \end{aligned}$$

Definition 6.10. The (estimated or) **fitted two-factor interactions of pairs of factors A, B, and C** in a complete three-way factorial are

$$ab_{ij} = \bar{y}_{ij.} - (\bar{y}_{...} + a_i + b_j),$$
$$ac_{ik} = \bar{y}_{i.k} - (\bar{y}_{...} + a_i + c_k), \quad \text{and}$$
$$bc_{jk} = \bar{y}_{.jk} - (\bar{y}_{...} + b_j + c_k).$$

A main effect is in some sense the difference between what exists (in terms of an average response) and what is explainable in terms of only a grand mean. A two-factor interaction is similarly a difference between what exists (in terms of an average response) and what can be accounted for by considering a grand mean and the factors acting individually.

Just as interactions in two-way factorials sum to zero across rows or columns, it is a consequence of the form of Definitions 6.9 and 6.10 that two-factor interactions in p-way factorials also sum to 0 over levels of either factor involved. In symbols,

$$\sum_i ab_{ij} = \sum_j ab_{ij} = 0, \quad \sum_i \alpha\beta_{ij} = \sum_j \alpha\beta_{ij} = 0,$$
$$\sum_i ac_{ik} = \sum_k ac_{ik} = 0, \quad \sum_i \alpha\gamma_{ik} = \sum_k \alpha\gamma_{ik} = 0, \quad \text{and}$$
$$\sum_j bc_{jk} = \sum_k bc_{jk} = 0, \quad \text{and} \quad \sum_j \beta\gamma_{jk} = \sum_k \beta\gamma_{jk} = 0.$$

One important consequence of these relationships is that for cases where factors have only two levels, one needs to calculate only one of the four interactions for a given pair of factors. The other three can then be obtained by appropriate changes of sign.

Example 6.4 continued. Turning once again to the alumina powder density study, consider the calculation of AB two-factor interactions. Averaging front to back on the cube plot of Figure 6.15 produces the values

$$\bar{y}_{11.} = 2.218, \quad \bar{y}_{21.} = 2.121, \quad \bar{y}_{12.} = 2.264, \quad \text{and} \quad \bar{y}_{22.} = 2.110.$$

An AB interaction plot of these is shown in Figure 6.16 and there is some lack of parallelism in evidence. The size of this lack of parallelism can be measured by computing

$$ab_{22} = \bar{y}_{22.} - (\bar{y}_{...} + a_2 + b_2) = 2.110 - (2.178 + (-.063) + .009) = -.014.$$

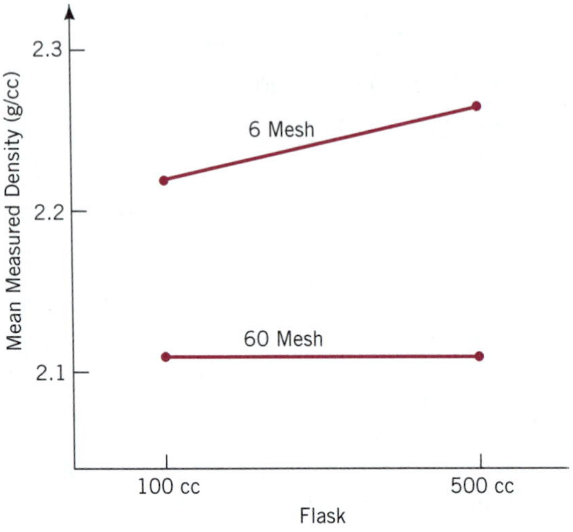

Figure 6.16
Interaction plot for alumina powder density after averaging over vibration conditions.

Then, since $ab_{21}+ab_{22} = 0$, $ab_{21} = .014$. Since $ab_{12}+ab_{22} = 0$, $ab_{12} = .014$. And finally, since $ab_{11} + ab_{21} = 0$, $ab_{11} = -.014$.

Similar calculations can be done to find the fitted two-way interactions of A and C and of B and C. The reader should verify that (except possibly for roundoff error)

$$ac_{22} = .067 \quad \text{and} \quad bc_{22} = .018.$$

Others of the AC and BC two-factor interactions in this 2^3 study can be obtained by making appropriate sign changes.

Main effects and two-factor interactions do not tell the whole story about a p-way factorial set of means. For example, in a three-factor context there are many quite different sets of means having a given set of main effects and two-factor interactions. One must go further in defining effects to distinguish between these different possibilities. The next logical step beyond two-factor interactions would seem to be some kind of **three-factor interactions**. To see what these might be, it is helpful to remember that (1) a main effect is the difference between what exists and what is explainable in terms of only a grand mean and (2) a two-factor interaction is the difference between what exists and what can be accounted for by considering a grand mean and main effects. This suggests that one define a three-factor interaction to be the difference between what exists in terms of an average response and what is explainable in terms of a grand mean, main effects, and two-factor interactions. That is, one has Definitions 6.11 and 6.12 for the case of $p = 3$. (The reader can reason by analogy to produce definitions of three-factor interactions in higher way studies by adding some dot subscripts to the three-way factorial expressions.)

Definition 6.11. The (theoretical) **three-factor interactions of factors A, B, and C** in a complete three-way factorial are

$$\alpha\beta\gamma_{ijk} = \mu_{ijk} - (\mu_{...} + \alpha_i + \beta_j + \gamma_k + \alpha\beta_{ij} + \alpha\gamma_{ik} + \beta\gamma_{jk}).$$

Definition 6.12. The (estimated or) **fitted three-factor interactions of factors A, B, and C** in a complete three-way factorial are

$$abc_{ijk} = \bar{y}_{ijk} - (\bar{y}_{...} + a_i + b_j + c_k + ab_{ij} + ac_{ik} + bc_{jk}).$$

Three-factor interactions sum to 0 over any of their three indices. That means that for the case of $p = 3$ factors

$$\sum_i abc_{ijk} = \sum_j abc_{ijk} = \sum_k abc_{ijk} = 0 \quad \text{and}$$
$$\sum_i \alpha\beta\gamma_{ijk} = \sum_j \alpha\beta\gamma_{ijk} = \sum_k \alpha\beta\gamma_{ijk} = 0.$$

So, in the case of 2^p studies it again suffices to compute only one fitted interaction for a set of three factors and then obtain all others of that type by appropriate choice of signs.

Example 6.4 continued. In the powder density study, the fitted three-factor interaction for the "all high levels" combination is

$$\begin{aligned} abc_{222} &= \bar{y}_{222} - (\bar{y}_{...} + a_2 + b_2 + c_2 + ab_{22} + ac_{22} + bc_{22}) \\ &= 2.334 - (2.178 + (-.063) + .009 + .123 + (-.014) + .067 + .018) \\ &= .016. \end{aligned}$$

Because the fitted interactions in this 2^3 study must add to 0 over levels of any one of the factors, it is straightforward to see that those abc_{ijk} with an even number of subscripts equal to 1 are .016, while those with an odd number of subscripts equal to 1 are −.016.

Hopefully, the pattern of how factorial effects are defined is by now clear. To define an interaction effect involving a particular set of q factors, one first averages means over all levels of all other factors, and then takes a difference between such an average mean and the sum of a grand mean, main effects, and interactions of order less than q. The result is a quantity that in some sense measures how much of the system response is explainable only in terms of what the factors do "q at a time." The object of a p-way factorial analysis is to hopefully identify some few effects that taken together both account for most of the variation in response and also have a simple interpretation. This, of course, need not always be possible. But when it is, a factorial analysis can provide important insight into how p factors impact the response.

Objective of a p-Way Analysis

6.3.2 Judging the Detectability of 2^p Factorial Effects in Studies with Replication

Although the examples used in this section have been ones where every factor has only two levels, the definitions of effects have been perfectly general, applicable to any full factorial. But from this point on in this section, the methods introduced are going to be specifically 2^p factorial methods. Of course there are data analysis tools for the more general case (that can be found in intermediate-level statistical methods texts). The methods that follow, however, are particularly simple and cover what is with little doubt the most important part of full factorial experimentation for modern process improvement.

We have noted that all effects of a given type in a 2^p factorial differ from each other by at most a sign change. This makes it possible to concentrate on the main effects and interactions for the "all factors at their high levels" treatment combination and still have a complete description of how the factors impact the response. In fact, people sometimes go so far as to call $a_2, b_2, ab_{22}, c_2, ac_{22}, bc_{22}, abc_{222}$, and so on "the" fitted effects in a 2^p factorial (slurring over the fact that there are effects corresponding to low levels of the factors). This subsection considers the issue of identifying those fitted effects that are big enough to indicate that the corresponding effect is detectable above the baseline experimental variation, under the assumption that there is some replication in the data.

The most effective tool of inference for 2^p factorial effects is the relevant specialization of expression (6.8). As it turns out, every effect in a 2^p factorial is a linear combination of means (an L) with coefficients that are all $\pm 1/2^p$. The corresponding fitted effect is the corresponding linear combination of sample means (the corresponding $\hat{L}$). So under the constant variance normal distributions model assumptions (6.1), if E is a generic 2^p factorial effect and $\hat{E}$ is the corresponding fitted effect, formula (6.8) can be specialized to give confidence limits for E of the form

$$\hat{E} \pm t s_{\mathrm{P}} \frac{1}{2^p} \sqrt{\frac{1}{n_{(1)}} + \frac{1}{n_{\mathrm{a}}} + \frac{1}{n_{\mathrm{b}}} + \frac{1}{n_{\mathrm{ab}}} + \cdots} . \qquad (6.31)$$

Confidence Limits for a 2^p Factorial Effect

When the plus or minus figure prescribed by this formula is larger in magnitude than a fitted effect, the real nature (positive, negative, or 0) of the corresponding effect is in doubt.

Example 6.4 continued. The fitted effects corresponding to the 60 mesh/500cc/ vibrated flask conditions in the alumina powder density study (the "all high levels" combination) have already been calculated to be

$$a_2 = -.063, b_2 = .009, ab_{22} = -.014, c_2 = .123,$$
$$ac_{22} = .067, bc_{22} = .018, \quad \text{and} \quad abc_{222} = .016.$$

Formula (6.31) allows one to address the question of whether any of these empirical values provides clear evidence of the nature of the corresponding long-run effect.

In the first place, using the sample standard deviations given in Table 6.8, one has

$$s_{\mathrm{P}} = \sqrt{\frac{(5-1)(.030)^2 + (5-1)(.040)^2 + \cdots + (5-1)(.044)^2}{(5-1) + (5-1) + \cdots + (5-1)}} = .059 \text{ g/cc}.$$

Actually, before going ahead to use s_{P} and formula (6.31) one should apply the methods of Section 6.1 to check on the plausibility of the basic one-way normal model assumptions. The reader can verify that a normal plot of the residuals is fairly linear. But, in fact, a test like "Bartlett's test" applied to the sample standard deviations in Table 6.8 draws into serious question the appropriateness of the "constant σ" part of the usual model assumptions. For the time being, the fact that there is nearly an order of magnitude difference between the smallest and largest sample standard deviations in Table 6.8 will be ignored. The rationale for doing so is as follows. The t intervals (6.8) are generally thought to be fairly "robust" against moderate departures from the constant σ model assumption (meaning that nominal confidence levels, while not exactly correct, are usually not ridiculously wrong either). So rather than just "give up and do nothing in the way of inference" when it seems there may be a problem with the model assumptions, it is better to go ahead with caution. One should then remember that the confidence levels cannot be trusted completely and agree to avoid making "close calls" of large engineering or financial impact based on the resulting inferences.

Then, assuming for the moment that the one-way model is appropriate, note that s_P has associated with it $\nu = (5-1)+(5-1)+\cdots+(5-1) = 32$ degrees of freedom. So since the .975 quantile of the t_{32} distribution is 2.037, 95% two-sided confidence limits for any one of the 2^3 factorial effects E are

$$\hat{E} \pm 2.037(.059)\frac{1}{2^3}\sqrt{\frac{1}{5}+\frac{1}{5}+\cdots+\frac{1}{5}}, \text{ i.e., } \hat{E} \pm .019 \text{ g/cc.}$$

By this standard, only the A main effects, C main effects, and AC two-factor interactions are clearly detectable. It is comforting here (especially in light of the caution necessitated by the worry over appropriateness of the constant variance assumption) that all of a_2, c_2, and ac_{22} are not just larger than .019 in magnitude, but substantially so. It seems pretty safe to conclude that mesh size (Factor A) and vibration condition (Factor C) have important effects on the mean measured density of this powder, but that the size of the flask used (Factor B) does not affect mean measured density in any way that can be clearly delineated on the basis of these data. (No main effect or interaction involving B is visible above the experimental variation.)

Notice, by the way, that the fact that the AC interaction is nonnegligible says that one may *not* think of changing mesh size as doing the same thing to mean measured density when the flask is vibrated as when it is not. Figure 6.17 shows an interaction plot for the average sample means $\bar{y}_{i.k}$ obtained by averaging out over the two flask sizes. Six mesh material consists of (a mixture of both coarse and finely ground) material that will pass through a fairly coarse screen. Sixty mesh material is that (only relatively fine material) that will pass through a fine screen. It is interesting that Leigh and Taylor's original motivation for their experimentation

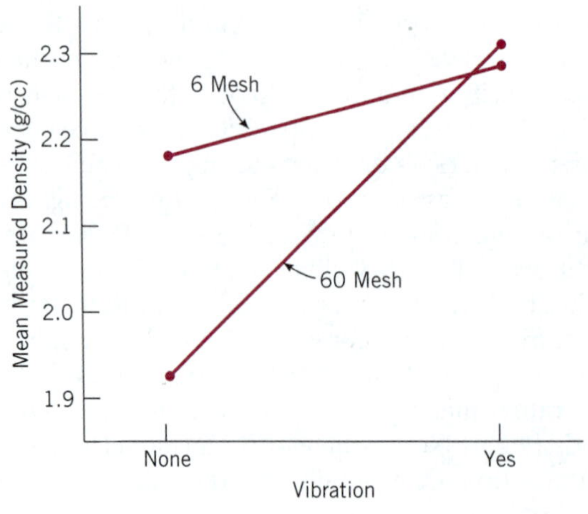

Figure 6.17
Interaction plot for alumina powder density after averaging over levels of flask size.

was to determine if their density measurement system was capable of detecting changes in material particle size mix on the basis of measured density. The form of Figure 6.17 suggests strongly that to detect mix changes on the basis of measured density, their system should be operated with unvibrated samples.

Confidence limits of the form (6.31) are individual limits. Since one will typically simultaneously investigate the detectability of all $2^p - 1$ effects in a 2^p factorial, it is of interest to see what can be easily done in the way of making simultaneous intervals for all of these effects. As it turns out, a simple method akin to the P-R method for estimating mean responses introduced in Section 6.1 can be used for balanced data situations. That is, when all 2^p sample sizes are m, one can replace t in formula (6.31) with an appropriate quantile of a studentized maximum modulus distribution and obtain simultaneous two-sided confidence intervals for all of $\alpha_2, \beta_2, \alpha\beta_{22}, \ldots$. In precise terms, if all 2^p sample sizes are m, then simultaneous two-sided 95% confidence limits for all factorial effects are

Balanced Data Simultaneous Confidence Limits for All Effects

$$\hat{E} \pm \kappa_2^* s_P \sqrt{\frac{1}{m2^p}} \tag{6.32}$$

where the constant κ_2^* is obtained from Table 6.12.

Example 6.4 continued. As an alternative to the individual 95% confidence limits already computed for the effects in the powder density study, consider the calculation of simultaneous 95% limits using display (6.32). Since the data in Table 6.8 are balanced 2^3 data with $m = 5$, Table 6.12 shows that $\kappa_2^* = 2.854$ should be used. That is, for simultaneous 95% confidence in estimating all of

Table 6.12 Factors κ_2^* for Simultaneous 95% Two-Sided Confidence Limits for All Effects in a Balanced 2^p Factorial

	p				
m	3	4	5	6	7
2	3.454	3.377	3.410	3.502	3.627
3	3.039	3.145	3.276	3.425	3.584
4	2.914	3.071	3.232	3.400	3.569
5	2.854	3.034	3.211	3.387	3.561
6	2.819	3.013	3.198	3.379	3.556

$\alpha_2, \beta_2, \alpha\beta_{22}, \gamma_2, \alpha\gamma_{22}, \beta\gamma_{22}$, and $\alpha\beta\gamma_{222}$, one should associate with each of the seven corresponding fitted effects a plus or minus value of

$$2.854(.059)\sqrt{\frac{1}{(5)2^3}} = .027 \text{ MPa}$$

instead of the .019 MPa value found earlier. This value simply reinforces the engineering conclusions already drawn from the data. The A and C main effects and the AC interaction are still clearly detectable above this level of uncertainty, while it is even more clear that the other effects (involving Factor B) are not.

6.3.3 Judging the Detectability of 2^p Factorial Effects in Studies Lacking Replication

The use of formula (6.31) to make confidence limits for 2^p factorial effects depends upon the existence of some replication somewhere in a data set, so that s_P can be calculated. It ought to be an axiom of life for any engineer with even the smallest amount of statistical background that experiments need to include some replication. Most fundamentally, replication allows one to verify that experimental results are to some degree repeatable and to establish the limits of that repeatability. Without it, there is no completely honest way to tell whether changing levels of experimental factors is what causes observed changes in y, or if instead changes one observes in y amount only to random variation. But having said all this, one is sometimes forced to make the best of completely unreplicated data. And it is thus appropriate to consider what can be done to analyze data like those in Table 6.9 that include no replication.

The best existing method of detecting factorial effects in unreplicated 2^p data is one suggested by Cuthbert Daniel. His method depends upon the **principle of effect sparsity** and makes use of probability plotting. The principle of effect sparsity is a kind of Pareto principle for experiments. It says that often a relatively few factors account for most of the variation seen in experimental data. And when the principle governs a physical system, the job of detecting real effects amounts only to picking out the largest few. Daniel's logic for identifying those then goes as follows. Any fitted effect $\hat{E}$ is related to its corresponding effect E as

$$\hat{E} = E + noise.$$

When σ (and therefore the noise level) is large, one might thus expect a normal plot of the $2^p - 1$ fitted effects (for the all-high combination) to be roughly linear. On the other hand, should σ be small enough that one has a chance of seeing the few important effects, they ought to lead to points that plot "off the line" established by the majority (that themselves represent small effects plus noise). These exceptional values thereby identify themselves as more than negligible effects plus experimental variation.

Actually, Daniel's original suggestion for the plotting of the fitted effects was slightly different than that just described. There is an element of arbitrariness associated with the exact appearance of a normal plot of a_2, b_2, ab_{22}, c_2, ac_{22}, ab_{22}, $abc_{222}, \ldots$. This is because the signs on these fitted effects depend on the arbitrary designation of the "high" levels of the factors. (And the choice to plot fitted effects associated with the all-high combination is only one of 2^p possible choices.) Daniel reasoned that plotting with the absolute values of fitted effects would remove this arbitrariness.

Now if Z is standard normal and has quantile function $Q_z(p)$, then $|Z|$ has quantile function

"Half-Normal" Quantile Function

$$Q(p) = Q_z\left(\frac{1+p}{2}\right). \tag{6.33}$$

Rather than plotting standard normal quantiles against fitted-effect quantiles as in Section 5.1, Daniel's original idea was to plot values of the quantile function $Q(p)$ specified in equation (6.33) versus absolute-fitted-effect quantiles, to produce a **half-normal plot**. This text will use both normal plotting of fitted effects and half-normal plotting of absolute fitted effects. The first is slightly easier to describe and usually quite adequate, but as Daniel pointed out, the second is somewhat less arbitrary.

Example 6.5 continued. Return again to Zwickl's data on pull-outs on dual in-line packages given in Table 6.9 early in this section. As it turns out, the fitted effects for the counts of Table 6.9 are:

$$\begin{aligned} a_2 &= 11.5, b_2 = -6.0, ab_{22} = -4.6, c_2 = -.6, ac_{22} = -1.5, bc_{22} = -2.3,\\ abc_{222} &= -1.6, d_2 = -10.3, ad_{22} = -7.1, bd_{22} = 2.9, abd_{222} = 2.5,\\ cd_{22} &= 3.8, acd_{222} = 3.6, bcd_{222} = -.6, \quad \text{and} \quad abcd_{2222} = -1.3. \end{aligned}$$

Table 6.13 then gives the coordinates of the 15 points that one plots to make a normal plot of the fitted effects. And Table 6.14 gives the coordinates of the points that one plots to make a half-normal plot of the absolute fitted effects. The corresponding plots are given in Figures 6.18 and 6.19.

The plots in Figures 6.18 and 6.19 are not as definitive as one might have hoped. None of the fitted effects or absolute fitted effects stand out as tremendously larger than the others. But it is at least clear from the half-normal plot in Figure 6.19 that no more than four, and probably at most two of the effects should be judged "detectable" on the basis of this data set. There is some indication in these

Table 6.13 Coordinates of Points for a Normal Plot of the Pull-Out Fitted Effects

i	i th smallest $\hat{E}$	$p = \frac{i-.5}{15}$	$Q_z(p)$
1	$d_2 = -10.3$	.033	-1.83
2	$ad_{22} = -7.1$	.100	-1.28
3	$b_2 = -6.0$	.167	$-.97$
4	$ab_{22} = -4.6$	.233	$-.73$
5	$bc_{22} = -2.3$	.300	$-.52$
6	$abc_{222} = -1.6$	.367	$-.34$
7	$ac_{22} = -1.5$	.433	$-.17$
8	$abcd_{2222} = -1.3$	.500	0
9	$bcd_{222} = -.6$	.567	.17
10	$c_2 = -.6$	.633	.34
11	$abd_{222} = 2.5$	.700	.52
12	$bd_{22} = 2.9$	.767	.73
13	$acd_{222} = 3.6$	.833	.97
14	$cd_{22} = 3.8$	.900	1.28
15	$a_2 = 11.5$	.967	1.83

Table 6.14 Coordinates of Points for a Half-Normal Plot of the Pull-Out Absolute Fitted Effects

i	i th smallest $\lvert\hat{E}\rvert$	$p = \frac{i-.5}{15}$	$Q(p) = Q_z\left(\frac{1+p}{2}\right)$
1	$\lvert bcd_{222}\rvert = .6$	.033	.04
2	$\lvert c_2\rvert = .6$	.100	.13
3	$\lvert abcd_{2222}\rvert = 1.3$	.167	.21
4	$\lvert ac_{22}\rvert = 1.5$	.233	.30
5	$\lvert abc_{222}\rvert = 1.6$	.300	.39
6	$\lvert bc_{22}\rvert = 2.3$	.367	.48
7	$\lvert abd_{222}\rvert = 2.5$	.433	.57
8	$\lvert bd_{22}\rvert = 2.9$	.500	.67
9	$\lvert acd_{222}\rvert = 3.6$	.567	.78
10	$\lvert cd_{22}\rvert = 3.8$	.633	.90
11	$\lvert ab_{22}\rvert = 4.6$	.700	1.04
12	$\lvert b_2\rvert = 6.0$	.767	1.19
13	$\lvert ad_{22}\rvert = 7.1$	.833	1.38
14	$\lvert d_2\rvert = 10.3$	.900	1.65
15	$\lvert a_2\rvert = 11.5$	.967	2.13

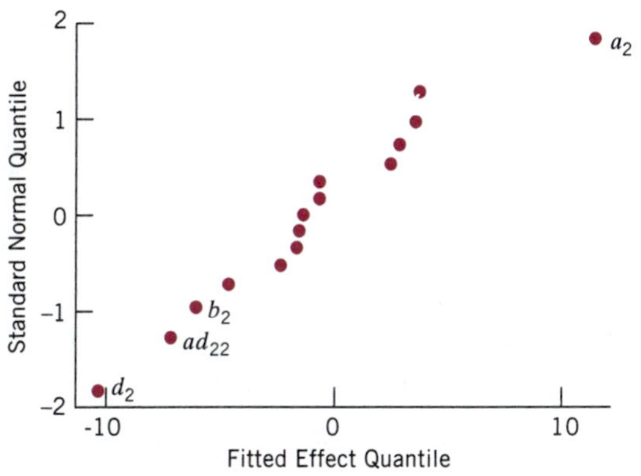

Figure 6.18
Normal plot of fitted effects for pull-outs on dual in-line packages.

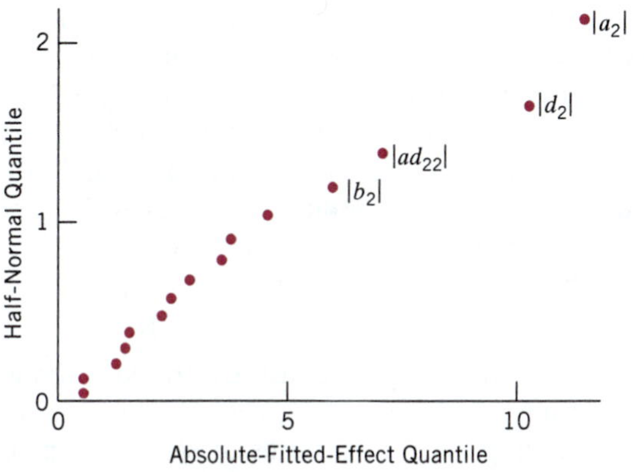

Figure 6.19
Half-normal plot of absolute fitted effects for pull-outs on dual in-line packages.

data that the A and D main effects are important in determining bond strength, but the conclusion is unfortunately clouded by the lack of replication.

Example 6.4 continued. As a second (and actually more satisfying) application of the notion of probability plotting in the analysis of 2^p data, return again to the powder density data and the issue of the *variability* in response. We noted earlier that the eight sample standard deviations in Table 6.8 do not really look as if they could all have come from distributions with a common σ. To investigate this matter further, one might look for 2^3 factorial effects *on the standard deviation of response.* A common way of doing this is to compute and plot fitted effects for the natural logarithm of the sample standard deviation, $\log(s)$. (The logarithm is used because it tends to make the theoretical distribution of the sample standard deviation look more symmetric and Gaussian than it is in its raw form.) The reader is invited to verify that the logarithms of the standard deviations in Table 6.8 produce fitted effects

$$a_2 = -.16, b_2 = -.26, ab_{22} = -.24, c_2 = .57,$$
$$ac_{22} = -.07, bc_{22} = .15, \quad \text{and} \quad abc_{222} = -.00.$$

For example,

$$a_2 = \frac{1}{4}(\log(.040) + \log(.011) + \log(.079) + \log(.044))$$
$$- \frac{1}{8}(\log(.030) + \log(.040) + \cdots + \log(.104) + \log(.044)) = -.16.$$

A half-normal plot of the absolute values of these is given in Figure 6.20.

It seems clear from Figure 6.20 that the fitted main effect of Factor C is more than just experimental noise. Since $c_2 = .57$, $c_2 - c_1 = 1.14$. One would judge from this analysis that for any mesh size and flask size, the logarithm of the standard deviation of measured density for vibrated flasks is 1.14 more than that for unvibrated flasks. This means that the standard deviation itself is about $\exp(1.14) = 3.13$ times as large when vibration is employed in density determination as when it is not. Not only does the lack of vibration fit best with the researchers' original goal of detecting mix changes via density measurements, but it provides more consistent density measurements than are obtained with vibration.

The kind of analysis of sample standard deviations just illustrated in Example 6.4 is an important rough-and-ready approach to the problem of seeing how several factors affect σ. Strictly speaking, even logarithms of sample standard deviations from normal populations are not really legitimately treated like sample means. But often, at least for balanced 2^p factorial data with m of at least 4 or 5, this kind of crude analysis will draw attention to interpretable structure in

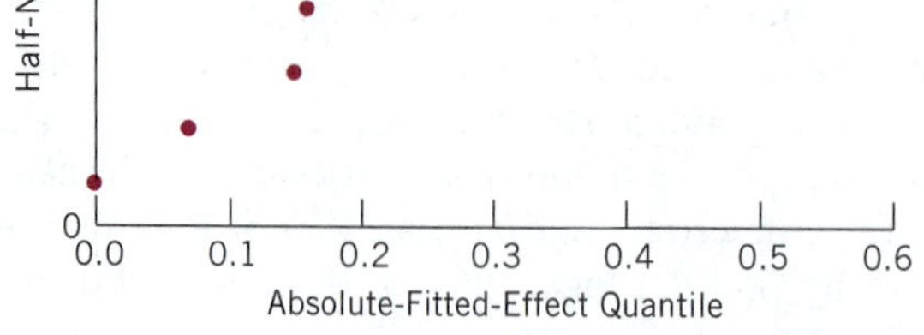

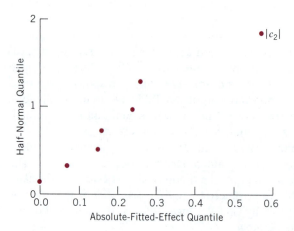

Figure 6.20
Half-normal plot of absolute fitted effects on $\log(s)$ in the solder joint strength study.

observed values of s. And in light of the fact that variation is the enemy of quality of conformance, this can be very important to quality engineering efforts.

6.3.4 The Yates Algorithm for Computing Fitted 2^p Effects

Computing fitted effects directly from definitions like those given in this section is no fun. Of course there are commercial computer programs that will do the work. (And the reader is also referred to the Fortran code provided in the paper "An Interactive Program for the Analysis of Data from Two Level Factorial Experiments via Probability Plotting," by Crowder, Jensen, Stephenson, and Vardeman, published in *Journal of Quality Technology* in 1988.) But it is also very helpful to have an efficient means of computing 2^p fitted effects more or less by hand. Happily, Frank Yates discovered such a means many years ago. His algorithm will produce, all at once and with minimum pain, the average sample mean and the fitted effects for the all-high combination.

Yates Algorithm for 2^p Fitted Effects

One sets up the **Yates algorithm** by first listing 2^p sample means in a column in what is called "Yates standard order." This order is easily remembered by beginning with combination (1) (the "all-low" combination) and then combination a, and then "multiplying by b" to get b and ab, then "multiplying by c" to get c, ac, bc, and abc, and so on. One then creates a second column of numbers by adding the numbers in the first column in pairs and then subtracting them in pairs (*the first value in a pair being subtracted from the second*). The additions and subtractions are applied to the new column, and so on, until a total of p successive new columns have been generated from the original list of $\bar{y}$'s. Finally, the entries in the last column generated by additions and subtractions are divided by 2^p to produce the fitted effects (themselves listed in Yates standard order applied to effects rather than combinations).

Example 6.4 continued. The means listed in Table 6.11 are in fact in Yates standard order. Table 6.15 shows the use of the Yates algorithm to quickly obtain the fitted effects computed earlier by much more laborious means.

The final column of Table 6.15 is really a very compact and helpful summary of what the data of Leigh and Taylor say about mean density measurements. And now that the hard work of defining (and understanding) the fitted effects has been done, the computations needed to produce them turn out to be fairly painless. It should be evident that 2^p factorial analyses are quite doable by hand for as many as four factors.

Example 6.4 and Table 6.15 illustrate the Yates algorithm computations for $p = 3$ factors. There, three cycles of additions and subtractions are done and the final division is by $2^3 = 8$. For the case of $p = 4$, 16 means would be listed, four

Table 6.15 Use of the Yates Algorithm on the Alumina Powder Density Means

Combination	$\bar{y}$	cycle #1	cycle #2	cycle #3	cycle #3 ÷ 8
(1)	2.164	4.130	8.222	17.426	$2.178 = \bar{y}_{...}$
a	1.966	4.092	9.204	−.502	$-.063 = a_2$
b	2.206	4.548	−.518	.070	$.009 = b_2$
ab	1.886	4.656	.016	−.114	$-.014 = ab_{22}$
c	2.272	−.198	−.038	.982	$.123 = c_2$
ac	2.276	−.320	.108	.534	$.067 = ac_{22}$
bc	2.322	.004	−.122	.146	$.018 = bc_{22}$
abc	2.334	.012	.008	.130	$.016 = abc_{222}$

cycles of additions and subtractions are required, and the final division is by $2^4 = 16$. And so on.

6.3.5 Fitting Simplified Models for Balanced 2^p Data

Just as in the case of two-way factorials, after identifying effects in a 2^p factorial that seem to be detectable above the experimental variation, it is often useful to fit to the data a model that includes only those effects. Residuals can then be computed and examined for indications that something important has been missed in the data analysis. So the question is how one accomplishes the fitting of a simplified model to 2^p data. The general answer involves (as in the case of two-way data) the use of a regression program and dummy variables and cannot be adequately discussed here. What can be done is to consider the case of balanced data.

Fitted Values for Balanced Data

For balanced 2^p data sets, fitted or predicted values for (constant σ, normal) models containing only a few effects can be easily generated by adding to a grand mean only those fitted effects that correspond to effects one wishes to consider. These fitted values, $\hat{y}$, then lead to residuals in the usual way,

Residuals

$$e = y - \hat{y}.$$

Example 6.4 continued. Return yet again to the alumina powder density data of Leigh and Taylor and consider fitting a model of the form

$$y_{ijkl} = \mu + \alpha_i + \gamma_k + \alpha\gamma_{ik} + \epsilon_{ijkl} \tag{6.34}$$

to their data. This says that mean density depends only on level of Factor A and level of Factor C and (assuming that the ϵ have constant standard deviation) that variation in density measurements is the same for all four AC combinations. The four different corresponding fitted or predicted values are

$$\begin{aligned}
\hat{y}_{1j1l} &= \bar{y}_{...} + a_1 + c_1 + ac_{11} = 2.178 + .063 + (-.123) + .067 = 2.185, \\
\hat{y}_{1j2l} &= \bar{y}_{...} + a_1 + c_2 + ac_{12} = 2.178 + .063 + .123 + (-.067) = 2.297, \\
\hat{y}_{2j1l} &= \bar{y}_{...} + a_2 + c_1 + ac_{21} = 2.178 + (-.063) + (-.123) + (-.067) \\
&= 1.926, \quad \text{and} \\
\hat{y}_{2j2l} &= \bar{y}_{...} + a_2 + c_2 + ac_{22} = 2.178 + (-.063) + .123 + .067 = 2.305.
\end{aligned}$$

Residuals for the simple model (6.34) can then be obtained by subtracting these four fitted values from the corresponding data of Table 6.8. And plots of the residuals can again make clear the lack of constancy in the variance of response. For example, Figure 6.21 is a plot of residuals against level of Factor C and shows the increased variation in response that comes with vibration of the material in the density measurement process.

The brute force additions just used to produce fitted values in Example 6.4 are effective enough when there are a very few factorial effects not assumed to be 0. But where there are more than four different $\hat{y}$'s, such additions become tedious and prone to contain errors. It is thus helpful that a modification of the Yates algorithm can be used to produce 2^p fitted values all at once. The modification is called the **reverse Yates algorithm**.

Reverse Yates Algorithm for Balanced Data Fitted Values

To use the reverse Yates algorithm to produce fitted values, one writes down a column of 2^p effects (including a grand mean) in Yates standard order *from bottom to top*. Then, p normal cycles of the Yates additions and subtractions applied to the column (with no final division) produce the fitted values (listed in reverse

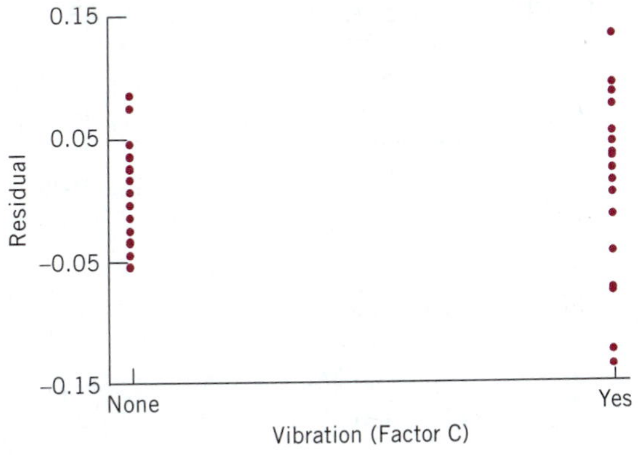

Figure 6.21
Plot of residuals from model (6.34) versus level of Factor C in the powder density study.

Yates order). In setting up the initial column of effects, one sets to 0 all those assumed to be negligible and uses fitted effects for those believed from earlier analysis to be nonnegligible.

Example 6.4 continued. As an illustration of the reverse Yates calculations, consider the fitting of model (6.34) to the powder density data. Table 6.16 shows the calculations (to more decimal places than are really justified, in order to avoid roundoff).

The calculations in Table 6.16 involve $p = 3$ factors and therefore begin with a column of $2^3 = 8$ effects. They involve 3 cycles of the Yates additions and subtractions and end with 8 fitted means. An example involving $p = 4$ factors would start with a column of $2^4 = 16$ effects, involve 4 cycles of the Yates additions and subtractions, and yield 16 fitted means.

As a final point in this chapter, it should be said that while the full factorial analyses presented here are important and useful in their own right, they are rarely the starting point for real engineering experimentation on systems with many potentially important factors. Rather, the fractional factorial methods presented first in the next chapter are often used to "screen" a large number of factors down to what look like they may be the most important few before full factorial experimentation is done. It is, however, necessary to introduce the material in the order that has been employed here, since it is impossible to understand the principles of intelligent fractional factorial experimentation without really understanding complete factorial design and analysis.

Table 6.16 Use of the Reverse Yates Algorithm to Fit a Model with Only A and C Effects to the Alumina Powder Density Data

Effect	Fitted Value	cycle #1	cycle #2	cycle #3
$\alpha\beta\gamma_{222}$	0	0	.1895	$2.305 = \hat{y}_{abc} = \hat{y}_{222l}$
$\beta\gamma_{22}$	0	.1895	2.1155	$2.297 = \hat{y}_{bc} = \hat{y}_{122l}$
$\alpha\gamma_{22}$	.06675	0	.0560	$2.305 = \hat{y}_{ac} = \hat{y}_{212l}$
γ_2	.12275	2.1155	2.2410	$2.297 = \hat{y}_{c} = \hat{y}_{112l}$
$\alpha\beta_{22}$	0	0	.1895	$1.926 = \hat{y}_{ab} = \hat{y}_{221l}$
β_2	0	.0560	2.1155	$2.185 = \hat{y}_{b} = \hat{y}_{121l}$
α_2	−.06275	0	.0560	$1.926 = \hat{y}_{a} = \hat{y}_{211l}$
$\mu_{...}$	2.17825	2.2410	2.2410	$2.185 = \hat{y}_{(1)} = \hat{y}_{111l}$

6.4 CHAPTER SUMMARY

Once a process has been brought to physical stability and is behaving predictably, further improvements typically require fundamental process changes. Intelligent experimentation can be used to guide those changes. This chapter has begun a discussion of relevant methods of experimental planning and data analysis. It opened with a presentation of general tools for comparing r experimental conditions that can be used without regard to any special structure in the conditions. Then, statistical methods for two-factor studies with complete factorial structure were considered. Finally, the long final section of the chapter addressed (full factorial) p factor studies, giving primary attention to cases where all p factors each have two levels.

6.5 CHAPTER 6 EXERCISES

6.1. An engineer and a material scientist are interested in a process for making a synthetic material. Contamination of the material during production is thought to be a real possibility. To find manufacturing conditions that produce minimal contamination, they select three different curing times and two different coating conditions (none versus some) for producing the material. Suppose two pieces of the synthetic material are produced for each combination of time and coating condition. The order of production is randomized and the raw material is randomly assigned to combinations of time and coating. The amount of iron (an important contaminant) in the final product is measured.

a. What type of (designed) experiment is this?

b. How many treatments (or treatment combinations) are there in this experiment?

c. To what future production circumstances can conclusions based on this experiment be extended?

d. Write the model equation for this experiment. Let "time" be Factor A and "coating condition" be Factor B. Say what each model term means, both in mathematical terms and in the context of the physical experiment.

e. What are the degrees of freedom for the estimated variance in response (for pieces made under a given set of process conditions)?

f. How many factors are there in this experiment? Identify the numbers of levels for all the factors.

g. Write your answer to (b) as $I \times J \times \cdots$, in terms of the numbers of levels you gave in answer to (f).

6.2. Refer to problem (6.1).

a. Let i represent the level of Factor A and j the level of Factor B ($j = 1$ indicating no coating). Write out the 6 equations for treatment means μ_{ij} that follow from the model in part (d) of problem (6.1).

b. For each of the three levels of Factor A (curing time) average your answers to (a) across levels of Factor B, and apply the facts in displays (6.23) and (6.24) to write the averages in terms of $\mu_{..}$ and the curing time main effects.

c. Find the difference between your mean in (a) for the second level of curing time and no coating and your mean in (a) for the first level of curing time and no coating.

d. Find the difference between your mean in (a) for the second level of curing time and some coating and your mean in (a) for the first level of curing time and some coating.

e. Find the difference between your answer to (b) for the second level of curing time and your answer to (b) for the first level of curing time.

f. Reflect on (c), (d), and (e). Would you use your answer in (e) to represent the quantities in (c) or (d)? Why or why not?

g. What must be true for the two differences identified in (c) and (d) to both be equal to the difference in (e)?

6.3. Refer to problems (6.1) and (6.2).

a. Average your answers to part (a) of problem (6.2) across time levels and use the facts in displays (6.23) and (6.24) to express the no coating and coating means in terms of $\mu_{..}$ and the coating main effects.

b. Find the difference between your mean in part (a) of problem (6.2) for the third level of curing time and no coating and your answer to part (a) of problem (6.2) for the third level of curing time and some coating.

c. Find the difference between your mean in part (a) of problem (6.2) for the first level of curing time and no coating and your answer to part (a) of problem (6.2) for the first level of curing time and some coating.

d. Reflect on (a), (b), and (c). Would you use the difference of your two answers in (a) to represent the quantities requested in (b) and (c)? Why or why not?

e. What must be true for the two differences identified in (b) and (c) to both be equal to the difference of your two answers in (a)?

6.4. Refer to problem (6.1). Suppose the following table gives mean iron contents (mg) for the set of six different treatment combinations.

	Low Time	Medium Time	High Time
No Coating	8	11	7
Coating	2	5	7

a. Plot the mean responses versus level of time. Use line segments to connect successive points for means corresponding to "no coating." Then separately connect the points for means corresponding to "coating." (You will have then created an interaction plot.)

b. Find the average (across levels of time) mean for "no coating" pieces. Do the same for "coating" pieces. Find the difference in these two average means (coated minus no coating).

c. Find the following three differences in means: (low time/coating) minus (low time/no coating), (medium time/coating) minus (medium time/no coating), and (high time/coating) minus (high time/no coating).

d. Compare the three differences found in (c) to the difference in (b). What is implied about the $\alpha\beta_{ij}$ values? (Hint: Use equation (6.24) and Definition 6.5.)

e. What feature of your graph in (a) reflects your answers in (d)?

f. If asked to describe the effect of coating on iron content, will your answer depend on the time level? Why or why not?

6.5. Repeat problem (6.4) with the table of means given below.

	Low Time	Medium Time	High Time
No Coating	6	8	7
Coating	1	10	4

6.6. Repeat problem (6.4) with the table of means given below.

	Low Time	Medium Time	High Time
No Coating	7	14	22
Coating	1	5	8

6.7. NASA Polymer. In their article "Statistical Design in Isothermal Aging of Polyimide Resins" that appeared in *Journal of Applied Polymer Science* in 1995, Sutter, Jobe, and Crane reported on their study of polymer resin weight loss at high temperatures. (The work is part of the NASA Lewis Research Center HITEMP program in polymer matrix composites. The focus of the HITEMP program is the development of high-temperature polymers for advanced aircraft engine fan and compressor applications.) The authors' efforts centered on evaluating the thermal oxidative stability of various polymer resins at high temperatures. The larger the weight loss in a polymer resin specimen for a given temperature/time combination, the less attractive (more unstable) the polymer resin. Maximum jet engine temperatures of interest are close to 425°C (700°F) and the experiment designed by the researchers exposed specimens of selected polymer resins to a temperature of 371°C for 400 hours. The specimens were initially essentially all the same size. Upon completion of the 400 hour exposures, percent weight loss was recorded for each specimen. The following data are from two of the polymer resins included in the study, Avimid-N, a polymer resin developed by DuPont, and VCAP-75, a polymer resin investigated in earlier engine component development programs.

Avimid-N	VCAP-75
10.6576	25.8805
9.1014	23.5876
9.0909	26.4873
10.2299	30.5398

a. How many experimental factors were there in the study described above? What were the levels of that factor or factors?
b. Give a model equation for a "random samples from normal distributions with a common variance" description of this scenario. Say what each term in your model means in the context of the problem. Also, give the numeric range for each subscript.
c. How many treatments are there in this study? What are they?
d. Find the two estimated treatment means (the two sample means) and the eight residuals.
e. Find the estimated (common) standard deviation.
f. Normal plot the eight residuals you found in (d). Does this plot indicate any problems with the "samples from two normal distributions with a common variance" model? Explain.

6.8. Refer to the **NASA Polymer** case in problem (6.7). Suppose that the "random samples from two normal distributions with a common variance" model is appropriate for this situation.
a. Find simultaneous 95% two-sided confidence intervals for the two mean percent weight losses using the Bonferroni approach.
b. Find simultaneous 95% two-sided confidence intervals for the two mean percent weight losses using the P-R method.
c. Which set of intervals, the one from (a) or the one from (b), is best? Why?
d. Find a 99% two-sided confidence interval for the difference in mean percent weight losses for VCAP-75 and Avimid-N. (VCAP-75 minus Avimid-N.)
e. If you were to choose one of the two polymer-resins on the basis of the data from problem (6.7), which one would you choose? Why?

6.9. Refer to the **NASA Polymer** case in problems (6.7) and (6.8).

a. Transform each of the percent weight loss responses, y, to $\ln(y)$. Find the sample means for the transformed responses.

b. Find the new residuals for the transformed responses.

c. Make a normal probability plot of the residuals found in (b). Does normal theory appear appropriate for describing the log of percent weight loss? Does it seem more appropriate for the log of percent weight loss than for percent weight loss itself? Why or why not?

Henceforth, in this problem, assume that usual normal theory is an appropriate description of log percent weight loss.

d. Make simultaneous 95% two-sided confidence intervals for the two average $\ln(y)$'s using the Bonferroni approach.

e. Repeat (d) using the P-R method.

f. Which set of intervals, the one from (d) or the one from (e), do you prefer and why?

g. If $\ln(y)$ is normal, the mean of $\ln(y)$ (say $\mu_{\ln(y)}$) is also the .5 quantile of $\ln(y)$. What quantile of the distribution of y is $\exp(\mu_{\ln(y)})$?

h. If end points of the intervals produced in (e) are transformed using the exponential function, $\exp(\cdot)$, what parameters of the original y distributions does one hope to bracket?

6.10. Refer to the **NASA Polymer** case in problem (6.7). Sutter, Jobe, and Crane experimented with three other polymer resins in addition to the two mentioned in problem (6.7). (Specimens of all five polymer resins were exposed to a temperature of 371°C for 400 hours.) Percent weight losses for specimens of the three additional polymer resins N-CYCAP, PMR-II-50, and AFR700B are given below. All specimens were originally of approximately the same size.

N-CYCAP	PMR-II-50	AFR700B
25.2239	27.5216	28.4327
25.3687	29.1382	28.9548
24.5852	29.8851	24.7863
25.5708	28.5714	24.8217

a. Use both the data above and the data in problem (6.7) and answer (a)–(f) of problem (6.7) for the study including all five polymer resins.

b. Answer (a)–(c) from problem (6.8) using data from all five polymer resins.

c. How many different pairs of polymer resins can be made from the five in this study?

d. Use the Tukey method to produce 95% simultaneous two-sided confidence intervals for all differences in pairs of weight loss means.

e. The smaller the weight loss, the better from an engineering/strength perspective. Which polymer resin is best? Why?

6.11. Refer to the **NASA Polymer** case in problems (6.7), (6.8), and (6.10).

a. Find the five sample standard deviations (of percent weight loss) for the five different polymer resins (the data are given in problems (6.7) and (6.10)).

b. Transform the data for the five polymer resins given in problems (6.7) and (6.10) by taking the natural logarithm of each percent weight loss. Then find the five sample standard deviations for the logged values.

c. Consider the two sets of standard deviations in (a) and (b). Which set is more consistent? What relevance does this comparison have to the use of the methods in Section 6.1?

d. Find the five different sample means of log percent weight loss. Find the five sets of residuals for log percent weight loss (by subtracting sample means from individual values).

e. Plot the residuals found in (d) versus the sample means from (d).

f. Plot the residuals found in part (a) of problem (6.10) (calculated as in problem (6.7d)) versus the sample means (of percent weight loss). Compare the plot in (e) to this plot.

g. Which set of data, the original one or the log transformed one, better satisfies the assumption of a common response variance for all five polymer resins?

6.12. Refer to the **NASA Polymer** case in problems (6.7), (6.8), (6.10), and (6.11).

a. Normal plot the residuals found in part (d) of problem (6.11). Does the normal theory model of Section 6.1 seem appropriate for transformed percent weight loss (for all five resins)? Why or why not?

b. Answer (a)–(c) from problem (6.8) using the log transformed data from all five polymer resins. (The requested intervals are for the means of the log transformed percent weight loss.)

c. If one exponentiates the end points of the intervals from (b) (plugs the values into the function $\exp(\cdot)$) to produce another set of intervals, what will the new intervals be estimating?

d. Use the Tukey method and make simultaneous 99% confidence intervals for all differences in pairs of mean log percent weight losses for different polymer resins.

e. If one exponentiates the end points of the intervals from (d) (plugs the values into the function $\exp(\cdot)$) to produce another set of intervals, what will the new intervals be estimating? (Hint: $\exp(x - y) = \exp(x)/\exp(y)$.)

6.13. Refer to the **NASA Polymer** case in problems (6.7) and (6.10). The data given in those problems were obtained from one region or position in the oven used to bake the specimens. The following (percent weight loss) data for the same five polymer resins came from a second region or position in the oven.

Avimid-N	VCAP-75	N-CYCAP	PMR-II-50	AFR700B
9.3103	24.6677	26.3393	25.5882	23.2168
9.6701	23.7261	23.1563	25.0346	24.8968
10.9777	22.1910	25.6449	24.9641	23.8636
9.3076	22.5213	23.5294	25.1797	22.4114

Suppose the investigators were not only interested in polymer resin effects, but also in the possibility of important position effects and position/polymer resin interactions.

a. How many experimental factors are there in the complete experiment as presented in problems (6.7), (6.10), and above?

b. How many levels does each factor have? Name the levels of each factor.

c. Give a model equation for a "random samples from normal distributions with a common variance" description of this multifactor study. Say what each term in your model represents in the context of the problem and define each one in terms of the $\mu_{..}$, $\mu_{i.}$, $\mu_{.j}$, and μ_{ij}.

d. Find the complete set of 10 sample average percent weight losses and plot them in an interaction plot format. Plot oven position on the horizontal axis and use two-sided individual 95% confidence limits to establish error bars around the sample means on your plot. Does your plot suggest that there are strong interactions between position and resin? Why or why not?

e. Find and plot the ten sets of residuals. (Make and label ten tick marks on a horizontal axis corresponding to the ten combinations of oven position and polymer resin and plot the residuals against treatment combination.)

f. Find the sample standard deviation for each oven position/polymer resin combination.

g. Does it appear from (e) and (f) that variability in percent weight loss is consistent from one oven position/polymer resin combination to another? Why or why not?

h. Make a normal probability plot for the 40 residuals. Does the plot suggest any problems with the "normal distributions with a common variance" model for percent weight loss? Why or why not?

i. Possibly ignoring your answer to (g), find an estimated standard deviation of percent weight loss (supposedly) common to all oven position/polymer resin combinations. What degrees of freedom are associated with your estimate?

j. Continuing to possibly ignore your answer to (g), find individual 99% two-sided confidence intervals for the five oven position/polymer resin interaction effects for position 1. Should one be 99% confident that all five intervals are correct? Use the Bonferroni inequality to place a minimum simultaneous confidence on this whole set of intervals.

k. In terms of the model from (c), what is the position 1/Avimid-N mean minus the position 1/VCAP-75 mean? Again in terms of the model from (c), what is the difference between mean percent weight loss from Avimid-N averaged over positions 1 and 2 and mean percent weight loss from VCAP-75 averaged over positions 1 and 2? (Use fact (6.24) in answering this second question.) What must be true for the two differences here to be the same?

l. Continuing with (k), what condition will guarantee that the difference between any two different polymer resin means from the same position will be the same as the difference between means for the same two polymer resins averaged across positions?

6.14. Refer to the **NASA Polymer** case of problems (6.7), (6.10), and (6.13). Use a "normal distributions with common variance" description of percent weight loss.

a. Do the intervals found in part (j) of problem (6.13) suggest that the condition in part (l) of problem (6.13) is satisfied?

b. Consider the difference in sample mean percent weight losses for position 1/Avimid-N and position 1/VCAP-75. Find the estimated standard deviation of this difference.

c. Consider the average over positions 1 and 2 of the Avimid-N sample mean percent weight losses minus the average over positions 1 and 2 of the VCAP-75 sample mean weight losses. Find the estimated standard deviation of this difference.

d. Reflect on your answers to parts (j)–(l) of problem (6.13) and (a) through (c) of this problem. If you wished to estimate a difference between the mean percent weight losses for two of the polymer resins, would you report two intervals, one for position 1 and one for position 2, or one interval for the difference averaged across positions? Defend your answer. What would be the width of your interval(s) (using a 90% confidence level)?

e. Use the Tukey method with simultaneous 95% confidence and make a set of intervals for all possible differences between polymer resin means averaged across positions (all possible differences between resin main effects). Which polymer resin seems to be best? Why? (See the paragraph immediately before Section 6.2.3.)

6.15. Refer to the **NASA Polymer** case in problems (6.7), (6.10), and (6.13). Transform the 40 observed percent weight losses by taking their natural logarithms.

a. Answer parts (d)–(j) of problem (6.13) using the transformed data.

b. Answer parts (a)–(e) of problem (6.14) using the transformed data.

c. Should conclusions from the transformed data or original data be presented? Defend your answer.

d. Consider the final set of intervals made in (b) above (the ones corresponding to part (e) of problem (6.14)). Exponentiate the end points of these intervals (i.e., plug them into the exponential function $\exp(\cdot)$) to get new ones. What do these resulting intervals estimate? (Hint: Consider the original percent weight loss distributions and recall that $\exp(x - y) = \exp(x)/\exp(y)$.)

6.16. NASA Polymer II. (See problems (6.7) and (6.13) for background.) In a preliminary investigation, Sutter, Jobe, and Crane designed a weight loss study that was to be balanced (with an equal number of specimens per polymer resin/oven position combination). Polymer resin specimens of a standard size were supposed to be randomly allocated (in equal numbers) to each of two positions in the oven. The following percent weight loss values were in fact obtained.

	Position 1	Position 2
Avimid-N	8.9	10.2, 9.1, 9.1, 8.5
PMR-II-50	29.8, 29.8	27.4, 25.5, 25.7

a. The lack of balance in the data set above resulted from a misunderstanding of how the experiment was to be conducted. How would you respond to a colleague who says of this study "Well, since an equal number of specimens were not measured for each polymer resin/oven position combination, we don't have a balanced experiment so a credible analysis cannot be made"?

b. How many experimental factors were there in this study? Identify the numbers of levels for each of the factors.

c. Write a model equation for the "normal distributions with a common variance" description of this study. Give the numeric ranges for all subscripts.

d. Find the fitted main effect for each level of each factor.

e. Find the fitted interactions for all combinations of polymer resin and oven position.

f. Use the model from (c) and find a 99% two-sided confidence interval for the interaction effect for Avimid-N and position 1. Are intervals for the other interaction effects needed? Why or why not?

g. Use the model from (c) and find 95% two-sided confidence intervals for each of the main effects.

h. Use the model from (c) and find a 90% two-sided confidence interval for the difference in oven position main effects.

i. Use the model from (c) and find a 90% confidence interval for the difference in polymer resin main effects.

6.17. Refer to the **NASA Polymer II** case in problem (6.16).

a. Find the residuals and plot them against the cell means (that serve as fitted values in this context). Does it appear that response variability is consistent from treatment combination to treatment combination? Why or why not?

b. Normal plot the residuals found in (a). Does this plot suggest problems with the basic "normal distributions with a common variance" model? Why or why not?

c. Transform each response using $y' = \arcsin(y/100)$, where y is a percent (between 0 and 100) as given in problem (6.16). Compute the four sample means for the transformed values.

d. Find residuals for the transformed data from (c). Plot these against the sample means from (c). Does it appear that variability in transformed response is consistent from treatment combination to treatment combination? Normal plot these residuals. Does this plot suggest problems with the basic "normal distributions with a common variance" model for the transformed response? Why or why not?

e. Transform each response in problem (6.16) using $y' = \ln(y)$. Answer (c) and (d) with the newly transformed data.

f. Which version of the data (the original, the arcsin transformed, or the log transformed) seems best described by the "normal distributions with a common variance" model? Why?

6.18. Refer to the **NASA Polymer II** case in problems (6.16) and (6.17). Consider the original data and use the "normal distributions with a common variance" model.

a. Plot the sample means in interaction plot format, placing "oven position" on the horizontal axis and using two-sided individual 90% confidence limits to establish error bars around each of the sample means. (Note that since the sample sizes vary, the error bars will not be the same length mean to mean.) Does the line segment connecting the two Avimid-N points cross that connecting the two PMR-II-50 points?

b. Find a 95% two-sided confidence interval for the difference in average percent weight losses at position 1 (Avimid-N minus PMR-II-50).

c. Find a 95% two-sided confidence interval for the difference in Avimid-N and PMR-II-50 main effects.

d. Which interval, the one from (b) or the one from (c), is better for estimating the difference in mean percent weight losses (Avimid-N minus PMR-II-50) at oven position 1? Why? (See part (f) of problem (6.16) and (a) above.)

e. Find a 90% two-sided confidence interval for the difference in oven position mean percent weight losses (position 2 minus position 1) for Avimid-N.

f. Find a 90% two-sided confidence interval for the difference in the two oven position main effects (position 2 minus position 1).

g. Which interval, the one from (e) or the one from (f), is better for estimating the difference in mean percent weight losses at the two oven positions for Avimid-N? Why? (See part (f) of problem (6.16) and (a) above.)

6.19. NASA Percent Weight Loss. Sutter, Jobe, and Ortiz designed and conducted an experiment as part of NASA Lewis Research Center's efforts to evaluate the effects of two-level factors kapton, preprocessing time at 700°F, 6F dianhydride type, and oven position on percent weight loss of PMR-II-50 specimens baked at 600°F for 936 hours. Two specimens were produced for each of the possible combinations of kapton, preprocessing time, and dianhydride type. One specimen of each type was randomly selected for baking at position 1 (and the other was assigned to position 2). Exact baking locations of the 8 specimens at each position were randomized within that region. The oven was set at 600°F and the specimens were exposed to this temperature for 936 hours. All specimens initially had about the same mass. Percent weight losses (y) similar to those below were observed.

Oven Position	Kapton	Preprocessing Time	Dianhydride Type	y
1	1	1	1	4.5
2	1	1	1	5.0
1	2	1	1	4.7
2	2	1	1	5.3
1	1	2	1	4.4
2	1	2	1	5.0
1	2	2	1	4.8
2	2	2	1	5.2
1	1	1	2	3.9
2	1	1	2	3.8
1	2	1	2	3.9
2	2	1	2	3.8
1	1	2	2	4.0
2	1	2	2	3.8
1	2	2	2	3.4
2	2	2	2	3.9

The levels of the factors "kapton," "preprocessing time," and "dianhydride type" were:

Kapton	no kapton (1) vs. with kapton (2)
Preprocessing Time	15 minutes (1) vs. 2 hours (2)
Dianhydride Type	polymer grade (1) vs. electronic grade (2)

a. How many experimental factors were there in this study? Identify them and say how many levels of each were used.

b. Describe the factorial structure of the treatment combinations using a "base and exponent" notation.

c. How many treatment combinations were there in this experiment? Identify them.

d. Was there replication in this study? Why or why not?

e. Give a model equation for the "normal distributions with a common variance" description of percent weight loss that uses "factorial effects" notation.

f. Find the fitted effects corresponding to the "all-high" treatment combination.

g. Find the $(i - .5)/15$ standard normal quantiles (for $i = 1, 2, \ldots, 15$).

h. Use equation (6.33) and find the $(i - .5)/15$ half-normal quantiles (for $i = 1, 2, \ldots, 15$).

i. Using your answers to (f)–(h), make a full normal plot of the fitted effects and a half normal plot for the absolute fitted effects corresponding to the all-high treatment combination.

j. Does it appear from your plots in part (i) that there are any statistically detectable effects on mean percent weight loss? Defend your answer.

6.20. Refer to the **NASA Percent Weight Loss** case in problem (6.19). Recall that single specimens of all eight different combinations of kapton, preprocessing time, and dianhydride type were randomized within position 1 and another set of eight were randomized within position 2. The factor "position" was not really one of primary interest. Its levels were, however, different, and its systematic contribution to response variability could have been nontrivial. (This type of experimental design is sometimes referred to as a factorial arrangement of treatments (FAT) in a randomized block.) Problem (6.19) illustrates one extreme of what is possible in the way of analysis of such studies, namely that where the main effects and all interactions with the "blocking" factor are considered potentially important. The other extreme (possibly adopted on the basis of the results of an analysis like that in problem (6.19)) is that where the main effects of the blocking variable and all interactions between it and the other experimental factors are assumed to be negligible. (Intermediate possibilities also exist. For example, one might conduct an analysis assuming that there are possibly main effects of "blocks" but that all interactions involving "blocks" are negligible.)

In this problem consider an analysis of the data from problem (6.19) that completely ignores the factor "oven position." If one treats the main effects and all interactions with oven position as negligible, the data in problem (6.19) can be thought of as three-factor factorial data (in the factors kapton, preprocessing time, and dianhydride type) with $m = 2$.

a. Redo parts (a)–(e) of problem (6.19) taking this new point of view regarding oven position.

b. Find the fitted effects for the all-high treatment combination in this three-factor study.

c. Is normal plotting the only possible means of judging the statistical detectability of the factorial effects you estimated in (b)? Why or why not?

d. Find the residuals for this study by subtracting sample means from individual percent weight loss measurements.

e. Plot the residuals versus the sample means. Does it appear the assumption of a common variance across treatment combinations is reasonable? Why or why not?

f. Find the sample standard deviation, s, for each treatment combination. Let $y^* = \ln(s)$ and find fitted effects for the log sample standard deviations, y^*.

g. Use equation (6.33) and find the $(i - .5)/7$ half-normal quantiles (for $i = 1, 2, \ldots, 7$).

h. Make a half-normal plot of the absolute values of the fitted effects found in (f) using the quantiles found in (g). Do you see evidence in this plot of any kapton, preprocessing time, or dianhydride type effects on the *variability* of percent weight loss? Why or why not?

6.21. Refer to the **NASA Percent Weight Loss** case in problems (6.19) and (6.20). Continue the analysis begun in problem 6.20 where the factor "oven position" is ignored.

a. How many pairs can be formed from the set of eight different treatment combinations considered by Jobe, Sutter, and Ortiz?

b. Find s_{pooled} based on the eight "samples" of size $m = 2$. What degrees of freedom are associated with this estimate of σ?

c. Use Tukey's method to find a value Δ, so that if one attaches an uncertainty of $\pm\Delta$ to every possible difference in the eight percent weight loss sample means, one may be 95% confident of the whole set of resulting intervals (as estimates of the corresponding differences in mean weight losses).

d. Using the P-R method, find a set of (eight) two-sided interval estimates for the treatment mean weight losses that have an associated 95% simultaneous confidence level.

e. Consider the all-high treatment combination. Find individual 95% confidence intervals for all three two-factor interactions and the three-factor interaction corresponding to this combination. How "sure" are you that all four intervals bracket their corresponding theoretical interactions? (Use the Bonferroni inequality.)

f. The investigators wanted a good estimate of the effect (averaged over the two time conditions) of kapton on the mean percent weight loss of electronic grade PMR-II-50 polymer resin. Find a 95% two-sided confidence interval for the mean percent weight loss of electronic grade PMR-II-50 polymer resin made with kapton, minus the mean for specimens made without kapton (both averaged over time). Find a 95% two-sided confidence interval for the difference in the two kapton main effects (with kapton minus without kapton). Which interval is better for the investigators' purposes? Defend your answer.

g. The investigators wanted a good estimate of the effect (averaged over the two time conditions) of kapton on the mean percent weight loss of polymer grade PMR-II-50 polymer resin. Find a 95% two-sided confidence interval for the mean percent weight loss of polymer grade PMR-II-50 polymer resin made with kapton minus the mean for specimens made without kapton (both averaged over time). Find a 95% two-sided confidence interval for the difference in the two kapton main effects (with kapton minus without kapton). Which interval is better for the investigators' purposes? Defend your answer.

h. The investigators wanted a good estimate of the effect of dianhydride type (grade) (averaged over the two time conditions) on the mean percent weight loss of PMR-II-50 polymer resin. Find a 95% confidence interval for the mean percent weight loss of polymer grade PMR-II-50 with kapton minus the mean for specimens of electronic grade PMR-II-50 with kapton (both averaged over time). Find a 95% interval for the difference in dianhydride type main effects (polymer grade minus electronic grade). Which interval is better for the investigators' purposes? Defend your answer.

6.22. **NASA Fe.** Refer to the **NASA Percent Weight Loss** case in problem (6.19). A portion of each PMR-II-50 specimen represented by the data in problem (6.19) was not exposed to 600°F temperature for 936 hours but instead was analyzed for iron (Fe) content. It is known that electronic grade dianhydride has small amounts of iron and polymer grade dianhydride has larger amounts of iron. NASA researchers Sutter and Ortiz were also aware of the possibility of iron transfer from the pressing mechanism used to form the polymer resin specimens. Thus, the protective kapton coating was used on the mechanism in pressing half of the PMR-II-50 polymer resin specimens and no kapton was used for the others. Data like the following sixteen responses (y) (in ppm Fe) were obtained.

Kapton	Preprocessing Time	Dianhydride Type	y
1	1	1	12.0, 2.5
2	1	1	9.0, 9.1
1	2	1	2.6, 2.5
2	2	1	25.0, 2.5
1	1	2	18.0, 34.0
2	1	2	7.0, 4.5
1	2	2	3.0, 14.0
2	2	2	2.5, 2.6

a. How many experimental factors were there in this study? Identify them and say how many levels of each were used.
b. Describe the factorial structure of this arrangement of treatment combinations using a "base and exponent" notation.
c. How many treatment combinations are there in this experiment? Identify them.
d. Does "oven position" play any role in the iron content measurements here?
e. Was there replication in this study? What are the eight sample sizes in this study?
f. Give a model equation in factorial effects notation for the "normal distributions with a common variance" description of Fe content.
g. Find the eight sample means.
h. Find the fitted effects corresponding to the all-high treatment combination using the Yates algorithm. Say what terms in your model from (f) each of these is meant to estimate.
i. Find the 16 residuals for this data set. Plot the residuals versus the sample means (that can function as fitted or predicted Fe responses). Does it appear that a constant variance model is reasonable? Why or why not?
j. Find the sample standard deviation of iron content, s, for each of the eight treatment combinations. Let $y^* = \ln(s)$ and find new fitted effects for the log standard deviations, y^*.
k. Use equation (6.33) and find the $(i - .5)/7$ half-normal quantiles (for $i = 1, 2, \ldots, 7$).
l. Make a half-normal plot of the absolute values of the fitted effects computed in (j). Do you see evidence in this plot of any kapton, preprocessing time, or dianhydride type effects on the *variability* of iron content? Explain.

6.23. Refer to the **NASA Fe** case in problem (6.22). Transform each Fe content response by taking its natural logarithm. Answer parts (g)–(l) of problem (6.22) based on the log iron contents. Should the data in problem (6.22) be analyzed in terms of the original y or in terms of $\ln(y)$? Why?

6.24. Refer to **NASA Fe** case in problems (6.22) and (6.23). In this problem, use the log transformed responses.
a. Find s_P for the log Fe contents. What degrees of freedom are associated with this estimate of σ?
b. Use the P-R method and make simultaneous 95% two-sided confidence intervals for the eight mean log iron contents.
c. If the end points of each interval in (b) are exponentiated (plugged into the function $\exp(\cdot)$), what parameters of the original iron distributions are estimated?
d. Use Tukey's method to find a value Δ, so that if one attaches an uncertainty of $\pm\,\Delta$ to every possible difference in the eight sample mean log Fe contents, one may be 95% confident of the whole set of resulting

intervals (as estimates of the corresponding differences in theoretical mean log iron contents).

e. If the end points of each interval in (d) are exponentiated (plugged into the function exp(·)), what does a given interval estimate in terms of parameters of the original iron distributions?

f. Consider the all-high treatment combination. Find individual 99% confidence intervals for all three two-factor interactions and the three-factor interaction corresponding to this combination. How "sure" are you that all four intervals bracket their corresponding theoretical interactions? (Use the Bonferroni inequality.)

g. The investigators wanted a good estimate of the effect (averaged over the two time conditions) of kapton on the mean log Fe content of electronic grade PMR-II-50 polymer resin. Find a 95% two-sided confidence interval for the mean log Fe content of electronic grade PMR-II-50 polymer resin made with kapton minus the mean for specimens made without kapton (both averaged over time). Find a 95% two-sided confidence interval for the difference in the two kapton main effects (with kapton minus without kapton). Which interval is better for the investigators' purposes? Defend your answer.

h. The investigators wanted a good estimate of the effect of dianhydride type (grade) (averaged over the two time conditions) on the mean log Fe content of PMR-II-50 polymer resin. Find a 95% confidence interval for the mean log Fe content of polymer grade PMR-II-50 with kapton minus the mean for electronic grade PMR-II-50 with kapton (both averaged over time). Find a 95% confidence interval for the difference in dianhydride type main effects (polymer grade minus electronic grade). Which interval is better for the investigators' purposes? Defend your answer.

i. Using the fitted effects computed in part (h) of problem (6.23), find a set of interval estimates for the factorial effects corresponding to the all-high treatment combination. Make 99% two-sided individual confidence intervals. What can be said concerning the confidence that one should have that each one of these intervals simultaneously contains its corresponding theoretical effect? (Use the Bonferroni inequality.)

j. If an interval in (i) includes 0, set that fitted effect to 0, otherwise use the value from part (h) of problem (6.23) and find fitted values for all eight treatment combination mean log Fe contents using the reverse Yates algorithm. Show your work in table form.

6.25. Refer to the **NASA Percent Weight Loss** and **NASA Fe** cases in problems (6.19), (6.20), (6.21), (6.22), (6.23), and (6.24). Reflect on your responses to these problems. Does it appear Fe content is related to percent weight loss? Why or why not?

6.26. Refer to the **Brush Ferrules** case in problem (1.15). Adams, Harrington, Heemstra, and Snyder identified several factors that potentially affected ferrule thickness. Ultimately, they were able to design and conduct an experiment with the two factors "crank position" and "slider position." Two levels (1.625 inch and 1.71875 inch) were selected for crank position and two levels (1.75 inch and 2.25 inch) were selected for slider position. Four new ferrules were produced for every combination of crank position and slider position. The resulting ferrule thicknesses (in inches) are given below.

Crank Position	Slider Position	Thickness
1.625	1.75	.421, .432, .398, .437
1.71875	1.75	.462, .450, .444, .454
1.625	2.25	.399, .407, .411, .404
1.71875	2.25	.442, .451, .439, .455

a. Give a model equation for the "normal distributions with a common variance" description of ferrule thickness that uses "factorial effects" notation. Say what each term in your model represents in the context of the problem and define each one in terms of the $\mu_{..}$, $\mu_{i.}$, $\mu_{.j}$, and μ_{ij}.

b. Find the four sample means.

c. Plot the sample means versus crank position in interaction plot format. (Connect means having the same slider level with line segments.) Does it appear there are strong interaction effects? Why or why not?

d. Find the residuals. Plot these versus sample means. Does it appear that a common variance (across treatment combinations) assumption is reasonable? Why or why not?

e. In Example 6.4, in particular the portion of it appearing just before the beginning of Section 6.3.4, half-normal plotting of absolute fitted effects of two-level factors on logged sample standard deviations is used as a means of looking for factor effects on response *variability*. Why is that method likely to provide little insight here? (Hint: How many points would you end up plotting? How decidedly "nonlinear" could such a plot possibly look?)

f. Use the Yates algorithm and find fitted effects for the thickness data corresponding to the all-high treatment combination.

g. Find s_P for the thickness data.

h. Enhance your plot from (c) with the addition of error bars around the sample means derived from P-R simultaneous 95% two-sided confidence intervals for the four treatment combination means.

i. Find a 95% two-sided confidence interval for the interaction effect for the all-high treatment combination. Interpret this interval and point out in what ways it is consistent with the enhanced plot from (h).

j. Find a 95% two-sided confidence interval for the difference in slider position main effects (2.25-inch setting minus 1.75-inch setting). Would you use this interval to estimate the difference in mean thicknesses for 1.625-inch crank position ferrules (2.25-inch slider setting minus 1.75-inch slider setting)? Why or why not? (Consider your answer to part (i).)

k. Find a 95% two-sided confidence interval for the difference in crank position main effects (1.71875-inch setting minus 1.625-inch setting). Would you use this interval to estimate the difference in mean thicknesses for 2.25-inch slider position ferrules (1.71875-inch crank setting minus 1.625-inch crank setting)? Why or why not? (Consider your answer to part (i).)

6.27. Refer to the **Brush Ferrules** case in problems (1.15) and (6.26). The slider mentioned in problems (1.15) and (6.26) controls the first stage of a forming process and the crank controls the second.

a. Make individual 95% two-sided confidence intervals for the two main effects and the two-factor interaction corresponding to the 1.71875 crank position and 2.25 slider position (the all-high) treatment combination. Why is it sufficient to consider only these factorial effects in this 2^2 factorial study? (For instance, why is nothing additional gained by considering the effects corresponding to the "all-low" treatment combination?)

b. Replace by 0 any effect whose interval in (a) includes zero and use the reverse Yates algorithm to find fitted means for a model of ferrule thickness that involves only those effects judged (via the intervals from part (a)) to be statistically detectable.

c. Plot the fitted thickness values from part (b) versus crank position. Connect the two points for the 1.75 slider position with a line segment and then connect the two points for the 2.25 slider position with another line segment.

d. The desired ferrule thickness was .4205 inches. Assuming the linearity of your plot in (c) is appropriate, give several (at least two) different combinations of crank and slider positions that would produce the desired ferrule thickness.

6.28. Refer to the **Cut-Off Machine** case in problem (1.18). The focus of project team efforts was improving the tool life of carbide cutting inserts. Carbide cutting inserts are triangular shaped pieces of titanium-coated carbide about 3/16-inch thick. All three corners of a given insert can be used to make cuts. A crater, break, or poor quality dimension of a cut part are typical indicators of a "failed" corner, and the objective was to improve (increase) the number of cuts that could be made before failure of an insert.

Stop delay and feed rate were seen as factors potentially affecting tool life. Stop delay is the time required to insert raw material (during which the insert cools). Feed rate is the rate at which the cutting insert is forced into the tubing being cut. Tubing RPM was also identified as a factor possibly affecting tool life, but because of time constraints this factor was held fixed during the team's study. The team considered two stop delay settings (low, high) and four different feed rate settings (coded here as 1 through 4). For each combination of stop delay and feed rate, a new carbide insert was used to cut 304 stainless steel. The number of tubes cut (until a failure occurred) was recorded for each corner of the insert. Thus, three responses (one from each corner of an insert) were recorded for each combination of stop delay and feed rate. The resulting data are below.

Stop Delay	Feed Rate	Number of Tubes Cut
1	1	125, 129, 146
1	2	135, 130, 176
1	3	194, 183, 166
1	4	176, 187, 204
2	1	136, 141, 149
2	2	169, 155, 177
2	3	162, 207, 198
2	4	163, 195, 224

a. How many treatment combinations were there in this study?

b. Find the eight "sample" means.

c. Plot the sample means from (b) in interaction plot format, placing feed rate on the horizontal axis. Does there seem to be serious interaction between stop delay and feed rate? Why or why not?

d. Find all the fitted factorial effects for this study. (Find two stop delay main effects, four feed rate main effects and eight two-factor interaction effects.)

e. How many carbide inserts were used in this experiment? Do you think that this experiment was equivalent to one in which only one corner is used from each insert (three inserts per treatment combination)? Would you expect to see more variation or less variation in response than that in the data above, if three inserts (one corner of each) had been used?

f. Probably the safest analysis of the data above would simply use the averages from (b) as responses, admitting that there was no real replication in the study and that one really has eight samples of size $m = 1$. If this route were taken, could you go beyond the computations in (d) to make confidence intervals for main effects and interactions? Explain.

Henceforth in this problem suppose that the data in the table above actually represent total cuts (for all corners) for three different inserts per treatment combination (so that it makes sense to think of the data as eight samples of size $m = 3$).

g. Give a model equation for the "normal distributions with a common variance" description of number of tubes cut that uses "factorial effects" notation. Say what each term in your model represents in the context of the problem and define each one in terms of the $\mu_{..}$, $\mu_{i.}$, $\mu_{.j}$, and μ_{ij}.

h. Find the 24 residuals by subtracting sample means from observations.

i. Plot the residuals found in (h) versus the sample means. Does it appear that the constant variance feature of the model in (g) is appropriate for number of tubes cut? Why or why not?

j. Find s_P, the pooled estimate of the supposedly common standard deviation, σ.

k. Normal plot the 24 residuals found in (h). What insight does a plot of this type provide?

6.29. Refer to the **Cut-Off Machine** case in problems (1.18) and (6.28). As in the last half of problem (6.28), treat the data given in that problem as if they had been obtained using three different inserts per treatment combination. Use the "normal distributions with a common variance" model for number of cuts per insert in the following analysis.

a. Use the P-R method to make simultaneous 95% lower confidence bounds for the eight mean numbers of cuts.

b. Use Tukey's method to find a value Δ, so that if one attaches an uncertainty of $\pm\Delta$ to every possible difference in the eight sample mean numbers of cuts, one may be 99% confident of the whole set of resulting intervals (as estimates of the corresponding differences in theoretical mean numbers of cuts).

c. On the basis of your analysis in (b), is there a combination of stop delay and feed rate that is clearly better than the others? Why or why not?

d. Find a value Δ so that an interval with end points $ab_{ij} \pm \Delta$ can be used as a 95% two-sided confidence interval for the corresponding interaction $\alpha\beta_{ij}$.

e. Make a 99% two-sided confidence interval for the difference in the mean numbers of cuts for feed rate 2 with stop delay 1 and feed rate 2 with stop delay 2.

f. Make a 99% two-sided confidence interval for the difference in the mean numbers of cuts for feed rate 4 with stop delay 1 and feed rate 3 with stop delay 1.

g. Find a 99% two-sided confidence interval for the difference in stop delay main effects (stop delay 1 minus stop delay 2). Would you use this interval in place of the one from (e)? Why or why not?

h. Find a 99% two-sided confidence interval for the difference in feed rate 4 and feed rate 3 main effects. Would you use this interval in place of the one from (f)? Why or why not?

6.30. Refer to the **Tablet Hardness** case in problem (4.22). Tablet hardness is measured in Standard Cobb Units (SCUs) and specifications for hardness are 17 ± 5 SCUs. After analyzing the tablet production process, engineers concluded that tablet press compression level and powder moisture level had large effects on final tablet hardness. Management was using a trial-and-error method to adjust compression and moisture levels. The engineering team decided to adopt a more systematic approach to finding a good combination of compression and moisture. Low and high settings were identified for both factors for purposes of experimentation. The following summaries were obtained from an experiment where the four different combinations of compression and moisture content were used to produce tablets. Given in the table are means and standard deviations of hardness for a number of batches made under each treatment combination. (Batch hardness, y, was defined by testing and averaging test results for ten tablets from the batch. The n's in the following table are numbers of batches and the $\bar{y}$'s and s's were obtained by averaging and finding the standard deviations of the batch hardnesses, y.)

Compression	Moisture	n	$\bar{y}$	s
low	low	29	17.59	1.22
high	low	17	16.75	.71
low	high	27	17.53	.78
high	high	23	17.60	1.02

a. How many experimental factors were there in this study? Name them.

b. Give a model equation for the "normal distributions with a common variance" description of batch hardness that uses "factorial effects" notation.

c. Find the pooled estimate of the (supposedly common) variance of batch hardness under a fixed set of processing conditions. What are the associated degrees of freedom?

d. Plot the sample means in interaction plot format. Put compression on the horizontal axis. Does it appear that there are strong interactions between compression and moisture? Why or why not?

e. Find simultaneous 95% two-sided confidence intervals for the four treatment combination means.

f. Enhance your plot in (d) by drawing error bars around the means based on your confidence limits from (e). What is suggested by these about the statistical detectability of interaction (lack of parallelism) in this study?

g. Find the fitted effects for each effect in the model given in (b).

h. Find a 99% two-sided confidence interval for the two factor interaction at the high levels of both factors.

i. Find a 99% two-sided confidence interval for the high moisture main effect. Find a 99% two-sided confidence interval for the high compression main effect.

j. Using the Bonferroni inequality, find a set of 94% simultaneous confidence intervals for the three effects estimated in (h) and (i).

k. Based on your intervals from (j) and the enhanced interaction plot from (f) discuss what has been learned about how compression and moisture impact tablet hardness.

l. Which combination of compression and moisture seems best? Defend your answer.

m. Using the information given in this problem estimate the fraction of high compression/low moisture batches that have measured "batch hardness" within 1 SCU of the target value of 17 SCU. Why don't you have appropriate information here to estimate the fraction of *individual* tablets made under these same conditions with tested hardness inside the 17 ± 5 SCU specifications (for individuals)?

6.31. Refer to the **Tablet Hardness** case in problems (4.22) and (6.30). One might apply the ideas of Section 5.5 to the situation described in problem (6.30) *for a particular compression/moisture combination.* In the notation of that section, for σ_α^2 a between batch variance component and σ^2 a within batch variance component, s_P^2 from part (c) of problem (6.30) is sensibly thought of as estimating $\sigma_\alpha^2 + \sigma^2/10$. (Remember that sample "batch" hardness was computed based on 10 tablets.) On the other hand, the variance of the hardness of a single tablet taken from a single batch is $\sigma_\alpha^2 + \sigma^2$. Suppose that (in addition to the information given in problem (6.30)) you are now told that tests on any single batch have a standard deviation of approximately .88 SCUs.

a. Find an estimate of $\sqrt{\sigma_\alpha^2 + \sigma^2}$.

b. Use your answer to (a) and the sample mean from problem (6.30) and estimate the fraction of individual high compression/low moisture tablets that have hardnesses inside the 17 ± 5 specifications.

c. Redo your calculation from (b) for each of the other three treatment combinations.

d. Discuss how you would proceed to estimate fractions of individuals inside specifications if "batch" hardnesses were based on the average measurements for only 3 (not 10) tablets. (How would the analysis of this problem change?)

6.32. Consider a situation where fatigue life testing of steel bar stock is to be done. Bar stock can be ordered from several different vendors and it can be ordered to several different sets of specifications with regard to each of the factors "dimensions," "hardness," and "chemical composition." There are several different testing machines in a lab and several different technicians could be assigned to do the testing. The response variable will be the number of cycles to failure experienced by a given test specimen.

a. Describe one three-factor full factorial study that might be carried out in this situation. Make out a data table that could be used to record the results. For each "run" of the experiment, specify the levels of each of the 3 factors to be used. What constitutes "replication" in your plan?

b. Suppose that attention is restricted to steel bar stock from a single vendor, ordered to a single set of specifications, tested on a single machine by a single technician. Suppose further, that either 10 specimens from a single batch or 1 specimen from each of 10 different batches can be used. Under which of the two scenarios would you expect to get the larger variation in observed fatigue life? Under what circumstances would the first test plan be most appropriate? Under what circumstances would the second be most appropriate?

c. Suppose that five specimens from each of three different batches (say I, II, and III) ordered to a single set of specifications from a single vendor are to be tested on a single machine by a single technician. However, the time required to make a test is such that only three specimens can be tested on a given day. Suppose that there is some concern that lab conditions may vary day to day in ways that could possibly impact the observed fatigue lives. Develop a plan for doing the required testing. (Specify for each of five days which batches will be tested and in what order.) Carefully describe the rationale behind your plan. If, for data analysis purposes, one simply computes three sample standard deviations for the specimens from a given batch, do you expect these values to overstate or to understate "within batch variability when testing on a single day"? Explain.

6.33. Resistance Measurements. Anderson, Koppen, Lucas, and Schotter made some resistance measurements on five nominally 1000 Ω resistors. The measurements were made with an analog meter, an old digital meter and a new digital meter. As it turns out, the analog readings and a set of new digital readings were made on one day and the old digital readings and a second set of new digital readings were made on another day. The students' data are given on page 344. We will assume that the readings were all made by the same person, so that the matter of reproducibility is not an issue here.

a. Discuss why it doesn't necessarily make sense to treat the two measurements for each resistor made with the new digital meter as a "sample" of size $m = 2$ from a single population, and to then derive a measure of digital meter repeatability from these five samples of size two. If the students had wanted to evaluate measurement repeatability how should they have collected some data?

b. It is possible to make a judgment as to how important the "day" effect is for the new digital meter by doing the following. For each resistor, subtract the day 1 new digital reading from the day 2 new digital reading to produce five differences d. Then apply the formula from elementary statistics, $\overline{d} \pm ts_d/\sqrt{5}$ to make a two-sided confidence interval for the mean difference, μ_d. Do this using 95% confidence. Does your interval include 0? Is there a statistically detectable "day" effect in the new digital readings? What does this analysis say about the advisability of ignoring "day," treating the two measurements on each resistor as a sample of size $m = 2$ and attempting to thereby estimate repeatability for the new digital meter?

Resistor	Day	Meter	Measured Resistance
1	1	analog	999
1	1	new digital	994
2	1	analog	1000
2	1	new digital	1001
3	1	analog	999
3	1	new digital	992
4	1	analog	999
4	1	new digital	987
5	1	analog	1000
5	1	new digital	1001
1	2	old digital	981
1	2	new digital	993
2	2	old digital	988
2	2	new digital	1000
3	2	old digital	979
3	2	new digital	992
4	2	old digital	974
4	2	new digital	987
5	2	old digital	988
5	2	new digital	1000

c. Differencing the day 1 analog and new digital readings to produce five differences (say analog minus new digital), gives a way of looking for systematic differences between these two meters. One may apply the formula from elementary statistics, $\bar{d} \pm ts_d/\sqrt{5}$ to make a confidence interval for the mean difference, μ_d. Do this, making a 95% two-sided interval. Does your interval include 0? Is there a statistically detectable systematic difference between how these meters read?

d. Redo (c) for the day 2 old digital and new digital readings.

e. A means of comparing the analog and old digital meters looking for a systematic difference in readings is the following. For each resistor one may compute

$$\begin{aligned} y &= (\textit{analog} - \textit{new digital day} \, 1) \\ &\quad - (\textit{old digital} - \textit{new digital day} \; 2) \end{aligned}$$

and then apply the formula from elementary statistics, $\bar{y} \pm ts_y/\sqrt{5}$ to make a confidence interval for the mean μ_y. Do this, making a 95% interval. Does your interval include 0? Is there a statistically detectable systematic difference between how these meters read?

f. Carefully describe a complete factorial study with the factors "meters," "resistors," and "days" that includes some replication and would allow more straightforward use of the material of this chapter in assessing the effects of these factors. Does your plan allow for the estimation of the repeatability variance component discussed in Section 2.2?

6.34. Consider the situation of Example 6.2. Suppose a colleague faced with a similar physical problem says "We don't need to change levels of both factors at once. The scientific way to proceed is to experiment one factor at a time. We'll hold the cooling method fixed and change antimony level to see the antimony effect. Then we'll hold antimony level fixed and change cooling method in order to see the cooling method effect." What do you have to say to this person?

6.35. **Heat Treating Steel.** Bockenstedt, Carrico, and Smith investigated the effects of steel formula (1045 and 1144), austenizing temperature (800°C and 1000°C), and cooling rate (furnace and oil quench) on the hardness of heat-treated steel.

a. If all possible combinations of levels of the factors mentioned above are included in an experiment, how many treatment combinations total will be studied?

b. If only a single specimen of each type alluded to in (a) was tested, how would one go about judging the importance of the main effects and interactions of the three factors?

In fact, a 2^3 full factorial with $m = 3$ steel specimens per treatment combination was run and hardness data like those below were obtained.

Steel	Temperature	Cooling Rate	Hardness, y
1045	800	furnace cool	186.0, 191.0, 187.0
1144	800	furnace cool	202.5, 204.0, 202.0
1045	1000	furnace cool	146.0, 153.0, 147.0
1144	1000	furnace cool	154.0, 156.0, 156.5
1045	800	oil quench	222.5, 230.0, 221.5
1144	800	oil quench	239.5, 248.5, 249.0
1045	1000	oil quench	268.0, 278.0, 272.5
1144	1000	oil quench	297.5, 296.0, 299.0

c. Compute the pooled standard deviation here, s_P.

d. Find the eight sample means and apply the Yates algorithm to find the fitted 2^3 factorial effects corresponding to the all-high (1144/1000/oil quench) treatment combination.

e. Apply the basic formula (6.31) to make seven individual 95% two-sided confidence intervals for the main effects, two-factor interactions and three-factor interaction for the all-high treatment combination here.

f. Apply formula (6.32) to make seven two-sided intervals with simultaneous 95% confidence level for estimating the main effects, two-factor interactions and three-factor interaction for the all-high treatment combination.

g. Based on the intervals from (f), which effects do you judge to be statistically detectable?

h. In light of your answer to (g), if maximum hardness is desired, but hardness being equal the preferable levels of the factors (perhaps for cost reasons) are 1045, 800, and oil quench, how do you recommend setting levels of these factors? Explain.

6.36. Valve Airflow. In their "Quality Quandaries" article in the 1996, volume 8, number 2 issue of *Quality Progress*, Bisgaard and Fuller further developed an example due originally to Moen, Nolan, and Provost in their *Improving Quality Through Planned Experimentation*. The emphasis of the Bisgaard and Fuller analysis was to consider the effects of four experimental factors on both mean *and standard deviation* of a response variable, y, measuring airflow through a solenoid valve used in an auto air pollution control device. The factors in a 2^4 factorial study with $m = 4$ were length of armature (A) (.595 inch vs. .605 inch), spring load (B) (70 g vs. 100 g), bobbin depth (C) (1.095 inch vs. 1.105 inch), and tube length (D) (.500 inch vs. .510 inch). Mean responses and sample standard deviations are given below.

A	B	C	D	$\bar{y}$	s
−	−	−	−	.46	.04
+	−	−	−	.42	.16
−	+	−	−	.57	.02
+	+	−	−	.45	.10
−	−	+	−	.73	.02
+	−	+	−	.71	.01
−	+	+	−	.70	.05
+	+	+	−	.70	.01
−	−	−	+	.42	.04
+	−	−	+	.28	.15
−	+	−	+	.60	.07
+	+	−	+	.29	.06
−	−	+	+	.70	.02
+	−	+	+	.71	.02
−	+	+	+	.72	.02
+	+	+	+	.72	.01

a. As in the part of Example 6.4 just preceding Section 6.3.4, take the natural logarithms of the sample standard deviations and then use the Yates algorithm and normal plotting (or half-normal plotting) to look for statistically detectable effects on the *variability* of airflow. Which factors seem to have significant effects? How would you recommend setting levels of these factors if the only object were consistency of airflow?

b. Your analysis in part (a) should suggest some problems with a "constant (across treatment combinations) variance" model for airflow. Nevertheless, at least as a preliminary or rough analysis of *mean* airflow, compute s_P, apply the Yates algorithm to the sample means, and apply the simultaneous 95% confidence limits from display (6.32) to judge the detectability of the 2^4 factorial effects on mean airflow. Which variables seem to have the largest influence on this quantity?

6.37. Collator Machine Stoppage Rate. Klocke, Tan, and Chai worked with the ISU Press on a project aimed at reducing jams or stoppages on a large collator machine. They considered the two factors "bar tightness" and "air pressure" in a 2×3 factorial study. For each of six different treatment combinations, they counted numbers of stoppages and recorded machine running time in roughly 5 minutes of machine operation. (The running times didn't include downtime associated with fixing jams. As such, every instant of running time could be thought of as providing opportunity for a new stoppage.) The table below summarizes their data.

Bar Tightness	Air Pressure	Number of Jams, X	Running Time, k (sec)	$\hat{u} = X/k$
tight	low	27	295	.0915
tight	medium	21	416	.0505
tight	high	33	308	.1071
loose	low	15	474	.0316
loose	medium	6	540	.0111
loose	high	11	498	.0221

This situation can be thought of as a "mean nonconformities per unit" scenario, where the "unit of product" is a 1-second time interval.

a. A crude method of investigating whether there are any clear differences in the six different operating conditions is provided by the retrospective u chart material of Section 3.3.2. Apply that material to the six $\hat{u}$ values given in the table above and say whether there is clear evidence of some differences in the operating conditions (in terms of producing stoppages). (Note that a total of 113 stoppages were observed in a total of 2531 seconds of running time, so that a pooled estimate of a supposedly common λ is $113/2531 = .0446$.)

b. Plot the $\hat{u}$ values in interaction plot format, placing levels of air pressure on the horizontal axis.

c. As in Section 3.3.2, a Poisson model for X (with mean $k\lambda$) produces a standard deviation for $\hat{u}$ of $\sqrt{\lambda/k}$. This in turn suggests estimating the standard deviation of $\hat{u}$ with

$$\hat{\sigma}_{\hat{u}} = \sqrt{\hat{u}/k}.$$

Then, very crude approximate confidence limits for λ might then be made as

$$\hat{u} \pm z\,\hat{\sigma}_{\hat{u}}$$

(for z a standard normal quantile). For each of the six treatment combinations, make approximate 99% two-sided limits for the corresponding stoppage rates. Use these to place error bars around the rates plotted in (b).

d. Based on the plot from (b) and enhanced in (c), does it appear that there are detectable bar tightness/air pressure interactions? Does it appear that there are detectable bar tightness or air pressure main effects? Ultimately, how do you recommend running the machine?

e. The same logic used in (c) says that for r different conditions leading to r values $\hat{u}_1, \ldots, \hat{u}_r$ and r constants $c_1, \ldots, c_r$, the linear combination

$$\hat{L} = c_1\hat{u}_1 + \cdots + c_r\hat{u}_r$$

has a standard error

$$\hat{\sigma}_{\hat{L}} = \sqrt{\sum_{i=1}^{r} \frac{c_i^2 \hat{u}_i}{k_i}},$$

that can be used to make approximate confidence intervals for

$$L = c_1\lambda_1 + \cdots + c_r\lambda_r$$

as

$$\hat{L} \pm z\,\hat{\sigma}_{\hat{L}}.$$

Use this method and make an approximate 95% two-sided confidence interval for

$$\alpha_1 - \alpha_2 = \frac{1}{3}(\lambda_{11} + \lambda_{12} + \lambda_{13}) - \frac{1}{3}(\lambda_{21} + \lambda_{22} + \lambda_{23}),$$

the difference in tight and loose bar main effects.

6.38. Refer to the **Collator Machine Stoppage Rate** case in problem (6.37). The analysis in problem (6.37) is somewhat complicated by the fact that the standard deviation of $\hat{u}$ depends not only on k, but on λ as well. A way of somewhat simplifying the analysis is to replace $\hat{u}$ with $y = g(\hat{u})$ where g is chosen so that (at least approximately) the variance of y is independent of λ. The Freeman-Tukey suggestion for g is

$$y = g(\hat{u}) = \frac{\sqrt{\hat{u}} + \sqrt{\hat{u} + \frac{1}{k}}}{2},$$

and unless λ is very small,

$$\text{Var } y \approx \frac{1}{4k}.$$

This problem considers the application of this idea to simplify the analysis of the collator machine data.

a. Compute the six y values corresponding to the different observed jam rates in problem (6.37).

b. Plot the y values in interaction plot format, placing levels of air pressure on the horizontal axis.

c. Approximate confidence limits for the mean of y ($\text{E}y = \mu_y$) are

$$y \pm z \frac{1}{2\sqrt{k}},$$

for z a standard normal quantile. Make individual two-sided 99% confidence limits for each of the six different means of y. Use these to enhance the plot in (b) with error bars.

d. Based on the plot from (b) and enhanced in (c), does it appear that there are detectable bar tightness/air pressure interactions? Does it appear that there are detectable bar tightness or air pressure main effects? Ultimately, how do you recommend running the machine?

e. If r different conditions lead to r different values $y_1, \ldots, y_r$ and one has in mind r constants $c_1, \ldots, c_r$, the linear combination

$$\hat{L} = c_1 y_1 + \cdots + c_r y_r$$

has an approximate standard deviation

$$\sigma_{\hat{L}} \approx \frac{1}{2}\sqrt{\sum_{i=1}^{r} \frac{c_i^2}{k_i}}.$$

The quantity

$$L = c_1\mathrm{E}y_1 + \cdots + c_r\mathrm{E}y_r$$

then has approximate confidence limits

$$\hat{L} \pm z\sigma_{\hat{L}},$$

and these provide means for judging the statistical detectability of factorial effects (on the mean of the transformed variable). Use this method and make an approximate 95% two-sided confidence interval for

$$\alpha_1 - \alpha_2 = \frac{1}{3}(\mathrm{E}y_{11} + \mathrm{E}y_{12} + \mathrm{E}y_{13}) - \frac{1}{3}(\mathrm{E}y_{21} + \mathrm{E}y_{22} + \mathrm{E}y_{23}),$$

the difference in tight and loose bar main effects. Does this method show that there is a clear difference between the bar tightness main effects?

CHAPTER 7

Experimental Design and Analysis for Process Improvement Part 2: Advanced Topics

The basic tools of experimental design and analysis provided in Chapter 6 form a foundation for effective multifactor experimentation. But as with any foundation, a corresponding superstructure is needed before there is a whole building. This chapter provides some of the superstructure of statistical methods for process improvement experiments.

The first section of the chapter provides an introduction to the important topic of fractional factorial experimentation. The 2^{p-q} designs and analyses presented there give engineers effective means of screening a large number of factors, looking for a few that need more careful subsequent scrutiny. Then there are two sections concerned with using regression analysis as a tool in optimizing a process with quantitative inputs. The first of these considers the issue of "response surface methods" in systems where the process variables can be changed independently of each other. The second concerns the more specialized situation of "mixture studies" where process variables are composition fractions of some blend of material, and are thus constrained to total to 1.0. Finally, there is a section that discusses a number of important qualitative issues in experimentation for process improvement, including some given recent emphasis by G. Taguchi and his followers.

7.1 2^{p-q} FRACTIONAL FACTORIALS

It is common for engineers engaged in process-improvement activities to be initially faced with many more factors/"process knobs" than can be studied in a practically feasible full factorial experiment. For example, even with just two levels for each of $p = 10$ factors (which is not that many by the standards of real industrial processes) there are already $2^{10} = 1{,}024$ different combinations to be considered. And few real engineering experiments are run in environments where there are both time and resources sufficient to collect 1,000 or more data points.

If one cannot afford a full factorial experiment in many factors, the alternatives are two. One must either hold the levels of some factors fixed (effectively eliminating them from consideration in the experiment), or find some way to vary all of the factors over some appropriate **fraction** of a full factorial (and then make a sensible analysis of the resulting data). This section concerns methods for this second approach. The thinking here is that it is best in early stages of experimentation to run fractional factorial experiments in many factors, letting data (rather than educated guessing alone) help *screen* those down to a smaller number that can subsequently be studied more carefully.

The discussion begins with some additional motivation for the section and some preliminary insights into what can and cannot possibly come out of a fractional factorial study. Then specific methods of design and analysis are provided for half-fractions of 2^p factorials. Finally, the methods for half-fractions are generalized to provide corresponding tools for studies involving only a fraction $1/2^q$ of all possible combinations from a full 2^p factorial.

7.1.1 Motivation and Preliminary Insights

Table 7.1 lists two levels of 15 factors from a real industrial experiment discussed by C. Hendrix in his article, "What Every Technologist Should Know About Experimental Design," which appeared in *Chemtech* in 1979. The object of experimentation was to determine what factors were principal determiners of the cold crack resistance of an industrial product. Now $2^{15} = 32{,}768$ and there is clearly no way that plant experimentation could be carried out in a full 2^{15} factorial fashion in a situation like this. Something else had to be done.

Rather than just guessing at which of the 15 factors represented in Table 7.1 might be most important and varying only their levels in an experiment, Hendrix and his colleagues were able to conduct an effective fractional factorial experiment varying all $p = 15$ factors in only 16 experimental runs. (A 1/2048th fraction of all possible combinations of levels of these factors was investigated!) Methods of experimental design and analysis for problems like this are the subject of this section. But before jumping headlong into technical details, it is best to begin with some qualitative/common sense observations about what will ultimately be possible. (And the reader is encouraged to return to this subsection after wrestling

Table 7.1 15 Process Variables and Their Experimental Levels

Factor	Process Variable	Levels
A	Coating Roll Temperature	115° (−) vs. 125° (+)
B	Solvent	Recycled (−) vs. Refined (+)
C	Polymer X-12 Preheat	No (−) vs. Yes (+)
D	Web Type	LX-14 (−) vs. LB-17 (+)
E	Coating Roll Tension	30 (−) vs. 40 (+)
F	Number of Chill Rolls	1 (−) vs. 2 (+)
G	Drying Roll Temperature	75° (−) vs. 80° (+)
H	Humidity of Air Feed to Dryer	75% (−) vs. 90% (+)
J	Feed Air to Dryer Preheat	Yes (−) vs. No (+)
K	Dibutylfutile in Formula	12% (−) vs. 15% (+)
L	Surfactant in Formula	.5% (−) vs. 1% (+)
M	Dispersant in Formula	.1% (−) vs. .2% (+)
N	Wetting Agent in Formula	1.5% (−) vs. 2.5% (+)
O	Time Lapse Before Coating Web	10 min (−) vs. 30 min (+)
P	Mixer Agitation Speed	100 rpm (−) vs. 250 rpm (+)

with the technical details that will follow, in an effort to avoid missing the forest for the trees.)

To begin with, there is no Santa Claus or Good Fairy and one cannot hope to learn from a small fractional factorial experiment all that could be learned from the corresponding full factorial. In the 15-factor situation represented by Table 7.1, there are potentially 32,768 effects of importance in determining cold crack resistance (from a grand mean and 15 main effects through a 15-way interaction). Data from only 16 different combinations cannot possibly be used to detail all of these. In fact, intuition should say that y's from 16 different conditions ought to let one estimate at most 16 different "things." Anyone who maintains that there is some magic by which it is possible to learn all there is to know about a p variable system from a small fractional factorial study is selling snake oil. (The reader is hereby warned that there *are* high-priced industrial "consultants" who promise essentially this if their "system" of experimentation is followed.)

In fact, unless the principle of effect sparsity is strongly active, small fractions of large factorials are doomed to provide little useful information for process improvement efforts. That is, in complicated systems the methods of this section can fail, in spite of the fact that they are indeed the best ones available. So from the outset, the reader needs to understand that although they are extremely important and can be instrumental in producing spectacular process-improvement results, the methods of this section have unavoidable limitations that are simply inherent in the problem they address.

Having sounded this note of warning, it is important next to point out that *if* one is to do fractional factorial experimentation there are more effective and less effective ways of doing it. Take, for example, the (completely unrealistic but instructive) case of two factors A and B, both with two levels, supposing that one can afford to conduct only half of a full 2^2 factorial experiment. In this artificial context, consider the matter of experimental design, the choice of which of the four combinations (1), a, b, and ab to include in one's study. Temporarily supposing that the "all-low" combination, (1), is one of the two included in the experiment, it is obvious that combination ab should be the other. Why? The two combinations a and b can be eliminated from consideration because using one of them together with combination (1) produces an experiment where the level of only one of the two factors is varied. This kind of reasoning shows that the only two sensible choices of half-fractions of the 2^2 factorial are those consisting of "(1) and ab" or of "a and b." And while it is simple enough to reason to these choices of half of a 2^2 factorial, how in general to address the choice of a $1/2^q$ fraction of a 2^p factorial is much less obvious.

The artificial situation of a half-fraction of a 2^2 factorial can be used to make several other points as well. To begin with, suppose that one uses either the "(1) and ab" or the "a and b" half-fraction as an experimental plan and sees a huge change in response between the two sets of process conditions. How is that outcome to be interpreted? After all, *both* the A "knob" and the B "knob" are changed as one goes from one set of experimental conditions to the other. Is it the A main effect that causes the change in response? Or is it the B main effect? Or is it perhaps both? There is inevitable ambiguity of interpretation inherent in this example. Happily, in more realistic problems (at least under the assumption of effect sparsity) the prospects of sensibly interpreting the results of a 2^{p-q} experiment are not so bleak as they seem in this artificial example.

Finally, in the case of the half-fraction of the 2^2 factorial, it is useful to consider what information about the 2^2 factorial effects $\mu_{..}$, α_2, β_2, and $\alpha\beta_{22}$ can be carried by, say, $\bar{y}_{(1)}$ and $\bar{y}_{ab}$ alone. Clearly, $\bar{y}_{(1)}$ tells one about $\mu_{(1)}$ and $\bar{y}_{ab}$ tells one about μ_{ab}, but what information about the factorial effects do they provide? As it turns out, the story is this. On the basis of $\bar{y}_{(1)}$ and $\bar{y}_{ab}$, one can estimate two *sums of effects*, namely $\mu_{..} + \alpha\beta_{22}$ and $\alpha_2 + \beta_2$, but cannot further separate these four effects. The jargon typically used in the world of experimental design is that the A and B main effects are **confounded** or **aliased** (as are the grand mean and the AB interaction). And the fact that such basic quantities as the two main effects are aliased (and in some sense indistinguishable) in even this "best" half fraction of the 2^2 factorial, pretty much lays to final rest the possibility of any practical application of this pedagogical example.

With this motivation and qualitative background, it is hopefully clear that there are three basic issues to be faced in developing tools for 2^{p-q} fractional factorial experimentation. One must know how to

1. choose wisely a $1/2^q$ fraction of all 2^p possible combinations of levels of p two-level factors,
2. determine exactly which effects are aliased with which other effects for a given choice of the fractional factorial, and

3. do intelligent data analysis in light of the alias structure of the experiment.

The following two subsections address these matters, first for the case of half-fractions and then for the general $1/2^q$ fraction situation.

7.1.2 Half Fractions of 2^p Factorials

Consider now a situation where there are p two-level experimental factors and for some reason one wishes to include only half of the 2^p possible combinations of levels of these factors in an experiment. Standard notation for this kind of circumstance is that a **2^{p-1} fractional factorial** is contemplated. (The p exponent identifies the number of factors involved and the -1 exponent indicates that a half fraction is desired, one that will include 2^{p-1} different treatment combinations.)

The following is an algorithm for identifying a best possible half-fraction of a 2^p factorial:

Algorithm for Identifying a Best Half-Fraction

> Write down for the "first" $p-1$ factors, a $2^{p-1}\times(p-1)$ table of plus and minus signs specifying all possible combinations of levels of these, columns giving the levels for particular factors in combinations specified by rows. Then multiply together the "signs" in a given row (treating negative signs as -1's and positive signs as $+1$'s) to create an additional column. This new (product) column specifies levels of the "last" factor to be used with the various combinations of levels of the first $p-1$ factors.

Example 7.1 A Hypothetical 2^{4-1} Design. The preceding prescription can be followed to produce a good choice of 8 out of 16 possible combinations of levels of factors A, B, C, and D. Table 7.2 on page 356 shows that the combinations following from the algorithm are those that have an even number of factors set at their high levels.

Example 7.2 An Unreplicated 2^{5-1} Chemical Process Improvement Study. The article "Experimenting With a Large Number of Variables" by R. Snee, which appeared in the 1985 ASQC Technical Supplement *Experiments in Industry*, describes a $p = 5$ factor experiment aimed at improving the consistency of the color of a chemical product. The factors studied and their levels were:

Factor A—Solvent/Reactant	low $(-)$	vs.	high $(+)$
Factor B—Catalyst/Reactant	.025 $(-)$	vs.	.035 $(+)$
Factor C—Temperature	150°C $(-)$	vs.	160°C $(+)$
Factor D—Reactant Purity	92% $(-)$	vs.	96% $(+)$
Factor E—pH of Reactant	8.0 $(-)$	vs.	8.7 $(+)$

Table 7.2 Construction of a Best Half-Fraction of a 2^4 Factorial

A	B	C	(D) Product	Combination
−	−	−	−	(1)
+	−	−	+	ad
−	+	−	+	bd
+	+	−	−	ab
−	−	+	+	cd
+	−	+	−	ac
−	+	+	−	bc
+	+	+	+	abcd

Snee's unreplicated 2^{5-1} fractional factorial data are given here in Table 7.3, in a way that makes it clear that the algorithm for producing a best 2^{p-1} fractional factorial was followed. (The reader might also notice that for $p = 5$ factors, the prescription for constructing a good half-fraction picks out those combinations with an odd number of factors set at their high levels.)

Table 7.3 Observed Color Index for 16 Combinations of Levels of Five Two-Level Factors

A	B	C	D	(E) Product	Combination	Color Index, y
−	−	−	−	+	e	−.63
+	−	−	−	−	a	2.51
−	+	−	−	−	b	−2.68
+	+	−	−	+	abe	−1.66
−	−	+	−	−	c	2.06
+	−	+	−	+	ace	1.22
−	+	+	−	+	bce	−2.09
+	+	+	−	−	abc	1.93
−	−	−	+	−	d	6.79
+	−	−	+	+	ade	6.47
−	+	−	+	+	bde	3.45
+	+	−	+	−	abd	5.68
−	−	+	+	+	cde	5.22
+	−	+	+	−	acd	9.38
−	+	+	+	−	bcd	4.30
+	+	+	+	+	abcde	4.05

Having identified which 2^{p-1} combinations of levels of the factors one is going to include in a fractional factorial study, the next issue is understanding the implied **alias structure**, the pattern of what is confounded or aliased with what. As it turns out, 2^p factorial effects in a half-fraction are aliased in 2^{p-1} different *pairs*. One can hope to estimate 2^{p-1} sums of two effects, but cannot further separate the aliased effects. Exactly which pairs are confounded can be identified using a **system of formal multiplication** as follows:

Method for Determining the Alias Structure for a Best Half-Fraction

One begins by writing down the relationship

$$\text{name of the last factor} \leftrightarrow \text{product of names of the first } p-1 \text{ factors} \quad (7.1)$$

called the **generator** of the design, in that it specifies how the column of signs is made up for the last factor. Then one can multiply both sides of the relationship (7.1) by any letter under the rules that

1. any letter times itself produces the letter I, and
2. I times any letter is that letter.

The relationships that then arise identify pairs of effects that are aliased.

For example, it follows from display (7.1) and these rules of multiplication that

$$\text{I} \leftrightarrow \text{product of the names of all } p \text{ factors.} \quad (7.2)$$

Identifying I with the grand mean, this relationship says that the grand mean is aliased with the p factor interaction. Relationship (7.2), having I on the left side, is often called the **defining relation** for the fractional factorial. This is because one can easily multiply through by any string of letters of interest and see what is confounded with the corresponding main effect or interaction. For example, multiplying both sides of a defining relation like (7.2) by A, one sees that the main effect of A is aliased with the $(p-1)$-way interaction of all other factors.

Example 7.1 continued. Return to the hypothetical 2^{4-1} situation introduced earlier. In making up the combinations listed in Table 7.2, levels of Factor D were chosen using products of signs for Factors A, B, and C. Thus the generator for the design in Table 7.2 is

$$\text{D} \leftrightarrow \text{ABC}.$$

From this (multiplying through by D and remembering that D·D is I), the defining relation for the design is

$$\text{I} \leftrightarrow \text{ABCD},$$

and all aliases can be derived from this relationship. To begin with, the grand mean is aliased with the four-factor interaction. That is, based on data from the eight combinations listed in Table 7.2, one can estimate $\mu_{....} + \alpha\beta\gamma\delta_{2222}$ but cannot further separate the summands. Or, multiplying through by A, one has

$$A \leftrightarrow BCD,$$

and the A main effect and BCD three-factor interaction are aliases. One can estimate $\alpha_2 + \beta\gamma\delta_{222}$ but cannot further separate these effects. Or, multiplying both sides of the defining relation by both A and B, one has

$$AB \leftrightarrow CD.$$

The combinations in Table 7.2 produce data leaving the AB two-factor interaction confounded with the CD two-factor interaction. One can estimate $\alpha\beta_{22} + \gamma\delta_{22}$ but cannot further separate these two-factor interactions on the basis of half-fraction data alone.

Example 7.2 continued. In Snee's color index study, levels of Factor E were set using a column of signs derived as products of signs for Factors A, B, C, and D. That means that the generator of the design used in the study is

$$E \leftrightarrow ABCD,$$

so the defining relation is (upon multiplying through by E)

$$I \leftrightarrow ABCDE.$$

This implies that the grand mean is aliased with the five-factor interaction. Then, for example, multiplying through by A, one has

$$A \leftrightarrow BCDE$$

and the A main effect is confounded with the four-way interaction of B, C, D, and E. On the basis of the data in Table 7.3, one can estimate $\alpha_2 + \beta\gamma\delta\epsilon_{2222}$ but neither of the summands separately.

Examples 7.1 and 7.2 are instructive in terms of showing what happens to the alias structure of best 2^{p-1} designs with increasing p. For the 2^{4-1} case, main effects are aliased with three-factor interactions, while for five factors, main effects are confounded with four-factor interactions. If one expects high-order interactions to typically be negligible, this is a comforting pattern. It says that for moderate to large p, an estimate of a main effect plus the aliased $(p-1)$-way interaction may often be thought of as essentially characterizing the main effect alone. And this kind of thinking suggests what is standard design doctrine for 2^{p-q} studies. One wants to set things up so that (often important) low-order effects (main effects and low-order interactions) are aliased only with high-order interactions (that in simple systems are small). The virtue of the prescription for half-fractions given in this section is that it produces the best alias structure possible in this regard.

The final issue needing attention in this discussion of half-fractions is the matter of **data analysis**. How does one make sense out of 2^{p-1} fractional factorial data like those given in Table 7.3? This question has a simple answer:

Data Analysis Method for a 2^{p-1} Study

> To analyze 2^{p-1} fractional factorial data, one first ignores the existence of the last factor, treating the data as if they were complete factorial data in the first $p-1$ factors. "Fitted effects" are computed and judged exactly as in Section 6.3. Then the statistical inferences are interpreted in light of the alias structure, remembering that one actually has estimates of not single effects, but sums of pairs of 2^p effects.

Example 7.3 **An Artificial 2^{3-1} Data Set.** For purposes of illustrating the meaning of the instructions for data analysis just given when there is some replication in a 2^{p-1} data set, consider the artificial figures in Table 7.4.

The means in Table 7.4 are listed in Yates standard order as regards the first two factors. The reader should check that (ignoring Factor C) application of the Yates algorithm (two cycles and then division by $2^2 = 4$) produces the "fitted effects" (listed in Yates order for A and B) 5.0, 2.0, .5, and .1.

The pooled sample variance here is

$$s_P^2 = \frac{(2-1)1.5 + (3-1)1.8}{(2-1)+(3-1)} = 1.7,$$

with $n - r = 7 - 4 = 3$ associated degrees of freedom. So then formula (6.31) can be used to make confidence intervals for judging the statistical detectability of the

Table 7.4 Some Summary Statistics from a Hypothetical 2^{3-1} Study

A	B	C	Combination	n	$\bar{y}$	s^2
−	−	+	c	1	2.6	
+	−	−	a	1	6.4	
−	+	−	b	2	3.4	1.5
+	+	+	abc	3	7.6	1.8

(sums of) effects estimated by the output of the Yates algorithm. The "p" appropriate in formula (6.31) is 2, since one is computing as if the last factor does not exist. Then, since the .975 quantile of the t_3 distribution is 3.182, using individual two-sided 95% confidence limits, a plus or minus value of

$$3.182\sqrt{1.7}\left(\frac{1}{2^2}\right)\sqrt{\frac{1}{1}+\frac{1}{1}+\frac{1}{2}+\frac{1}{3}} = 1.7$$

should be associated with the values 5.0, 2.0, .5, and .1. By this standard, only the first two represent effects visible above the experimental variation.

This statistical analysis has to this point ignored the existence of Factor C. But now in interpreting the results, it is time to remember that the experiment was not a 2^2 factorial in Factors A and B, but rather a 2^{3-1} fractional factorial in Factors A, B, and C. Note that the signs in Table 7.4 show that the generator for this hypothetical study was

$$\mathrm{C} \leftrightarrow \mathrm{AB},$$

so that the defining relation is

$$\mathrm{I} \leftrightarrow \mathrm{ABC}.$$

Now if the means in Table 7.4 were from a 2^2 factorial, the value 5.0 appearing on the first line of the Yates calculations would be an estimate of a grand mean. But in the 2^{3-1} fractional factorial, the grand mean is aliased with the three-factor interaction, that is,

$$5.0 \quad \text{estimates} \quad \mu_{...} + \alpha\beta\gamma_{222}.$$

Similarly, if the means were from a 2^2 factorial, the value 2.0 appearing on the second line of the Yates calculations would be an estimate of the A main effect. But in the 2^{3-1} fractional factorial, the A main effect is aliased with the BC two-factor interaction, that is,

$$2.0 \quad \text{estimates} \quad \alpha_2 + \beta\gamma_{22}.$$

Of course, the simplest (and possibly quite wrong) interpretation of the fact that both $\mu_{...} + \alpha\beta\gamma_{222}$ and $\alpha_2 + \beta\gamma_{22}$ are statistically detectable would follow from an assumption that the two interactions are negligible and the estimated sums of effects are primarily measuring the overall mean and the main effect of Factor A.

Example 7.2 continued. As an example of how the analysis of a half fraction proceeds in the absence of replication, consider what can be done with Snee's color index data given in Table 7.3. The observations in Table 7.3 are listed in

Table 7.5 The Output of the Yates Algorithm Applied to the 16 Color Indices

Combination	Color Index, y	(Yates Cycle 4) ÷ 16	Sum Estimated
e	−.63	2.875	$\mu_{.....} + \alpha\beta\gamma\delta\epsilon_{22222}$
a	2.51	.823	$\alpha_2 + \beta\gamma\delta\epsilon_{2222}$
b	−2.68	−1.253	$\beta_2 + \alpha\gamma\delta\epsilon_{2222}$
abe	−1.66	.055	$\alpha\beta_{22} + \gamma\delta\epsilon_{222}$
c	2.06	.384	$\gamma_2 + \alpha\beta\delta\epsilon_{2222}$
ace	1.22	.064	$\alpha\gamma_{22} + \beta\delta\epsilon_{222}$
bce	−2.09	.041	$\beta\gamma_{22} + \alpha\delta\epsilon_{222}$
abc	1.93	.001	$\alpha\beta\gamma_{222} + \delta\epsilon_{22}$
d	6.79	2.793	$\delta_2 + \alpha\beta\gamma\epsilon_{2222}$
ade	6.47	−.095	$\alpha\delta_{22} + \beta\gamma\epsilon_{222}$
bde	3.45	−.045	$\beta\delta_{22} + \alpha\gamma\epsilon_{222}$
abd	5.68	−.288	$\alpha\beta\delta_{222} + \gamma\epsilon_{22}$
cde	5.22	−.314	$\gamma\delta_{22} + \alpha\beta\epsilon_{222}$
acd	9.38	.186	$\alpha\gamma\delta_{222} + \beta\epsilon_{22}$
bcd	4.30	−.306	$\beta\gamma\delta_{222} + \alpha\epsilon_{22}$
abcde	4.05	−.871	$\alpha\beta\gamma\delta_{2222} + \epsilon_2$

Yates standard order as regards Factors A, B, C, and D. One may apply the (4-cycle, final division by $2^4 = 16$) Yates algorithm to these data and arrive at the fitted sums of effects listed in Table 7.5.

Snee's data include no replication. So the only method presented in this text that is helpful for judging the detectability of effects in his study is probability plotting. Figures 7.1 and 7.2 show respectively a normal plot of the last 15 estimates listed in Table 7.5 and a half-normal plot of their magnitudes. (Since one is probably willing to grant *a priori* that the mean response is other than 0, there is no

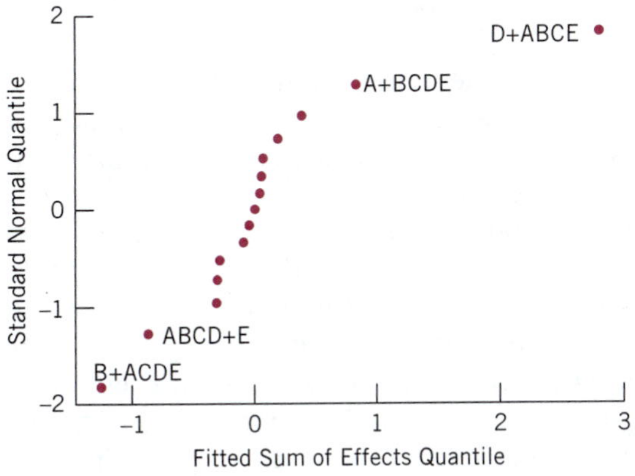

Figure 7.1 Normal plot of fitted sums of effects for Snee's color index data.

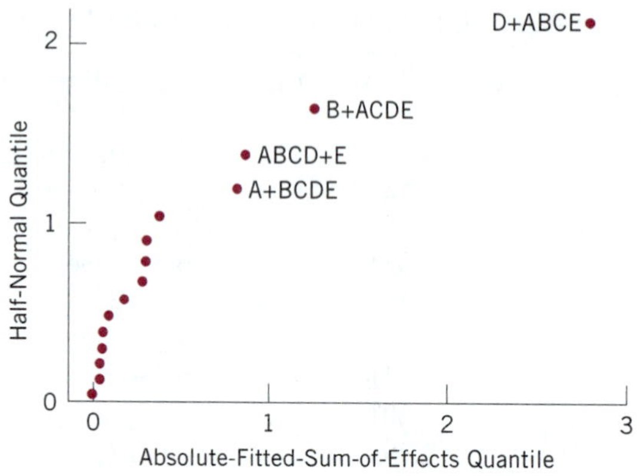

Figure 7.2 Half-normal plot of absolute-fitted-sums-of-effects for Snee's color index data.

reason to include the first estimate listed in Table 7.5 in the plot.) Especially from Figure 7.2 it is clear that even if the bulk of the estimates in Table 7.5 consist of nothing more than experimental variation, those corresponding to "D+ABCE," "B+ACDE," "ABCD+E," and "A+BCDE" do not. The four sums of effects

$$\delta_2 + \alpha\beta\gamma\epsilon_{2222},\ \beta_2 + \alpha\gamma\delta\epsilon_{2222},\ \alpha\beta\gamma\delta_{2222} + \epsilon_2 \quad \text{and} \quad \alpha_2 + \beta\gamma\delta\epsilon_{2222}$$

are statistically detectable.

The simplest possible interpretation of the judgment that the four largest (in magnitude) of the last 15 estimates in Table 7.5 correspond to detectable sums of effects is that only the main effects of Factors D, B, E, and A are important. This interpretation says

1. that (in decreasing order of importance) reactant purity, catalyst/reactant ratio, reactant pH, and solvent/reactant ratio affect color index for this product,
2. that temperature has no appreciable impact on product color, and
3. that the important factors act separately on the color index.

Since the original motivation for the experimentation was to find a way of improving color *consistency*, this result would guide engineers to the very careful control of process inputs. Reactant purity would deserve first attention, catalyst/reactant ratio would deserve second attention, and so on.

It should be said that these tentative conclusions about color index are so clean and intuitively appealing that there would seem to be little reason to doubt that they are the right ones for the color index problem. But it must be kept in mind that they are based on a fractional factorial study and an assumption of simple structure for the chemical system. If, in fact, the chemical system is not simple (and, for example, there are important four-factor interactions), then they could lead one in wrong directions when looking for a way to improve color consistency.

7.1.3 $1/2^q$ Fractions of 2^p Factorials

The tools just presented for half-fractions of 2^p factorials all have their natural extensions to the general case of $1/2^q$ fractions. To begin with, the problem of choosing 2^{p-q} out of 2^p possible combinations to include in a fractional factorial study can be addressed using the **product of signs** idea introduced for half-fractions. That is, one may follow this prescription:

Algorithm for Producing a 2^{p-q} Fractional Factorial

> Write down for the first $p-q$ factors a $2^{p-q} \times (p-q)$ table of plus and minus signs specifying all possible combinations of levels of these, columns giving the levels for particular factors in combinations specified by rows. Then make up q (different) additional columns of signs as products (a row at a time) of signs in q (different) groups of the first $p-q$ columns. These new product columns specify levels of the last q factors to be used with the various combinations of levels of the first $p-q$ factors.

This set of instructions is somewhat ambiguous in that it does not specify exactly *which* product columns one ought to construct. The fact is that some choices are better than others in terms of the alias structures that they produce. But discussion of this must wait until the matter of actually *finding* an alias structure has been considered.

Example 7.4 **A 2^{5-2} Catalyst Development Study.** In a paper presented at the 1986 National Meeting of the American Statistical Association, Hanson and Best described an experimental program for the development of an effective catalyst for the production of ethyleneamines by the amination of monoethanolamine. One part of that program involved a quarter-fraction of a 2^5 factorial with the following factors and levels:

Factor A—Ne/Re Ratio	2/1 (−) vs. 20/1 (+)
Factor B—Precipitant	$(NH_4)_2CO_3$ (−) vs. none (+)
Factor C—Calcining Temperature	300° (−) vs. 500° (+)
Factor D—Reduction Temperature	300° (−) vs. 500° (+)
Factor E—Support Used	alpha-alumina (−) vs. silica-alumina (+)

The response variable of interest was y = the percent water produced.

A quarter-fraction of a full 2^5 factorial involves 8 out of 32 possible combinations of levels of Factors A, B, C, D, and E. To follow the prescription just given for choosing such a 2^{5-2} fractional factorial, one begins by writing down an 8×3 table of signs specifying all eight combinations of levels of the factors A, B, and C. Then one must make up $q = 2$ product columns to use in choosing corresponding levels of Factors D and E. The particular choice made by Hanson and Best was to use ABC products to choose levels of Factor D, and BC products to choose levels of E. Table 7.6 shows the construction used, the 2^5 names of the eight combinations selected, the raw data, and some (sample-by-sample) summary statistics.

Table 7.6 % H_2O Values Produced in Runs of Eight Combinations of Five Factors in the Catalyst Development Study and Some Summary Statistics

A	B	C	(D) ABC Product	(E) BC Product	Combination	% H_2O, y	$\bar{y}$	s^2
−	−	−	−	+	e	8.70, 11.60, 9.00	9.767	2.543
+	−	−	+	+	ade	26.80	26.80	
−	+	−	+	−	bd	24.88	24.88	
+	+	−	−	−	ab	33.15	33.15	
−	−	+	+	−	cd	28.90, 30.98	29.940	2.163
+	−	+	−	−	ac	30.20	30.20	
−	+	+	−	+	bce	8.00, 8.69	8.345	.238
+	+	+	+	+	abcde	29.30	29.30	

The choice of product columns made by the engineers in this study was by no means the only one possible. Different choices would have led to different confounding patterns (that in other circumstances might have seemed preferable on the basis of engineering considerations).

The way one finds the alias structure of a general 2^{p-q} fractional factorial is built on the same **formal multiplication** idea used for half-fractions. The only new complication is that one must now work with q generators, and these lead to a defining relation that says that the grand mean is aliased with $2^q - 1$ effects. (So in the end, effects are aliased in 2^{p-q} different groups of 2^q.) To be more precise:

Method for Determining the Alias Structure for a 2^{p-q} Study

One begins by writing down the ***q* generators** of the design of the form

$$\text{name of one of the last } q \text{ factors} \leftrightarrow \text{product of names of some of the first } p - q \text{ factors} \quad (7.3)$$

that specify how the columns of signs are made up for choosing levels of the last q factors. Then each of these q relationships (7.3) is multiplied through by the letter on the left to produce a relationship of the form

$$\text{I} \leftrightarrow \text{product.}$$

These are taken individually, multiplied in pairs, multiplied in triples, and so on to produce a **defining relation** of the form

$$\text{I} \leftrightarrow \text{product 1} \leftrightarrow \text{product 2} \leftrightarrow \cdots \leftrightarrow \text{product}(2^q - 1) \quad (7.4)$$

specifying $2^q - 1$ aliases for the grand mean. This defining relation (7.4) and the formal multiplication scheme are then used to find all aliases of any 2^p effect of interest.

Example 7.4 continued. Consider again the 2^{5-2} catalyst study of Hanson and Best. Table 7.6 shows the two generators for the eight combinations used in the study to be

$$D \leftrightarrow ABC \quad \text{and} \quad E \leftrightarrow BC.$$

Multiplying the first of these through by D and the second by E, one has the two relationships

$$I \leftrightarrow ABCD \quad \text{and} \quad I \leftrightarrow BCE.$$

But then, multiplying the left sides and the right sides of these two together, one also has

$$I \cdot I \leftrightarrow (ABCD) \cdot (BCE), \quad \text{that is,} \quad I \leftrightarrow ADE.$$

Finally, combining the three "I↔product" statements into a single string of aliases, one has the defining relation for this study

$$I \leftrightarrow ABCD \leftrightarrow BCE \leftrightarrow ADE.$$

This relationship shows immediately that one may estimate the sum of effects $\mu_{.....} + \alpha\beta\gamma\delta_{2222} + \beta\gamma\epsilon_{222} + \alpha\delta\epsilon_{222}$ but may not separate the summands. Or, multiplying through the defining relation by A, one has

$$A \leftrightarrow BCD \leftrightarrow ABCE \leftrightarrow DE,$$

and sees that the A main effect is aliased with the BCD three-factor interaction, the ABCE four-factor interaction, and the DE two-factor interaction. It should be easy for the reader to verify that (as expected) the 2^5 factorial effects are aliased in $8 = 2^{5-2}$ sets of $4 = 2^2$ effects. With data from eight different combinations, one can estimate "eight things," the corresponding eight different sums of four effects.

Example 7.5 Defining Relations for Two Different 2^{6-2} Plans. Consider the choice of generators for a 2^{6-2} plan. Two different possibilities are

$$E \leftrightarrow ABCD \quad \text{and} \quad F \leftrightarrow ABC, \tag{7.5}$$

and

$$E \leftrightarrow BCD \quad \text{and} \quad F \leftrightarrow ABC. \tag{7.6}$$

The reader should do the work (parallel to that in the previous example) necessary to verify that the defining relation corresponding to the set of generators (7.5) is

$$\text{I} \leftrightarrow \text{ABCDE} \leftrightarrow \text{ABCF} \leftrightarrow \text{DEF}.$$

And the defining relation corresponding to the set of generators (7.6) is

$$\text{I} \leftrightarrow \text{BCDE} \leftrightarrow \text{ABCF} \leftrightarrow \text{ADEF}.$$

Notice that this second defining relation is arguably better than the first. The first shows that the choice of generators (7.5) leaves some main effects aliased with *two-factor* interactions, fairly low-order effects. In contrast, the choice of generators (7.6) leads to the main effects being aliased with only three-factor (and higher-order) interactions. This example shows that not all choices of q generators in a 2^{p-q} study are going to be equally attractive in terms of the alias structure they produce.

Example 7.6 Finding the Defining Relation for a 1/8th Fraction of a 2^6 Study. As an example of what must be done to find the defining relation for fractions of 2^p factorials smaller than 1/4th, consider the set of generators of a 2^{6-3} study

$$\text{D} \leftrightarrow \text{AB}, \quad \text{E} \leftrightarrow \text{AC}, \quad \text{and} \quad \text{F} \leftrightarrow \text{BC}.$$

These immediately produce

$$\text{I} \leftrightarrow \text{ABD}, \quad \text{I} \leftrightarrow \text{ACE}, \quad \text{and} \quad \text{I} \leftrightarrow \text{BCF}.$$

Then multiplying these in pairs, one has

$$\text{I} \cdot \text{I} \leftrightarrow (\text{ABD}) \cdot (\text{ACE}), \quad \text{i.e., I} \leftrightarrow \text{BCDE},$$

$$\text{I} \cdot \text{I} \leftrightarrow (\text{ABD}) \cdot (\text{BCF}), \quad \text{i.e., I} \leftrightarrow \text{ACDF},$$

and

$$\text{I} \cdot \text{I} \leftrightarrow (\text{ACE}) \cdot (\text{BCF}), \quad \text{i.e., I} \leftrightarrow \text{ABEF}.$$

And finally multiplying *all three* of these together, one has

$$\text{I} \cdot \text{I} \cdot \text{I} \leftrightarrow (\text{ABD}) \cdot (\text{ACE}) \cdot (\text{BCF}), \quad \text{i.e., I} \leftrightarrow \text{DEF}.$$

So then, stringing together all of the aliases of the grand mean, one has the defining relation

$$\text{I} \leftrightarrow \text{ABD} \leftrightarrow \text{ACE} \leftrightarrow \text{BCF} \leftrightarrow \text{BCDE} \leftrightarrow \text{ACDF} \leftrightarrow \text{ABEF} \leftrightarrow \text{DEF},$$

and it is evident that 2^6 effects are aliased in eight groups of eight effects.

Armed with the ability to choose 2^{p-q} plans and find their confounding structures, the only real question remaining is how **data analysis** should proceed. Again, essentially the same method introduced for half-fractions is relevant. That is:

Data Analysis Method for a 2^{p-q} Study

> To analyze 2^{p-q} fractional factorial data, one first ignores the existence of the last q factors, treating the data as if they were complete factorial data in the first $p - q$ factors. "Fitted effects" are computed and judged exactly as in Section 6.3. Then the statistical inferences are interpreted in light of the alias structure, remembering that one actually has estimates of not single effects, but sums of 2^p factorial effects.

Example 7.4 continued. The sample means in Table 7.6 are listed in Yates standard order as regards Factors A, B, and C. The reader may do the arithmetic to verify that the (three-cycle, final division by $2^3 = 8$) Yates algorithm applied directly to these means (as listed in Table 7.6) produces the eight fitted sums of effects listed in Table 7.7.

There is replication in the data listed in Table 7.6, so one may use the confidence interval approach to judge the detectability of the sums corresponding to the estimates in Table 7.7. First, the pooled estimate of σ must be computed from the sample variances in Table 7.6,

$$s_{\text{P}}^2 = \frac{(3-1)2.543 + (2-1)2.163 + (2-1).238}{(3-1) + (2-1) + (2-1)},$$

$$= 1.872,$$

Table 7.7 Fitted Sums of Effects from the Catalyst Development Data

A	B	C	D	E	$\bar{y}$	Estimate (8 Divisor)	Sum Estimated
−	−	−	−	+	9.767	24.048	$\mu_{.....}$ + *aliases*
+	−	−	+	+	26.80	5.815	α_2 + *aliases*
−	+	−	+	−	24.88	−.129	β_2 + *aliases*
+	+	−	−	−	33.15	1.492	$\alpha\beta_{22}$ + *aliases*
−	−	+	+	−	29.94	0.399	γ_2 + *aliases*
+	−	+	−	−	30.20	−.511	$\alpha\gamma_{22}$ + *aliases*
−	+	+	−	+	8.345	−5.495	$\beta\gamma_{22}$ + *aliases*
+	+	+	+	+	29.30	3.682	$\alpha\beta\gamma_{222}$ + *aliases*

so that

$$s_P = \sqrt{1.872} = 1.368 \text{ \% water.}$$

Then the $p = 3$ version of formula (6.31) says that using individual 95% two-sided confidence limits, a plus or minus value of

$$2.776(1.368)\frac{1}{2^3}\sqrt{\frac{1}{3}+\frac{1}{1}+\frac{1}{1}+\frac{1}{1}+\frac{1}{2}+\frac{1}{1}+\frac{1}{2}+\frac{1}{1}} = 1.195 \text{ \% water}$$

should be associated with each of the estimates in Table 7.7. By this criterion, the estimates on the first, second, fourth, seventh, and eighth lines of Table 7.7 are big enough to force the conclusion that they represent more than experimental error.

How then does one interpret such a result? Ignoring the sum involving the grand mean, the sums of 2^5 effects corresponding to the four largest (in absolute value) fitted sums are (using the defining relation for this study)

$$\alpha_2 + \beta\gamma\delta_{222} + \alpha\beta\gamma\epsilon_{2222} + \delta\epsilon_{22} \quad \text{estimated as} \quad 5.815,$$
$$\beta\gamma_{22} + \alpha\delta_{22} + \epsilon_2 + \alpha\beta\gamma\delta\epsilon_{22222} \quad \text{estimated as} \quad -5.495,$$
$$\alpha\beta\gamma_{222} + \delta_2 + \alpha\epsilon_{22} + \beta\gamma\delta\epsilon_{2222} \quad \text{estimated as} \quad 3.682, \quad \text{and}$$
$$\alpha\beta_{22} + \gamma\delta_{22} + \alpha\gamma\epsilon_{222} + \beta\delta\epsilon_{222} \quad \text{estimated as} \quad 1.492.$$

Concentrating initially on the first three of these, one can reason to at least four different, relatively simple interpretations. First, if the largest components of these sums are the main effects appearing in them, one might have an "A, E, and D main effects only" explanation of the pattern of responses seen in Table 7.6. But picking out from the first sum the A main effect, from the second the E main effect, and from the third the AE interaction, an explanation involving only Factors A and E would be "A and E main effects and interactions only." And that doesn't end the relatively simple possibilities. One might also contemplate an "A and D main effects and interactions only" or a "D and E main effects and interactions only" description of the data. Which of these possibilities is the most appropriate for the catalyst system cannot be discerned on the basis of the data in Table 7.7 alone. One needs either more data or the guidance of a subject matter expert who might be able to eliminate one or more of these possibilities on the basis of some physical theory. (As it turns out, in the real situation, additional experimentation confirmed the usefulness of an "A, E, and D main effects only" model for the chemical process.)

Exactly what to make of the detectability of the sum corresponding to the 1.492 estimate is not at all clear. Thankfully (in terms of ease of interpretation of the experimental results) its estimate is less than half of any of the others. While it may be statistically detectable, it does not appear to be of a size to rival the other effects in terms of physical importance.

This 2^{5-2} experiment does what any successful fractional factorial does. It gives some directions to go in further experimentation and hints as to which factors may *not* be so important in determining the response variable. (The Factors B and C do not enter into any of the four simple candidates for describing mean response.) But definitive conclusions await either confirmation by actually "trying out" tentative recommendations/interpretations based on fractional factorial results, or further experimentation aimed at removing ambiguities and questions left by the study.

Example 7.7 **Tentative Conclusions in a 2^{15-11} Fractional Factorial Study.** As a final (and fairly extreme) example of what is possible in the way of fractional factorial experimentation, return to the scenario represented by the factors and levels in Table 7.1. The study actually conducted by Hendrix and his associates was a 2^{15-11} study with the 11 generators

$$E \leftrightarrow ABCD,\ F \leftrightarrow BCD,\ G \leftrightarrow ACD,\ H \leftrightarrow ABC,\ J \leftrightarrow ABD,\ K \leftrightarrow CD,\ L \leftrightarrow BD,$$
$$M \leftrightarrow AD,\ N \leftrightarrow BC,\ O \leftrightarrow AC, \quad \text{and} \quad P \leftrightarrow AB.$$

The combinations run and the cold crack resistances observed are given in Table 7.8.

Table 7.8 16 Experimental Combinations and Measured Cold Crack Resistances

Combination	y
eklmnop	14.8
aghjkln	16.3
bfhjkmo	23.5
abefgkp	23.9
cfghlmp	19.6
acefjlo	18.6
bcegjmn	22.3
abchnop	22.2
dfgjnop	17.8
adefhmn	18.9
bdeghlo	23.1
abdjlmp	21.8
cdehjkp	16.6
acdgkmo	16.7
bcdfkln	23.5
abcdefghjklmnop	24.9

Notice that in this scenario it is practically infeasible to write out the whole defining relation. Since only 16 out of the 32,768 possible combinations are involved in this 2^{15-11} study, every 2^{15} factorial effect is aliased with 2,047 other effects. But at least the generators make it clear which main effects are aliased with the effects involving only Factors A, B, C, and D.

The combinations and means in Table 7.8 are listed in Yates standard order as regards the first $15 - 11 = 4$ factors, A, B, C, and D. Applying the (four-cycle and 16 divisor) Yates algorithm to them (in the order listed), one obtains the estimates given in Table 7.9.

The business of judging the detectability of the sums of 2^{15} factorial effects in this study is complicated by the lack of replication. All that can be done is to probability plot the estimates of Table 7.9. Figures 7.3 and 7.4 are respectively a normal plot of the last 15 estimates in Table 7.9 and a half-normal plot of the absolute values of these. The plots show clearly that even if all other estimates really represent only experimental variation, the ones corresponding to the B main effect plus its aliases and the F main effect plus its aliases do not. Of course, the simplest possible interpretation of this outcome is that only the factors B and F impact cold crack resistance in any serious way, the factors acting independently on the response and the high levels of both factors leading to the largest values of y. In light of the very small fraction involved here, however, the wise engineer would treat such conclusions as very tentative. They are intriguing and are perhaps even absolutely correct. But it would be foolhardy to conclude such on the basis of 2^{15-11} data. Of course, if one's object is only to find a good combination of

Table 7.9 Estimates of Sums of Effects for the Data of Table 7.8

Sum of Effects Estimated	Estimate
grand mean + aliases	20.28
A main + aliases	.13
B main + aliases	2.87
P main + aliases (including AB)	−.08
C main + aliases	.27
O main + aliases (including AC)	−.08
N main + aliases (including BC)	−.19
H main + aliases (including ABC)	.36
D main + aliases	.13
M main + aliases (including AD)	.03
L main + aliases (including BD)	.04
J main + aliases (including ABD)	−.06
K main + aliases (including CD)	−.26
G main + aliases (including ACD)	.29
F main + aliases (including BCD)	1.06
E main + aliases (including ABCD)	.11

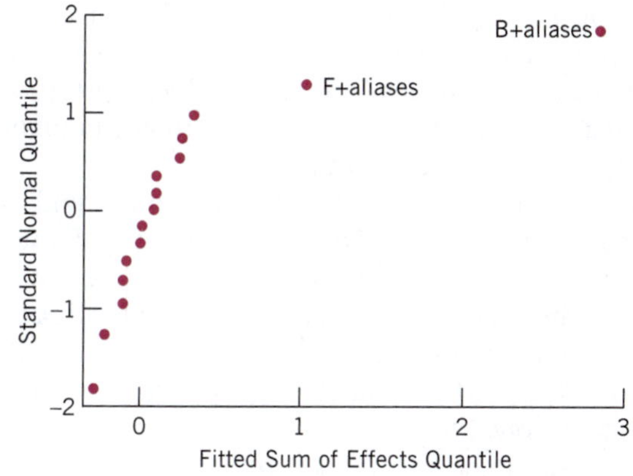

Figure 7.3
Normal plot of fitted sums of effects for the cold crack resistance data.

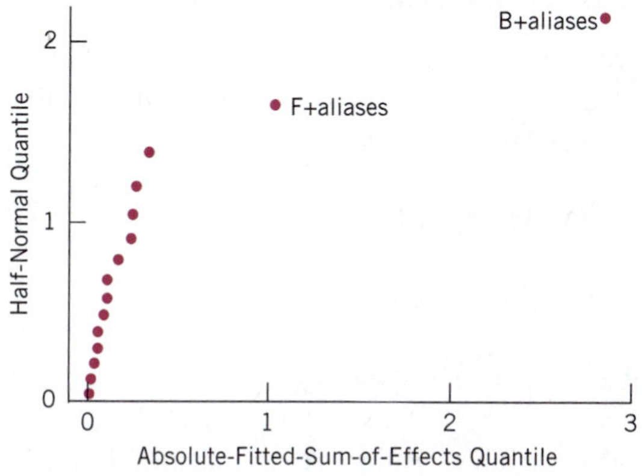

Figure 7.4
Half-normal plot of absolute-fitted-sums-of-effects for the cold crack resistance data.

levels of these 15 factors, this analysis points out what is in retrospect completely obvious about the data in Table 7.8. It is those four combinations with both of B and F at their high levels that have the largest responses.

7.2 RESPONSE SURFACE STUDIES

When p process variables $x_1, x_2, \ldots, x_p$ are all quantitative, there exists the possibility of doing some experimentation and then using the resultant data to produce an equation describing how (at least in approximate terms) a response y depends upon the process variables. That equation can then, for example, provide direction as one tries to optimize (maximize or minimize) y by choice of settings of the process variables. This section primarily concerns methods for using such equations in the exploration of the approximate relationship between y and $x_1, x_2, \ldots, x_p$.

Standard methodology for turning n data points $(x_1, x_2, \ldots, x_p, y)$ into an equation for y as a function of the x's is the **multiple regression analysis** that is a main topic in most introductions to engineering statistics. This section does not repeat basic regression material, but instead assumes that the reader is already familiar with the subject and shows how it is useful in process improvement.

The section opens with a discussion of using graphical means to aid understanding of equations fit to $(x_1, x_2, \ldots, x_p, y)$ data. Then the topic of quadratic response functions is introduced, along with the experimental design issues associated with their use. There follows a discussion of analytical tools for interpreting fitted quadratic functions. Finally, the section concludes with a discussion of search strategies that can be used when one's goal is to optimize a response and issues of process modeling are not of particular concern.

7.2.1 Graphics for Understanding Fitted Response Functions

The primary output of a standard regression analysis based on n data points $(x_{11}, x_{21}, \ldots, x_{p1}, y_1), (x_{12}, x_{22}, \ldots, x_{p2}, y_2), \ldots, (x_{1n}, x_{2n}, \ldots, x_{pn}, y_n)$ is an equation that we will temporarily represent in generic terms as

$$\hat{y} = f(x_1, x_2, \ldots, x_p). \tag{7.7}$$

A typical specific version of equation (7.7) is, of course,

$$\hat{y} = b_0 + b_1 x_1 + b_2 x_2 + \cdots + b_p x_p, \tag{7.8}$$

an equation linear in all of the process variables. But more complicated equations are possible and, in many cases, are necessary to really adequately describe the relationship of y to the process variables.

Example 7.8 A Two-Variable Drilling Experiment. The paper "Design of a Metal-Cutting Drilling Experiment: A Discrete Two-Variable Problem" by E. Mielnik appeared in *Quality Engineering* in 1993 and concerns the drilling of 7075–T6 aluminum alloy. The two process variables

$$x_1 = \text{feed rate (ipr)} \quad \text{and}$$
$$x_2 = \text{drill diameter (inches)}$$

were varied in (800 rpm) drilling of aluminum specimens and both

$$y_1 = \text{thrust (lbs)} \quad \text{and}$$
$$y_2 = \text{torque (ft-lbs)}$$

Table 7.10 Thrust and Torque Measurements for Nine Feed Rate/Drill Diameter Combinations

Feed Rate, x_1 (ipr)	Diameter, x_2 (in)	Thrust, y_1 (lbs)	Torque, y_2 (ft-lbs)
.006	.250	230	1.0
.006	.406	375	2.1
.013	.406	570	3.8
.013	.250	375	2.1
.009	.225	280	1.0
.005	.318	255	1.1
.009	.450	580	3.8
.017	.318	565	3.4
.009	.318	400, 400, 380, 380	2.2, 2.1, 2.1, 1.9

were measured. A total of nine different (x_1, x_2) combinations were studied and 12 data points (x_1, x_2, y) were collected for both thrust and torque. Mielnik's data are given in Table 7.10.

Apparently on the basis of established drilling theory, Mielnik found it useful to express all variables on a log scale and fit linear regressions for $y_1' = \ln(y_1)$ and $y_2' = \ln(y_2)$ in terms of the variables $x_1' = \ln(x_1)$ and $x_2' = \ln(x_2)$. Multiple regression analysis then provides fitted equations

$$\widehat{y_1'} = 10.0208 + .6228x_1' + .9935x_2'$$

and

$$\widehat{y_2'} = 6.8006 + .8927x_1' + 1.6545x_2'.$$

Notice that by exponentiating, one gets back to equations for the original responses y_1 and y_2 in terms of x_1 and x_2, namely

$$\widehat{y_1} = 22{,}489x_1^{(.6228)}x_2^{(.9935)}$$

and

$$\widehat{y_2} = 898.39x_1^{(.8927)}x_2^{(1.6545)}.$$

Example 7.9 Lift-to-Drag Ratio for a Three-Surface Configuration. P. Burris studied the effects of the placement of a canard (a small forward "wing") and a tail, relative to the main wing of a model aircraft. He measured the lift/drag ratio for nine different configurations. With

$$x_1 = \text{the canard placement in inches above the main wing}$$

and

$$x_2 = \text{the tail placement in inches above the main wing,}$$

part of his data are given in Table 7.11.

Notice that Burris's data set has the unfortunate feature that it contains no replication. It is a real weakness of the study that there is no pooled (or "pure error") sample standard deviation against which to judge the appropriateness of a fitted equation. However, making the best of the situation, a multiple regression analysis described on pages 133 through 136 of Vardeman's *Statistics for Engineering Problem Solving* leads to the equation

$$\hat{y} = 3.9833 + .5361x_1 + .3201x_2 - .4843x_1^2 - .5042x_1x_2$$

as a plausible description of lift-to-drag ratio in terms of canard and tail positions. (Equations simpler than this one turn out to have obvious deficiencies when one plots residuals.)

Notice that in Example 7.8, the equations for log thrust and log torque are linear in the (logs of the) process variables. That is, they are exactly of the form in display (7.8). But the equation for lift-to-drag ratio obtained from Burris's data in

Table 7.11 Lift/Drag Ratio for Nine Different Canard and Tail Configurations (Positions in Inches Above the Main Wing)

Canard Position, x_1	Tail Position, x_2	Lift/Drag Ratio
−1.2	−1.2	.858
−1.2	0	3.156
−1.2	1.2	3.644
0	−1.2	4.281
0	0	3.481
0	1.2	3.918
1.2	−1.2	4.136
1.2	0	3.364
1.2	1.2	4.018

Example 7.9 (while still linear in the fitted parameters) is not linear in the variables x_1 and x_2. (It is, in fact, a kind of quadratic equation in the predictor variables x_1 and x_2. Much more will be said about such equations later in this section.)

After one has fit an equation for y in the variables $x_1, x_2, \ldots, x_p$ to data from a process-improvement experiment, the question of how to interpret that equation must be faced. Several possibilities exist for helpfully representing the generic fitted relationship (7.7) in graphical terms. The most obvious is to simply make plots of y against a single process variable, x_i, for various combinations of the remaining variables that are of particular interest. This amounts to viewing **slices** of the fitted response function.

A second possibility is to make **contour plots** of y against a pair of the process variables $(x_i, x_{i'})$, for various combinations of any remaining variables that are of particular interest. Such plots function as "topographic maps" of the response surface. They can be especially helpful when there are several more or less competing responses and one must find a compromise setting of the process variables that balances off process performance on one response against performance on another.

And a final method of representing a fitted response function is to make use of modern graphics software and produce **surface plots/perspective plots** of y against a pair of the process variables $(x_i, x_{i'})$, again for various combinations of any remaining variables that are of particular interest. These plots attempt to give a "3-D" rendering of the relationship between y and $(x_i, x_{i'})$ (with values of the remaining variables held fixed).

Example 7.8 continued. Figure 7.5 is a plot of the fitted log torque in the drilling study as a function of log feed rate for drills of several diameters. Figure 7.6 is the corresponding plot where both torque and feed rate are in their original units (rather than being portrayed on logarithmic scales).

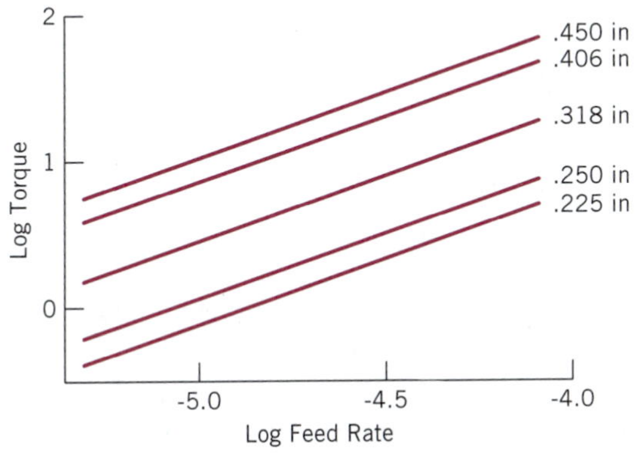

Figure 7.5
Fitted log torque as a function of log feed rate for five different drill diameters.

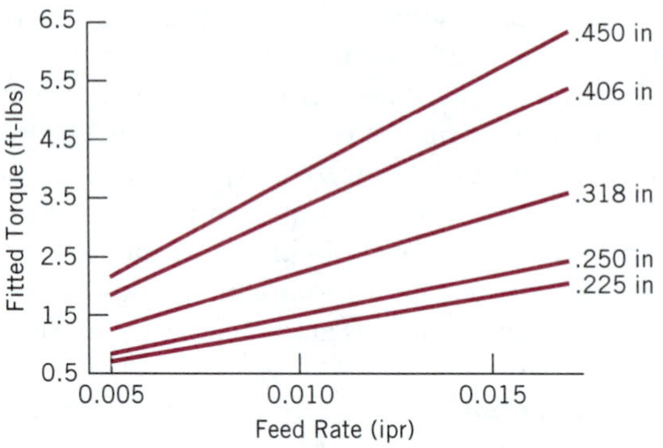

Figure 7.6
Fitted torque as a function of feed rate for five different drill diameters.

Notice that on the logarithmic scales of Figure 7.5 the response changes linearly in the feed rate variable. In addition, the linear traces on Figure 7.5 are parallel. In the language of factorial analysis, on the logarithmic scales there are no interactions between feed rate and drill diameter. On the logarithmic scales, the fitted relationship between feed rate, diameter, and torque is a very simple one. The relationship on the original scales of measurement represented by Figure 7.6 is not impossibly complicated, but neither is it as simple as the one in Figure 7.5.

Figures 7.7 and 7.8 are respectively a contour plot and a surface plot of the fitted relationship between log torque and the logarithms of feed rate and drill diameter. They illustrate clearly that on the logarithmic scales, the fitted equation for torque defines a plane in three-space.

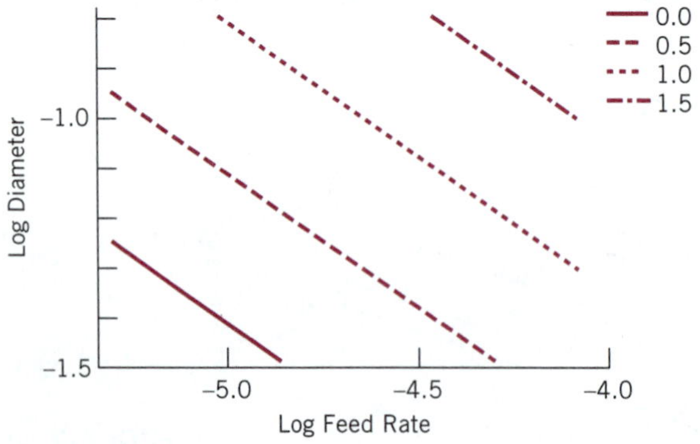

Figure 7.7
Contour plot of fitted log torque as a function of log feed rate and log drill diameter.

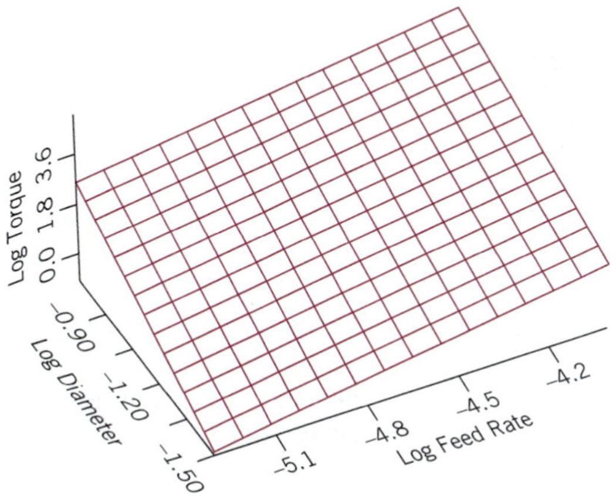

Figure 7.8
Surface plot of fitted log torque as a function of log feed rate and log drill diameter.

Example 7.9 continued. Figures 7.9 and 7.10 are respectively a contour plot and surface plot for the equation for lift-to-drag ratio fit to the data of Table 7.11. They show a geometry that is substantially more complicated than the planar geometry in Figures 7.7 and 7.8. It is evident from either examining the fitted equation itself or viewing the surface plot that, for a fixed tail position (x_2), the fitted lift-to-drag ratio is quadratic in the canard position (x_1). And even though for fixed canard position (x_1) the fitted lift-to-drag ratio is linear in tail position (x_2), the relevant slope depends upon the canard position. That is, there are canard position by tail position interactions.

Having raised the issue of interactions in the examples, it is worth pointing out how one can tell from the form of a fitted equation whether or not it implies the

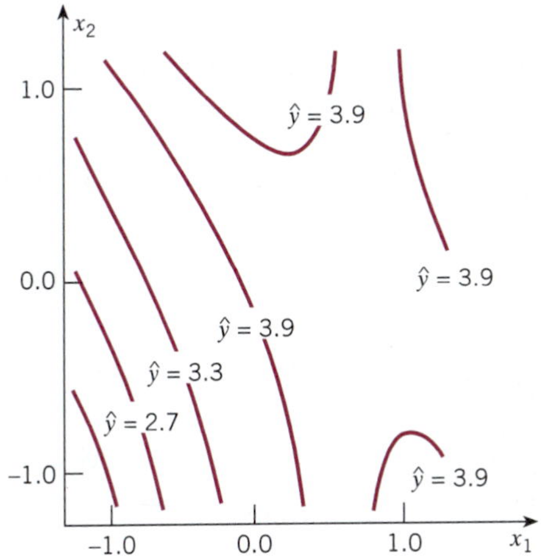

Figure 7.9
Contour plot for fitted lift/drag ratio as a function of canard and tail positions.

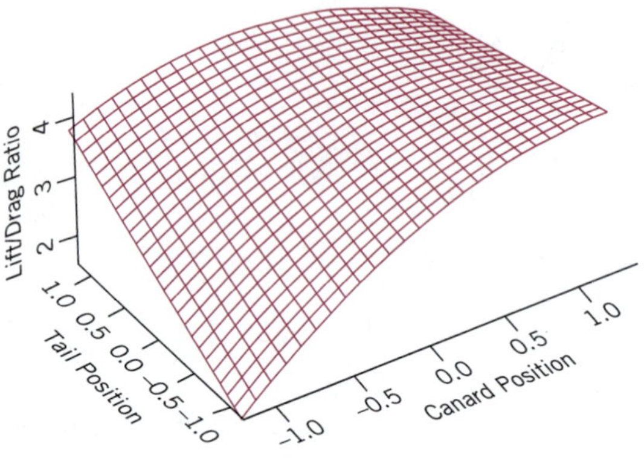

Figure 7.10
Surface plot for fitted lift/drag ratio as a function of canard and tail positions.

existence of interactions between the process variables. The general story is this. If the function $f(x_1, x_2, \ldots, x_p)$ in equation (7.7) can be written as a *sum* of two functions, the first of which has as its arguments $x_1, x_2, \ldots, x_l$ and the second of which has as its arguments $x_{l+1}, x_{l+2}, \ldots, x_p$, then the fitted equation implies that there are no interactions between any variables in the first set of arguments and any variables in the second set of arguments. Otherwise there *are* interactions. For example, in the case of the lifting surface study, the existence of the cross product term $-.5042x_1x_2$ in the fitted equation makes it impossible to write it in terms of a sum of a function of x_1 and a function of x_2. So the interactions noted on Figure 7.10 are to be expected.

Example 7.10 **Yield and Cost Associated with a Chemical Process.** The data in Table 7.12 are from the paper "More on Planning Experiments to Increase Research Efficiency" by Hill and Demler, which appeared in *Industrial and Engineering Chemistry* in 1970. The responses are a yield (y_1) and a filtration time

Table 7.12 Yields and Filtration Times for Nine Combinations of Condensation Temperature and Amount of B

Temperature, x_1 (°C)	Amount of B, x_2 (cc)	Yield, y_1 (g)	Time, y_2 (sec)
90	24.4	21.1	150
90	29.3	23.7	10
90	34.2	20.7	8
100	24.4	21.1	35
100	29.3	24.1	8
100	34.2	22.2	7
110	24.4	18.4	18
110	29.3	23.4	8
110	34.2	21.9	10

(y_2) for a chemical process run under nine different combinations of the process variables condensation temperature (x_1) and amount of boron (x_2). Notice that the filtration time is a cost factor, and ideally one would like large yield and small filtration time. But chances are that these two responses more or less work against each other, and some compromise is necessary.

The study represented in Table 7.12 suffers from a lack of replication, making it impossible to do a completely satisfactory job of judging the appropriateness of regressions of the responses on the process variables. But as far as is possible to tell given the inherent limitations of the data, the fitted equations (derived using multiple regression analysis)

$$\hat{y}_1 \approx -113.2 + 1.254x_1 + 5.068x_2 - .009333x_1^2 - .1180x_2^2 + .01990x_1x_2,$$

and

$$\widehat{\ln(y_2)} \approx 99.69 - .8869x_1 - 3.348x_2 + .002506x_1^2 + .03375x_2^2 + .01196x_1x_2$$

do a reasonable job of summarizing the data of Table 7.12. Figure 7.11 goes a long way toward both making the nature of the fitted response surfaces clear and providing guidance in how compromises might be made in the pursuit of good process performance. The figure shows contour plots for yield and log filtration time overlaid on a single set of (x_1, x_2)-axes.

The three examples used thus far in this section are all ones involving only two process variables, x_1 and x_2. As such, the figures shown here provide reason-

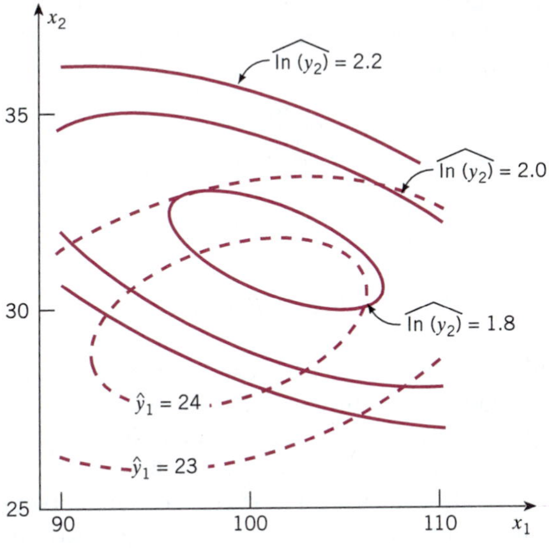

Figure 7.11
Overlaid contour plots for yield and filtration time.

ably straightforward and complete pictures of the fitted response functions. When more than two variables are involved, it becomes harder to make helpful graphics. In the case of three process variables, one can make plots like those shown here involving x_1 and x_2 for several different values of x_3. And in the case of four process variables, one can make plots involving x_1 and x_2 for several different (x_3, x_4) combinations. But unless the fitted equation is such that the function $f(x_1, x_2, \ldots, x_p)$ in equation (7.7) can be written as a sum of several functions of small and disjoint sets of process variables (that can be individually represented using the kind of plots illustrated in the examples), the plots become less and less helpful as p increases.

7.2.2 Using Quadratic Response Functions

Equation (7.8) represents the simplest kind of response function that can be easily fit to $(x_1, x_2, \ldots, x_p, y)$ data. As illustrated in Example 7.8, this kind of equation represents a p dimensional plane in $(p + 1)$ dimensional space. As useful as it is, it does not allow for any **curvature** in the response. This means, for example, that when used to describe how y varies over some region in $(x_1, x_2, \ldots, x_p)$-space, an equation of the type in display (7.8) will always predict that optimum y occurs on the boundary of the region. And that is not appropriate for many situations where one is *a priori* fairly certain that optimum settings for the process variables occur in the interior of a region of experimentation. But to allow for this kind of circumstance, one needs alternatives to relationship (7.8) that provide for curvature.

The simplest convenient alternative to the linear relationship (7.8) with the ability to portray curvature is the **general quadratic relationship** between process variables $x_1, x_2, \ldots, x_p$ and a response y. This involves a constant term, p linear terms in the process variables, p pure quadratic terms in the process variables, and $\binom{p}{2}$ cross product terms in the process variables. In the simple case of $p = 2$, this means fitting the approximate relationship

General Quadratic Relationship Between Two Process Variables and a Response

$$y \approx \beta_0 + \beta_1 x_1 + \beta_2 x_2 + \beta_3 x_1^2 + \beta_4 x_2^2 + \beta_5 x_1 x_2 \tag{7.9}$$

via least squares to obtain the fitted equation

$$\hat{y} = b_0 + b_1 x_1 + b_2 x_2 + b_3 x_1^2 + b_4 x_2^2 + b_5 x_1 x_2.$$

Use of quadratic equations like that in display (7.9) can be motivated on a number of grounds. For one, such equations turn out to be convenient to fit to data (multiple linear regression programs can still be used). For another, they can provide a variety of shapes (depending on the values of the coefficients β) and thus can be thought of as a kind of "mathematical French curve" for summarizing empirical data. And they can also be thought of as natural approximations to more complicated theoretical relationships between process variables and a response. That is, if there is some general relationship of the form

$$y \approx h(x_1, x_2, \ldots, x_p)$$

at work, making a second-order Taylor approximation of h about any relevant base point (i.e., finding a function with the same first- and second-order partial derivatives as h at the point of interest, but all partials of higher order equal to 0 at that point) will produce a relationship like that in equation (7.9).

Quadratic response functions have already proved helpful in Examples 7.9 and 7.10, and once one realizes that (taking the coefficients on the pure quadratic and cross product terms to be 0) linear equations of the form (7.8) are just special quadratics, it is evident that even Example 7.8 can be thought of as using them. The potential usefulness of quadratic equations then prompts the question of what **data requirements** are in order to be able to fit them. As it turns out, not every set of n data points $(x_1, x_2, \ldots, x_p, y)$ will be adequate to allow the fitting of a quadratic response function. At a minimum, one needs at least as many different $(x_1, x_2, \ldots, x_p)$ combinations as coefficients β. So one must have $n \geq 1 + 2p + \binom{p}{2}$. But in addition, the set of $(x_1, x_2, \ldots, x_p)$ combinations must be "rich" enough, sufficiently spread out in p-space.

One kind of pattern in the $(x_1, x_2, \ldots, x_p)$ combinations that is sufficient to support the fitting of a quadratic response function is a full 3^p factorial design. For example, the data sets in Examples 7.9 and 7.10 are (unreplicated) 3^2 factorial data sets. But as p increases, 3^p grows very fast. For example, for $p = 4$, a full 3^4 factorial requires a minimum of 81 observations. And it turns out that for $p > 2$, a full 3^p factorial is really "overkill" in terms of what is needed. On the other hand, a 2^p factorial *will not* support the fitting of a quadratic response function.

There are two directions to go. In the first place, one might try to "pare down" a 3^p factorial to some useful (small) fraction that will still allow the use of a full quadratic response function. One popular kind of fraction of a 3^p of this type is due to Box and Behnken and is discussed, for example, in Section 15.4 of *Empirical Model-Building and Response Surfaces* by Box and Draper.

A second route to finding experimental designs that allow the economic (and precise) fitting of second-order surfaces is to begin with a 2^p factorial in the p process variables, and then augment it until it provides enough information for the effective fitting of a quadratic. This route has the virtue that it suggests how the designs can be used in practice in a **sequential** manner. One can begin with a 2^p factorial experiment, fitting a linear equation of form (7.8), and augment it only when there is evidence that a response function allowing for curvature is really needed to adequately describe a response. This kind of experimental strategy is important enough to deserve a thorough exposition.

Sequential Experimental Strategy

So suppose that in a study involving the process variables $(x_1, x_2, \ldots, x_p)$ one desires to center experimentation about the values $x_1^*, x_2^*, \ldots, x_p^*$. Then for $\Delta_1, \Delta_2, \ldots, \Delta_p$ positive constants, one can let the low level of the ith process variable be $x_i^* - \Delta_i$ and the high level be $x_i^* + \Delta_i$ and begin experimentation with a 2^p factorial. In fact, it is a good idea to initially also collect a few responses at the center point of experimentation $(x_1^*, x_2^*, \ldots, x_p^*)$. Data from such a **$2^p$ plus repeated center point** experimental design provide a good basis for checking the adequacy of a linear equation like the one in display (7.8). If (applying regression techniques) one finds important lack of fit to the linear equation and/or (applying the Yates algorithm to the 2^p part of the data) important interactions among the process variables, the need for more data collection is indicated.

Then, a clever way of choosing additional points at which to collect data is to augment the 2^p with **star or axial points**. For some constant α (usually taken to be at least 1) these are $2p$ points of the form

$$(x_1^*, x_2^*, \ldots, x_{l-1}^*, x_l^* \pm \alpha\Delta_l, x_{l+1}^*, \ldots, x_p^*)$$

for $l = 1, 2, \ldots, p$. Geometrically, if one thinks of setting up a coordinate system with origin at the center of the 2^p factorial part of the experimental design, these are points on the axes of the system, α times as far from the center as the "faces" of the 2^p design. Figure 7.12 shows $p = 2$ and $p = 3$ versions of this "2^p factorial plus center points plus star points" arrangement of combinations of process variables. In practice, it is also wise to make some additional observations at the center point in the second round of experimentation. (These repeats of a condition from the first round of experimentation allow one to investigate whether experimental conditions appear to be different in the second round than they were in the first.)

When the whole "2^p plus center points plus star points" experimental program is completed, the jargon typically employed is that one has used a **central composite** experimental design. Allowing for the fact that only certain settings of the drilling machine were possible and only certain drill diameters were available, the

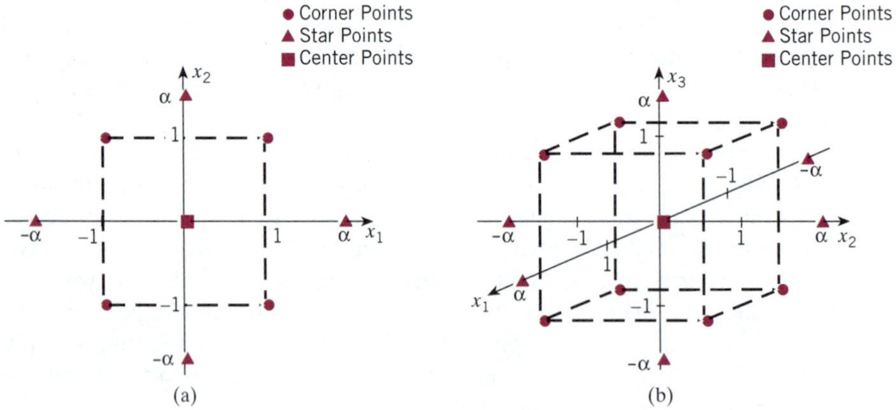

Figure 7.12
(a) 2^2 factorial plus center points and star points, (b) 2^3 factorial plus center points and star points.

data set in Table 7.10 came approximately from a $p = 2$ central composite design. The next example involves a $p = 3$ central composite study.

Example 7.11 **A Seal Strength Optimization Study.** The article "Sealing Strength of Wax-Polyethylene Blends" by Brown, Turner, and Smith appeared in *Tappi* in 1958 and contains an early and instructive application of a central composite design and a quadratic response function. Table 7.13 contains the data from the article. Three process variables,

x_1 = seal temperature (°F),

x_2 = cooling bar temperature (°F), and

x_3 = polyethylene content (%)

were considered regarding their effects on

y = the seal strength of a paper breadwrapper stock (g/in).

The data in Table 7.13 compose a central composite design with center point (255, 55, 1.1), $\Delta_1 = 30$, $\Delta_2 = 9$, $\Delta_3 = .6$, and $\alpha = 1.682$. The first eight rows of Table 7.13 represent a 2^3 design in the three process variables. The ninth row represents six observations taken at the center point of the experimental region. And the last six rows represent the axial or star points of the experimental design. It is

Table 7.13 Seal Strengths Under 15 Different Combinations of $p = 3$ Process Variables

x_1	x_2	x_3	y
225	46	.5	6.6
285	46	.5	6.9
225	64	.5	7.9
285	64	.5	6.1
225	46	1.7	9.2
285	46	1.7	6.8
225	64	1.7	10.4
285	64	1.7	7.3
255	55	1.1	10.1, 9.9, 12.2, 9.7, 9.7, 9.6
204.54	55	1.1	9.8
305.46	55	1.1	5.0
255	39.862	1.1	6.9
255	70.138	1.1	6.3
255	55	.0908	4.0
255	55	2.1092	8.6

important to note that the repeated center point provides an honest estimate of experimental variability. The sample standard deviation of the six y's at the center of the design is $s_P = 1.00$ g/in to two decimal places.

It is instructive to consider the analysis of the data in Table 7.13 in two stages, corresponding to what could have first been known from the "2^3 factorial plus repeated center point" part of the data, and then from the whole data set. To begin with, the reader can verify that fitting a linear regression

$$y \approx \beta_0 + \beta_1 x_1 + \beta_2 x_2 + \beta_3 x_3$$

to the data on the first nine lines of Table 7.13 produces the prediction equation

$$\hat{y} = 13.0790 - .0292x_1 + .0306x_2 + 1.2917x_3,$$

$R^2 = .273$, $s_{SF} = 1.75$ a regression-based (or "surface fitting") estimate of σ, and residuals that when plotted against any one of the process variables have a clear "up then back down again pattern." There are many indications that this fitted equation is not a good one. The R^2 value is small by engineering standards, the regression-based estimate of σ shows signs of being inflated (when compared to $s_P = 1.00$) by lack of fit (in fact, a formal test of lack of fit has a p-value of .048), and the fitted equation underpredicts the response at the center of the experimental region, while overpredicting at all but one of the corners of the 2^3 design. (This last fact is what produces the clear patterns on plots of residuals against x_1, x_2, and x_3.)

A way of formally showing the discrepancy between the mean response at the center of the experimental region and what an equation linear in the process variables implies, is based on the fact that if a mean response is linear in the process variables, the average of the means at the 2^p corner points is the same as the mean at the center point. This implies that the average of the y's on the first eight lines of Table 7.13 should be about the same as the average of the y's on the ninth line. But this is clearly not the case. Using the obvious notation for corners and center points,

$$\bar{y}_{\text{center}} - \bar{y}_{\text{corners}} = 10.20 - 7.65 = 2.55 \text{ g/in.}$$

Then formula (6.8) can be applied to make 95% two-sided confidence limits (based on the 5-degree-of freedom s_P computed from the repeated center point) for the linear combination of nine μ's corresponding to this combination of $\bar{y}$'s. Doing this, an uncertainty of only

$$\pm 2.571(1.00)\sqrt{\frac{(1)^2}{6} + \frac{8\left(-\frac{1}{8}\right)^2}{1}} = \pm 1.39 \text{ g/in}$$

can be associated with the 2.55 g/in figure. That is, it is clear that the 2.55 figure is more than just noise. A linear equation simply does not adequately describe the first part of the data set. There is some curvature evident in how seal strength changes with the process variables.

So based on the first part of the data set, one would have good reason to collect data at the star points and fit a quadratic model to the whole data set. The reader is invited to verify that using multiple regression to fit a quadratic to all 20 observations represented in Table 7.13 produces

$$\begin{aligned}\hat{y} &= -104.82 + .49552x_1 + 1.72530x_2 + 14.27454x_3 - .00084x_1^2 - .01287x_2^2 \\ &\quad - 3.19013x_3^2 - .00130x_1x_2 - .02778x_1x_3 + .02778x_2x_3,\end{aligned}$$

$R^2 = .856$, $s_{SF} = 1.09$, and residuals that look much better than those for the linear fit to the first part of the data. This fitted equation is admittedly more complicated than the linear equation, but it is also a much better empirical description of how seal strength depends upon the process variables.

As a means of aiding understanding of the nature of the quadratic response function, Figure 7.13 on page 386 shows a series of contour plots of this function versus x_1 and x_2, for $x_3 = 0, .5, 1.0, 1.5$, and 2.0. This series of plots suggests that optimum (maximum) seal strength may be achieved for x_1 near 225°F, x_2 near 57°F, and x_3 near 1.5%, and that a mean seal strength exceeding 11 g/in may be possible by proper choice of values for the process variables.

It is worth noting that on the basis of some theoretical optimality arguments, it is common to recommend values of α for constructing central composite designs that are larger than 1.0 (thereby placing the star points outside of the region in $(x_1, x_2, \ldots, x_p)$-space with corners at the 2^p design points). There are cases, however, where other considerations may come into play and suggest smaller choices for α. For example, there are experimental scenarios where one really wants to minimize the number of different levels of a given factor that one uses. And the choice $\alpha = 1$ places the star points on the "faces" of the 2^p design and makes the resulting central composite a fraction of the 3^p design (so that only three different levels of each factor are used, rather than five).

7.2.3 Analytical Interpretation of Quadratic Response Functions

For p much larger than 2 or 3, graphical interpretation of a fitted quadratic response function becomes difficult at best and impossible at worst. Happily (at least for those with some familiarity with matrix manipulations) there are some analytic tools that can help. Those tools are the subject of this subsection. Readers without a background in matrix algebra can skim this material (reading for main points and not technical detail) without loss of continuity.

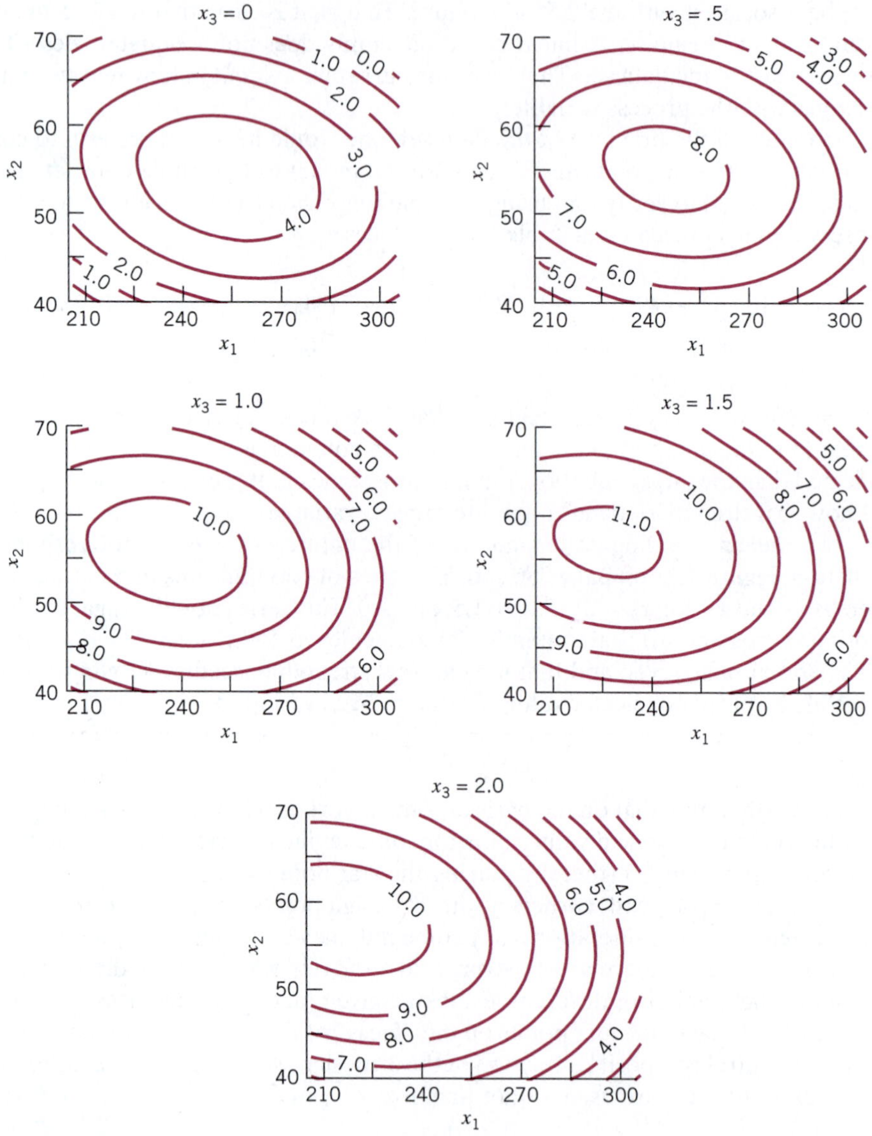

Figure 7.13
Series of five contour plots of seal strength for polyethelene contents x_3 between 0 and 2.0 (%).

For the quadratic function of $p = 1$ variable,

$$y \approx \beta_0 + \beta_1 x + \beta_2 x^2,$$

it is the coefficient β_2 that governs the basic nature of the relationship between x and y. For $\beta_2 > 0$, the equation graphs as a parabola opening up, and y has a minimum at $x = -\beta_1/2\beta_2$. For $\beta_2 < 0$, the equation graphs as a parabola opening down, and y has a maximum at $x = -\beta_1/2\beta_2$. And for $\beta_2 = 0$, the function is actually linear in x and (if β_1 is not 0) has neither a maximum nor a minimum value when x is allowed to vary over all real numbers. Something similar to this story is true for p larger than 1.

It is the coefficients of the pure quadratic and mixed terms of a multivariate quadratic relationship like that in display (7.9) which govern the nature of the response function. In order to detail the situation, some additional notation is required. Suppose that based on n data points $(x_1, x_2, \ldots, x_p, y)$ one arrives at a quadratic regression equation with fitted coefficients

b_i = the fitted coefficient of the linear term x_i,
b_{ii} = the fitted coefficient of the pure quadratic term x_i^2, and
$b_{ii'}$ = the fitted coefficient of the mixed term $x_i x_{i'}$ (note that $b_{ii'} = b_{i'i}$).

Then using these, define the $p \times 1$ vector $\boldsymbol{b}$ and $p \times p$ matrix $\boldsymbol{B}$ by

$$\boldsymbol{b} = \begin{bmatrix} b_1 \\ b_2 \\ \vdots \\ b_p \end{bmatrix} \quad \text{and} \quad \boldsymbol{B} = \begin{bmatrix} b_{11} & \frac{1}{2}b_{12} & \cdots & \frac{1}{2}b_{1p} \\ \frac{1}{2}b_{21} & b_{22} & \cdots & \frac{1}{2}b_{2p} \\ \vdots & \vdots & & \vdots \\ \frac{1}{2}b_{p1} & \frac{1}{2}b_{p2} & \cdots & b_{pp} \end{bmatrix}. \tag{7.10}$$

Vector and Matrix of Coefficients for a p Variable Quadratic

It is the **eigenvalues** of the matrix $\boldsymbol{B}$ that govern the shape of the fitted quadratic, and $\boldsymbol{b}$ and $\boldsymbol{B}$ together determine where (if at all) the quadratic has a minimum or maximum.

The eigenvalues of the matrix $\boldsymbol{B}$ are the p solutions λ to the equation

$$\det(\boldsymbol{B} - \lambda \boldsymbol{I}) = 0. \tag{7.11}$$

Equation Solved by Eigenvalues

The fitted surface has a stationary point (i.e., a point where all first partial derivatives with respect to the process variables are 0) at the point

$$\boldsymbol{x} = \begin{bmatrix} x_1 \\ x_2 \\ \vdots \\ x_p \end{bmatrix} = -\frac{1}{2}\boldsymbol{B}^{-1}\boldsymbol{b}. \tag{7.12}$$

Location of the Stationary Point for a p Variable Quadratic

When all p eigenvalues are positive, the fitted quadratic has a **minimum** at the point defined by relationship (7.12). When all p eigenvalues are negative, the fitted quadratic has a **maximum** at the point defined in display (7.12). When some eigenvalues are positive and the rest are negative, the fitted response function has a **saddle** geometry. (Moving away from the point defined in display (7.12) in some directions causes an increase in fitted y, while moving away from the point in other directions produces a decrease in fitted y.) And when some eigenvalues are 0 (or in practice, nearly 0), the fitted quadratic has a **ridge** geometry.

Example 7.10 continued. The fitted equation for yield in the Hill and Demler chemical process study is a quadratic in $p = 2$ variables, and the corresponding vector and matrix defined in display (7.10) are

$$\boldsymbol{b} = \begin{bmatrix} 1.254 \\ 5.068 \end{bmatrix} \quad \text{and} \quad \boldsymbol{B} = \begin{bmatrix} -.009333 & \frac{1}{2}(.01990) \\ \frac{1}{2}(.01990) & -.1180 \end{bmatrix}.$$

Figure 7.11 indicates that the fitted yield surface has a mound-shaped geometry, with maximum somewhere near $x_1 = 99$ and $x_2 = 30$. This can be confirmed analytically by using equation (7.12) to find the stationary point and examining the eigenvalues defined in equation (7.11).

To begin with, note that

$$-\frac{1}{2}\boldsymbol{B}^{-1}\boldsymbol{b} = -\frac{1}{2}\begin{bmatrix} -.009333 & \frac{1}{2}(.01990) \\ \frac{1}{2}(.01990) & -.1180 \end{bmatrix}^{-1}\begin{bmatrix} 1.254 \\ 5.068 \end{bmatrix} = \begin{bmatrix} 99.09 \\ 29.79 \end{bmatrix},$$

so that the stationary point has $x_1 = 99.09$ and $x_2 = 29.79$. Then note that equation (7.11) for the eigenvalues is

$$0 = \det\left(\begin{bmatrix} -.009333 & \frac{1}{2}(.01990) \\ \frac{1}{2}(.01990) & -.1180 \end{bmatrix} - \lambda\begin{bmatrix} 1 & 0 \\ 0 & 1 \end{bmatrix}\right),$$

that is,

$$0 = (-.009333 - \lambda)(-.1180 - \lambda) - \frac{1}{4}(.01990)^2.$$

This is a quadratic equation in the variable λ, and the quadratic formula can be used to find its roots

$$\lambda = -.0086 \quad \text{and} \quad \lambda = -.1187.$$

The fact that these are both negative is analytical confirmation that the stationary point provides a maximum of the fitted yield surface.

Example 7.9 continued. Figures 7.9 and 7.10 portraying the fitted equation for Burris's lift/drag ratio data indicate the saddle surface nature of the fitted response. This can be confirmed analytically by noting that $\boldsymbol{B}$ corresponding to the fitted equation is

$$\boldsymbol{B} = \begin{bmatrix} -.4843 & \frac{1}{2}(-.5042) \\ \frac{1}{2}(-.5042) & 0 \end{bmatrix},$$

and that in this instance equation (7.11) for the eigenvalues,

$$0 = \det(\boldsymbol{B} - \lambda \boldsymbol{I}) = (-.4843 - \lambda)(-\lambda) - \frac{1}{4}(-.5042)^2,$$

has roots

$$\lambda = -.592 \qquad \text{and} \qquad \lambda = .107,$$

one of which is positive and one of which is negative.

For p larger than two, the calculations indicated in equations (7.11) and (7.12) are reasonably done only using some kind of statistical or mathematical software that supports matrix algebra. Printout 7.1 illustrates a Minitab session for finding the eigenvalues of $\boldsymbol{B}$ and the stationary point identified in display (7.12) for the quadratic response function fit to the seal strength data of Table 7.13. It illustrates that as suggested by Figure 7.13, the quadratic fit to the data has a maximum when $x_1 = 226°\text{F}$, $x_2 = 57.2°\text{F}$, and $x_3 = 1.50\%$.

Printout 7.1

```
MTB > Read 3 3 M1.
DATA> -.00084 -.00065 -.01389
DATA> -.00065 -.01287 .01389
DATA> -.01389 .01389 -3.19013
      3 rows read.
MTB > Eigen M1 c1.
MTB > Print c1

Data Display
C1
 -3.19025   -0.01285   -0.00074
```

```
MTB > Read 3 1 M2.
DATA> .49552
DATA> 1.72530
DATA> 14.27454
      3 rows read.
MTB > Invert M1 M3.
MTB > Multiply M3 M2 M4.
MTB > Multiply M4 -.5 M4
MTB > Print M4

Data Display
 Matrix M4

  225.794
   57.247
    1.503
```

A final caution needs to be sounded before leaving this discussion of interpreting fitted response surfaces. This concerns the very real possibility of overinterpreting a fitted relationship between p process variables and y. One needs to be sure that a surface is really fairly well identified by data in hand before making conclusions based on the kind of calculations illustrated here (or, for that matter, based on the graphical tools of this section).

One useful rule of thumb for judging whether a surface is well enough determined to justify its use is due to Box, Hunter, and Hunter (see their *Statistics for Experimenters*) and goes as follows: If a response function involving l coefficients b (including a constant term where relevant) is fit to n data points via multiple regression, producing n fitted values $\hat{y}$ and an estimate of σ (based on surface fitting) s_{SF}, then one checks to see whether

Criterion for Judging Whether a Response is Adequately Determined

$$\max \hat{y} - \min \hat{y} > 4\sqrt{\frac{l s_{\text{SF}}^2}{n}} \tag{7.13}$$

before interpreting the surface. The difference on the left of inequality (7.13) is a measure of the movement of the surface over the region of experimentation. The fraction under the root on the right is an estimate of average variance of the n values $\hat{y}$. The check is meant to warn its user if the shape of the fitted surface is really potentially attributable completely to random variation.

7.2.4 Response Optimization Strategies

The tools of this section are primarily tactical devices, useful for understanding the "local terrain" of a response surface. What remains largely unaddressed in this section is the broader, more strategic issue of how one finds a region in $(x_1, x_2, \ldots, x_p)$-space deserving careful study, particularly where the ultimate objective is to optimize one or more responses y. Many sensible strategies are possible and this subsection discusses two. The first is something called evolutionary operation (or EVOP for short) and the second is an empirical optimization strategy that uses the linear and quadratic response surface tools just discussed.

EVOP is a strategy for conservative ongoing experimentation on a working production process that aims to simultaneously produce good product and also provide information for the continual improvement (the "evolution") of the process. The notion was first formally discussed by Box and Wilson and is thoroughly detailed in the book *Evolutionary Operation* by Box and Draper. Various authors have suggested different particulars for implementation of the general EVOP strategy, including a popular "simplex" empirical hill-climbing algorithm put forth by Spendley, Hext, and Himsworth. As the present treatment must be brief, only the simplest (and original) factorial implementation will be discussed, and that without many details.

Outline of an EVOP Strategy

So consider a situation where p process variables $x_1, x_2, \ldots, x_p$ are thought to affect l responses (or quality characteristics) $y_1, y_2, \ldots, y_l$. If $x_1^*, x_2^*, \ldots, x_p^*$ are current standard operating values for the process variables, an EVOP program operates in the following set of steps.

Step 1. An EVOP Committee (consisting of a broad group of experts with various kinds of process knowledge and interests in process performance) chooses a few (two or three) of the process variables as most likely to provide improvements in the response variables. For sake of concreteness, suppose that x_1 and x_2 are the variables selected. A "two-level factorial plus center point" experimental design is set up in the selected variables with (x_1^*, x_2^*) as the center point. The high and low levels of the variables are chosen to be close enough to x_1^* and x_2^* so that any change in any mean response across the set of experimental conditions is expected to be small.

Step 2. Holding variables $x_3, x_4, \ldots, x_p$ at their standard values $x_3^*, x_4^*, \ldots, x_p^*$, in normal process operation (x_1, x_2) is cycled through the experimental design identified in Step 1, and values for all of $y_1, y_2, \ldots, y_l$ are recorded. This continues until for every response, enough data have been collected so that the changes in mean as one moves from the center point to corner points are estimated with good precision (relative to the sizes of the changes).

Step 3. If, in light of the estimated changes in the means for all of $y_1, y_2, \ldots, y_l$, the EVOP Committee finds no corner point of the design to be preferable to the center point, the program returns to Step 1 and a different set of process variables is chosen for experimentation.

Step 4. If there is a corner of the two-level factorial design that the EVOP Committee finds preferable to the center point, *new* standard values of

x_1 and x_2, x_1^* and x_2^*, are established between the previous ones and the values for the superior corner point. A new two-level factorial plus center point experimental design is set up in variables x_1 and x_2 with the new (x_1^*, x_2^*) as the center point. The high and low levels of the variables are chosen to be close enough to the new x_1^* and x_2^* so that any change in any mean response across the experimental conditions is expected to be small. The EVOP program then returns to Step 2.

Evolutionary Operation is intended to be a relatively cautious program that succeeds in process improvement because of its persistence. Only a few variables are changed at once and only small moves are made in the standard operating conditions. Nothing is done in an "automatic" mode. Instead the EVOP Committee considers potential impact on all important responses before authorizing movement. The caution exercised in the ideal EVOP program is in keeping with the fact that the intention is to run essentially no risk of noticeably degrading process performance in any of the experimental cycles.

A Second Response Optimization Strategy

A more aggressive posture can be adopted when experimentation on a process can be done in a "get in, make the improvement, and get out" mode and there is no serious pressure to assure that every experimental run produces acceptable product. The following steps outline a common kind of strategy that makes use of the linear and quadratic response function ideas of this section in the optimization of a single response y, where p process variables are at work and initial standard operating values are $x_1^*, x_2^*, \ldots, x_p^*$.

Step 1. A "2^p plus center points" experimental design is run (with center at the point $(x_1^*, x_2^*, \ldots, x_p^*)$) and a linear response function for y is fit to the resulting data and examined for its adequacy using the tools of regressions analysis.

Step 2. In the event that the linear equation fit to the 2^p factorial plus center points data is adequate, a sequence of observations is made along a ray in $(x_1, x_2, \ldots, x_p)$-space beginning at the center point of the design and proceeding in the direction of steepest ascent (or steepest descent depending upon whether the object is to maximize or to minimize y). That is, if the fitted values of the parameters of the linear response function are $b_0, b_1, b_2, \ldots, b_p$, observations are made at points of the form

$$(x_1^* + ub_1, x_2^* + ub_2, \ldots, x_p^* + ub_p)$$

for positive values of u if the object is to maximize y (or for negative values of u if the object is to minimize y). If the object is to maximize y, the magnitude of u is increased until the response seems to cease to increase (or seems to cease to decrease if the object is to minimize u). Polynomial regression of y on u can be helpful in seeing the pattern of response to these changes in u. The point of optimum response (or optimum fitted response if one smoothes the y values using regression on u) along the ray becomes a new point $(x_1^*, x_2^*, \ldots, x_p^*)$ and the algorithm returns to Step 1.

Step 3. If the linear surface fit in Step 1 is not an adequate description of the 2^p plus center point data, star points are added and a quadratic surface is fit. The quadratic is examined for adequacy as a local description of mean response and the location of the best fitted mean within the experimental region is identified. This point becomes a new center of experimental effort $(x_1^*, x_2^*, \ldots, x_p^*)$ and the algorithm returns to Step 1.

Fairly obviously, at some point this algorithm typically "stalls out" and ceases to provide improvements in mean y. At that point, engineering attention can be turned to some other process or project.

7.3 MIXTURE STUDIES

Behind the response surface material of the previous section stands an implicit assumption that the p process variables can be adjusted completely independently of each other. There is, however, an important class of applications of response surface ideas where this is not the case. **Mixture studies** are carried out to determine the effects of changing the component proportions of some material made by mixing p pure ingredients. The physical properties of gasoline, concrete, various kinds of plastic, inks, metals, and so on depend upon the proportions of their constituents. And one cannot increase the proportion of cement in a concrete mixture without simultaneously decreasing the proportion of sand, and/or water, and/or other additives in the mixture.

This fact (that mixture proportions cannot be changed completely independently of each other) ultimately requires that some special methods of data analysis and experimental design be developed for mixture studies. Entire large books, most notably Cornell's *Experiments With Mixtures: Designs, Models and the Analysis of Mixture Data,* have been written on the subject. So, the goal here is to provide only the most rudimentary introduction to the issues that arise in the study of mixtures and some elementary examples of effective use of mixture techniques. The reader faced with a serious mixture problem in more than a few components will need to do further reading elsewhere (and likely obtain some specialized software if anything but the most elementary experimental design is contemplated). But this section should at least serve to make readers aware of the kinds of things that are possible.

The section begins with a discussion of the impact on regression analysis of the linear dependence of mixture proportions, and of the forms usually used for linear and quadratic response functions in mixture studies. Then some graphical devices useful in mixture studies are presented. Finally, there is a brief discussion of some issues and elementary methods in experimental design for mixture studies.

7.3.1 Linear Dependence of Mixture Proportions and Fitting Linear and Quadratic Response Functions

In this section, the process variables $x_1, x_2, \ldots, x_p$ are the (nonnegative) proportions of p pure components in a mixture and the response y is some measured physical property of the mixture. The fundamental fact that makes mixture studies special is that the proportions must satisfy the relationship

Basic Mixture Constraint

$$x_1 + x_2 + \cdots + x_p = 1. \tag{7.14}$$

This constraint on the values of $x_1, x_2, \ldots, x_p$ has a number of implications for the analysis and planning of mixture studies.

Consider first the apparently simple problem of fitting a linear relationship

$$y \approx \beta_0 + \beta_1 x_1 + \beta_2 x_2 + \cdots + \beta_p x_p \tag{7.15}$$

to data from a mixture study. Relationship (7.14) has the unpleasant implication that when trying to fit equation (7.15), two sets of fitted coefficients

$$b_0, b_1, \ldots, b_p \qquad \text{and} \qquad b'_0, b'_1, \ldots, b'_p$$

related by

$$b'_0 = b_0 + c \qquad \text{and} \qquad b'_i = b_i - c \text{ for } i = 1, 2, \ldots, p$$

(for any constant c) will have the same set of predicted responses, and are thus indistinguishable in terms of least squares fitting. That is, in the mixture context relationship (7.15) is simply overparameterized. If used naively, standard regression programs will either refuse to fit the equation or will make some preprogrammed choice to eliminate one of the process variables (and corresponding parameter) from the equation in order that there be a single best set of fitted parameters.

But rather than fitting a version of equation (7.15) with, say, the last x eliminated, that is,

$$y \approx \beta_0 + \beta_1 x_1 + \beta_2 x_2 + \cdots + \beta_{p-1} x_{p-1}, \tag{7.16}$$

it is typically preferable in mixture contexts to drop the constant term from relationship (7.15) and fit instead the equation

Linear Response Function for a Mixture Study

$$y \approx \beta_1 x_1 + \beta_2 x_2 + \cdots + \beta_p x_p. \tag{7.17}$$

Most multiple regression programs have a **no-constant option**, and the coefficients in relationship (7.17) have a more straightforward interpretation than do the ones in relationship (7.16). That is, in form (7.16) β_0 is the mean response when x_1

through x_{p-1} are all 0 and x_p is therefore equal to 1, while β_i for $i = 1, 2, \ldots, p-1$ is the rate of change of mean response as x_i increases, all other x's appearing in the equation are held constant, *but x_p decreases*. These relationships are both unpleasantly asymmetric and also somewhat unnatural.

On the other hand, each coefficient β_i in the form (7.17) has the straightforward interpretation as the mean response when x_i is 1 and the proportions of all other components are 0. That is, when equation (7.17) is a good description of y for all $(x_1, x_2, \ldots, x_p)$ with nonnegative entries satisfying the constraint (7.14), the coefficients in equation (7.17) are the mean responses for the pure components. (Note that it will not always be the case that this makes physical sense. Some materials simply cannot be made when any x_i is too extreme. The possibility of "pure water" concrete is absurd enough to illustrate this point in memorable fashion.)

If one's regression program does not have a no-constant option, it is possible to simply fit equation (7.16), getting

$$\hat{y} = b_0 + b_1x_1 + b_2x_2 + \cdots + b_{p-1}x_{p-1}, \tag{7.18}$$

and then translate the results to a fitted version of relationship (7.17). That is, substituting constraint (7.14) into equation (7.18), one has

$$\begin{aligned}\hat{y} &= b_0(x_1 + x_2 + \cdots + x_p) + b_1x_1 + b_2x_2 + \cdots + b_{p-1}x_{p-1} \\ &= (b_1 + b_0)x_1 + (b_2 + b_0)x_2 + \cdots + (b_{p-1} + b_0)x_{p-1} + b_0x_p\end{aligned}$$

and it is evident how to get fitted parameters for equation (7.17) from the fitted equation (7.18). Probably the biggest advantage of fitting equation (7.17) directly is not so much the small bit of arithmetic saved in translating the results (7.18), but rather the fact that a no-constant regression output will usually give standard errors for the estimates of the parameters β in equation (7.17) that otherwise would not be directly available.

Example 7.12 **A $p = 5$ Component Gasoline Blending Study.** The article "Developing Blending Models for Gasoline and Other Mixtures" by R. Snee appeared in *Technometrics* in 1981 and contains a mixture data set relating

y = measured research octane at 2.0 g/gal Pb

for 16 different gasoline blends, to the mixture proportions

x_1 = fraction of Butane in the mixture,

x_2 = fraction of Alkylate in the mixture,

x_3 = fraction of Light Straight Run in the mixture,

x_4 = fraction of Reformate in the mixture, and

x_5 = fraction of Cat Cracked in the mixture.

Table 7.14 Research Octane for 16 Different Gasoline Blends

x_1	x_2	x_3	x_4	x_5	y
.10	.10	.05	.20	.55	95.1
.10	.00	.15	.20	.55	93.4
.00	.10	.15	.20	.55	93.3
.10	.10	.15	.20	.45	94.1
.00	.00	.05	.40	.55	91.8
.10	.00	.05	.40	.45	91.8
.00	.10	.05	.40	.45	92.5
.00	.00	.15	.40	.45	90.5
.00	.00	.05	.35	.60	92.7
.10	.10	.15	.25	.40	93.5
.10	.00	.05	.25	.60	94.8
.00	.10	.05	.25	.60	93.7
.00	.00	.15	.25	.60	92.5
.10	.10	.05	.35	.40	93.1
.10	.00	.15	.35	.40	91.8
.00	.10	.15	.35	.40	91.6

Snee's data are given here in Table 7.14.

Printout 7.2 shows two Minitab regressions for Snee's data, the first fitting relationship (7.16) and the second fitting relationship (7.17).

Printout 7.2

```
MTB > Name c7 = 'QUAD1'
MTB > RSReg 'y' = 'x1' 'x2' 'x3' 'x4';
SUBC>    Quadratic 'QUAD1';
SUBC>    Brief 3.

Response Surface Regression

Estimated Regression Coefficients for y

Term            Coef       Stdev    t-ratio      p
Constant       97.59      0.5203    187.564  0.000
x1              4.82      1.6854      2.861  0.015
x2              3.07      1.6854      1.822  0.096
x3            -12.43      1.6854     -7.374  0.000
x4            -12.86      1.1917    -10.789  0.000

s = 0.3153        R-sq = 95.2%     R-sq(adj) = 93.5%
```

Analysis of Variance for y

Source	DF	Seq SS	Adj SS	Adj MS	F	P
Regression	4	21.6839	21.6839	5.42098	54.53	0.000
Linear	4	21.6839	21.6839	5.42098	54.53	0.000
Residual Error	11	1.0936	1.0936	0.09942		
Total	15	22.7775				

Obs.	y	Fit	Stdev.Fit	Residual	St.Resid
1	95.100	95.189	0.181	-0.089	-0.35
2	93.400	93.639	0.181	-0.239	-0.93
3	93.300	93.464	0.181	-0.164	-0.64
4	94.100	93.946	0.160	0.154	0.57
5	91.800	91.829	0.160	-0.029	-0.11
6	91.800	92.311	0.181	-0.511	-1.98
7	92.500	92.136	0.181	0.364	1.41
8	90.500	90.586	0.181	-0.086	-0.33
9	92.700	92.471	0.160	0.229	0.84
10	93.500	93.304	0.160	0.196	0.72
11	94.800	94.239	0.181	0.561	2.17R
12	93.700	94.064	0.181	-0.364	-1.41
13	92.500	92.514	0.181	-0.014	-0.06
14	93.100	93.261	0.181	-0.161	-0.62
15	91.800	91.711	0.181	0.089	0.35
16	91.600	91.536	0.181	0.064	0.25

R denotes an obs. with a large st. resid.

```
MTB > Mixreg 'y' = 'x1' 'x2' 'x3' 'x4' 'x5' ;
SUBC>   Brief 3.
```

Regression for Mixtures

Estimated Regression Coefficients for y

Term	Coef	Stdev	t-ratio	p
x1	102.41	1.5148	67.610	0.000
x2	100.66	1.5148	66.455	0.000
x3	85.16	1.4304	59.539	0.000
x4	84.74	0.7448	113.765	0.000
x5	97.59	0.5203	187.564	0.000

s = 0.3153   R-sq = 95.2%   R-sq(adj) = 93.5%

Analysis of Variance for y

Source	DF	Seq SS	Adj SS	Adj MS	F	P
Regression	4	21.6839	21.6839	5.42098	54.53	0.000
Linear	4	21.6839	21.6839	5.42098	54.53	0.000
Residual Error	11	1.0936	1.0936	0.09942		
Total	15	22.7775				

Obs.	y	Fit	Stdev.Fit	Residual	St.Resid
1	95.100	95.189	0.181	-0.089	-0.35
2	93.400	93.639	0.181	-0.239	-0.93
3	93.300	93.464	0.181	-0.164	-0.64
4	94.100	93.946	0.160	0.154	0.57
5	91.800	91.829	0.160	-0.029	-0.11
6	91.800	92.311	0.181	-0.511	-1.98
7	92.500	92.136	0.181	0.364	1.41
8	90.500	90.586	0.181	-0.086	-0.33
9	92.700	92.471	0.160	0.229	0.84
10	93.500	93.304	0.160	0.196	0.72
11	94.800	94.239	0.181	0.561	2.17R
12	93.700	94.064	0.181	-0.364	-1.41
13	92.500	92.514	0.181	-0.014	-0.06
14	93.100	93.261	0.181	-0.161	-0.62
15	91.800	91.711	0.181	0.089	0.35
16	91.600	91.536	0.181	0.064	0.25

R denotes an obs. with a large st. resid.

The reader should note that the two regressions described on Printout 7.2 are essentially equivalent. Fitted responses are the same, as are the estimated standard deviations of y (for a fixed blend) and the R^2 values. And as has been indicated, estimated coefficients for one of the fitted equations translate directly to fitted coefficients for the other. The second regression does have the virtue of providing standard errors for the estimated coefficients for relationship (7.17), but otherwise there is little to choose between the two outputs.

The lack of any replication prevents a completely thorough evaluation of the appropriateness of the linear fit to Snee's data. But as far as is possible to tell, it seems to be an effective summarization of the octane values. Further, looking at the table of fitted coefficients for the second run, it is evident that the proportions of all five components are important in determining research octane, and that (by virtue of the sizes of the fitted coefficients) the proportions of Butane and then Alkylate and Cat Cracked in the mixture have the strongest effects on research octane. But it is probably unwise to attempt to interpret the fitted coefficients as octanes for the pure components, since the data of Table 7.14 cover a fairly limited set of mixtures. So such interpretations would amount to extrapolations far beyond the data set. (Note in the data of Table 7.14 that $x_1 \leq .10$, $x_2 \leq .10, .05 \leq x_3 \leq .15, .20 \leq x_4 \leq .40$, and $.40 \leq x_5 \leq .60$.)

One point that is not obvious from Printout 7.2 but deserves some comment, concerns R^2 and no-constant regressions. R^2 only makes sense for regressions that either explicitly or implicitly include a constant term. So no-constant options on standard regression programs do not usually print R^2 values, even when used with mixture data. However, since Minitab's special "Mixreg" command was used for Printout 7.2 and since in light of constraint (7.14) mixture data implicitly allow for the shifting of all fitted responses by any constant value in the process of producing a least squares fit, an R^2 value is printed for the no-constant regression. If the reader's statistical software does not have special mixture routines, it may be necessary to fit a linear response in form (7.16) in order to automatically obtain a value for R^2.

To allow for **curvature** in a response in a mixture study, one must be prepared to consider relationships more complicated than those indicated in displays (7.16) and (7.17). To see what a general quadratic response function should look like in a mixture context, one can use the device of writing out a general quadratic (from Section 7.2) in $p - 1$ variables and then substituting in constraint (7.14). In the case of $p = 3$ components, this gives

$$\begin{aligned} y &\approx \beta_0 + \beta_1 x_1 + \beta_2 x_2 + \beta_3 x_1^2 + \beta_4 x_2^2 + \beta_5 x_1 x_2 \\ &= \beta_0(x_1 + x_2 + x_3) + \beta_1 x_1 + \beta_2 x_2 + \beta_3 x_1(1 - x_2 - x_3) + \beta_4 x_2(1 - x_1 - x_3) + \beta_5 x_1 x_2 \\ &= (\beta_0 + \beta_1 + \beta_3)x_1 + (\beta_0 + \beta_2 + \beta_4)x_2 + \beta_0 x_3 + (-\beta_3 - \beta_4 + \beta_5)x_1 x_2 \\ &\quad + (-\beta_3)x_1 x_3 + (-\beta_4)x_2 x_3. \end{aligned}$$

And this bit of algebra correctly suggests that in a mixture context, the general form of a quadratic response function that explicitly displays dependence upon all p component proportions is one without a constant term that includes a linear term for each component and all possible cross product terms. For $p = 3$, this is

$$y \approx \beta_1 x_1 + \beta_2 x_2 + \beta_3 x_3 + \beta_4 x_1 x_2 + \beta_5 x_1 x_3 + \beta_6 x_2 x_3. \tag{7.19}$$

$p = 3$ Quadratic Response Function for a Mixture Study

In equation (7.19) only the coefficients of the linear terms have obvious direct interpretations (as mean responses for the pure components). But, as it treats all components symmetrically, the form (7.19) is usually preferred to one that displays only $p - 1 = 2$ of the component proportions. Of course the mechanics of fitting a quadratic equation can be accomplished either by eliminating one of the variables x and applying the material of the previous section, or by fitting an equation like (7.19) directly, using a no-constant regression.

Example 7.13 Quadratic Regressions in a $p = 3$ Component Pipe Compound Mixture Study. The article "Design and Analysis of an ABS Pipe Compound Experiment" by Koons and Wilt appeared in the 1985 ASQC Technical Supplement *Experiments in Industry* and contains an interesting mixture data set. The physical properties of acrylonitrile-butadiene-styrene (ABS) pipe material

Table 7.15 Physical Properties of 10 Different ABS Pipe Compound Formulations

x_1	x_2	x_3	y_1	y_2	y_3
.45	.55	.00	6.1	213	6,080
.45	.425	.125	2.1	191	5,740
.45	.30	.25	.8	183	5,065
.50	.45	.05	4.9	202	5,615
.50	.35	.15	2.5	186	5,385
.535	.38	.085	4.4	197	5,410
.575	.425	.00	7.3	204	5,280
.575	.30	.125	3.9	188	5,080
.60	.35	.05	6.1	192	5,035
.70	.30	.00	7.4	205	4,350

y_1 = Izod impact strength (ft-lbs/in),
y_2 = deflection temperature under load (°F), and
y_3 = yield strength (psi)

were measured for 10 different mixtures of $p = 3$ components. With

x_1 = fraction of grafted polybutadiene in the mixture,
x_2 = fraction of styrene-acrylonitrile copolymer in the mixture, and
x_3 = fraction of coal tar pitch in the mixture,

the data from the study are given in Table 7.15.

Printout 7.3 then shows the result of using Minitab to fit the relationship (7.19) to each of the three response variables. Unfortunately, the pipe compound data set contains no replication, and so efforts to judge the appropriateness of the quadratic fits are accordingly hindered. But as far as is possible to tell, the three prediction equations indicated by the printout do a decent job of describing the physical properties of ABS pipe compound. (There is some hint that a linear equation for y_2 would be adequate, but the curvature provided by the quadratic form is definitely important in the description of both y_1 and y_3.)

Printout 7.3

```
MTB > Mixreg 'y1' 'y2' 'y3' = 'x1' 'x2' 'x3';
SUBC>   Quadratic;
SUBC>   Brief 2.
```

Regression for Mixtures

Estimated Regression Coefficients for y1

Term	Coef	Stdev	t-ratio	p
x1	2.35	1.724	1.365	0.244
x2	-9.22	3.204	-2.877	0.045
x3	16.76	5.774	2.902	0.044
x1*x2	40.74	9.874	4.126	0.015
x1*x3	-7.67	9.834	-0.780	0.479
x2*x3	-84.74	9.874	-8.582	0.001

s = 0.1411 R-sq = 99.8% R-sq(adj) = 99.6%

Analysis of Variance for y1

Source	DF	Seq SS	Adj SS	Adj MS	F	P
Regression	5	45.2453	45.24532	9.049064	454.26	0.000
Linear	2	43.4475	0.98192	0.490960	24.65	0.006
Quadratic	3	1.7978	1.79782	0.599273	30.08	0.003
Residual Error	4	0.0797	0.07968	0.019921		
Total	9	45.3250				

Estimated Regression Coefficients for y2

Term	Coef	Stdev	t-ratio	p
x1	235.5	39.83	5.913	0.004
x2	312.4	74.00	4.221	0.013
x3	339.0	133.37	2.542	0.064
x1*x2	-260.2	228.08	-1.141	0.318
x1*x3	-359.1	227.16	-1.581	0.189
x2*x3	-350.2	228.08	-1.535	0.199

s = 3.260 R-sq = 95.0% R-sq(adj) = 88.7%

Analysis of Variance for y2

Source	DF	Seq SS	Adj SS	Adj MS	F	P
Regression	5	802.385	802.3851	160.4770	15.10	0.011
Linear	2	734.291	79.2548	39.6274	3.73	0.122
Quadratic	3	68.094	68.0940	22.6980	2.14	0.238
Residual Error	4	42.515	42.5149	10.6287		
Total	9	844.900				

```
Estimated Regression Coefficients for y3

Term         Coef      Stdev    t-ratio      p
x1           1887      462.7      4.078  0.015
x2           8404      859.7      9.776  0.001
x3          -7644     1549.4     -4.933  0.008
x1*x2        2430     2649.6      0.917  0.411
x1*x3       22989     2638.9      8.712  0.001
x2*x3        9334     2649.6      3.523  0.024


s = 37.87        R-sq = 99.7%      R-sq(adj) = 99.4%

Analysis of Variance for y3

Source           DF     Seq SS     Adj SS     Adj MS       F      P
Regression        5    1991403    1991403     398280  277.67  0.000
  Linear          2    1860868     305982     152991  106.66  0.000
  Quadratic       3     130534     130534      43511   30.33  0.003
Residual Error    4       5737       5737       1434
Total             9    1997140
```

Many other types of equations (beyond linear and quadratic ones) have been suggested for use in modeling mixture response surfaces. For example, cubic generalizations of the relationships (7.17) and (7.19) have proved helpful in some studies. Where response behavior becomes extreme when any proportion x becomes too small, regression models with "$1/x$" terms have been suggested. And sometimes models involving ratio terms like x_1/x_2 are useful. But the simple linear and quadratic relationships discussed here indicate the basics of what is possible and what approaches must be taken to regression analysis where mixture data are involved.

7.3.2 Special Graphical Devices for Mixture Studies

It is possible (by dropping one of the mixture proportions from explicit consideration) to directly apply the kinds of graphical devices discussed in Section 7.2 to mixture studies. But this is not really completely satisfactory, and so it is natural to consider tools crafted specifically for mixture situations. This subsection thus presents the tools of plotting **response traces** for fitted mixture response functions, and of making various plots in a kind of **triangular or trilinear coordinate system**.

To begin, let

$$\hat{y} = f(x_1, x_2, \ldots, x_p) \tag{7.20}$$

be a generic fitted (mixture) response function for p components. In trying to understand the nature of a relationship (7.20), one faces the complication of constraint (7.14) that was not an issue in Section 7.2. One cannot, for example, simply hold $x_2, x_3, \ldots, x_p$ fixed and expect to look at meaningful plots of $\hat{y}$ versus x_1. If x_1 varies, so must $x_2, x_3, \ldots, x_p$. But exactly how they are varied will affect the appearance of the plot. Further, unless a fitted equation (7.20) is known to be appropriate for *all* possible mixture vectors, it is in general difficult to tell whether in varying x_1 (and making corresponding changes in the other x_i) one has moved outside the region where there are data and the fitted equation can be relied upon.

One circumstance where some sense can be made of this matter and there is a reasonably standard way of plotting curves based on a fitted equation is the case where the region over which the relationship (7.20) is valid can be defined in terms of p lower bounds on the component proportions. That is, suppose that the fitted equation is a good description of the response provided

$$x_i \geq l_i \geq 0 \qquad \text{for all} \qquad i = 1, 2, \ldots, p$$

and that the (possibly 0) lower bounds l_i are such that

$$\Lambda = \sum_i l_i < 1.$$

One can then define **pseudocomponents** corresponding to the original x_i by

$$x_i' = \frac{x_i - l_i}{1 - \Lambda}. \tag{7.21}$$

*i*th Pseudo-component Where There Are p Lower Bound Constraints

(In the event that the lower bounds l_i are all 0, pseudocomponents are all simply the corresponding component proportions. When some l_i are positive, pseudocomponents provide one with variables that satisfy the constraint (7.14), but otherwise each vary from 0 to 1 without the corresponding vector of component proportions leaving the region where the fitted equation is relevant.)

Then, a common choice for representing the impact of a mixture proportion x_i on $\hat{y}$ is to plot values of f as a function of t, for $x_i' = t$ and all pseudocomponents except x_i' equal to $(1-t)/(p-1)$. Other choices are possible, but this choice shows how f changes along a line segment in p-space that passes through both the "equal pseudocomponents point" and the "extreme pseudocomponent i" point (corresponding to the case where $t = 1$). Noting that equation (7.21) implies that

$$x_i = l_i + (1 - \Lambda)x_i', \tag{7.22}$$

it is then evident that one should plot (as a function of t) f evaluated where

($x_1, x_2, \ldots, x_p$) Points Where f Is Evaluated to Produce a Response Trace for x_i

$$x_i = l_i + (1 - \Lambda)t \quad \text{and} \quad x_{i'} = l_{i'} + (1 - \Lambda)\left(\frac{1-t}{p-1}\right) \quad \text{for all } i' \neq i. \tag{7.23}$$

Example 7.13 continued. The (x_1, x_2, x_3) vectors in Table 7.15 turn out to be spread across the set of possible mixture proportion vectors with

$$x_1 \geq .45 \quad \text{and} \quad x_2 \geq .30.$$

So using $l_1 = .45$, $l_2 = .30$, and $l_3 = 0$ in display (7.23) and the fitted quadratic response function for y_1 shown in Printout 7.3, it is possible to develop response traces for the ABS pipe compound Izod impact strength. For example, corresponding to the variable x_1 (the grafted polybutadiene fraction), one plots as a function of t the fitted strengths

$$\begin{aligned}\hat{y}(t) = {} & 2.35(.45 + .25t) - 9.22\left(.30 + .25\left(\frac{1-t}{2}\right)\right) \\ & + 16.76\left(.25\left(\frac{1-t}{2}\right)\right) + 40.74(.45 + .25t)\left(.30 + .25\left(\frac{1-t}{2}\right)\right) \\ & - 7.67(.45 + .25t)\left(.25\left(\frac{1-t}{2}\right)\right) - 84.74\left(.30 + .25\left(\frac{1-t}{2}\right)\right)\left(.25\left(\frac{1-t}{2}\right)\right).\end{aligned}$$

Figure 7.14 shows Izod impact strength traces plotted on the same set of axes for all three of the components. The figure illustrates the slight curvature of the

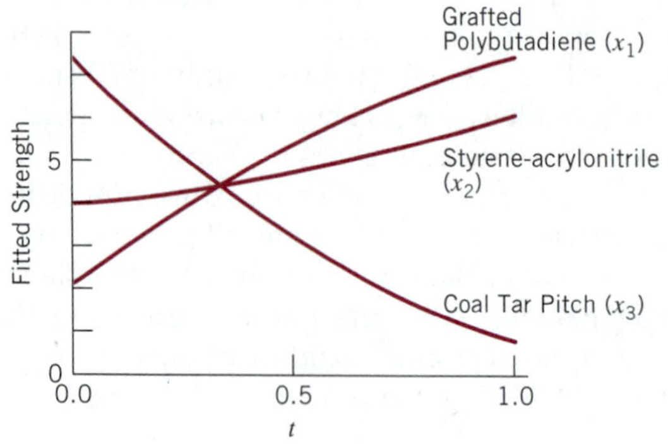

Figure 7.14
Response traces for Izod impact strength.

response function (traces for linear fitted responses are themselves all linear, unlike the curves in Figure 7.14) and the fact that over the region used to develop the fitted equation, the third component of the mixture seems to have the strongest effect on impact strength.

The response traces just introduced are the mixture analogs of the "regular" response surface device of plotting curves representing how $\hat{y}$ changes as a function of one of the process variables (when the others are held fixed). There is also a mixture version of the contour plotting idea. The new wrinkle in the mixture case is only that instead of plotting in a rectangular coordinate system involving x_1 and x_2, one usually plots in a triangular or trilinear coordinate system. That is, for $p = 3$ component proportions, instead of eliminating x_3 and plotting on standard graph paper with coordinate axes labeled with x_1 and x_2, it is common to plot in the kind of coordinate system pictured in Figure 7.15. In Figure 7.15, each vector of mixture coordinates (x_1, x_2, x_3) corresponds to a point whose distance from the "$x_1 = 0$" edge of the triangle is a fraction x_1 of that to the "$x_1 = 1$" corner, whose distance from the "$x_2 = 0$" edge of the triangle is a fraction x_2 of that to the "$x_2 = 1$" corner, and (therefore) whose distance from the "$x_3 = 0$" edge of the triangle is a fraction x_3 of that to the "$x_3 = 1$" corner. This kind of coordinate system has the virtue of treating each of $p = 3$ component proportions equally

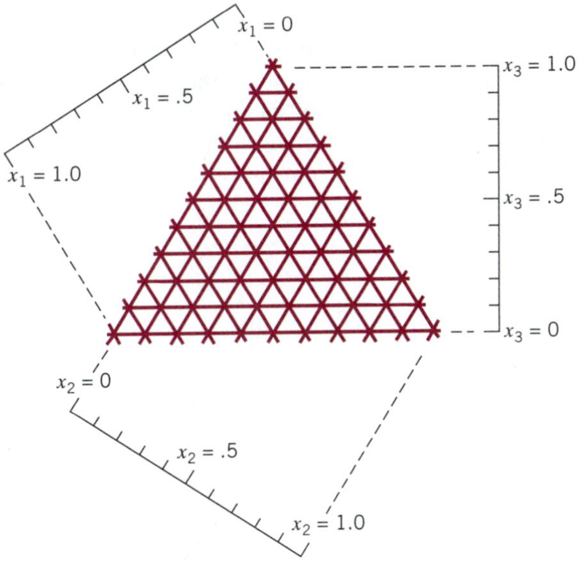

Figure 7.15
A triangular coordinate system for three-component mixtures.

(rather than forcing the elimination of one of them). It is useful both for helping visualize sets of points (x_1, x_2, x_3) and for contour plotting a fitted response as a function of x_1, x_2, and x_3.

Example 7.13 continued. Figure 7.16 shows the set of 10 triples (x_1, x_2, x_3) for which there are ABS pipe compound data in Table 7.15. It is evident that these points cover the (shaded) region defined by the two lower bound constraints

$$x_1 \geq .45 \quad \text{and} \quad x_2 \geq .30.$$

(This fact was earlier used to argue for the relevance of the response traces in Figure 7.14 for describing the nature of the fitted quadratic for y_1 over this region.)

Figure 7.17 is a blowup of the shaded region from Figure 7.16 with a contour plot for the quadratic regression fit to the Izod impact strength. Plots like this are typically made using specialized software, but it is possible to make such a plot by hand. When looking for points (x_1, x_2, x_3) producing a particular value of $\hat{y}_1$, say v, one uses the fact that $x_3 = 1 - x_1 - x_2$ and arrives at an equation

$$v = \hat{y}_1 = f(x_1, x_2, 1 - x_1 - x_2)$$

that for fixed x_1 is quadratic in x_2, and can thus be solved (using the quadratic formula) for x_2 as a function of x_1.

It is useful to contemplate the relationship between Figures 7.17 and 7.14. The ith trace on Figure 7.14 tells the elevations on the "topographic map" in Figure 7.17 as one begins on the "$x_i = l_i$ edge" of the triangle and moves in a straight line through the center point of the triangle to the "$x_i = 1 - \sum_{i' \neq i} l_{i'}$ corner" of the triangle. It is exactly in this sense that response traces summarize the nature of a fitted surface.

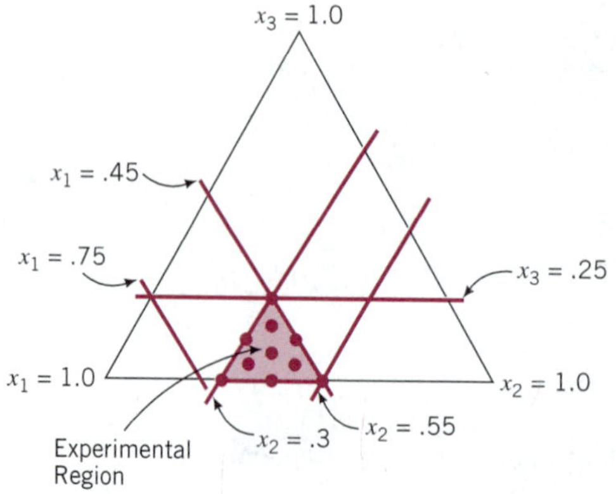

Figure 7.16
Ten triples (x_1, x_2, x_3) for the ABS pipe compound data.

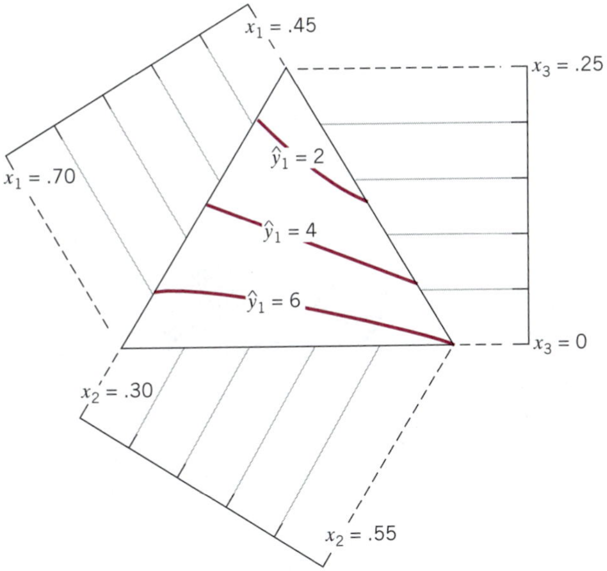

Figure 7.17
Contour plot for the quadratic regression fit to Izod impact strength.

The notion that overlaying contour plots for different responses can be helpful is just as relevant in the mixture context as it is where the process variables are completely unconstrained. In the pipe compound study, engineers had goals for all three of the fitted responses based on ASTM lower specifications for those material properties plus some heuristic "safety factors." (Completely proper lower bounds could have been based on tolerance bounds related to those discussed in Section 5.3 but derived from regression analysis.) But in any case, the engineers wanted

$$\hat{y}_1 > 1.2, \quad \hat{y}_2 > 195 \quad \text{and} \quad \hat{y}_3 > 5{,}050.$$

Figure 7.18 shows a blowup of the experimental region with the $\hat{y}_1 = 1.2$, $\hat{y}_2 = 195$, and $\hat{y}_3 = 5050$ contours marked on it and the portion of the region where the engineering goals are satisfied shaded. Those pipe compound formulas corresponding to points in the shaded region meet the engineering requirements for impact strength, deflection temperature, and yield strength.

7.3.3 Experimental Design for Mixture Studies

General goals for experimentation with mixtures are very much the same as those for "regular" response surface studies. One needs to

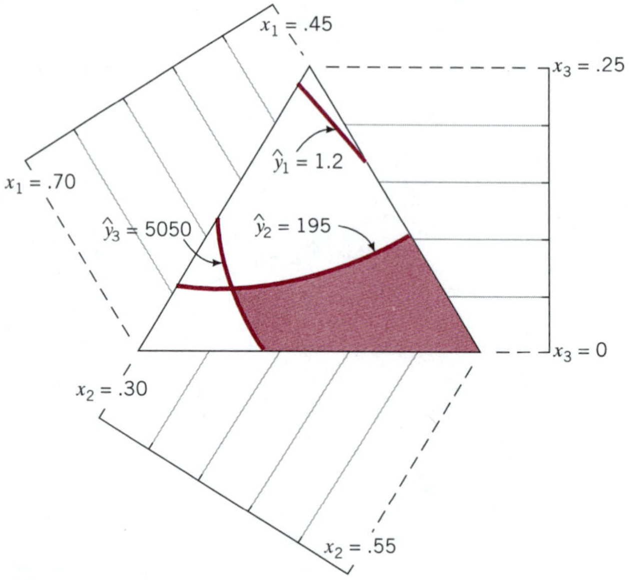

Figure 7.18
Blowup of the experimental region, shaded where engineering goals are satisfied.

1. collect data spread across the $(x_1, x_2, \ldots, x_p)$ region of interest,
2. provide enough variety in the $(x_1, x_2, \ldots, x_p)$ combinations to assure that equations of interest can be fit and that departures from fitted forms can be detected, and
3. provide for replication somewhere in the data set, so as to produce an honest estimate of baseline experimental variation.

The main difficulty to be faced when planning a mixture study is that the basic constraint (7.14) and others that frequently arise in mixture studies (sometimes involving individual component proportions and other times involving several at once) can make it next to impossible to visualize a region of experimentation for p any larger than 3. As a result, the design of mixture studies is an area where specialized computer-implemented mathematical optimization algorithms play an important role, and only the very simplest design ideas are reasonably discussed in a general book like this one. But even these will serve to illustrate the kind of issues that arise. With them understood, the reader will be ready to consult other sources when more specialized information is needed.

The simplest design problem met in mixture contexts is that where the region of interest is either the entire set of $(x_1, x_2, \ldots, x_p)$ vectors satisfying the basic constraint (7.14) or is a subset that "has the same shape" as the whole $(x_1, x_2, \ldots, x_p)$-space. Now it turns out that "having the same shape as the whole

$(x_1, x_2, \ldots, x_p)$-space" means exactly that the region of interest is defined by p lower bounds l_i on the component proportions, and that pseudocomponents (7.21) can be defined (that sum to 1 and each can vary from 0 to 1 just like mixture proportions that are unconstrained except for the basic relationship (7.14)).

Simplex Lattice Experimental Design

Two well-known methods for spreading design points out over a region defined by p lower bound constraints are the simplex lattice and the simplex centroid methods. A **(p, q) simplex lattice design** takes observations at all $\binom{p+q-1}{q}$ possible $(x_1, x_2, \ldots, x_p)$ vectors (or corresponding pseudocomponent vectors), such that each x_i (or each x_i') is a nonnegative integer multiple of $\frac{1}{q}$. Figure 7.19 shows the locations of design points for (3, 2), (3, 3) and (3, 4) simplex lattice designs.

Figure 7.19 correctly suggests that unless $q \geq p$, the set of simplex lattice design points has the deficiency of being confined to the boundary of the experimental region. (And when $q = p$, there is only one design point in the interior.) This suggests that if one is going to use only simplex lattice points, $q > p$ should be employed in order to adequately cover the interior of the experimental region. But the drawback to this possibility is that for $q > p$, the number of simplex lattice points is *much* larger than the minimum needed to fit even a quadratic response function. So even before allowing for some important replication, taking data at all (p, q) simplex lattice design points for $q > p$ calls for more experimentation than is typically practical.

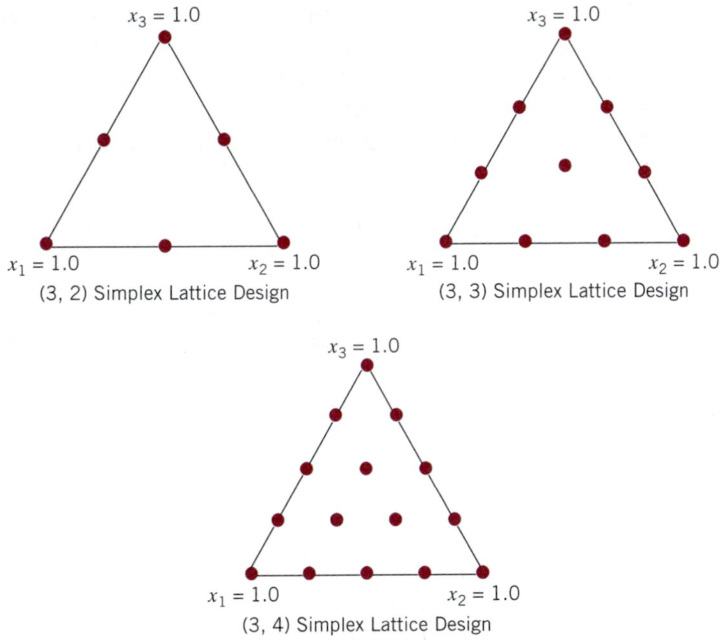

Figure 7.19
Design points for three simplex lattice designs.

Simplex Centroid Experimental Design

A **simplex centroid design** in p mixture variables takes observations at all $2^p - 1$ vectors $(x_1, x_2, \ldots, x_p)$ (or corresponding pseudocomponent vectors), such that

1. one x_i is 1 and all others are 0,
2. two x_i's are $\frac{1}{2}$ and all others are 0,
3. three x_i's are $\frac{1}{3}$ and all others are 0,
4. $p - 1$ x_i's are $\frac{1}{p-1}$ and all others are 0, and
5. all p mixture proportions x_i are $\frac{1}{p}$.

Figure 7.20 shows the set of design points for a $p = 3$ simplex centroid design.

Figure 7.20 correctly suggests that while doing a good job of covering the boundary of the experimental region, the set of simplex centroid design points has the unfortunate property that it places only 1 point in the interior (at the very center). This means that using only simplex centroid design points, one will be in the rather precarious position of interpolating response behavior in the interior of an experimental region almost completely on the basis of behavior on the region's boundary.

A useful means of "fixing" the deficiencies of "small q" simplex lattice designs and the simplex centroid designs is to **augment** them with some additional points in the interior of the experimental region. One way of doing this is to add the center point (with all $x_i = \frac{1}{p}$) to a "small q" simplex lattice design (it is already part of a simplex centroid design) and then place additional points at an appropriate fraction u of the distance from the center point to some or all of the points in the simplex lattice or simplex centroid set. For a point $(x_1, x_2, \ldots, x_p)$ in the simplex lattice or simplex centroid set of design points, this means the addition of the point

$$u \cdot (x_1, x_2, \ldots, x_p) + (1 - u)\left(\frac{1}{p}, \frac{1}{p}, \ldots, \frac{1}{p}\right). \tag{7.24}$$

Leading candidates for points $(x_1, x_2, \ldots, x_p)$ for use in expression (7.24) are the pure component points where one x_i is 1 and all others are 0.

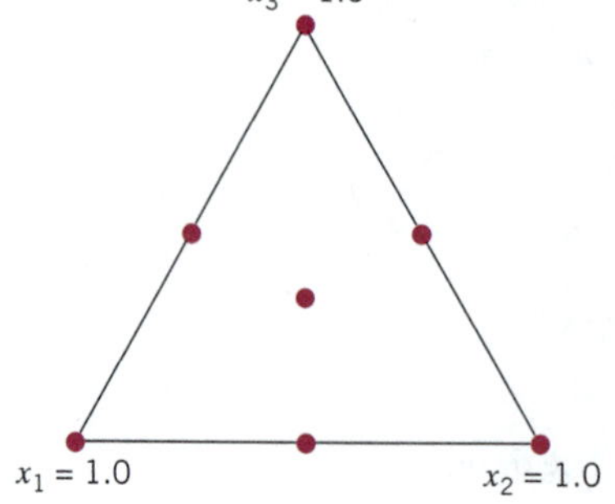

Figure 7.20
Design points for a $p = 3$ simplex centroid design.

In addition to the matter of augmentation with interior points to fully cover the experimental region, the issue of **replication** needs to be explicitly considered. In order to provide a solid basis for estimating the size of basic experimental variation, mixture studies (like almost all data-based studies) should employ some replication. If only a few formulations are going to be replicated, the logical place to begin is with the center point of the experimental region and other interior points.

Example 7.13 continued. The 10 locations (x_1, x_2, x_3) of data collection in the pipe compound study were plotted in Figure 7.16. By using formula (7.21) and computing pseudocomponents for the data of Table 7.15, the reader can verify that (at least approximately) the design used by Koons and Wilt is a $p = 3$ simplex centroid design augmented with three additional interior points, each a fraction $u = .39$ of the way from the center point of the design toward a corner of the experimental region. As noted earlier, unfortunately the Koons and Wilt data do not include any replication.

Where experimental regions with shapes more complicated than the simple "triangular" regions specified by p lower bound constraints are appropriate, some of the general motivation provided here can still be relevant. There are, for example, algorithms for finding "corners," "edges," "faces," and so on for more complicated geometries, and the set of corners and centers of such edges and faces provides a starting point for building mixture designs much like the simplex centroid points do in the simple situation discussed here. And the simplex lattice idea (applied to the raw x_i) is really independent of the shape of an experimental region. (The tricks become finding an appropriate q so that the experimental region contains enough, but not too many so as to be impractical, simplex lattice points, and doing the sorting necessary to determine which simplex lattice design points fall into the experimental region.) But ultimately, for $p > 3$ the mixture design problem is hard enough to think about that the reader faced with an important project is well advised to do some more reading on the subject and probably obtain some special software written to aid the design process.

7.4 QUALITATIVE CONSIDERATIONS IN EXPERIMENTING FOR QUALITY IMPROVEMENT

The discussion of experimental design and analysis in Chapter 6 and thus far in Chapter 7 has concentrated primarily on technical statistical tools. These are essential. But there are also a number of more qualitative matters that must be handled intelligently if experimentation for process improvement is to be successful.

This section discusses some of these. It begins by considering the implications of some "classical" experimental design issues for process improvement studies. Then some matters given recent emphasis by Genichi Taguchi and his followers are discussed.

7.4.1 "Classical" Issues

It is a truism that the world is highly multivariate. Accordingly, almost any process response variable that a person would care to name is potentially affected by *many, many* factors. And successful quality-improvement experimentation requires that those myriad factors be handled in intelligent ways. Section 2.1 has already presented some simple aids for the identification/naming of the most important of those factors. But there is also the matter of how to treat them during an experiment.

There are several possibilities in this regard. For factors of primary interest, it is obvious that one will want to vary their levels during experimentation, so as to learn how they impact the response variable. That is, one will want to manipulate them, treating them as **experimental variables.**

For factors of little interest (usually because outside of the experimental environment they are not under the direct influence of those running a process) but that nevertheless may impact the response, one approach is to **control** them in the sense of holding their levels fixed. This is the laboratory/pilot plant approach, where one tries to cut down the background noise and concentrate on seeing the effects of a few experimental variables. Scientists doing basic research often think of this approach as "the" scientific method of experimentation. But for technologists, it has its limitations. Results produced in carefully controlled laboratory/pilot plant environments are notoriously hard to reproduce in full-scale facilities where many extraneous variables are not even close to constant. That is, when a variable is controlled during experimentation, there is in general no guarantee that the kind of responses one sees at the single level of that variable will carry over to other levels met in later practice. Controlling a factor can increase experimental sensitivity to the effects of primary experimental variables, *but it also limits the scope of application of experimental results.*

Another approach to the handling of extraneous factors (those not of primary interest) is to include them as experimental factors, purposefully varying them in spite of the fact that in later operation one will have little or no say in their values. If a full-scale chemical process is going to be run in a building where humidity and temperature change dramatically over the course of a year, technological experimentation with the basic process chemistry might well be carried out in a variety of temperatures and humidities. Or if a metal-cutting process is affected by the hardness of steel bar stock that itself cannot be guaranteed to be completely constant in plant operations, it may be wise to purposely vary the hardness in a process-improvement experiment. This is in spite of the fact that what may be of primary interest are the effects of tooling and cutting methods. This notion of possibly including as experimental variables some whose precise effects are of at most secondary interest is related to two concepts that need further discussion. The first is the notion of **blocking** from classical experimental design and the second is Taguchi's notion of **noise variables** that will be discussed later in the section.

There are situations in which experimental units or times of experimentation can be segmented into groups that are likely to be relatively homogeneous in terms of their impact on a response variable, but when looked at as a whole are quite heterogeneous. In such cases, it is common (and useful) to break the experimental units or times up into homogeneous **blocks** and to essentially conduct an independent experiment in each of these blocks. If one has in mind two types of tooling and two cutting methods, it makes sense to conduct a 2^2 factorial experiment on a block of "soft" steel bar stock specimens and another one on a block of "hard" steel bar stock specimens. Of course, a way to then think about the study as a whole is to recognize that one has done a complete 2^3 factorial, one factor of which is "blocks." Blocking has the effect of allowing one to account for variation in response attributable to differences between the groups (blocks) rather than, so to speak, lumping it into an omnibus measure of experimental error. It amounts to a sort of local application of the notion of control and provides several relatively homogeneous environments in which to view the effects of primary experimental variables.

Of course there are limits to the size of real-world experiments, and it is not possible to purposely vary *every* variable that could potentially affect a response. (Indeed, one can rarely even be aware of all factors that could possibly affect it!) So there need to be means of protecting the integrity of experimental results against effects of variables not included as experimental factors (and perhaps not even explicitly recognized). Where one is aware of a variable that will not be controlled or purposely varied during data collection and yet might influence experiment results, it makes sense to at least record values for that **concomitant variable**. It is then often possible to account for the effects of such variables during data analysis by, for example, treating them as additional explanatory variables in regression analyses.

In addition, whether or not one is going to try to account for effects of nonexperimental variables in data analysis, it is a good idea to take steps to try and balance their effects between the various experimental conditions using the notion of **randomization.** Randomization is the use of a table of random digits (or other randomizing device) to make those choices of experimental protocol not already specified by other considerations. For example, if one is doing a machining study with two types of tooling and two cutting methods and is planning to apply each of the 2^2 combinations to three steel specimens, one might initially divide 12 specimens up into four groups of three using a randomizing device. (The hope is that the randomization will treat all four combinations "fairly" and, to the extent possible, average between them the effects of all named and unnamed extraneous factors like hardness, surface roughness, microstructure, and so on. This is, of course, a different approach to handling an extraneous factor than treating it as a blocking variable.)

And one might well also randomize the order of cutting in such a study, particularly if there is some concern that ambient conditions might change over the course of experimentation and impact a response of interest. If, for example, one were to make all experimental runs involving tooling type 1 before making the runs for tooling type 2, any factor that changed over time and had an impact on the response would contribute to what the experimenter perceives as the effects of tool type. This is a very unhappy possibility. In terms of understanding what is really affecting a response, it would be far better to either completely randomize the order of the 12 experimental runs or to create three (time) blocks of size four and run a full 2^2 factorial within each of the blocks of experimental runs.

This matter of randomizing order in an engineering experiment brings up two additional qualitative issues. The first is that engineers frequently argue that there are cases where some experimental factors have levels that cannot be easily changed, and in those cases it is often much more economical to make runs in an order that minimizes the number of changes of levels of those factors. It is impossible to say that this argument should never be allowed to stand. Physical and economic realities in engineering experimentation are just that, and the object is not to stick to inflexible rules (like "You must always randomize") but rather to artfully make the best of a given set of circumstances *and recognize the implications of the choices one makes* in experimental design. Where it is absolutely necessary to make all experimental runs with level 1 of Factor A before those with level 2 of Factor A, so be it. But the wise analyst will recognize that what looks like an effect of Factor A is really the effect of A *plus* the effects of any other important but unnamed factors whose levels change over time.

The second matter is the whole issue of what one really means by the word **replication.** Not all methods of obtaining multiple observations associated with a given set of process conditions are equal. Consider, for example, an injection molding process where one is studying the effects of two raw material mixes, two shot sizes, and two temperatures on some property of parts produced by the process. To set up a machine for a given material mix, shot size, and temperature combination and to make and measure five consecutive parts is not at all the same as making that setup and measuring one part, going to some other combination(s) and then returning later to make the second part at that setup, and so on. Five consecutive parts could well look much more alike than five manufactured with intervening time lapses and changes in setup. And it is probably variation of the second (typically larger) magnitude that should be viewed as the baseline against which one correctly judges the effects of material mix, shot size, and temperature. A way of describing this concern is to contrast **repetition** or **remeasurement** with "true" **replication.** With the former, one learns a lot about a particular setup, but not nearly so much about all setups of that type as with the latter.

This issue is also sometimes called the **unit of analysis problem**. If one wishes to compare physical properties for two steel formulas, the logical unit of analysis is a heat of metal. Making one heat of the new formula and one heat from the old, pouring these into ingots and measuring some property of each of 10 ingots produces 20 measurements. But in effect one has two (highly multivariate) samples of size *one*, not two samples of size 10. Why? The object is almost surely to compare the formulas, not the two specific heats. And that being the case, the unit of analysis is a heat, and there is but one of these for each formula. Replication means repeating the unit of analysis, not just remeasuring a single one. Engineers who are constrained by economics to build and tweak a single prototype of a new product need to be very conscious that they have but a single unit of analysis, whose behavior will almost certainly exhibit much less variation than would several "similar" prototypes.

A final matter that should be mentioned in this discussion of classical issues in experimental design concerns resource allocation. It is implicit in much of what has been said in this chapter, but needs to be clearly enunciated, that experimentation for process improvement is almost always a **sequential** business. One typically collects some data, rethinks one's view of process behavior, goes back for a second round of experimentation, and so on. This being true, it is important to spend resources wisely, and not all on round one. A popular rule of thumb (traceable to Box, Hunter, and Hunter) is that not more than 25% of an initial budget for process experimentation ought to go into a first round of data collection.

7.4.2 "Taguchi" Emphases

Genichi Taguchi is a Japanese statistician and quality consultant whose ideas about "offline quality control" have been extremely popular in some segments of U.S. manufacturing over the last decade. Some of the "Taguchi methods" are repackaged versions of well-established classical statistical tools like those presented in this text. Others are, in fact, less than reliable and should be avoided. But the general philosophical outlook and emphases brought by Taguchi and his followers are important and deserve some discussion. The reader who finds his or her interest piqued by this subsection is referred to the article "Taguchi's Parameter Design: A Panel Discussion," edited by V. Nair, which appeared in *Technometrics* in 1992, for more details and an entry into the statistical literature on the subject.

Taguchi's offline quality control ideas have to do with engineering experimentation for the improvement of products and processes. Important points that have been emphasized by him and his followers are

1. small variation in product and process performance is important,
2. the world of engineered products and processes is not a constant variance world,
3. not all means of producing small variation are equal ... some are cheaper and more effective than others, and
4. it is experimental design and analysis that will enable one to discover effective means of producing products and processes with consistent performance.

The goal of consistent product and process performance in a variety of environments has become known as the goal of **robust** (product and process) **design**.

As an artificial but instructive example, consider a situation where, unbeknownst to design engineers, some system response variable y is related to three other variables x_1, x_2, and x_3 via

$$y = 2x_1 + (x_2 - 5)x_3 + \epsilon, \tag{7.25}$$

where x_1 and x_2 are system variables that one can pretty much set at will (and expect to remain where set), x_3 is a random input to the system that has mean 0 and standard deviation η, and ϵ has mean 0 and standard deviation σ and is independent of x_3. Suppose further that it is important that y have a distribution tightly packed about the number 7.0, and that if necessary η can be made small, but only at huge expense.

Having available the model (7.25), it is pretty clear what one should do in order to achieve desired results for y. Thinking of x_1 and x_2 as fixed and x_3 and ϵ as independent random variables,

$$\mathrm{E}y = 2x_1 \quad \text{and} \quad \mathrm{Var}\,y = \sqrt{(x_2 - 5)^2\eta^2 + \sigma^2}.$$

So taking $x_2 = 5$, one achieves minimum variance for y without the huge expense of making η small, and then choosing $x_1 = 3.5$, one gets an ideal mean for y. The object of Taguchi's notions of experimentation is to enable an empirical version of this kind of optimization, applicable where one has no model like that in display (7.25) but does have the ability to collect and analyze data.

Notice, by the way, what it is about the model (7.25) that allows one to meet the goal of robust design. In the first place, the variable x_2 interacts with the system input x_3 and it is possible to find a setting of x_2 that makes the response "flat" as a function of x_3. Then, x_1 interacts with neither of x_2 nor x_3 and is such that it can be used to put the mean of y on target (without changing the character of the response as a function of x_2 and x_3).

How common it is for real products and processes to have a structure like this hypothetical example just waiting to be discovered and exploited is quite unknown. But the Taguchi motivation to pay attention to response variance as well as response mean is clearly sound and in line with classical quality control philosophy (which, for example, has always advised "Get the R chart under control before worrying about the $\bar{x}$ chart"). Statistical methods for plotting residuals after fitting "mean-oriented" models can be used to look for indications of change in response variance. And the final analysis of Example 6.4 presented in this book shows explicitly (using a logged standard deviation as a response variable) what can be done in the way of applying methods for response mean to the analysis of patterns in response variance. The genuinely important reminder provided by Taguchi is that nonconstant variance is not only to be discovered and noted, but is to be *exploited* in product- and process-design.

There is some special terminology and additional experimental philosophy used by Taguchi and his followers that needs to be mentioned here. In a product or process design experiment, those factors whose levels are under the designer's control both in the experiment and in later use are termed **control factors**. The designer gets to choose levels of those variables as parameters of the product or process. Other factors which may affect performance but will not be under the designer's influence in normal use are then called **noise factors**. And Taguchi and followers emphasize the importance of including *both* control variables and noise variables as factors in product- and process-development experiments, which they term **parameter design experiments.**

In the synthetic example represented by display (7.25), Taguchi would call the variables x_1 and x_2 control variables, and the variable x_3 would be termed a noise variable. The objects of experimentation on a system like that modeled by equation (7.25) then become finding settings of some control variables that minimize sensitivity of the response to changes in the noise variables, and finding other control variables that have no interactions with the noise variables and that can be used to bring the mean response on target.

There has been a fair amount of discussion in the statistical literature about exactly how one should treat noise factors in product- and process-development experimentation. The "Taguchi" approach has been to develop separate experimental designs involving first the control factors and then the noise factors. These are sometimes referred to as respectively the **inner array** and the **outer array.** Then each combination of control factors is run with every combination of the noise factors, and summary measures (like a mean and "standard deviation" taken over all combinations of the noise factors) of performance are developed for each combination of control factors. These summary measures are (for better or worse) sometimes called **signal-to-noise ratios** and serve as responses in analyses of experimental results in terms of the "effects" of only the control variables.

This approach to handling noise variables has received serious criticism on a number of practical grounds. In the first place, there is substantial evidence that this Taguchi **product array** approach ultimately leads to very large experiments ... much larger than are really needed to understand the relationships between experimental factors and the response. (For more on this point, the reader is referred to the panel discussion mentioned at the beginning of this section and to the paper "Are Large Taguchi-Style Experiments Necessary? A Reanalysis of Gear and Pinion Data" by Miller, Sitter, Wu, and Long, which appeared in *Quality Engineering* in 1993.) Further, there seems to be little to guarantee that any pattern of combinations of noise variables set up for purposes of experimentation will necessarily mimic how those variables will fluctuate in later product or process use. And the whole notion of "summarizing out" the influence of noise variables before beginning data analysis in earnest seems misguided. It appears to preclude the possibility of completely understanding the original response in terms of main effects and interactions of all experimental variables or in terms of the kind of response function ideas discussed in Sections 7.2 and 7.3. The growing current statistical literature indicates that it is both more economical and more effective to simply treat control variables and noise variables on equal footing in setting up a single **combined array** experimental design, applying the methods of design and analysis already discussed in Chapter 6 and the present chapter of this book.

If nothing else is obvious on a first reading of this section, hopefully it is clear that the tools of experimental design and analysis presented in this book are just that, tools. They are not substitutes for consistent, careful, and clear thinking. They can be combined in many different and helpful ways, but no amount of cleverness in their use can salvage a study whose design has failed to take into account one or more of the important qualitative issues discussed here. Experimentation for process improvement cannot be reduced to something that can be looked up in a cookbook. Instead, it is a subtle but also tremendously interesting and rewarding enterprise that can repay an engineer's best efforts with order of magnitude quality improvements.

7.5 CHAPTER SUMMARY

Chapter 6 introduced the basics of experimental design and analysis for process improvement, covering full factorial studies. This chapter has filled in the picture of process experimentation sketched in Chapter 6. It opened with a discussion of the design and analysis of fractional factorial studies, which are widely used to efficiently screen a large number of factors looking for plausible descriptions of process behavior in terms of a few factorial effects. The second section considered the use of regression analysis and response surface methods in problems where all experimental factors are quantitative. The third section presented response surface methods for studies where the experimental variables are the proportions of pure components in a mixture. Finally, the last section discussed a number of qualitative issues in experimental planning/design that must be thoughtfully addressed if one hopes to achieve process improvement.

7.6 CHAPTER 7 EXERCISES

7.1. Tile Manufacturing. Taguchi and Wu (*Introduction to Off-Line Quality Control,* 1980) discussed a tile manufacturing experiment. The experiment involved seven factors, each having two levels. A current operating condition and a newly suggested value were used as levels for each of the factors. A set of 100 tiles was produced for each treatment combination included in the study, and the number of nonconforming tiles from each set of 100 was recorded.

a. If all possible treatment combinations had been included in the study, how many different experimental runs would have been required (at a minimum)? (An experimental run consists of all items produced under a single set of process conditions.) How many tiles would that have required?

b. In fact, a 1/16th fraction of the full factorial set of possible treatments was actually applied. How many different treatment combinations were studied? How many total tiles were involved?

Suppose the seven factors considered in the experiment were A, B, C, D, E, F, and G, where the current operating conditions are (arbitrarily) designated as the "high" or "+" levels of the factors.

c. How many generators are needed for specifying which combinations are to be run in this 2^{7-4} study?

d. Using the multiplication-of-signs convention introduced in Section 7.1, suppose

i. levels of factor D are chosen by making products of signs for levels of factors A and B,

ii. levels of factor E are chosen by making products of signs for levels of factors A and C,

iii. levels of factor F are chosen by making products of signs for levels of factors B and C, and

iv. levels of factor G are chosen by making products of signs for levels of factors A, B, and C.

Make the table of signs that indicates which eight treatment combinations will be run in the experiment. (List the combinations in Yates standard order as regards factors A, B and C.)

e. The grand mean will be aliased with how many other 2^7 factorial effects? Give the defining relation for this experiment.

7.2. Refer to the **Tile Manufacturing** case in problem (7.1). The data actually collected are given in the table below in a format close to that used by the original authors. (Note that the real experiment was *not* run according to the hypothetical set of generators used for exercise in problem (7.1.d).) The response variable, y, is the fraction of tiles that were nonconforming.

	Factor							
Run	A	B	C	D	E	F	G	y
1	+	+	+	+	+	+	+	.16
2	+	+	+	−	−	−	−	.17
3	+	−	−	+	+	−	−	.12
4	+	−	−	−	−	+	+	.06
5	−	+	−	+	−	+	−	.06
6	−	+	−	−	+	−	+	.68
7	−	−	+	+	−	−	+	.42
8	−	−	+	−	+	+	−	.26

a. Suppose that levels of factors C, E, F, and G were chosen by multiplication of signs for factors A, B, and D. (The table above is in reverse Yates order for the three factors D, B, and A.) What four generators were then used? (Hint: Find the four possible "product" columns of signs and identify which factors correspond to each of these.)

b. Find the defining relation from the four generators in (a).

c. Find the 15 effects that are aliased with the A main effect.

d. Rearrange the rows of the table to produce the Yates standard order for Factors A, B, and D.

e. Apply the Yates algorithm to the rearranged list of y's and find the eight fitted sums of effects.

f. Normal plot the last seven fitted sums from (e).

g. Make a half-normal plot of the absolute values of the last seven fitted values from (e).

h. What conclusions do you draw from the plots in (f) and (g) regarding significant effects on the fraction nonconforming?

7.3. Refer to the **Tile Manufacturing** case in problems (7.1) and (7.2). The analysis of the data suggested in problem (7.2) ignores the fact that the responses "y" were really fractions nonconforming "$\hat{p}$" and that if p is the long-run fraction nonconforming for a given set of conditions, reasoning as in Section 3.3.1, Var $\hat{p} = p(1-p)/n$. This in turn suggests that here (where each sample size is $m = 100$ tiles) for $\hat{E}$ a fitted sum of effects, one might estimate the standard deviation for $\hat{E}$ as

$$\hat{\sigma}_{\hat{E}} = \frac{1}{2^3\sqrt{100}}\sqrt{\sum \hat{p}(1-\hat{p})},$$

where the sum is over the 2^3 treatment combinations included in the study. Then, based on such a standard error for fitted sums of effects, it is further possible to make crude approximate confidence intervals for the corresponding sum of effects, E, using the end points

$$\hat{E} \pm z\hat{\sigma}_{\hat{E}}.$$

(In this equation, z is standing for an appropriate standard normal quantile. For example, a two-sided interval made with $z = 1.96$ would have associated approximate confidence of 95%.) Make intervals such as this (with approximate individual confidence levels of 95%) for the eight sums of effects based on the data of problem (7.2). Are your intervals consistent with your conclusions in problem (7.2)? Explain.

7.4. Nickel-Cadmium Cells. In a 1988 *Journal of Quality Technology* paper, Ophir, El-Gad, and Snyder reported results of quality improvement efforts focusing on finding optimal process conditions for the making of nickel-cadmium cells. (Their study resulted in a process producing almost no defects, annual monetary savings of many thousands of dollars, and an improved workplace atmosphere.) In the production of nickel-cadmium batteries, sometimes contact between the two electrodes occurs. This causes an internal short and the shorted cell must be rejected.

The authors used Pareto and Ishikawa diagrams in their efforts to identify factors possibly influencing battery shorts. Seven factors, each at two levels, were selected as most likely to control the production of shorts. A full factorial design would have involved 128 treatments. Instead, the levels of the four factors "rolling order," "rolling direction," "nylon sleeve on edge," and "side margin of the plate" were held constant at plant standard levels. The three experimental factors considered were:

A—Method of Sintering	old (−) vs. new (+)
B—Separator	thin (−) vs. thick (+)
C—Rolling Pin	thin (−) vs. thick (+)

Data like the following resulted.

	Factor				
Run	**Sintering Method**	**Separator**	**Rolling Pin**	**Number Short**	**Number Tested**
1	new	thick	thick	0	50
2	new	thick	thin	0	50
3	new	thin	thick	1	50
4	new	thin	thin	1	50
5	old	thick	thick	1	50
6	old	thick	thin	1	50
7	old	thin	thick	1	50
8	old	thin	thin	2	41

a. Reorder the rows in the table above to put it in Yates standard order for the factors A, B, and C. Compute the sample fraction nonconforming, $\hat{p}$, for each treatment combination.

b. The most natural way to think of this study is as a full factorial in the three factors A, B, and C. Is it possible to think of it as a fractional factorial in the original seven factors? Why or why not? What can potentially be learned from the data above about the effects of the four factors "rolling order," "rolling direction," "nylon sleeve on edge," and "side margin of the plate"?

c. Find the fitted effects of the three factors corresponding to the all-high treatment combination by applying the Yates algorithm to the sample fractions $\hat{p}$ calculated in (a).

d. Make a normal probability plot for the fitted effects found in (c).

e. Make a half-normal plot for the absolute values of the fitted effects found in (c).

f. Are there factors that seem to be important? Why? (Notice that your conclusions here apply to operation at the standard levels of the factors "rolling order," "rolling direction," "nylon sleeve on edge," and "side margin of the plate.")

g. A rough-and-ready method of investigating whether there is evidence of *any* real differences in the fractions nonconforming across the eight treatment combinations is to apply the retrospective p chart ideas from Section 3.3.1. Do that here (temporarily entertaining the possibility that one has samples from eight equivalent sets of process conditions). Is there clear evidence of a difference somewhere in this factorial set of treatment combinations?

7.5. Refer to the **Nickel-Cadmium Cells** case in problem (7.4) and the discussion in problem (7.3). Since the sample sizes in problem (7.4) were not all the same, the estimated standard deviation of a fitted effect $\hat{E}$ is a bit messier than that suggested in problem (7.3). In place of the $m = 100$ formula given in problem (7.3), one has

$$\hat{\sigma}_{\hat{E}} = \frac{1}{2^3}\sqrt{\sum \frac{\hat{p}(1-\hat{p})}{n}},$$

where again the sum is over the 2^3 treatment combinations included in the study. Use the formula

$$\hat{E} \pm z\hat{\sigma}_{\hat{E}}$$

and make approximate 90% two-sided individual confidence intervals for the 2^3 factorial effects on fraction of cells nonconforming. What do these say about the importance of the factors in determining fraction nonconforming? How do your conclusions here compare to those in problem (7.4)?

7.6. Refer to the **Nickel-Cadmium Cells** case in problems (7.4) and (7.5). The investigators decided that the new sintering methodology combined with a thick separator produced a minimum of shorts and that rolling pin thickness had no appreciable influence on this quality measure. Using the new sintering method, a thick separator, and thin rolling pin thickness, half of a full 2^4 factorial experiment in the factors originally held constant was designed and carried out. Data like the following were obtained.

Factor					
A–Nylon Sleeve	**B–Rolling Direction**	**C–Rolling Order**	**D–Margin**	**Number of Shorts**	n
no	lower edge first	negative first	narrow	1	80
no	lower edge first	positive first	wide	8	88
no	upper edge first	negative first	wide	0	90
no	upper edge first	positive first	narrow	2	100
yes	lower edge first	negative first	wide	0	90
yes	lower edge first	positive first	narrow	1	90
yes	upper edge first	negative first	narrow	0	90
yes	upper edge first	positive first	wide	0	90

a. Designate "yes," "upper edge first," "positive first," and "wide" as the high levels of factors A through D. Rearrange the rows of the table above to place the combinations in Yates standard order as regards factors A, B, and C.

b. Is this design a fractional factorial of the type discussed in Section 7.1? Why or why not? If your answer is yes, describe the structure in terms of 2 raised to some "$p - q$" power.

c. Find the defining relation for the design.

d. What (single) 2^4 factorial effects are aliased with each of the main effects in this study?

e. Using the Yates algorithm, find the estimated sums of effects on the long-run fractions nonconforming.

f. Normal plot the last seven estimates from (e).

g. Make a half-normal plot of the magnitudes (absolute values) of the last seven estimates from (e).

h. Based on your plots from (f) and (g), do any of the sums of effects appear to be significantly different from 0? Why or why not?

i. A rough-and-ready method of investigating whether there is evidence of any real differences in the fractions nonconforming across the eight treatment combinations is to apply the retrospective p chart ideas from Section 3.3.1. Do that here (temporarily entertaining the possibility that one has samples from eight equivalent sets of process conditions). Is there clear evidence of a difference somewhere in this fractional factorial set of treatment combinations?

7.7. Refer to the **Nickel-Cadmium Cells** case in problem (7.6) and the discussion in problem (7.5). Since there are $2^{4-1} = 2^3 = 8$ treatment combinations represented in the study in problem (7.6), the formulas in problem (7.5) can be used to make rough judgments of statistical detectability of sums of 2^4 effects on the fraction of cells with shorts.

Use the formulas in problem (7.5) and make an approximate 95% two-sided individual confidence interval for each sum of 2^4 effects that can be estimated based on the data given in problem (7.6). What do these say about the importance of the factors in determining fraction nonconforming? How do your conclusions here compare to those in problem (7.6)?

7.8. **Paint Coat Thickness**. In an article that appeared in the *Journal of Quality Technology* in 1992, Eibl, Kess, and Pukelsheim reported on experimentation with a painting process. Prior to the experimentation, observed paint thickness varied between 2 mm and 2.5 mm, clearly exceeding the target value of .8 mm. The team's goal was to find levels of important process factors that would yield the desired target value without substantially increasing the cost of production. Pre-experimental discussions produced the following six candidate factors and corresponding experimental levels.

A—Belt Speed	low (−) vs. high (+)
B—Tube Width	narrow (−) vs. wide (+)
C—Pump Pressure	low (−) vs. high (+)

D—Paint Viscosity	low (−) vs. high (+)
E—Tube Height	low (−) vs. high (+)
F—Heating Temperature	low (−) vs. high (+)

The (−) or low levels of each of the experimental factors were in fact the same number of units below standard operating conditions as the (+) or high levels were above standard operating conditions. For some purposes, it is then useful to think of variables $x_A, x_B, \ldots, x_F$ giving (coded) values of the factors in the range −1 to 1, −1 corresponding to the (−) level of the factor, 1 corresponding to the (+) level, and 0 corresponding to a level half-way between the experimental ones (and corresponding to standard plant conditions).

Since an experiment including all possible combinations of even two levels of all six factors was judged to be infeasible, 1/8 of the possible treatment combinations were each applied to $m = 4$ different work pieces. The resulting paint thicknesses, y, were measured and are given below (in mm).

Combination	A	B	C	D	E	F	y
1	+	−	+	−	−	−	1.09, 1.12, .83, .88
2	−	−	+	−	+	+	1.62, 1.49, 1.40, 1.59
3	+	+	−	−	−	+	.88, 1.29, 1.04, 1.31
4	−	+	−	−	+	−	1.83, 1.65, 1.70, 1.76
5	−	−	−	+	−	+	1.46, 1.51, 1.59, 1.40
6	+	−	−	+	+	−	.74, .98, .79, .83
7	−	+	+	+	−	−	2.05, 2.17, 2.36, 2.12
8	+	+	+	+	+	+	1.51, 1.46, 1.42, 1.40

a. Is this experiment a fractional factorial of the type discussed in Section 7.1? Why or why not?

b. Describe the experiment in terms of an appropriate base and exponent. Say what the "base" and "exponent" correspond to in the context of the problem.

c. Rearrange the rows of the table above to put the treatment combinations into Yates standard order as regards factors A, B, and C.

d. Find the sample averages and standard deviations.

e. Find the defining relation for this study.

f. Name the seven 2^6 factorial effects aliased with the A main effect, α_2. Do the same for the B main effect, β_2.

g. Use the Yates algorithm and find the estimated sums of effects corresponding to the all-high treatment combination.

h. Use equation (6.31) and make 90% individual two-sided confidence intervals for the sums of effects on mean paint thickness. Which of these intervals fail to include 0 and thus indicate statistically detectable sums? What is the simplest possible interpretation of these results (in terms of 2^6 factorial effects)?

i. Suppose that only main effects of the factors A through F (and no interactions) are important and that any main effect not found to be detectable by the method of part (h) is judged to be ignorable. Find a predicted (or estimated mean) coating thickness for the all-high treatment combination. Find a predicted coating thickness for the "all-low" treatment combination. (Notice that this is *not* one included in the original study.)

j. Use a multiple regression program to fit the equation

$$y \approx \beta_0 + \beta_A x_A + \beta_B x_B + \beta_C x_C + \beta_D x_D + \beta_E x_E + \beta_F x_F$$

to the data. (You will have $n = 32$ data points $(x_A, x_B, x_C, x_D, x_E, x_F, y)$ to work with.) How do the estimates of the coefficients β compare to the results from part (g)?

k. Drop from your fitted equation in (j) all terms x with corresponding fitted coefficients that are not "statistically significant." (Drop all terms x corresponding to β's whose 90% two-sided confidence intervals include 0.) Use the reduced equation to predict coating thickness when all x's are 1. Then use the reduced equation to predict coating thickness when all x's are -1. How do these predictions compare to the ones from (i)?

l. The factorial-type analysis in parts (g) through (i) only presumes to infer mean paint thickness at combinations of $(-)$ and $(+)$ levels of the factors. On the other hand, the linear regression analysis in parts (j) and (k) could be used (with caution) for extrapolation or interpolation to other levels of the factors. What, for example, does your equation from part (k) predict for paint thickness when all x's are .5 (half-way between standard operating conditions and the high levels of the factors)?

7.9. Speedometer Cable. In a 1993 *Journal of Quality Technology* paper, Schneider, Kasperski, and Weissfeld discussed and reanalyzed an experiment conducted by Quinlan and published in the *American Supplier Institute News* in 1985. The objective of the experiment was to reduce postextrusion shrinkage of speedometer casing. Fifteen factors each at two levels were considered in 16 experimental trials. Four measurements were made on pieces of a single (3000 ft) cable segment for each different treatment combination. (A different cable segment was made using each of the different combinations of process conditions.) The fifteen factors were A—liner tension, B—liner line speed, C—liner die, D—liner outside diameter (OD), E—melt temperature, F—coating material, G—liner temperature, H—braiding tension, J—wire braid type, K—liner material, L—cooling method, M—screen pack, N—coating die type, O—wire diameter, P—line speed. The log (ln) transformation was applied to each of the 4×16 shrinkage responses and "sample" means for the 16 combinations included in the study were computed. Estimated sums of effects (on the log scale) for the all-high treatment combination were as follows.

Sum of Effects Estimated	Estimate
grand mean + aliases	−1.430
A main + aliases	.168
B main + aliases	.239
C main + aliases	−.028
D main + aliases	.222
E main + aliases	−.119
F main + aliases	.046
G main + aliases	−.084
H main + aliases	.212
J main + aliases	−.882
K main + aliases	−.317
L main + aliases	−.102
M main + aliases	−.020
N main + aliases	.309
O main + aliases	−.604
P main + aliases	−.025

a. If one objective of the experiment was to say what would happen for different 3,000-ft segments of cable made under a given set of process conditions, why would it be a bad idea to treat the four measurements made for a given treatment combination as "replicates" and base inferences on formulas for situations where a common sample size is $m = 4$? What is the real "sample size" in this study?

b. Normal plot the last 15 estimated sums of effects listed above.

c. Make a half-normal plot of the absolute values of the last 15 estimated sums of effects.

d. Based on your plots from (b) and (c), which (if any) sums appear to be clearly larger than background noise?

e. If one adopts the tentative conclusion that any sums of effects judged in (d) to be important consist primarily of main effects, it might make sense to follow up the initial 2^{15-11} fractional factorial study with a replicated full factorial in the seemingly important factors (holding the other, apparently unimportant, factors fixed at some standard levels). Describe such a 2^p study with, say, $m = 3$.

f. What will constitute "replication" in your new study from (e)?

7.10. **Bond Strength.** Grego (in a 1993 *Journal of Quality Technology* paper) and Lochner and Matar (in the 1990 book *Designing for Quality*) analyzed the effects of four two-level factors on the bond strength of an integrated circuit mounted on a metallized glass substrate. The four factors identified by engineers as potentially important determiners of bond strength were:

A—Adhesive Type	D2A (−) vs. H-1-E (+)
B—Conductor Material	Copper (−) vs. Nickel (+)

C—Cure Time at 90°C 90 min (−) vs. 120 min (+)
D—Deposition Material Tin (−) vs. Silver (+)

Half of all 2^4 possible combinations were included in an experiment. $m = 5$ observations were recorded for each treatment combination. Summary statistics from the experiment are given below.

A	B	C	D	$\bar{y}$	s^2
D2A	Cu	90	tin	73.48	2.452
D2A	Cu	120	Ag	87.06	.503
D2A	Ni	90	Ag	81.58	.647
D2A	Ni	120	tin	79.38	1.982
H-1-E	Cu	90	Ag	83.88	4.233
H-1-E	Cu	120	tin	79.54	8.562
H-1-E	Ni	90	tin	75.60	26.711
H-1-E	Ni	120	Ag	90.32	3.977

a. Describe the structure of this study in terms of a base raised to some power $p - q$. (Give numerical values for the base, p and q.)
b. The treatment combinations in the table are presently arranged in Yates standard order as regards factors C, B, and A. Rearrange the rows so that the table is in Yates standard order as regards factors A, B, and C.
c. Find the defining relation for this study.
d. For each factor A through D, find the effect aliased with that factor's main effect. (Use notation like A ↔ BCD.)
e. Write out (in terms of subscripted individual lowercase Greek letters and products of the same, like $\alpha_2 + \beta\gamma\delta_{222}$) the eight sums of 2^4 factorial effects that can be estimated based on the data from this study.
f. Use the Yates algorithm to find estimates of the eight sums identified in (e).
g. Find 95% two-sided individual confidence intervals for the sums identified in (e). (Use display (6.31).)
h. Use your intervals from (g) to identify statistically detectable sums of effects. What is it about an interval in (g) that says the associated sum of effects is statistically significant?

7.11. Refer to the **Bond Strength** case in problem (7.10). Factor screening is one motivation for using a fractional factorial experiment. Providing at least tentative suggestions as to which treatment combination(s) might maximize or minimize a response is another (regardless of whether such a combination was part of an original fractional factorial data set).
a. Reflect on the results of part (h) of problem (7.10). What is the simplest possible interpretation of your results in terms of the four original

factors? How many of the original four factors are involved in this interpretation? Which factors are not involved and might possibly be "inert," not affecting the response in any way?

b. Set up a data table for a full 2^p factorial in any factors that from part (a) you suspect of being "active," that is, affecting the response. (Suppose, for example, that $m = 3$ will be used.) Say which levels you will use for any factors in (a) that you suspect to be "inert."

c. Again, consider the results of part (h) of problem (7.10). What combination (or combinations) of levels of factors A through D do you suspect might maximize mean bond strength? Is such a combination in the original data set? (In general, it need not be.)

d. For the combination(s) in (c), what is the predicted bond strength?

7.12. Solder Thickness. Bisgaard, in a 1994 *Journal of Quality Technology* paper, discussed an experiment designed to improve solder layer mean thickness and thickness uniformity on printed circuit boards. (A uniform solder layer of a desired thickness provides good electrical contacts.) To stay competitive, it was important for a manufacturer to solve the problem of uneven solder layers. A team of engineers focused on the operation of a hot air solder leveler (HASL) machine. A 16-run screening experiment involving six two-level factors was run. A measure of solder layer uniformity was obtained from each of the 16 runs (y is a sample variance thickness based on 24 thickness measurements). The design generators were E ↔ ABC and F ↔ BCD. The 16 combinations of levels of factors A through D and corresponding values of the response variable are given below:

A	B	C	D	y
−	−	−	−	32.49
+	+	+	−	46.65
+	−	−	+	8.07
−	+	+	+	6.61
−	+	−	−	25.70
+	−	+	−	16.89
+	+	−	+	29.27
−	−	+	+	42.64
+	+	−	−	31.92
−	−	+	−	49.28
−	+	−	+	11.83
+	−	+	+	18.92
+	−	−	−	35.52
−	+	+	−	24.30
−	−	−	+	32.95
+	+	+	+	40.70

a. Rearrange the rows in the table to produce Yates standard order for factors A, B, C, and D. Then add two columns to the table giving the levels of factors E and F that were used.

b. Discuss why, even though there were 24 thickness measurements (probably from a single PC board) involved in the computation of "y" for a given treatment combination, this is really an experiment with no replication ($m = 1$).

c. Find the defining relation for this 2^{6-2} fractional factorial experiment.

d. The effects are aliased in 16 groups of four effects each. For each of the six main effects (α_2, β_2, γ_2, δ_2, ϵ_2 and ϕ_2), find the sums involving those which can be estimated using data from this experiment. (Use notation like $\alpha_2 + \beta\gamma\epsilon_{222} + \alpha\beta\gamma\delta\phi_{22222} + \delta\epsilon\phi_{222}$.)

e. Transform the responses y by taking natural logarithms, $y' = \ln(y)$. Then use the Yates algorithm to find the estimated sums of effects on y'.

f. Make a normal probability plot for the last 15 estimated sums of effects in (e). Then make a half-normal plot for the absolute values of these estimates.

g. Do you detect any important effects on solder layer uniformity? Why or why not? What is the simplest possible interpretation of the importance of these? Based on your results, what combination or combinations of levels of the six factors do you predict will have the best uniformity of solder layer? Why?

h. Would you be willing to recommend adoption of your projected best treatment combination(s) from (g) with no further experimentation? Explain.

7.13. A 1/2 replicate of a full 2^4 factorial experiment is to be conducted. Unfortunately, only four experimental runs can be made on a given day, and it is feared that important environmental variables may change from day to day and impact experimental results.

a. Make an intelligent recommendation of which ABCD combinations to run on each of 2 consecutive days. Defend your answer. (You should probably think of "day" as a fifth factor and at least initially set things up as a 1/4 replicate of a full 2^5 factorial.)

b. Assuming that environmental changes do not interact with the experimental factors A through D, list the eight sets of aliases associated with your plan from (a). (Your sets must include the main effects and interactions of A through D and the main effect (only) of E (day).)

c. Discuss how data analysis would proceed for your study.

7.14. Polymer Density. In a 1985 *Journal of Quality Technology* article, R. Snee discussed a study of the effects of annealing time and temperature on polymer density. The data were collected using a central composite design, except that no observation was possible at the highest level of temperature (because the polymer melted at that temperature), and an additional run was made at 170°C and 30 minutes. Snee's data are given below.

Temperature, x_1 (°C)	Time, x_2 (min)	Density, y
140	30	70
155	10	70
155	50	72
170	30	91
190	60	101
190	30	98
190	0	70
225	10	83
225	50	101

a. Plot the nine design points (x_1, x_2) in two-space, labeling each with the corresponding observed density, y.

b. Use a multiple regression program and fit the equation

$$y \approx \beta_0 + \beta_1 x_1 + \beta_2 x_2$$

to these data.

c. Make a contour plot for the fitted response from part (a) over the "region of experimentation." (This region is discernible from the plot in (a) by enclosing the nine design points with the smallest possible polygon.) Where in the experimental region does one have the largest predicted density? Where is the smallest predicted density?

d. Compute residuals from the fitted equation in (b) and plot them against both x_1 and x_2 and against $\hat{y}$. Do these plots suggest that an equation that allows for curvature of response might better describe these data?

e. Use a multiple regression program to fit the equation

$$y \approx \beta_0 + \beta_1 x_1 + \beta_2 x_2 + \beta_3 x_1^2 + \beta_4 x_2^2 + \beta_5 x_1 x_2$$

to these data.

f. Make a contour plot for the fitted response from (e) over the experimental region. Use this plot to locate a part of the experimental region that has the largest predicted density.

g. Compute residuals from the fitted equation in (e) and plot them against both x_1 and x_2 and against $\hat{y}$. Do these plots look "better" than the ones in (d)? Does it seem that the quadratic equation is a better description of density than the linear one?

h. Use equation (7.12) and find the stationary point for the fitted quadratic equation from (e). Is the stationary point the location of a maximum, a minimum, or a saddle point? Why?

i. Is the stationary point from (h) located inside the experimental region? Why might the answer to this question be important to the company producing the polymer?

j. The experiment was originally planned as a perfect central composite design, but when it became clear that no data collection would be possible at the set of conditions $x_1 = 240$ and $x_2 = 30$, the "extra" data point was added in its place. Snee suggests that when one suspects that some planned experimental conditions may be infeasible, those combinations should be run early in an experimental program. Why does this suggestion make good sense?

7.15. Surface Finish in Machining. The article "Including Residual Analysis in Designed Experiments: Case Studies" by Collins and Collins that appeared in *Quality Engineering* in 1994 contains discussions of several machining experiments aimed at improving surface finish for some torsion bars. The following is a description of part of those investigations. Engineers selected surface finish as a quality characteristic of interest because it is directly related to part strength and product safety, and because it seemed possible that its variation and production cost could simultaneously be reduced. Surface roughness was measured using a gage that records total vertical displacement of a probe as it is moved a standard distance across a specimen. (The same 1-inch section was gaged on each torsion bar included in the experiment.)

Speed rate (rate of spin during machining, in rpm) and feed rate of the machining (rate at which the cutting tool was moved across a bar, in inches per revolution) were the two factors studied in the experiment. Three levels of speed rate and three levels of feed rate were selected for study. $m = 2$ bars were machined at each combination of speed and feed. Speed rate levels were 2,500, 3,500 and 4,500 rpm. Feed rate levels were .001, .005 and .009 inches per revolution. In the following table, the speeds x_1 have been coded to produce values x_1' via the equation

$$x_1' = \frac{x_1 - 3{,}500}{1{,}000}$$

and the feed rates x_2 have similarly been coded to produce values x_2' using

$$x_2' = \frac{x_2 - .005}{.004}.$$

(Note that with this coding, low, medium, and high levels of both variables are respectively -1, 0, and 1.)

x_1'	x_2'	y
−1	−1	7, 9
−1	0	77, 77
−1	1	193, 190
0	−1	7, 9
0	0	75, 85
0	1	191, 191
1	−1	9, 18
1	0	79, 80
1	1	192, 190

a. Describe the factorial experimental design employed here in terms of a base and an exponent.

b. How many different (x_1, x_2) pairs were there in this study? Is this set sufficient to fit an equation like that given in display (7.8)?, like that in display (7.9)? Why or why not?

c. Use a multiple regression program to fit an equation for y that (like that in display (7.8)) is linear in x_1' and x_2'. (Note that there are 18 data points (x_1', x_2', y) indicated in the table above.)

d. For the equation from (c), plot contours in the (x_1', x_2')-plane corresponding to fitted values, $\hat{y}$, of 5, 55, 105, and 155.

e. Where in the experimental region (the square specified by $-1 \le x_1' \le 1$ and $-1 \le x_2' \le 1$) is $\hat{y}$ optimum (minimum)? What is the optimum predicted value? Do you find this value reasonable? (Can y be negative?)

f. Find the residuals for the equation fit in (c). Plot these against each of x_1', x_2' and $\hat{y}$.

g. Find R^2 for the equation fit in (c).

h. Based on (f) and (g), does it appear the fitted model in (c) fits the data well? Why or why not?

i. Fit a full quadratic equation in x_1',x_2' to the surface roughness data. (That is, fit an equation like (7.9) to the data.)

j. For the equation from (i), plot contours in the (x_1', x_2')-plane corresponding to fitted values, $\hat{y}$, of 5, 55, 105, and 155.

k. Based on the plot from (j), where in the experimental region does it seem that the predicted roughness is optimum? What is the optimum predicted value? Do you find this value reasonable? (Can y be negative?)

7.16. Refer to the **Surface Finish** case in problem (7.15). In problems like the surface finish case, where responses at different conditions vary over an order of magnitude, linear and quadratic fitted response functions often have difficulty fitting the data. It is often helpful to instead try to model the logarithm of the original response variable.

a. Redo the analysis of problem (7.15) using $y' = \ln(y)$ as a response variable. (Notice, for one thing, that modeling the logarithm of surface roughness deals with the embarrassment of negative predicted values of y.) Do the methods of Section 7.2 seem better suited to describing y or to describing y'?

b. For your quadratic description of y', find the stationary point of the fitted surface in both coded and raw units. Is the stationary point inside the experimental region? Is it a maximum, a minimum, or a saddle point? Why?

c. Use the rule of thumb summarized in display (7.13) to judge whether first the linear and then the quadratic fitted equations for y' are clearly "tracking more than experimental noise."

7.17. Refer to the **Surface Finish** case in problem (7.15). Suppose that in a similar situation, experimental resources are such that only 10 torsion bars can be machined and tested. Suppose further that the lowest possible machining speed is 2,200 rpm and the highest possible speed is 4,800 rpm. Further, suppose that the smallest feed rate of interest is .0002 inch/rev and the maximum one of interest is .0098 inch/rev.

a. Set up a central composite experimental plan for this situation. Use the four "corner points" of the design in problem (7.15) as "corner points" here. (These are the points with $x_1' = \pm 1$ and $x_2' = \pm 1$.) Base the "star points" on the minimum and maximum values of the process variables suggested above and allow for replication of the "center point." Make out a data collection form that could be used to record the 10 measured roughness values next to the values of the process variables that produce them.

b. What are α, Δ_1, and Δ_2 for your plan in (a)? (Answer this for both the raw (x) and coded (x') representations of your data collection plan.)

c. In what order would you suggest collecting the 10 data points specified in part (a)? Why is this a good order?

7.18. In Example 7.13, it was shown by using formula (7.21) and computing pseudocomponents for the data of Table 7.15 that (at least approximately) the design employed by Koons and Wilt was a $p = 3$ simplex centroid design augmented with three additional interior points, each a fraction $u = .39$ of the way from the center point of the design to a corner of the experimental region.

a. Find design points (for the same experimental constraints) that produce a $p = 3$ simplex centroid design augmented with three additional points, each a fraction $u = .50$ of the way from the center point of the design to a corner of the experimental region. Give your answer in terms of the original x_1, x_2, and x_3.

b. Find design points (for the same experimental constraints discussed in Example 7.13) that produce a (3, 4) simplex lattice design. Give your answer in terms of the original x_1, x_2, and x_3.

c. What do you see as the relative strengths and weaknesses of the designs in (a) and (b)?

7.19. Rocket Propellant. In the 1966 *Industrial Quality Control* article "Experiments with Mixtures of Components Having Lower Bounds," I. Kurotori discussed a rocket propellant mixture experiment. Physical constraints on the propellant components were that

$$\begin{aligned} x_1 &= \text{the proportion of binder in the mixture,} \\ x_2 &= \text{the proportion of oxidizer in the mixture, and} \\ x_3 &= \text{the proportion of fuel in the mixture} \end{aligned}$$

satisfy

$$x_1 \geq .20, \quad x_2 \geq .40 \quad \text{and} \quad x_3 \geq .20.$$

a. Find seven design points (x_1, x_2, x_3) for a simplex centroid design that satisfies the above constraints. (Hint: Use pseudocomponents.)

b. Augment the seven design points in (a) with six additional points, each 60% of the distance from the center point to one of the other six points from (a).

c. Suppose after considering the two experimental plans from (a) and (b), it was decided the design in (a) did not have enough points in the interior of the experimental region and the design in (b) required more experimentation than was economically feasible. Another possibility is the design from (a) augmented with three additional points, each half-way from the center to the corner points. Give the 10 design points for this plan.

d. The important issue of replication has been ignored in (a) through (c) above. Consider this matter particularly in reference to the design in part (c). If you could afford to replicate 1 of the 10 points from part (c), which one would you choose? If you could afford to make 13 experimental runs based on the plan in (c), how would you allocate the replication to the 10 design points?

7.20. Refer to the **Rocket Propellant** case in problem (7.19). The response variable of interest in this study was the modulus of elasticity of the propellant, y, and the data actually collected are given in the following table.

x_1	x_2	x_3	y
.4	.4	.2	2,350
.2	.6	.2	2,450
.2	.4	.4	2,650
.3	.5	.2	2,400
.3	.4	.3	2,750
.2	.5	.3	2,950
$.26\bar{6}$	$.46\bar{6}$	$.26\bar{6}$	3,000
$.33\bar{3}$	$.43\bar{3}$	$.23\bar{3}$	2,690
$.23\bar{3}$	$.53\bar{3}$	$.23\bar{3}$	2,770
$.23\bar{3}$	$.43\bar{3}$	$.33\bar{3}$	2,980

The goal of experimentation was to find a mixture of the components producing a modulus of elasticity of at least 3,000 with minimum x_1.

a. If the last three data points were not included, what type of experimental design would this be? (Hint: Use pseudocomponents.)

b. What fraction of the distance from the center point to an appropriate corner of the experimental region are each of the last three design points?

c. Fit a linear equation (with intercept) in the first two variables (x_1, x_2) to the data.

d. Fit a full quadratic (with intercept) in the first two variables (x_1, x_2) to the data.

e. Express your answer in (c) in terms of a no-intercept linear function of x_1, x_2, and x_3.

f. Express your answer in (d) in terms of a no-intercept second-order function of x_1, x_2, and x_3.

g. What advantages does your answer to (e) have over that in (c)?

h. What advantages does your answer to (f) have over that in (d)?

7.21. Refer to the **Rocket Propellant** case in problems (7.19) and (7.20).

a. Sketch the region in the (x_1, x_2)-plane (in regular rectangular coordinates) where the constraints on all of x_1, x_2, and $x_3 = 1 - x_1 - x_2$ given in problem (7.19) are satisfied.

b. On your graph from (a), plot contours for predicted modulus of elasticity of 2,950 and 3,000 based on the fitted quadratic equation from part (d) of problem (7.20).

c. Sketch in triangular coordinates the region where the constraints (given in problem (7.19)) on all three components are satisfied. Plot the 10 design points for the data from problem (7.20) on your graph.

d. On your graph from (c), plot contours for predicted modulus of elasticity 2,950 and 3,000 based on the fitted quadratic equation from part (d) (or (f)) of problem (7.20).

e. Using pseudocomponents and the fitted equation from part (f) of problem (7.20), plot a response trace for each of the three components (x_1, x_2, and x_3) all on the same set of axes. What do you conclude about the effects of these three variables on modulus of elasticity?

f. Using your answers to (a) through (e) above and problem (7.20), make a recommendation for (x_1, x_2, x_3) that will produce a predicted modulus of elasticity of at least 3,000 with minimum x_1.

g. It is important to bear in mind what a fitted equation attempts to portray, namely (estimated) mean response. Give two reasons why it would be unwise to expect your answer to part (f) to place all (or even most) measured moduli of elasticity above 3,000. (Hint: Consider the difference between a distribution parameter and an individual observation from that distribution and also the matter of uncertainty of estimation.)

7.22. Coating Opacity. J.A. Cornell (in a 1995 *Journal of Quality Technology* article) discussed a coating formulation experiment of Chau and Kelley (*Journal of Coatings Technology,* 1993). A coating used on identification labels and tags must be both absorbent (so that a printed image will remain on the coating without smudging or running) and opaque. Opacity, y, is known to depend upon

$$x_1 = \text{(weight) fraction of pigment 1,}$$
$$x_2 = \text{(weight) fraction of pigment 2,} \quad \text{and}$$
$$x_3 = \text{(weight) fraction of polymeric binder.}$$

(The binder is used to hold the pigments so that the coating makes a continuous film.)

Experience with similar coating products suggested the constraints on the proportions of pigments 1 and 2 and of the binder

$$.13 \leq x_1 \leq .45, \quad .21 \leq x_2 \leq .67,$$
$$.20 \leq x_3 \leq .34 \quad \text{and} \quad x_1 + x_2 + x_3 = 1.$$

The following data are similar to those reported for an experiment aimed at modeling opacity for a coating thickness of 10 μm.

x_1	x_2	x_3	y
.13	.67	.20	.710
.45	.35	.20	.802
.45	.21	.34	.823
.13	.53	.34	.698
.29	.51	.20	.772
.45	.28	.27	.818
.29	.37	.34	.772
.13	.60	.27	.700
.29	.44	.27	.861
.29	.44	.27	.818
.13	.67	.20	.680
.45	.35	.20	.822
.45	.21	.34	.798
.13	.53	.34	.711

a. Sketch the region in the (x_1, x_2)-plane (in regular rectangular coordinates) where the constraints on all of x_1, x_2, and $x_3 = 1 - x_1 - x_2$ given above are satisfied. Identify on your plot the design points for the experiment.

b. Sketch in triangular coordinates the region where the constraints on all of x_1, x_2, and $x_3 = 1 - x_1 - x_2$ given above are satisfied. Again, identify on your plot the design points for the experiment.

c. Was there replication in this study? Explain.

d. Is the experimental region defined by the original constraints on the proportions describable in terms of only lower bounds on the proportions (so that the pseudocomponent methodology described in section 7.3.2 can be applied here)? Explain.

e. Use a multiple regression program to fit the linear equation

$$y \approx \beta_0 + \beta_1 x_1 + \beta_2 x_2$$

to the data.

f. Use a multiple regression program to fit the quadratic function

$$\begin{aligned} y \approx \beta_0 &+ \beta_1 x_1 + \beta_2 x_2 \\ &+ \beta_3 x_1^2 + \beta_4 x_2^2 + \beta_5 x_1 x_2 \end{aligned}$$

to these data. Is this quadratic equation a better description of the relationship between components and opacity than the equation fit in (e)? Explain.

g. Express your equations from both (e) and (f) in no-intercept forms explicitly involving all of x_1, x_2, and x_3.

7.23. Refer to the **Coating Opacity** case in problem (7.22).

a. Use equation (7.12) and find the stationary point of the response surface fit in part (f) of problem (7.22). Is the stationary point inside the experimental region? Does the fitted surface have a minimum, a maximum, or a saddle point? Explain.

b. Plot (in rectangular (x_1, x_2) coordinates) contours corresponding to predicted opacities of .75, .775, and .80 based on the two equations fit in parts (e) and (f) of problem (7.22).

c. Plot the opacity contours from (b) in triangular coordinates.

d. At what points (x_1, x_2, x_3) does the equation from part (e) of problem (7.22) produce minimum and maximum opacities? (Parts (b) and (c) above may help identify these points.)

e. At about what points (x_1, x_2, x_3) does the equation from part (f) of problem (7.22) produce minimum and maximum opacities? (Parts (b) and (c) above may help identify these points.)

7.24. Tar Yield. In the article "Improving a Chemical Process Through Use of a Designed Experiment" that appeared in *Quality Engineering,* 1990–1991, J.S. Lawson discussed an experiment intended to find a set of production conditions that would produce minimum tars in a process that can be described in rough terms as

$$\mathrm{A} + \mathrm{B} \xrightarrow[\text{CATALYST}]{\text{SOLVENT}} \mathrm{PRODUCT} + \mathrm{TARS}\,.$$

(In the process studied, side reactions create tars that lower product quality. As more tars are produced, yields decrease and additional blending of finished product is needed to satisfy customers. Lower yields and increased blending raise the average production cost.)

High, medium, and low levels of the three process variables

$$x_1 = \text{temperature},$$
$$x_2 = \text{catalyst concentration}, \quad \text{and}$$
$$x_3 = \text{excess reagent B}$$

were identified and 15 experimental runs were made with high reagent A purity and low solvent stream purity. Then 15 additional runs were made with low reagent A purity and low solvent stream purity. In both sets of experimental runs, the response variable of interest was a measure of tars produced, y. Lawson's data are below in separate tables for the two different reagent A purity/solvent stream purity conditions. (The process variables have been coded by subtracting the plant standard operating values and then dividing by the amounts that the "high" and "low" values differ from the standard values.)

"High" Reagent A Purity and "Low" Solvent Stream Purity Data

x_1	x_2	x_3	y
−1	−1	0	57.81
1	−1	0	24.89
−1	1	0	13.21
1	1	0	13.39
−1	0	−1	27.71
1	0	−1	11.40
−1	0	1	30.65
1	0	1	14.94
0	−1	−1	42.68
0	1	−1	13.56
0	−1	1	50.60
0	1	1	15.21
0	0	0	19.62
0	0	0	20.60
0	0	0	20.15

"Low" Reagent A Purity and "Low" Solvent Stream Purity Data

x_1	x_2	x_3	y
−1	−1	0	37.29
1	−1	0	4.35
−1	1	0	9.51
1	1	0	9.15
−1	0	−1	20.24
1	0	−1	4.48
−1	0	1	18.40
1	0	1	2.29
0	−1	−1	22.42
0	1	−1	10.08
0	−1	1	13.19
0	1	1	7.44
0	0	0	12.29
0	0	0	11.49
0	0	0	12.20

a. Make a "three-dimensional" sketch of the cube-shaped region with $-1 \le x_1 \le 1$, $-1 \le x_2 \le 1$, and $-1 \le x_3 \le 1$. Locate the 13 different (x_1, x_2, x_3) design points employed in Lawson's experiment on that sketch.

b. Do the 13 design points in Lawson's study constitute a central composite design? Explain.

For the time being, consider only the first set of 15 data points, the ones for high reagent A purity.

c. Was there any replication in this study? Explain.

d. Use a multiple regression program to fit the linear equation

$$y \approx \beta_0 + \beta_1 x_1 + \beta_2 x_2 + \beta_3 x_3$$

to the data.

e. Use a multiple regression program to fit a full quadratic function in x_1, x_2, and x_3 to these data. (You will need a constant term, three linear terms, three pure quadratic terms, and three cross product terms.)

f. Is the quadratic equation in (e) a better description of the relationship between the process variables and tar production than the equation fit in (d)? Explain.

g. Find an "honest" estimate of σ, the experimental variability in y for a fixed set of process conditions.

Now consider the second set of 15 data points, the ones for low reagent A purity.

h. Redo parts (c) through (g) for this second situation.

7.25. Refer to the **Tar Yield** case in problem (7.24).

a. Consider the high reagent A purity data. Use the fitted quadratic equation from part (e) of problem (7.24). In succession, set x_1 at values -1, 0, and 1 to produce three quadratics for y in terms of x_2 and x_3. Make contour plots for each of these and find their associated (x_2, x_3) stationary points. Are these maxima, minima, or saddle points? Are these stationary points within the experimental region (i.e., do they satisfy $-1 \leq x_2 \leq 1$ and $-1 \leq x_3 \leq 1$)?

b. Consider the low reagent A purity data. Use the fitted quadratic equation from part (h) of problem (7.24). In succession, set x_1 at values $-1, 0$, and 1 to produce three quadratics for y in terms of x_2 and x_3. Make contour plots for each of these and find their associated (x_2, x_3) stationary points. Are these maxima, minima, or saddle points? Are these stationary points within the experimental region (i.e., do they satisfy $-1 \leq x_2 \leq 1$ and $-1 \leq x_3 \leq 1$)?

c. Find the stationary point for the quadratic equation in three process variables found in part (e) of problem (7.24). Is this a minimum, maximum, or a saddle point? Explain. Is the point inside the experimental region defined by $-1 \leq x_1 \leq 1$, $-1 \leq x_2 \leq 1$, and $-1 \leq x_3 \leq 1$?

d. Find the stationary point for the quadratic equation in three process variables found in part (h) of problem (7.24). Is this a minimum, maximum, or a saddle point? Explain. Is the point inside the experimental region defined by $-1 \leq x_1 \leq 1$, $-1 \leq x_2 \leq 1$, and $-1 \leq x_3 \leq 1$?

e. Reflect on your answers to problem (7.24) and (a), (b), (c), and (d) above. What combination of temperature, catalyst concentration, excess reagent B, reagent A purity, and solvent stream purity (inside the experimental region) seems to produce the minimum tar? Defend your answer.

7.26. Chemical Process Yield. The text, *Response Surface Methodology*, by Raymond Myers contains the results of a four-variable central composite experiment. The conversion of 1,2-propanediol to 2,5-dimethyl-piperazine is affected by

$$\begin{aligned} NH_3 &= \text{amount of ammonia (g)}, \\ T &= \text{temperature (°C)}, \\ H_2O &= \text{amount of water (g)}, \quad \text{and} \\ P &= \text{hydrogen pressure (psi)}. \end{aligned}$$

The response variable of interest was a measure of yield, y. For purposes of defining a central composite experimental plan, it was convenient to define the coded process variables

$$\begin{aligned} x_1 &= (NH_3 - 102)/51, \\ x_2 &= (T - 250)/20, \\ x_3 &= (H_2O - 300)/200, \quad \text{and} \\ x_4 &= (P - 850)/350. \end{aligned}$$

(The 2^4 factorial points of the design had $x_1 = \pm 1, x_2 = \pm 1, x_3 = \pm 1$, and $x_4 = \pm 1$.)

a. Find the raw (uncoded) high and low levels of each process variable in the 2^4 factorial part of the design.

b. α for this study was 1.4. Find the uncoded (NH_3, T, H_2O, P) coordinates of the "star points" in this study. What was the (uncoded) center point of this design?

c. How many design points total were there in this study (including the 2^4 factorial, the star points, and the center point)?

d. The quadratic equation fit to the data set was

$$\begin{aligned} \hat{y} = {} & 40.198 - 1.511x_1 + 1.284x_2 - 8.739x_3 \\ & + 4.995x_4 - 6.332x_1^2 - 4.292x_2^2 + .020x_3^2 \\ & - 2.506x_4^2 + 2.194x_1x_2 - .144x_1x_3 + 1.581x_1x_4 \\ & + 8.006x_2x_3 + 2.806x_2x_4 + .294x_3x_4. \end{aligned}$$

Find the stationary point for the fitted response surface. Is the stationary point a minimum, maximum, or saddle point? Is the stationary point within the experimental region? Why or why not?

e. Consider the following six different (x_3, x_4) ordered pairs

$$(-1.4, -1.4), (-1.4, 0), (-1.4, 1.4),$$
$$(-1.3, -1.4), (-1.3, 0), \quad \text{and} \quad (-1.3, 1.4).$$

Substitute each of these pairs into the equation in (d) and produce a quadratic equation for y in terms of x_1 and x_2. Find the stationary points for these six different equations and say whether they locate minima, maxima, or saddle points. What predicted responses are associated with these points?

f. Based on your analysis in this problem, recommend a point (x_1, x_2, x_3, x_4) that (within the experimental region) maximizes the fitted conversion yield. Explain your choice and translate the recommendation to raw (uncoded) values of process variables.

7.27. Turning and Surface Roughness. The article "Variation Reduction by Use of Designed Experiments" by Sirvanci and Durmaz appeared in *Quality Engineering* in 1993 and discusses a fractional factorial study on a turning operation and the effects of several factors on a surface roughness measurement, y. Below are the factors and levels studied in the experiment.

A—Insert	#5023 (−) vs. #5074 (+)
B—Speed	800 rpm (−) vs. 1000 rpm (+)
C—Feed Rate	50 mm/min (−) vs. 80 mm/min (+)
D—Material	Import (−) vs. Domestic (+)
E—Depth of Cut	.35 mm (−) vs. .55 mm (+)

Only $2^{5-2} = 8$ of the possible combinations of levels of the factors were considered in the study. These eight combinations were derived using the generators D ↔ AB and E ↔ AC. For each of these eight combinations, $m = 5$ different overhead cam block auxiliary drive shafts were machined and surface roughness measurements, y (in μ-inches), were obtained.

a. Finish the table below specifying which eight combinations of levels of the five factors were studied.

A	B	C	D	E	Combination Name
−	−	−			
+	−	−			
−	+	−			
+	+	−			
−	−	+			
+	−	+			
−	+	+			
+	+	+			

b. Use the two generators D ↔ AB and E ↔ AC and find the entire defining relation for this experiment. Based on that defining relation, determine which effects are aliased with the A main effect.

c. The experimenters made and measured roughness on a total of 40 drive shafts. If, in fact, the total number of shafts (and not anything connected with *which kinds of shafts*) was the primary budget constraint in this experiment, suggest an alternative way to "spend" 40 observations that might be preferable to the one the experimenters tried. (Describe an alternative and possibly better experimental design using 40 shafts.)

d. Below are the eight sample means ($\bar{y}$) and standard deviations (s) that were obtained in the experiment, listed in Yates standard order as regards factors A, B, and C, along with results of applying the Yates algorithm to the means.

$\bar{y}$	s	Estimate
74.96	36.84	65.42
57.92	3.72	−3.01
50.44	8.27	−6.80
49.16	4.19	2.27
80.04	8.25	7.30
75.96	3.69	1.57
68.28	5.82	1.52
66.60	6.32	−1.67

In the list of estimates, there are five that correspond to main effects and their aliases. Give the values of these.

e. The pooled sample standard deviation here is $s_P = 14.20$. For purposes of judging the statistical significance of the estimated sums of effects, one might make individual 95% two-sided confidence limits of the form $\hat{E} \pm \Delta$. Find Δ.

f. Based on your answers to parts (d) and (e), does it seem that the D main effect here might be tentatively judged to be statistically detectable? Explain.

g. What about the values of s listed in the table calls into question the appropriateness of the confidence interval analysis outlined in parts (e) and (f)? Explain.

7.28. Professor George Box is famous for saying that to find out what happens to a system when you interfere with it you have to interfere with it (not just passively observe it). Reflect on the roles (in modern quality improvement) of (1) control charting/process monitoring and (2) experimental design and analysis in the light of this maxim. What in the maxim corresponds to experimental design and analysis? What corresponds to control charting?

7.29. Consider a hypothetical 2^{3-1} experiment with generator $C \leftrightarrow AB$ and involving some replication that produces sample means $\bar{y}_c = 1$, $\bar{y}_a = 13$, $\bar{y}_b = 15$, and $\bar{y}_{abc} = 11$ and s_P small enough that all four of the sums of effects that can be estimated are judged to be statistically detectable. One possible simple interpretation of this outcome is that the grand mean $\mu_{...}$ and all of the main effects α_2, β_2, and γ_2 (and no interactions) are important.

a. If one adopts the above simple interpretation of the experimental results, what combination of levels of factors A, B, and C would you recommend as a candidate for producing maximum response? What mean response would you project?

b. Is the combination from (a) in the original experiment? Why would you be wise to recommend a "confirmatory run" for your combination in (a) before ordering a large capital expenditure to permanently implement your candidate from (a)?

7.30. Problems (7.3) and (7.5) offer one means of dealing with the fact that sample fractions nonconforming $\hat{p}$ have variances that depend not only on n, but on p as well, so that somewhat complicated formulas are needed for standard errors of estimated effects based on them. Another approach is to (for analysis purposes) transform $\hat{p}$ values to $y = g(\hat{p})$ where y has a variance that is nearly independent of p. The Freeman-Tukey transformation is

$$y = g(\hat{p}) = \frac{\arcsin\sqrt{\frac{n\hat{p}}{(n+1)}} + \arcsin\sqrt{\frac{n\hat{p}+1}{(n+1)}}}{2}.$$

Using this transformation, as long as p is not extremely small or extremely large

$$\text{Var}\, y \approx \frac{1}{4n}.$$

That means that if one computes fitted effects $\hat{E}$ based on the transformed values (using the k cycle Yates algorithm) then standard errors can be computed as

$$\hat{\sigma}_{\hat{E}} = \frac{1}{2^{k+1}}\sqrt{\sum \frac{1}{n}},$$

where the sum of reciprocal sample sizes is over the 2^k treatment combinations in the study. (Approximate confidence limits are $\hat{E} \pm z\hat{\sigma}_{\hat{E}}$ as in problems (7.3) and (7.5).)

a. Redo the analysis in part (e) through (h) of problem (7.2) using transformed values y.

b. Redo the analysis in problem (7.3) using the transformed values y.

7.31. A project team identifies three quantitative variables (x_1, x_2, and x_3) to be used in an experiment intended to find an optimal way to run a vibratory burnishing process. The project budget will allow 18 runs of the process to be made. Suppose that the process variables have been coded in such a way that it is plausible to expect optimum conditions to satisfy

$$-2 \leq x_1 \leq 2, \quad -2 \leq x_2 \leq 2, \quad \text{and} \quad -2 \leq x_3 \leq 2.$$

a. Make and defend a recommendation as to how the 18 runs should be allocated. That is, make up a data collection sheet giving 18 combinations (x_1, x_2, x_3) that you would use. List these in the order you recommend running the experiment and defend your proposed order.

Suppose that after running the experiment and fitting a quadratic response surface, $\hat{y}$ appears to be maximized at the point (x_1^*, x_2^*, x_3^*).

b. What are some circumstances in which you would be willing to recommend immediate use of the conditions (x_1^*, x_2^*, x_3^*) for production? What are some circumstances in which immediate use of the conditions (x_1^*, x_2^*, x_3^*) would be ill-advised?

CHAPTER 8

Sampling Inspection

The bulk of this book is concerned with *process*-oriented quality assurance methods. That is as it should be. A process orientation is essential to business success in a competitive world economy. But even best current practice and wise process management need not always produce perfect product. There remains, then, a legitimate need and place for *product*-oriented statistical quality assurance methods. This chapter provides an introduction to some of these, concentrating on tools based on *sampling* from lots or production streams.

The chapter begins with two sections on acceptance sampling. The first concerns the use of attributes data and the second discusses the use of measurements. The methods presented are aimed at wise decisions concerning disposal of individual lots of items where the quality of those lots is unknown. Then there is a section discussing the Military Standard 105 attributes acceptance sampling system (sometimes used in the routine sampling inspection of streams of lots). There follows a brief introduction to *continuous inspection plans* (that can be used more or less on-line in the manufacture of a stream of individual items). The chapter then concludes with a section that aims to put the technical material in perspective, discussing the proper role of product-oriented inspection in general and sampling inspection in particular.

8.1 ATTRIBUTES ACCEPTANCE SAMPLING

It sometimes happens that one faces the following generic problem. A decision on the disposal of a lot of N items is to be taken with the understanding that if the lot is "good enough" one action is most appropriate, while if it is not a second is better. The first possible action is given the generic name **acceptance** of the lot and the second is called **rejection,** although what they entail in actual practice can vary widely, depending upon the specific circumstances. When the decision to accept or reject is made on the basis of a sample from the lot, the methodology is known as **acceptance sampling.** This section treats elementary acceptance sampling where decisions are made on the basis of counts of either nonconforming items in a sample or nonconformities found on sampled items. That is, **attributes acceptance sampling** is considered.

The section begins with a discussion of single sampling plans for fraction nonconforming and mean nonconformities per unit, and the quantitative evaluation of their effectiveness using OC curves. Then the use of acceptance sampling in rectifying inspection is considered. Finally, the issue of choosing an attributes acceptance sampling plan is treated.

8.1.1 Attributes Single Sampling Plans and Their Operating Characteristics

Section 3.3 discussed process monitoring on the basis of attributes data and identified the "fraction nonconforming" and "mean nonconformities per unit" situations. The language and notation developed there will be continued in this section. That is, in a problem where items are individually either conforming or nonconforming and a (random) sample of n items is drawn from a lot of N, let

$$X = \text{the number of nonconforming items in the sample}\,. \tag{8.1}$$

And in a situation where multiple nonconformities/nonconformances can occur on inspection units and a sample of k units is drawn from a lot of N, let

$$X = \text{the total number of nonconformities/nonconformances on the } k \text{ units}\,. \tag{8.2}$$

Then, in both contexts, a sensible kind of decision criterion is to choose some integer c and

$$\text{accept a sampled lot if } X \le c\text{, and reject the lot if } X > c\,. \tag{8.3}$$

The criterion specified by either notation (8.1) or (8.2) together with display (8.3) is the classical **single sampling** (because the accept/reject decision is made on the basis of *one* sample of several units from the lot) attributes acceptance sampling criterion. The number c is usually called the **acceptance number** for the plan.

Example 8.1 **Sampling Utility Meters.** State regulation of a power utility requires that a lot of 5,000 meters (of a single model purchased at once by the company from a single supplier and installed at customer locations) be sampled and checked for calibration at regular intervals. If too many of the sampled meters fail a calibration test, then the entire lot of meters must be replaced or recalibrated. Current regulations call for a random sample of 200 of the meters to be taken and replacement ordered if more than 14 of the sampled meters fail the calibration test.

This can be thought of as a fraction nonconforming attributes acceptance sampling problem. Display (8.1) can be used to define X, the sample size is $n = 200$, and the acceptance number is $c = 14$.

Example 8.2 **Quality Control of Tomato Seed.** The quality control of seed is a major issue in agribusiness. Consider, for example, the case of tomato seed. Bacterial canker is an important problem for food processors who buy tomato seed for planting in contract fields. A means of checking an incoming lot of seed for canker is to select some part of the lot, grind it up, and do a lab analysis, counting the number of bacteria colonies in the sample. If too many colonies are found, the whole lot is rejected as unacceptable for planting.

This can be thought of as a mean nonconformities per unit attributes acceptance sampling scenario. If the lot contains N times as much seed as is taken for lab analysis, with $k = 1$, $X =$ the lab colony count (as in display (8.2)), and c the largest count that does not require rejection of the seed lot, a company can be thought of as doing classical attributes acceptance sampling.

Example 8.3 **Sampling Textiles.** If a garment manufacturer must decide whether to take receipt of a large lot of bolts of denim, a sensible quality control procedure is to select several bolts and thoroughly inspect these, counting important imperfections and demanding a discount from the producer if too many flaws are found. For example, one might inspect five bolts from a shipment of 250, demanding special consideration if more than 10 serious imperfections total are found in sampled bolts. This is a classical mean nonconformities per unit acceptance sampling problem. N here is 250, $k = 5$, X is as in display (8.2), and $c = 10$.

In each of the preceding examples there is an implicit assumption that one doesn't know the quality of the lot under consideration. The purpose of the (sampling) inspection is to make a (somewhat fuzzy because of the incomplete nature of sampling information) determination as to whether the lot in question is a "good one" or a "bad one." If lot quality is known, the sampling has no purpose. (More will be said about this in Section 8.5, but the point here is that under constant/known quality conditions, inspection on a sampling basis typically does not make sense.)

As a preliminary to discussing the quantitative properties of attributes acceptance sampling plans, it is important to note that there are two different perspectives that can be taken. Probably the most natural of the two is **perspective A**. This means that one thinks of a lot as an isolated entity and nonconforming items or nonconformities as pertaining to that lot alone. This is frequently the viewpoint of a consumer faced with the sentencing of a particular lot. On the other hand, it is also possible to adopt the point of view of a producer with a stable production process, who sees any single lot as simply a random output from that stable process. From this second viewpoint, **perspective B**, a lot in hand is taken as indicative of the (stable) process that created it.

In Example 8.1, there can be little argument that perspective A is most relevant. By the time the in-service inspection of meters is done (some years after installation), only the particular meters installed at customers' sites are relevant to the utility. But one might argue that in Example 8.3, both perspectives could be appropriate. The manufacturer of the denim might adopt the point of view that what happens to one lot at one garment maker's shipping dock reflects on what nonconformity rate is being run at the mill producing the cloth. But a garment manufacturer planning to use the cloth to meet a particular high-priority order might well see the results of acceptance sampling in more narrow terms.

Whether one adopts perspective A or perspective B, the primary tool for quantifying what an attributes acceptance sampling plan can do is the **operating characteristic (OC)** of the plan. This is the probability that a lot is accepted,

$$Pa = P[X \le c]. \tag{8.4}$$

How one actually computes the probability (8.4) depends upon whether one is doing fraction nonconforming or mean nonconformities per unit sampling, and upon whether perspective A or perspective B is adopted.

Consider first the case of fraction nonconforming acceptance sampling. If p is the actual fraction nonconforming in a lot of N items, the lot then contains $N(1-p)$ conforming items and Np nonconforming items, and the distribution of X from display (8.1) is hypergeometric. That is, for $p = 0, 1/N, 2/N, \ldots, 1$, the type A operating characteristic of the plan (8.3) is (as a function of p)

Type A Fraction Nonconforming Operating Characteristic

$$Pa(p) = \sum_{x=0}^{c} \frac{\binom{Np}{x}\binom{N(1-p)}{n-x}}{\binom{N}{n}}. \tag{8.5}$$

On the other hand, if p is the propensity for a stable process (that stands behind the lot) to produce a nonconforming item, then X from display (8.1) is binomially distributed for n trials and success probability p. That is, as a function of $p \in [0, 1]$, the type B operating characteristic of the plan (8.3) is

$$Pa(p) = \sum_{x=0}^{c} \binom{n}{x} p^x (1-p)^{n-x} . \tag{8.6}$$

Type B Fraction Nonconforming Operating Characteristic

Turning to the mean nonconformities per unit situation, one must compute under the assumption that nonconformities are generated by a Poisson process. This provides simple ways to evaluate the probability (8.4). To begin with, if a lot actually contains a total of T nonconformities and thus $\lambda = T/N$ is the realized nonconformity rate for the lot, the Poisson process assumption implies that the (conditional on T) distribution of X in display (8.2) is binomial for T trials and success probability k/N. That is, as a function of $\lambda = 0, 1/N, 2/N, \ldots$, the type A operating characteristic of the plan (8.3) is

$$Pa(\lambda) = \sum_{x=0}^{c} \binom{N\lambda}{x} \left(\frac{k}{N}\right)^x \left(1 - \frac{k}{N}\right)^{N\lambda - x} . \tag{8.7}$$

Type A Mean Nonconformities Per Unit Operating Characteristic

Finally, if λ is a theoretical mean nonconformities per unit, the Poisson process assumption implies that X defined by relationship (8.2) has the Poisson distribution with mean $k\lambda$. Thus, as a function of $\lambda \in [0, \infty)$, the type B operating characteristic of the plan (8.3) is

$$Pa(\lambda) = \sum_{x=0}^{c} \frac{\exp(-k\lambda)(k\lambda)^x}{x!} . \tag{8.8}$$

Type B Mean Nonconformities Per Unit Operating Characteristic

Plots of Pa versus either p or λ picture how the chance of accepting a lot varies with the rate of undesirables in the lot. Such plots are commonly called **OC curves**. Notice that this is something of misnomer when perspective A is taken, since the perspective A operating characteristic is defined only for values of p or λ that are multiples of $1/N$.

Example 8.4 Some Operating Characteristic Calculations. To illustrate the use of formulas (8.5) through (8.8), consider an attributes acceptance sampling plan with $c = 1$ and $n = 5$ or $k = 5$, applied in a situation where the lot size is $N = 100$. Suppose further that $p = .03$ or $\lambda = .03$.

First, in the fraction nonconforming context, $p = .03$ for perspective A means that $Np = 100(.03) = 3$ of the items in the lot are nonconforming. For such a situation, formula (8.5) shows that the probability of accepting the lot is

$$Pa(.03) = \frac{\binom{3}{0}\binom{97}{5}}{\binom{100}{5}} + \frac{\binom{3}{1}\binom{97}{4}}{\binom{100}{5}} = \frac{\left(\frac{3!}{0!3!}\right)\left(\frac{97!}{5!92!}\right)}{\left(\frac{100!}{5!95!}\right)} + \frac{\left(\frac{3!}{1!2!}\right)\left(\frac{97!}{4!93!}\right)}{\left(\frac{100!}{5!95!}\right)} = .856 + .138 = .994\,.$$

On the other hand, if one adopts perspective B in the fraction nonconforming situation, the operating characteristic when $p = .03$ is from equation (8.6)

$$Pa(.03) = \binom{5}{0}(.03)^0(.97)^5 + \binom{5}{1}(.03)^1(.97)^4 = .859 + .133 = .992\,.$$

Shifting to the mean nonconformities per unit setting, $\lambda = .03$ for perspective A means that there are $T = N\lambda = 100(.03) = 3$ nonconformities on the items in the lot. For such a case, formula (8.7) then shows that the probability of accepting the lot is

$$Pa(.03) = \binom{3}{0}\left(\frac{5}{100}\right)^0\left(1 - \frac{5}{100}\right)^3 + \binom{3}{1}\left(\frac{5}{100}\right)^1\left(1 - \frac{5}{100}\right)^2 = .8574 + .1354 = .993\,.$$

Finally, the perspective B mean nonconformities per unit operating characteristic when $\lambda = .03$ is from equation (8.8)

$$Pa(.03) = \frac{\exp(-.15)(.15)^0}{0!} + \frac{\exp(-.15)(.15)^1}{1!} = .861 + .129 = .990\,.$$

Calculations similar to these can be used to find operating characteristic values for other values of p or λ, and to produce the OC curves plotted in Figures 8.1 and 8.2. Figure 8.1 shows the two fraction nonconforming OC plots and Figure 8.2 provides the two plots for the mean nonconformities per unit scenario.

Example 8.4 is instructive on at least two accounts. For one thing, it brings into focus the fact that often the different types of OC curves for a given plan are quite similar. In retrospect, this is at least partially to be expected. It is a fact from elementary probability theory that for N large compared to n, the hypergeometric distributions for X can be approximated by appropriate binomial distributions. That is, in such circumstances expressions (8.5) and (8.6) give substantially the same numerical values. Similarly, a second fact from elementary probability theory says that for small p and large n, the binomial (n, p), distribution for X can

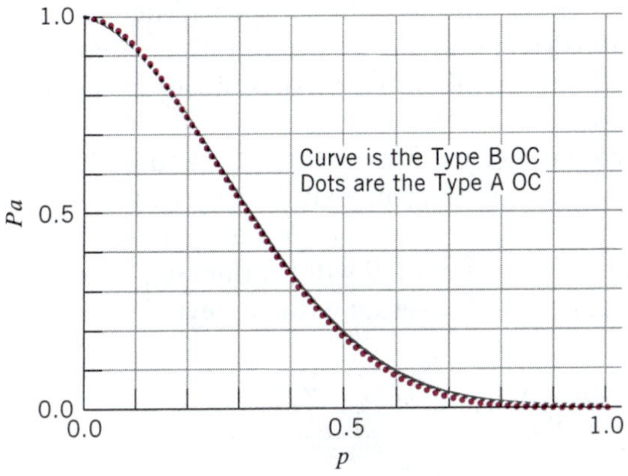

Figure 8.1
Fraction nonconforming OC for $n = 5, c = 1$, and $N = 100$.

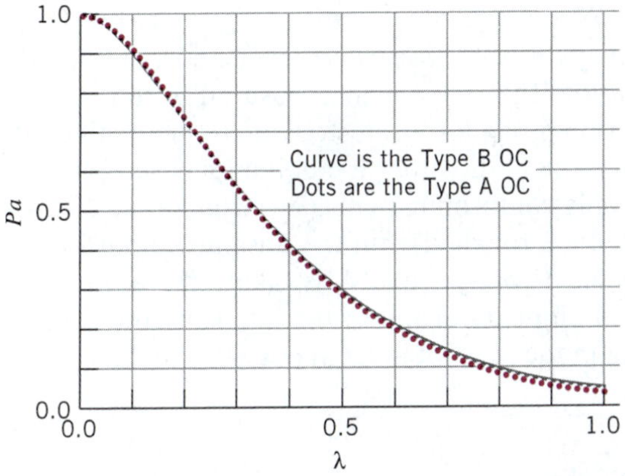

Figure 8.2
Mean nonconformities per unit OC for $k = 5, c = 1$, and $N = 100$.

be approximated by the Poisson distribution with mean np. That is, expressions (8.6) and (8.8) can give substantially the same numerical values.

Example 8.4 also illustrates that while elementary, the OC calculations indicated in formulas (8.5) through (8.8) can be tedious. It is thus relevant to note that commonly available statistical packages have built-in routines for computing cumulative binomial and Poisson (and sometimes even hypergeometric) probabilities. These can be used to compute OC values for attributes single sampling plans.

8.1.2 Rectifying Inspection and Fraction Nonconforming Acceptance Sampling

One use to which fraction nonconforming attributes acceptance sampling is sometimes put is as a screening device to determine whether lots should be placed directly into a production stream, or should instead be inspected item by item and nonconforming items either repaired or replaced with conforming items, that is, **rectified**. There are terminology and calculations that accompany this kind of application and these are considered next.

If "rejection" of a lot under the plan specified by displays (8.1) and (8.3) means that the unsampled portion of the lot will be inspected item by item, the total inspection load actually encountered is either n items or N items. It makes sense to ask what an average or expected inspection load will be, thinking of this as a function of p. And it should be clear that this **average total inspected** is

Average Total Inspected

$$ATI(p) = nPa(p) + N(1 - Pa(p)). \tag{8.9}$$

For small p, $Pa(p)$ is near 1 and the ATI is essentially n (a lot will be accepted with high probability). For large p, $Pa(p)$ is near 0 and the ATI is essentially N (a lot will be rejected with high probability).

A natural question is what rectifying inspection can be expected to accomplish in terms of the final quality of a lot. This is typically measured in terms of the expected final fraction nonconforming. If a lot is rejected due to the rectification the final fraction nonconforming is 0. On the other hand, if a lot is accepted, the sample is guaranteed to contain only conforming items (by virtue of the fact that all nonconforming items identified are rectified), but the unsampled portion can contain some nonconforming items. If perspective A is taken, there are $Np - X$ nonconforming items in the unsampled portion of the lot. Thus, the mean or expected fraction nonconforming in the final lot is, from perspective A,

Type A AOQ for Rectifying Inspection

$$AOQ(p) = \sum_{x=0}^{c} \left(\frac{Np - x}{N}\right) \frac{\binom{Np}{x}\binom{N(1-p)}{n-x}}{\binom{N}{n}}. \tag{8.10}$$

The acronym AOQ introduced in display (8.10) stands for the words **average outgoing quality**. (Average outgoing "unquality" might be more appropriate, but the AOQ language is standard.)

The calculation indicated in display (8.10) is almost never done. Instead, for better or worse, it is common to compute only type B AOQ's. If one adopts perspective B, the expected number of nonconforming items in the unsampled portion of a lot is $(N - n)p$. Thus, the mean fraction nonconforming in the final lot is, from perspective B,

Type B AOQ for Rectifying Inspection

$$AOQ(p) = 0(1 - Pa(p)) + \left(\frac{(N-n)p}{N}\right)Pa(p) = \left(1 - \frac{n}{N}\right)pPa(p). \tag{8.11}$$

This expression is simpler than the perspective A formula given in display (8.10). In fact, in those cases where the sampling fraction n/N is small, there is the even simpler approximation

$$AOQ(p) \approx pPa(p).$$

The AOQ defined in formula (8.10) or (8.11) behaves in a perfectly sensible manner. For small p, the rectifying inspection scheme can only improve the lot fraction nonconforming, so the AOQ is small. For large p, the chance of rejecting a lot (and thus ending with a rectified lot having no nonconforming items in it) is high and thus the AOQ is again small. That suggests that the average outgoing quality has a maximum (worst) value for some "moderate" value of p. This maximum value is called the **average outgoing quality limit**, the AOQL. That is,

$$AOQL = \max_p AOQ(p). \qquad (8.12)$$

Average Outgoing Quality Limit

The AOQL is the worst long-run fraction nonconforming that could exit the two-stage inspection scheme if a series of lots with a common p were submitted to it.

Example 8.4 continued. The attributes plan with $n = 5$ and $c = 1$ can be used to illustrate the meaning of formulas (8.9), (8.11) and (8.12). If, for example, $p = .03$ and $N = 100$, then formula (8.9) says that the type B average total inspected is

$$ATI(.03) = 5(.992) + 100(.008) = 5.76.$$

The corresponding average outgoing quality is from formula (8.11)

$$AOQ(.03) = \left(1 - \frac{5}{100}\right)(.03)(.992) = .028.$$

Further, Figure 8.3 is a plot of the AOQ for this scenario and shows that the average outgoing quality limit is approximately .152.

The calculations illustrated here are appropriate exactly when nonconforming items are rectified. A natural alternative to rectifying inspection is **removal inspection**. That is, one can simply cull or remove any nonconforming item located through inspection. And it is worth contemplating what formulas might describe the effects of removal inspection.

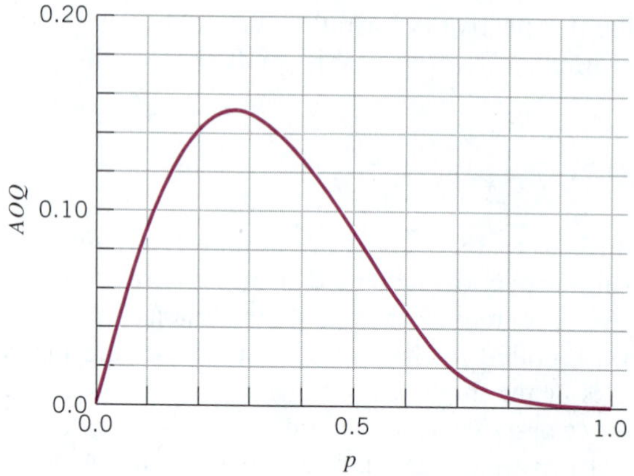

Figure 8.3
Type B AOQ for $n = 5, c = 1,$ and $N = 100$.

In the first place, an average total inspected for removal inspection is the same as for rectifying inspection. Formula (8.9) is still in force. But formulas for the AOQ are different from the ones for rectifying inspection. From perspective A, it turns out that

Type A AOQ for Removal Inspection

$$AOQ(p) = \sum_{x=0}^{c} \left(\frac{Np - x}{N - x}\right) \frac{\binom{Np}{x}\binom{N(1-p)}{n-x}}{\binom{N}{n}}. \tag{8.13}$$

And from perspective B, the average outgoing quality is

Type B AOQ for Removal Inspection

$$AOQ(p) = \sum_{x=0}^{c} \left(\frac{(N - n)p}{N - x}\right)\binom{n}{x} p^x (1-p)^{n-x}. \tag{8.14}$$

Values produced by formulas (8.13) and (8.14) are larger than the corresponding quantities specified for rectifying inspection in displays (8.10) and (8.11) because final lot sizes can be smaller than N.

A measure of lot shrinkage due to the removal of nonconforming items is the difference between N and the expected final lot size. And if M is the final lot size, from perspective A

Type A Expected Final Lot Size for Removal Inspection

$$\mathrm{EM} = N(1 - p(1 - Pa(p))) - \sum_{x=0}^{c} x \frac{\binom{Np}{x}\binom{N(1-p)}{n-x}}{\binom{N}{n}}, \tag{8.15}$$

while from perspective B

$$EM = N - np - p(N - n)(1 - Pa(p)) . \tag{8.16}$$

Type B Expected Final Lot Size for Removal Inspection

Example 8.4 continued. Consider now the use of an $n = 5$, $c = 1$ attributes plan as the basis for removal inspection of lots of size $N = 100$, supposing that $p = .03$. From perspective B

$$ATI\,(.03) = 5.76\,,$$

exactly as for rectifying inspection. The perspective B average outgoing quality is, from display (8.14)

$$AOQ = \frac{95(.03)}{100}(.859) + \frac{95(.03)}{99}(.133) = .028\,,$$

which to three decimal places is the same as for rectifying inspection. And from display (8.16) the mean size of a final lot under removal inspection is

$$EM = 100 - 5(.03) - .03(95)(.008) = 99.83\,,$$

nearly the original value of $N = 100$. For this value of p, there is little difference between rectifying and removal inspection. In fact, since $Pa(.03) = .992$, for $p = .03$, lots are rarely rejected and *neither* rectifying nor removal inspection has much practical effect. Of course, for p with a smaller corresponding acceptance probability, the differing impacts of rectifying and removal inspection will be more obvious and important.

8.1.3 Designing Attributes Acceptance Sampling Plans

The primary practical question yet unaddressed in this section is "How does one choose n (or k) and c?" This final subsection will provide some general comments on the problem, give some tools for "getting a ballpark answer," and end with the advice that one needs to look at the OC curves for candidate plans and tinker appropriately if they are not exactly what one wants.

First, some general observations: In rough terms, the sample size n (or k) governs how steeply an OC curve "drops" as a function of p (or λ). The larger the sample size the steeper the drop. And the ratio c/n (or c/k) governs the location of the drop. Roughly speaking, the drop is "centered" around $p = c/n$ (or $\lambda = c/k$). So (again in rough terms) if one has in mind a value θ that separates p's (or λ's) that are tolerable from those that are not, one might well set

Approximate Acceptance Number for Attributes Acceptance Sampling

$$c \approx n\theta \tag{8.17}$$

(or $c \approx k\theta$) and see what kind of OC curve is implied by various values of n (or k).

A crude calculation based on the normal approximation to the binomial distribution shows that if one has in mind a value θ for separating "OK" p's from those that are not, plans to use relationship (8.17), and wants a type B fraction nonconforming OC of about β at $p = \theta + \delta$, a relevant approximate sample size is

Approximate Sample Size for Attributes Acceptance Sampling

$$n \approx (Q_z(\beta))^2 \left(\frac{(\theta + \delta)(1 - \theta - \delta)}{\delta^2} \right). \tag{8.18}$$

($Q_z(\beta)$ is the standard normal β quantile.) (To the extent that mean nonconformity per unit acceptance probabilities approximate fraction nonconforming values, display (8.18) can also be used to obtain a relevant approximate k in a mean nonconformity per unit problem. In such a case, θ and $\theta + \delta$ of course refer to values of λ, not p.)

Example 8.5 Design for an Attributes Single Sampling Plan. As an example of using the design method just described, suppose that one wants an OC curve whose "drop" occurs in the vicinity of $p = \theta = .02$ and in fact wishes to have an acceptance probability of around $\beta = .10$ when $p = .025 = \theta + .005$. Formula (8.18) suggests a sample size of

$$n \approx (-1.282)^2 \left(\frac{(.025)(.975)}{(.005)^2} \right) \approx 1600\,,$$

and formula (8.17) prescribes a corresponding acceptance number

$$c \approx 1600(.02) = 32\,.$$

Table 8.1 gives some OC values (8.6) for this plan obtained using Minitab's facility for producing binomial cumulative probabilities.

Table 8.1 shows that the plan prescribed by the formulas (8.17) and (8.18) does a pretty good job of satisfying the original design criteria. That is because the sample size is fairly large, and therefore the normal approximation that stands behind display (8.18) is a good one. If for some reason one was not quite satisfied with the OC values in Table 8.1, the $n = 1{,}600$ and $c = 32$ plan could still serve as a starting point for minor "tinkering" to produce slightly adjusted acceptance

Table 8.1 Type B Fraction Nonconforming OC Values for an $n = 1{,}600$ and $c = 32$ Acceptance Sampling Plan

p	$Pa(p)$
.010	.9999
.015	.9546
.020	.5468
.025	.1123
.030	.0085

probabilities. For example, the reader can check that keeping $n = 1{,}600$ but reducing the acceptance number to $c = 31$ (not surprisingly) lowers all acceptance probabilities from those in Table 8.1. In particular, it gives $Pa(.020) = .4758$ and $Pa(.025) = .0828$, and this choice might be preferable to the $c = 32$ choice in some circumstances.

There are other somewhat more refined methods of making choices of n and c to produce desired fraction nonconforming OC properties. For example, H.R. Larson in the 1966 *Industrial Quality Control* article, "A Nomograph of the Cumulative Binomial Distribution," first presented a famous graph that can sometimes be used to find n and c meeting two-point (on a desired OC curve) design criteria.

Two-Point OC Design Method for Mean Nonconformities per Unit Plans

And there is a simple method that can be used to choose k and c to produce desired type B nonconformities per unit OC properties. (To the extent that the other OC curves match this one, the method produces corresponding properties for them as well.) It is based on a relationship between cumulative Poisson probabilities and cumulative probabilities for the χ^2 distributions that is discussed, for example, on pages 262–263 of *Statistical Process Control: Theory and Practice* by Wetherill and Brown.

Suppose that one has in mind acceptance probabilities Pa_1 and Pa_2 for two values of λ, say $\lambda_1 < \lambda_2$. If $Q_\nu(p)$ is the χ^2_ν quantile function, one can use a χ^2 table like Table A.7 to look at successively larger values of c until finding one such that

Equation for Mean Nonconformities per Unit Acceptance Number

$$\frac{Q_{2(c+1)}(1 - Pa_2)}{Q_{2(c+1)}(1 - Pa_1)} \approx \frac{\lambda_2}{\lambda_1} . \qquad (8.19)$$

If no $c \leq 19$ produces approximate equality in relationship (8.19), one runs out of entries in Table A.7 and needs to use the approximation to χ^2 quantiles given at the bottom of the table for larger c. As it turns out, plugging the approximation into relationship (8.19) produces an equation that can be solved explicitly for c. Writing

Abbreviations Used in Equation (8.20)

$$t = \left(\frac{\lambda_2}{\lambda_1}\right)^{1/3}, \quad q_1 = Q_z(1 - Pa_1) \text{ and } q_2 = Q_z(1 - Pa_2),$$

that solution is

Approximate Mean Nonconformities per Unit Acceptance Number

$$c \approx \left(\frac{\frac{2}{3}(t-1)}{tq_1 - q_2 + \sqrt{(tq_1 - q_2)^2 + 4(t-1)^2}} \right)^2 - 1. \tag{8.20}$$

Then using a c that solves equation (8.19) (possibly derived using equation (8.20) if no $c \leq 19$ works), one takes

Mean Nonconformities per Unit Sample Size

$$k \approx \frac{Q_{2(c+1)}(1 - Pa_1)}{2\lambda_1} \approx \frac{Q_{2(c+1)}(1 - Pa_2)}{2\lambda_2}. \tag{8.21}$$

(Again, if $c > 19$, the approximation given at the bottom of Table A.7 must be used to find the χ^2 quantiles in display (8.21).)

Example 8.6 **Design for Another Attributes Sampling Plan.** Suppose that for $\lambda_1 = .01$ one would like an acceptance probability of approximately $Pa_1 = .90$, while for $\lambda_2 = .05$ an acceptance probability of approximately $Pa_2 = .10$ is desired. Table 8.2 then shows some $.9 = 1 - Pa_2$ and $.1 = 1 - Pa_1$ quantiles of χ^2 distributions and their ratios.

Since $\lambda_2/\lambda_1 = 5.0$, Table 8.2 and relationship (8.19) suggest that $c = 2$ may be appropriate. Then applying relationship (8.21),

Table 8.2 Some .9 and .1 Quantiles of χ^2 Distributions and Their Ratios

c	$\nu = 2(c+1)$	$Q_\nu(.9)$	$Q_\nu(.1)$	$Q_\nu(.9)/Q_\nu(.1)$
0	2	4.605	.211	21.8
1	4	7.779	1.064	7.3
2	6	10.645	2.204	4.8
3	8	13.362	3.490	3.8

$$k \approx \frac{2.204}{2(.01)} = 110 \qquad \text{or} \qquad k \approx \frac{10.645}{2(.05)} = 106 \text{ is prescribed.}$$

Taking for sake of example $c = 2$ and $k = 108$, one can use formula (8.8) to verify that in fact

$$Pa(.01) = .904 \qquad \text{and} \qquad Pa(.05) = .095\,.$$

The story told in Example 8.6 is completely typical. As one requires the OC "drop" to occur over narrower and narrower ranges (as measured by the ratio λ_2/λ_1), one is forced to larger and larger c's and therefore larger and larger k's (or n's). Steep OC curves are bought at the expense of large sample sizes.

The two design methods represented by displays (8.17) and (8.18) and by displays (8.19) through (8.21) are effective means of "getting into the ballpark" of c and n (or k) combinations approximately meeting stated requirements on the location and steepness of the drop in an OC curve. But the design problem is not completely solved until one computes appropriate OC values via one of formulas (8.5) through (8.9), and either verifies that the values are acceptable or tries other combinations close to the original one until a satisfactory OC curve is obtained.

8.2 VARIABLES ACCEPTANCE SAMPLING FOR FRACTION NONCONFORMING

It sometimes happens that what makes an item conforming or nonconforming is definable in terms of a single measurement, x. When that is the case and one is somehow sure that the underlying distribution of measurements is normal, it is possible to make direct use of the measurements in acceptance sampling (instead of just counting each sampled item as conforming or nonconforming). This section discusses what can be easily done in this direction. The section begins with a discussion of "known σ" variables acceptance sampling plans and their OC curves. Then some "unknown σ" plans are presented and again the calculation of their OC properties considered. There follows a discussion of methods for designing a variables acceptance sampling plan. And the section closes with a short critique of the whole notion of variables acceptance sampling for fraction nonconforming.

8.2.1 Known σ Variables Acceptance Sampling Plans and Their Operating Characteristics

The starting point for variables acceptance sampling is the assumption that when some variable x is measured on an item and compared to an engineering specification or specifications U and/or L, one then knows whether the item is conforming or nonconforming. In a **single limit** problem with an **upper specification** U, an item is nonconforming if

$$x > U\,.$$

In a **single limit** problem with a **lower specification** L, an item is nonconforming if

$$x < L\,.$$

And in a **double limits** problem with both an upper specification U and a lower specification L, an item is nonconforming if

$$x > U \qquad \text{or if} \qquad x < L\,.$$

Example 8.7 **Testing Electric Meters (Example 8.1 Revisited).** Consider the situation of a power utility that must check the calibration of a lot of its electric meters installed at customer sites. For sake of illustration, further suppose that the test to which meters are put consists of measuring 1 kilowatt hour of energy. A possible calibration criteria is then that a meter is nonconforming if the measurement x it produces is less than .98 kW hr or larger than 1.02 kW hr. Such a situation could be termed a double limits case with upper specification $U = 1.02$ and lower specification $L = .98$.

On the other hand, it is possible to think that from a regulatory point of view, the case of meters that read too high is of primary concern. That is, it is possible to consider only a single (upper) engineering specification. If one specifies that a meter is nonconforming if the measurement it produces is above 1.02 kW hr, the situation could be termed a single limit case with upper specification $U = 1.02$.

If one next thinks of x as having a probability distribution, in the single lower limit case the probability an item is nonconforming is

$$p = P[x < L]\,, \tag{8.22}$$

in the single upper limit situation the probability an item is nonconforming is

$$p = P[x > U]\,, \tag{8.23}$$

and in the double limits situation, the probability that an item is nonconforming is

$$p = P[x < L] + P[x > U]\,. \tag{8.24}$$

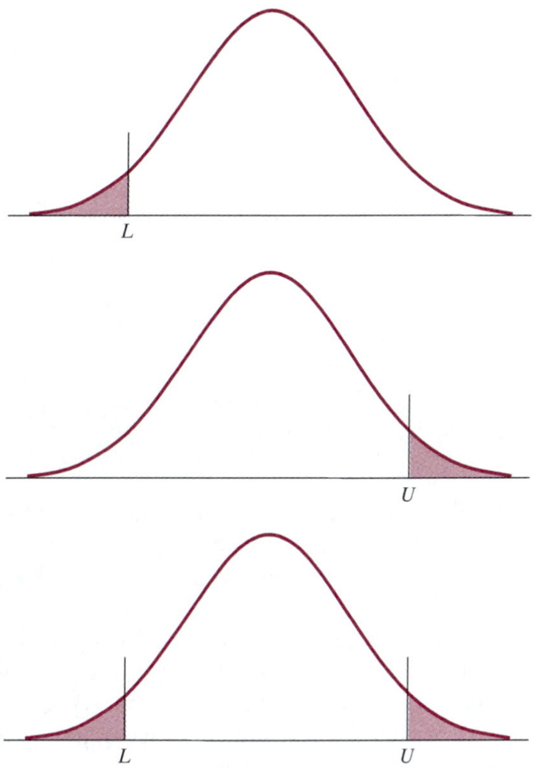

Figure 8.4
p (shaded) for a normal distribution in respectively a single lower limit case, a single upper limit case, and a double limits case.

These probabilities are each usually interpreted as the fraction of items nonconforming. Exactly how one computes them of course depends upon what one assumes about the distribution of x. Everything in this section depends upon a Gaussian/normal model for x. The shaded areas in the three parts of Figure 8.4 correspond to the probabilities (8.22) through (8.24) for x normally distributed.

Before going further, there must be some discussion of circumstances under which a normal model for measurements made on items from a lot can hope to make sense. One possibility is that perspective B is appropriate and it is sensible to assume that the *process standing behind a lot* produces normal x values. But if perspective A is most relevant, there are simply a finite number of values x corresponding to the N items in the lot. To then treat random draws from the lot essentially as if they produce independent normal random variables requires not only that the lot have a relative frequency distribution of x's that is fairly Gaussian looking, but also that the lot be *large*. Otherwise, both the fact that one is approximating a discrete relative frequency distribution by a continuous probability distribution and the lack of independence in random sampling from real finite populations will call into question the kind of mathematics used in variables acceptance sampling.

Then continuing on, if it is sensible to treat measurements x as normal, the probabilities (8.22) through (8.24) are respectively

Single Lower Limit Fraction Nonconforming

$$p = P[x < L] = P[Z < \frac{L - \mu}{\sigma}] = \Phi\left(\frac{L - \mu}{\sigma}\right), \quad (8.25)$$

Single Upper Limit Fraction Nonconforming

$$p = P[x > U] = P[Z > \frac{U - \mu}{\sigma}] = 1 - \Phi\left(\frac{U - \mu}{\sigma}\right), \quad (8.26)$$

and

Double Limits Fraction Nonconforming

$$p = P[x < L] + P[x > U] = 1 - \left(\Phi\left(\frac{U - \mu}{\sigma}\right) - \Phi\left(\frac{L - \mu}{\sigma}\right)\right), \quad (8.27)$$

for Z a generic standard normal random variable and Φ the standard normal cumulative probability function, $\Phi(z) = P[Z \leq z]$. The object of variables acceptance sampling is to use n measurements from sampled items to ensure that p defined in one of these equations is small.

This initial subsection considers variables acceptance sampling under the assumption that not only is it known that x is normal, but that (from considerations outside the acceptance sampling process) **σ is known.** This means that somehow (presumably from some kind of prior experience) one knows the variability of the measurements in the lot (perspective A) or of the process that produced the lot (perspective B). For fixed/known σ, it is evident from either the plots in Figure 8.4 or the formulas (8.25) through (8.27) that p is a function of μ alone. One can keep p small by ensuring that μ stays on the correct side of and far away from L and/or U.

Example 8.7 continued. For the time being, suppose that the standard deviation of laboratory energy measurements for the large lot of meters being checked by the utility is $\sigma = .008$ kW hr. If the two-sided specifications $L = .98$ and $U = 1.02$ are in force, some values of μ and corresponding values of p derived using formula (8.27) are given in Table 8.3.

Table 8.3 shows that p is minimum (but bigger than 0) when $\mu = 1.00 = (U + L)/2$. This is the way things always turn out in a double limits problem. The best position for μ is at the midpoint of the specifications, but even a perfect mean produces a nonzero fraction nonconforming. The business of double limits variables acceptance sampling is basically to ensure that μ is not too far from the ideal value.

Some values of μ and corresponding values of p derived using formula (8.26) are given in Table 8.4 for the single limit problem with upper specification $U = 1.02$, again assuming that $\sigma = .008$ kW hr.

Table 8.3 Some Means and Corresponding Fractions Nonconforming for a Double Limits Problem

μ	$z_L = \dfrac{.98 - \mu}{.008}$	$z_U = \dfrac{1.02 - \mu}{.008}$	$\Phi(z_U) - \Phi(z_L)$	$p = 1 - (\Phi(z_U) - \Phi(z_L))$
.97	1.25	6.25	.1056	.8944
.98	0	5.00	.5000	.5000
.99	−1.25	3.75	.8943	.1057
1.00	−2.50	2.50	.9876	.0124
1.01	−3.75	1.25	.8943	.1057
1.02	−5.00	0	.5000	.5000
1.03	−6.25	−1.25	.1056	.8944

Table 8.4 Some Means and Corresponding Fractions Nonconforming for a Single (Upper) Limit Problem

μ	$z_U = \dfrac{1.02 - \mu}{.008}$	$\Phi(z_U)$	$p = 1 - \Phi(z_U)$
.97	6.25	1.0000	.0000
.98	5.00	1.0000	.0000
.99	3.75	.9999	.0001
1.00	2.50	.9938	.0062
1.01	1.25	.8944	.1056
1.02	0	.5000	.5000
1.03	−1.25	.1056	.8944
1.04	−2.50	.0062	.9938
1.05	−3.75	.0001	.9999

It is evident from Table 8.4 that the larger μ becomes, the larger is the fraction nonconforming. More and more of the normal distribution is positioned to the right of the upper specification. The business of single limit variables sampling (with an upper specification) is basically to ensure that μ is comfortably below U.

When one assumes that measurements are normally distributed and σ is known, it is plausible that a variables acceptance sampling plan should depend upon n measurements only through their sample mean, $\bar{x}$. And in the single lower limit case (where small means μ produce large fractions nonconforming), it is natural to consider acceptance sampling criteria of the form

Single Lower Limit Known σ Decision Criteria

$$\text{accept a sampled lot if } \bar{x} \geq L + \Delta_1, \text{ and reject the lot if } \bar{x} < L + \Delta_1 \quad (8.28)$$

for an appropriate positive number Δ_1. The corresponding single upper limit variables acceptance sampling criterion is

Single Upper Limit Known σ Decision Criteria

$$\text{accept a sampled lot if } \bar{x} \leq U - \Delta_1, \text{ and reject the lot if } \bar{x} > U - \Delta_1 \quad (8.29)$$

for an appropriate positive number Δ_1. And a reasonably natural variables acceptance sampling criterion for the double limits case is

Double Limit Known σ Decision Criteria

$$\text{accept a sampled lot if } L + \Delta_2 \leq \bar{x} \leq U - \Delta_2, \text{ and reject the lot otherwise} \quad (8.30)$$

for an appropriate Δ_2 between 0 and $(U - L)/2$.

Each of the criteria (8.28) through (8.30) has its own corresponding formula for the acceptance probability. But all of these are based on the same very elementary probability fact. That is, the normal distribution assumption for x implies that $\bar{x}$ is normal as well, with mean $\mu_{\bar{x}} = \mu$ and standard deviation $\sigma_{\bar{x}} = \sigma/\sqrt{n}$. Corresponding to a given μ is not only a value of p specified in one of formulas (8.25) through (8.27), but also a value of Pa computed using the normal distribution of $\bar{x}$. For the single lower limit criterion (8.28),

Single Lower Limit Known σ Acceptance Probability

$$Pa = P[\bar{x} \geq L + \Delta_1] = P[Z \geq \frac{L + \Delta_1 - \mu}{\sigma/\sqrt{n}}] = 1 - \Phi\left(\frac{L + \Delta_1 - \mu}{\sigma/\sqrt{n}}\right). \quad (8.31)$$

For the single upper limit criterion (8.29),

Single Upper Limit Known σ Acceptance Probability

$$Pa = P[\bar{x} \leq U - \Delta_1] = P[Z \leq \frac{U - \Delta_1 - \mu}{\sigma/\sqrt{n}}] = \Phi\left(\frac{U - \Delta_1 - \mu}{\sigma/\sqrt{n}}\right). \quad (8.32)$$

And for the double limits criterion (8.30),

Double Limits Known σ Acceptance Probability

$$\begin{aligned} Pa &= P[L + \Delta_2 \leq \bar{x} \leq U - \Delta_2] \\ &= P\left[\frac{L + \Delta_2 - \mu}{\sigma/\sqrt{n}} \leq Z \leq \frac{U - \Delta_2 - \mu}{\sigma/\sqrt{n}}\right] \\ &= \Phi\left(\frac{U - \Delta_2 - \mu}{\sigma/\sqrt{n}}\right) - \Phi\left(\frac{L + \Delta_2 - \mu}{\sigma/\sqrt{n}}\right). \end{aligned} \quad (8.33)$$

By varying μ and using the appropriate one of equations (8.25) through (8.27) together with the corresponding one of equations (8.31) through (8.33), pairs (p, Pa) and an OC curve for any of the criteria (8.28) through (8.30) can be produced.

Example 8.7 continued. Consider again the utility meter scenario. And suppose that $n = 20$ meters are to be selected at random from those in the field and brought to the laboratory for testing. Considering initially the *double limits* case with $L = .98$ and $U = 1.02$, suppose that $\Delta_2 = .01$ kW hr is used in criterion (8.30). (One will reject the lot and require that all the meters be replaced or recalibrated if $\bar{x}$ fails to fall between .99 and 1.01.) Then if, for example, $\mu = 1.01$ (so that from Table 8.3 $p = .1057$), the corresponding acceptance probability is computed using formula (8.33). That is,

$$\begin{aligned} Pa &= \Phi\left(\frac{1.02 - .01 - 1.01}{.008/\sqrt{20}}\right) - \Phi\left(\frac{.98 + .01 - 1.01}{.008/\sqrt{20}}\right) \\ &= \Phi(0) - \Phi(-11.18) = .5000\,. \end{aligned}$$

Or, if $\mu = 1.00$, Table 8.3 shows that $p = .0124$ and again using formula (8.33) the corresponding acceptance probability is

$$\begin{aligned} Pa &= \Phi\left(\frac{1.02 - .01 - 1.00}{.008/\sqrt{20}}\right) - \Phi\left(\frac{.98 + .01 - 1.00}{.008/\sqrt{20}}\right) \\ &= \Phi(5.59) - \Phi(-5.59) = 1.000\,. \end{aligned}$$

Similar calculations (and the values of p from Table 8.3) yield the picture of the OC for this variables plan given in Table 8.5 and the kind of OC curve shown in Figure 8.5. Note that as expected, no points are plotted for p less than .0124, the fraction nonconforming corresponding to a perfect value of μ.

Table 8.5 Some OC Points for a Double Limits Problem

μ	p	Pa
.98	.5000	.0000
.99	.1057	.5000
1.00	.0124	1.000
1.01	.1057	.5000
1.02	.5000	.0000

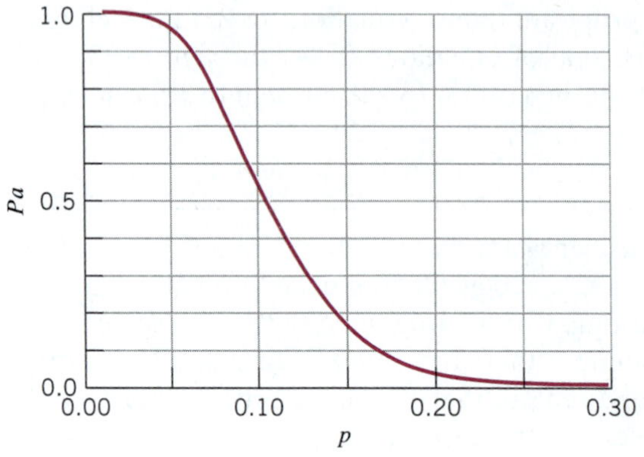

Figure 8.5
OC for the two-sided known σ variables acceptance sampling plan of Example 8.7.

To illustrate some *single limit* calculations, suppose that $U = 1.02$ and that $\Delta_1 = .015$ is used in criterion (8.29). (Rejection will occur if $\bar{x} > 1.005$.) Then if, for example, $\mu = 1.01$ (so that from Table 8.4 one sees that $p = .1056$), the corresponding acceptance probability is computed using formula (8.32). That is,

$$Pa = \Phi\left(\frac{1.02 - .015 - 1.01}{.008/\sqrt{20}}\right) = \Phi(-2.80) = .0026 .$$

Or, if $\mu = 1.00$, Table 8.4 shows that $p = .0062$, and again using formula (8.32), the corresponding acceptance probability is

$$Pa = \Phi\left(\frac{1.02 - .015 - 1.00}{.008/\sqrt{20}}\right) = \Phi(2.80) = .9974 .$$

Table 8.6 lists these two and a few more (p, Pa) pairs computed using formulas (8.26) and (8.32).

Table 8.6 Some OC Points for a Single (Upper) Limit Problem

μ	p	Pa
.995	.0009	1.000
1.000	.0062	.9974
1.005	.0380	.5000
1.010	.1056	.0026
1.015	.2660	.0000

8.2.2 Unknown σ Variables Acceptance Sampling Plans

When one cannot treat σ as a known constant, the business of variables acceptance sampling rapidly becomes more difficult. In the first place, while p is still given by the appropriate one of displays (8.25) through (8.27), it must be treated as a function of *both* μ and σ. And while one could *try* to use criteria like those given in displays (8.28) through (8.30) to decide lot disposal, where more than one σ is possible it is no longer the case that a given p has a single associated Pa. This makes the criteria worthless as a means of discriminating between lots with big p and those with small p.

Example 8.7 continued. Consider again the single (upper) limit plan for the meter testing scenario that uses $n = 20$ and rejects the lot if $\bar{x} > 1.005$. Previous calculations have shown that if $\sigma = .008$, $\mu = 1.00$ produces $p = .0062$ and $Pa = .9974$. If, however, this criterion is applied when $\sigma = .002$, the reader can verify that $\mu = 1.015$ produces $p = .0062$ but $Pa = .0000$. This is a very unhappy circumstance. If one wants to treat all lots with $p = .0062$ equally, rejecting the lot if $\bar{x} > 1.005$ is useless as a means of variables acceptance sampling.

Something new has to be done for **unknown σ** variables sampling. As it turns out, there is a fairly straightforward way to proceed in single limit cases. But the double limits situation is complicated beyond what is reasonable to address thoroughly in a book like this. So for the double limits case it will be possible only to point the interested reader to other sources for more complete discussions.

So consider the single upper limit problem. One wants to use data to somehow ensure that μ is far enough below U that the shaded area in the second part of Figure 8.3 is small. But "far below U" must be defined in terms of σ. That is, one needs to be sure of a large z-value corresponding to U,

$$z = \frac{U - \mu}{\sigma},$$

without having μ and σ to work with, but instead having only data from measurements.

Following this line of reasoning, it seems like a natural method of acceptance sampling in an unknown σ context might then be to compute a sample version of z, namely

$$\frac{U - \bar{x}}{s},$$

and accept the lot if this is big and reject it otherwise. That is (after a small bit of algebra), a possible acceptance sampling criterion for the single upper limit problem is

Single Upper Limit Unknown σ Decision Criteria

$$\text{accept a sampled lot if } \bar{x} \leq U - ks, \text{ and reject the lot if } \bar{x} > U - ks \tag{8.34}$$

for an appropriate positive number k. (This is acceptance when the sample version of the z-value for U is at least k.) The corresponding criterion for the single lower limit problem is

Single Lower Limit Unknown σ Decision Criteria

$$\text{accept a sampled lot if } \bar{x} \geq L + ks, \text{ and reject the lot if } \bar{x} < L + ks \tag{8.35}$$

for an appropriate positive number k. (This is acceptance when the sample version of the z-value for L is no more than $-k$.)

Acceptance sampling criteria (8.34) and (8.35) are not only intuitively reasonable, they have the happy property that all (μ, σ) pairs with a given p (computed via the appropriate one of (8.25) or (8.26)) turn out to have the same Pa. So one can sensibly use them to discriminate between lots with small p and lots with large p. But how to compute the acceptance probability for criterion (8.34) or criterion (8.35) is not completely elementary. Exact probabilities can be found using a statistical package that will evaluate "noncentral t" probabilities. That is, the exact acceptance probability corresponding to criterion (8.34) is

$$Pa = P[W \geq k\sqrt{n}],$$

for W a noncentral t random variable with $n-1$ degrees of freedom and noncentrality parameter $\sqrt{n}(U-\mu)/\sigma$. The corresponding exact acceptance probability for the lower limit criterion (8.35) is computed the same way, except that the noncentrality parameter is $\sqrt{n}(\mu - L)/\sigma$.

A good approximation to the exact acceptance probability for an unknown σ variables plan that involves only the normal distribution is due to W. A. Wallis. He noted that it is often adequate to treat the variable $\bar{x} \pm ks$ as if it were normal with mean $\mu \pm k\sigma$ and variance

$$\sigma^2\left(\frac{1}{n} + \frac{k^2}{2n}\right).$$

This leads to approximate acceptance probabilities. That is, for the single upper limit criterion (8.34),

Approximate Upper Limit Unknown σ Acceptance Probability

$$Pa \approx P\left[Z \leq \frac{U - (\mu + k\sigma)}{\sigma\sqrt{\frac{1}{n} + \frac{k^2}{2n}}}\right] = \Phi\left(\frac{\frac{U-\mu}{\sigma} - k}{\sqrt{\frac{1}{n} + \frac{k^2}{2n}}}\right). \tag{8.36}$$

And for the single lower limit criterion (8.35),

$$Pa \approx P\left[Z \geq \frac{L-(\mu-k\sigma)}{\sigma\sqrt{\frac{1}{n}+\frac{k^2}{2n}}}\right] = 1-\Phi\left(\frac{\frac{L-\mu}{\sigma}+k}{\sqrt{\frac{1}{n}+\frac{k^2}{2n}}}\right). \qquad (8.37)$$

Approximate Lower Limit Unknown σ Acceptance Probability

Example 8.7 continued. In the context of the meter calibration problem, suppose that with $U = 1.02$ the power company contemplates selecting a sample of $n = 20$ meters and applying criterion (8.34) with $k = 2.5$. Then further suppose that μ and σ are such that

$$\frac{1.02-\mu}{\sigma} = 1.645\,.$$

From formula (8.26),

$$p = 1-\Phi(1.645) = .05\,,$$

and using formula (8.36)

$$Pa \approx \Phi\left(\frac{1.645-2.5}{\sqrt{\frac{1}{20}+\frac{(2.5)^2}{2(20)}}}\right) = \Phi(-1.88) = .0301\,.$$

Similar calculations with this acceptance sampling plan but with different values of the ratio $(U-\mu)/\sigma$ give the (p, Pa) pairs listed in Table 8.7. The corresponding approximate OC curve for the variables plan is plotted in Figure 8.6.

As indicated earlier in this section, handling the double limits unknown σ variables sampling problem is not completely straightforward. No one has yet discovered a way to specify a useful acceptance criterion so that all (μ, σ) pairs having a given p defined by formula (8.27) have exactly the same Pa. Some reasonable criteria have been proposed that come close enough to this that they can be used. But they are complicated to describe, and finding the range of Pa's associated with a given p is well beyond the technical level of this text. So, if the reader is highly motivated to know details about what is possible, he or she is referred to the reading and references in Chapter 12 of Duncan's *Quality Control and Industrial Statistics* and Chapter 10 of Schilling's *Acceptance Sampling in Quality Control.*

Table 8.7 Some (p, Pa) Pairs for a Single Upper Limit Unknown σ Variables Acceptance Sampling Plan

$z_U = (1.02 - \mu)/\sigma$	$p = 1 - \Phi(z_U)$	$Pa \approx \Phi\left((z_U - 2.5)/\sqrt{\frac{1}{20} + \frac{(2.5)^2}{40}}\right)$
4.00	.00003	.9995
3.50	.0002	.9862
3.00	.0013	.8645
2.50	.0062	.5000
2.00	.0228	.1355
1.50	.0668	.0138
1.00	.1587	.0005

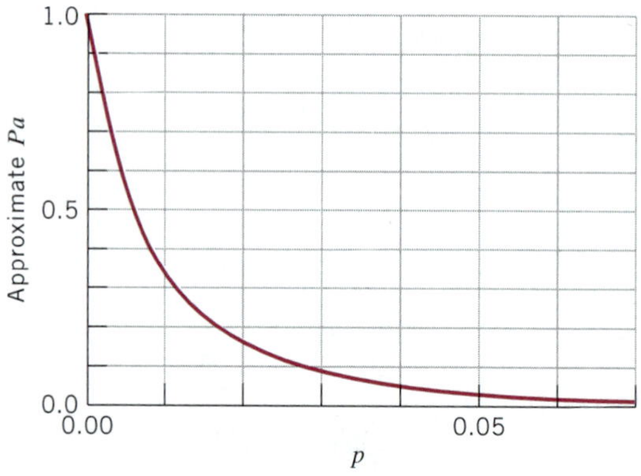

Figure 8.6
Approximate OC for the one-sided unknown σ variables acceptance sampling plan of Example 8.7.

8.2.3 Designing Variables Acceptance Sampling Plans

The variables acceptance sampling plans introduced in this section involve a constant to be chosen by the user (Δ_1, Δ_2, or k) and the sample size, n. In many cases, by properly choosing these one can produce a plan with an OC curve essentially meeting two-point design goals (i.e., having desired acceptance probabilities for two user-specified fractions nonconforming p_1 and p_2). Methods for accomplishing this are presented next.

Consider first the single limit known σ plans specified in displays (8.28) and (8.29). If one has in mind two fractions nonconforming $p_1 < p_2$ and corresponding acceptance probabilities $Pa_1 > Pa_2$, it is straightforward to set up and solve two equations in the unknowns Δ_1 and n. (This comes about by remembering that there are unique means μ_1 and μ_2 corresponding to p_1 and p_2 and that the distribution of $\overline{x}$ is normal with mean μ and standard deviation $\sigma/\sqrt{n}$.) The solutions to these equations are

$$n \approx \left(\frac{Q_z(Pa_1) - Q_z(Pa_2)}{Q_z(1-p_1) - Q_z(1-p_2)} \right)^2 , \tag{8.38}$$

Single Limit Known σ Sample Size

and

$$\Delta_1 \approx \sigma \left(\frac{Q_z(Pa_1)Q_z(1-p_2) - Q_z(Pa_2)Q_z(1-p_1)}{Q_z(Pa_1) - Q_z(Pa_2)} \right) . \tag{8.39}$$

Single Limit Known σ Critical Difference

Example 8.7 continued. Returning to the meter-sampling scenario, suppose that from substantial experience with the type of meter in the lot under investigation, engineers at the utility believe that a population standard deviation of $\sigma = .01$ kW hr for tests on meters from the lot is appropriate. Further, suppose that they would like to find an upper limit variables acceptance sampling plan with $Pa_1 = .95$ when $p_1 = .04$ and $Pa_2 = .05$ when $p_2 = .08$. Then, interpolating in the standard normal table, the quantities needed in equations (8.38) and (8.39) are approximately $Q_z(1-p_1) = 1.751$, $Q_z(1-p_2) = 1.405$, $Q_z(Pa_1) = 1.645$, and $Q_z(Pa_2) = -1.645$. So applying formula (8.38), a sample size of approximately

$$n \approx \left(\frac{1.645 - (-1.645)}{1.751 - 1.405} \right)^2 \approx 90$$

is prescribed, along with

$$\Delta_1 \approx .01 \left(\frac{1.645(1.405) - (-1.645)1.751}{1.645 - (-1.645)} \right) = .01578 .$$

The reader is invited to verify using formula (8.26), that with $U = 1.02$ and $\sigma = .01$, $\mu_1 = 1.00249$ produces $p_1 = .04$ and $\mu_2 = 1.00595$ produces $p_2 = .08$. The reader can then further verify that using $n = 90$ and $\Delta_1 = .01578$ in criterion (8.29), formula (8.32) produces $Pa(p_1) \approx .0486$ and $Pa(p_2) \approx .9514$, values close to the original design goals for the plan.

The (known σ) double limits two-point design problem is complicated by the fact that, in general, one must (for a given μ) worry about parts of the fraction nonconforming coming both from the possibility that $x < L$ and also from the possibility that $x > U$. When $U - L$ is large, say at least 6σ, for most practical purposes this problem disappears and one can use the single limit solution given in equations (8.38) and (8.39) (of course substituting Δ_2 for Δ_1 in display (8.39)). When $U - L$ is not large compared to σ, one must proceed more or less by trial and error. One must first find $\mu_1 < \mu_2$ both (say) larger than $(L + U)/2$ having corresponding fractions nonconforming $p_1 < p_2$ (defined in formula (8.27)). Then, the task is to set $Pa(p_1)$ (defined in formula (8.33) using μ_1) equal to a desired value Pa_1, to set $Pa(p_2)$ (defined in formula (8.33) using μ_2) equal to a desired value Pa_2, and to solve those equations simultaneously for n and Δ_2.

The unknown σ single limit design problem has a fairly clean approximate solution, based on the Wallis approximation alluded to as the basis of formulas (8.36) and (8.37). Again setting up and solving two simultaneous equations for k and n one arrives at approximate values

Single Limit Unknown σ Critical Sample z-Score, k

$$k \approx \frac{Q_z(Pa_1)Q_z(1-p_2) - Q_z(Pa_2)Q_z(1-p_1)}{Q_z(Pa_1) - Q_z(Pa_2)}, \tag{8.40}$$

and

Single Limit Unknown σ Sample Size

$$n \approx \left(1 + \frac{k^2}{2}\right)\left(\frac{Q_z(Pa_1) - Q_z(Pa_2)}{Q_z(1-p_1) - Q_z(1-p_2)}\right)^2 \tag{8.41}$$

for use in criterion (8.34) or (8.35). These equations are very similar to equations (8.39) and (8.38). Except that Δ_1 is in the original units and involves σ while k is unitless, essentially the same calculation needed to produce Δ_1 in a known σ case can be done to produce k in an unknown σ problem. And the sample size required in the unknown σ situation is that from the known σ case multiplied by $(1 + k^2/2)$. This last factor is sometimes thought of as a "penalty" incurred by lack of knowledge of σ.

Example 8.7 continued. Consider for a final time the meter sampling scenario. Suppose that the single upper limit design problem is faced with $Pa_1 = .95$ when $p_1 = .04$ and $Pa_2 = .05$ when $p_2 = .08$, now no longer assuming that σ is known. Recalling that $Q_z(1-p_1) = 1.751$, $Q_z(1-p_2) = 1.405$, $Q_z(Pa_1) = 1.645$, and $Q_z(Pa_2) = -1.645$, formula (8.40) prescribes

$$k \approx \left(\frac{1.645(1.405) - (-1.645)1.751}{1.645 - (-1.645)}\right) = 1.578,$$

while formula (8.41) then suggests that

$$n \approx \left(1 + \frac{(1.578)^2}{2}\right)\left(\frac{1.645 - (-1.645)}{1.751 - 1.405}\right)^2 \approx 203$$

is needed in this circumstance. The reader is invited to verify using equations (8.26) and (8.36) that indeed these values used in criterion (8.34) come close to satisfying the design criteria.

8.2.4 Some Cautions/Caveats Regarding Variables Acceptance Sampling

Before leaving the topic of variables acceptance sampling, it is important to offer some cautions about the whole enterprise. First, it is common to note that in order to meet a given set of OC curve design requirements, variables sampling plans typically require far smaller sample sizes than the attributes plans discussed in Section 8.1. Initially, this seems especially helpful in modern environments where parts-per-million fractions nonconforming are commonly desired. (A little experience with the attributes plan design tools of the last section will convince the reader that to effectively distinguish between lots with 10^{-6} nonconformity rates and those with 2×10^{-6} nonconformity rates, sample sizes of several million are needed for attributes sampling!) But there are good reasons why variables acceptance sampling plans should not be thought of as a panacea for large sample sizes.

It cannot be emphasized strongly enough that this entire section depends critically on the appropriateness of the normal distribution assumption. Equations (8.25) through (8.27) are essential to relating μ and σ to the fraction nonconforming. This fact is perhaps not so unnerving for moderate values of p, since many real relative frequency distributions look fairly bell-shaped in their "middles" and normal calculations can do pretty well at approximating observed relative frequencies. But when one begins to consider acceptance sampling where the object is to discriminate between very small fractions nonconforming, the situation is not so optimistic. For small p, one is relying on the appropriateness of the normal model *in the tails of the real distribution of measurements*. And the conventional wisdom is that if a normal model is going to go wrong, it will go wrong *first* in the tails of a real distribution.

One might think about trying to allay this worry over the shape of the distribution of measurements by comparing the shape of a sample to the normal distribution shape, for example via normal plotting. But that's not really much help where the extreme tails are concerned. In order to get a solid picture of the shape of the "10^{-6} tails" of a distribution of measurements, a sample of millions is needed. So one is faced with the proverbial "Catch 22." In order to really verify that one need not deal with sample sizes in the millions, one needs a sample of millions.

Where does that leave the reader on the matter of variables sampling for very small fractions nonconforming? Hopefully he or she is left fairly skeptical. While the methods of this section are surely better than nothing, your authors do not recommend taking very seriously the OC values provided here for very small values of p.

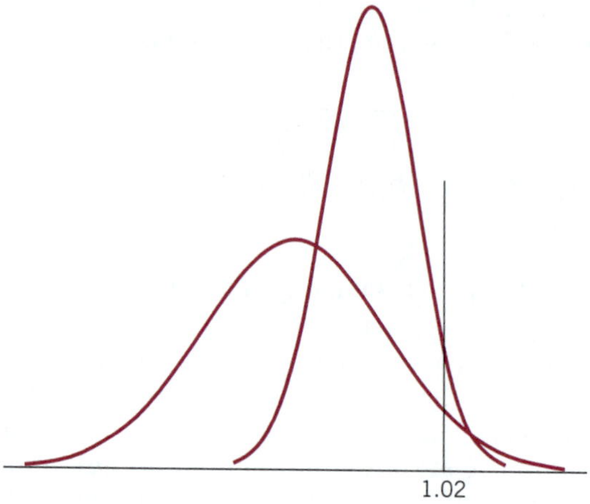

Figure 8.7
Two normal distributions with $p = .05$ for a single upper limit problem with $U = 1.02$.

And this point is not the only one that throws some cold water on the variables acceptance sampling enterprise. There is also the fact that variables acceptance sampling attempts to treat equally all (μ, σ) pairs with a single corresponding p. But it's not so clear how often that goal is rational. Take, for example, the power meter calibration problem. Figure 8.7 shows two quite different normal distributions, both having $p = .05$ for the single upper limit problem with $U = 1.02$. These are treated equally by variables acceptance sampling. But from a regulatory point of view, it seems like the distribution with the larger σ (and smaller μ) is preferable to the one with the smaller σ (and larger μ). Although both distributions indicate the same fraction of customers being overcharged, the distribution with the smaller σ leads to a much larger mean (and total) charge. (And the problem isn't even limited to the unknown σ situation. In double limits problems with *known* σ, it's not so clear that two means symmetrically placed about the mid-specification and thus producing the same p should typically be treated equivalently!)

There are no doubt cases where it does make sense to treat all (μ, σ) with a common p equally. But one cannot assume without careful thought that this is necessarily appropriate. The plea here is that before mechanically plugging into the formulas in this section and charging straight ahead, the reader consider whether the case he or she faces is really adequately described by the material that has been presented.

8.3 MILITARY STANDARD 105

There is nothing in the acceptance sampling material presented in the previous two sections that requires that any more than a single lot be under consideration. (Even the ATI and AOQ concepts, which have natural interpretations in terms of long-run average performance over a sequence of lots with the same p, can also be interpreted in terms of *expected* behavior of rectifying inspection on a single lot.) But this section explicitly considers situations where a (long) sequence of lots is to be subjected to sampling inspection.

The discussion here concerns an attributes **acceptance sampling system** originally developed for application to an incoming stream of lots in the context of military procurement. This system was spelled out in great detail in a document known as Military Standard 105. This document went through a number of revisions between its introduction in 1950 and its final "E" version issued in 1989. The standard has recently been withdrawn by the U.S. military, but there are various civilian versions of it still available, including ANSI/ASQC Z1.4 (which can be ordered from the American Society for Quality) and ISO 2859. These differ from the military standard in some qualitative details, but share most of the basic structure of the technical (statistical) part of the original.

The presentation here is limited to an introduction to the **switching system** prescribed by the standard and the main single sampling tables in the document, followed by some commentary on the limitations and proper application of the system. Fuller discussions of the quantitative rationale behind the standard and the more qualitative aspects of implementing it can be found in Chapter 10 of Duncan's *Quality Control and Industrial Statistics* and Chapter 11 of Schilling's *Acceptance Sampling in Quality Control.* (Both of these sources also have extensive chapter bibliographies for further reading in primary source material.)

8.3.1 The Mil. Std. 105 AQL, Switching System, and Single Sampling Tables

Military Standard 105 is structured around an **acceptable quality level** or AQL notion. Applications of the standard are made with the intention that if a supplier submits a series of lots with p (or λ) no more than an agreed-upon AQL, the sampling system will tend to operate with few rejections of lots or other negative (for the supplier) consequences. On the other hand, when there is evidence that p or λ has become larger than the agreed-upon AQL, the system prescribes actions by the receiving party (originally the U.S. military) that both makes it more difficult for lots to be accepted and eventually causes contractual consequences beyond mere rejections of individual lots.

Military Standard 105 allows that attributes acceptance sampling of lots be carried out either using single sampling as described in Section 8.1, or using more complicated (but sometimes more economical in terms of average sample size) "double sampling" or even "multiple sampling." There was no discussion of double or multiple sampling in Section 8.1, so this basic introduction to the military standard system will be confined to the use of single sampling. (The more complicated options are intended to provide statistical properties in terms of lot acceptance probabilities comparable to those encountered using single sampling.)

Under the standard, acceptance sampling of a stream of lots begins using an attributes plan (using n or k and c) intended to provide high probability of lot acceptance for $p < AQL$ (or $\lambda < AQL$). This initial plan (specified by the standard) is called the **normal inspection plan**. Should, however, two out of any five or less consecutive lots be rejected under normal inspection, the acceptance sampling plan is switched to a **tightened** plan that keeps the same sample size, but reduces the acceptance number (with the end effect of reducing *Pa* for a given p or λ). When on tightened inspection, return can be made to the normal acceptance sampling plan when five consecutive lots have been accepted. However, if 10 consecutive lots are inspected under the tightened plan without a return to normal inspection, one exits the standard's sampling inspection system and other measures (outside the standard) are taken to deal with the problem of quality that is evidently worse than the agreed-upon AQL.

The standard has an optional provision for reducing sampling intensity when it appears that indeed a supplier is more than living up to the agreed-upon AQL. This provision operates as follows: When 10 lots in a row have been accepted *under the normal sampling plan* and the total number nonconforming (or total number of nonconformities) seen in those 10 samples is small enough (and other qualitative requirements are met), then the change can be made to a **reduced inspection plan**. This plan has a smaller sample size and corresponding smaller acceptance number. And it is different from anything discussed in Section 8.1 in that for an **acceptance number** Ac, one does not reject for all counts bigger than Ac. Instead, a **rejection number** *Re* is also specified, and rejection is specified for counts bigger than or equal to this value, while for counts *between* the acceptance and rejection numbers, the lot is accepted but one returns to use of the normal plan.

Figure 8.8 is a schematic portrayal of the entire Military Standard 105 switching scheme.

Exactly which acceptance sampling plans are used for the normal, tightened, and reduced inspection, and exactly what it means to have a small number nonconforming (or small number of nonconformities) in a series of samples for purposes of authorizing a switch to reduced inspection, are all governed by a set of tables. The starting point for using these is the standard's Table I, reproduced here as Table 8.8. In order to enter Table I, one needs a lot size and inspection level. The commonly recommended inspection level is General Inspection Level II. General Inspection Level I calls for sample sizes (very roughly) half of those for General Inspection Level II, while General Inspection Level III (approximately) doubles the sample sizes over Level II. The Special Inspection Levels call for very small sample sizes (the S-1 Level having the very smallest ones) and are not recommended for common use. Based on the lot size and inspection level, Table I then prescribes a code letter (basically specifying sample sizes) for use in the other tables in the standard.

One then proceeds to Tables II-A, II-B, and II-C of the standard to find normal, tightened, and reduced acceptance sampling plans. These tables are reproduced here as Tables 8.9 through 8.11. They are entered on the row corresponding to the code letter taken from Table I, in the column under the agreed-upon AQL.

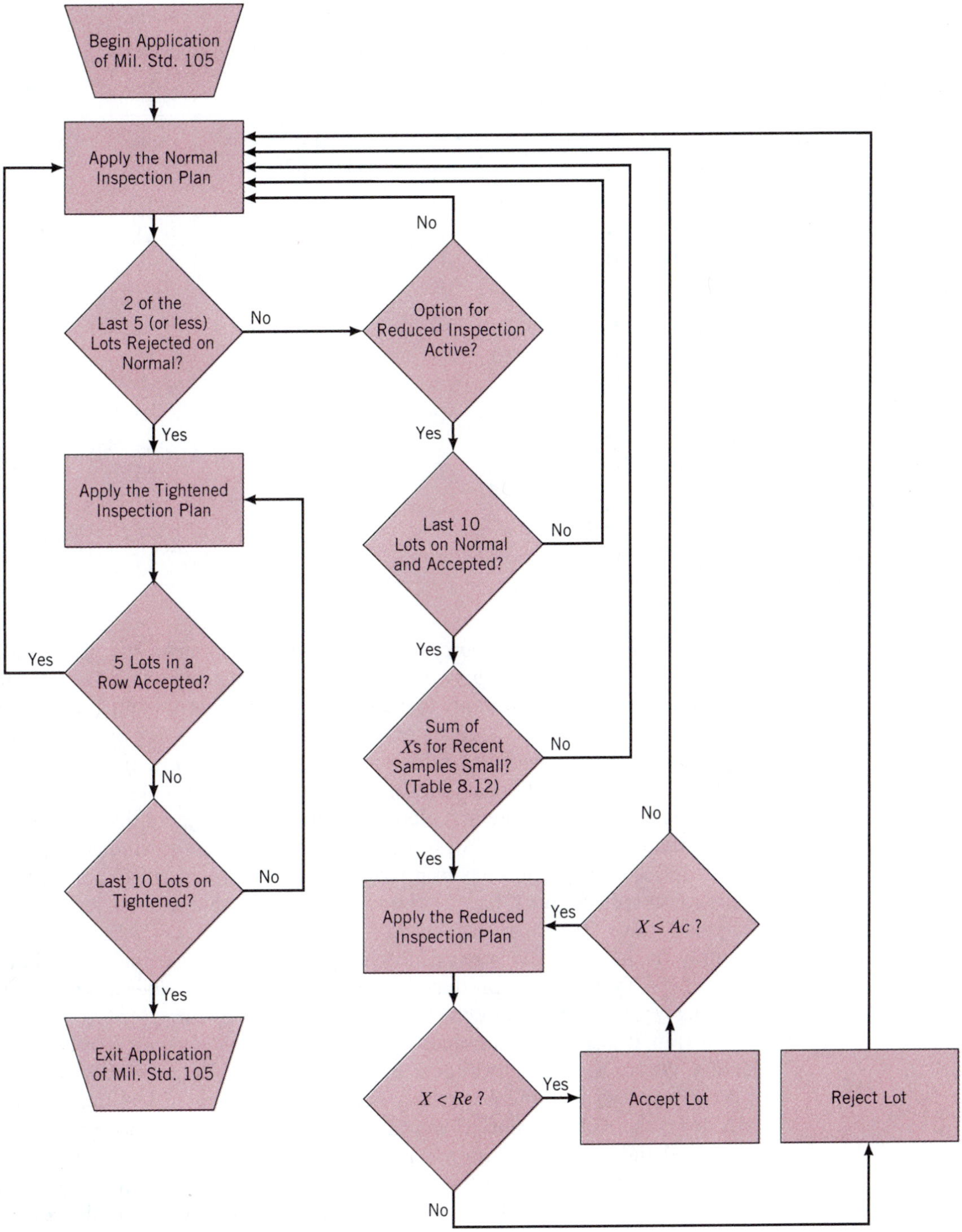

Figure 8.8
The Military Standard 105 switching scheme.

Table 8.8 Table I from Mil. Std. 105, Sample Size Code Letters

Lot or Batch Size			Special Inspection Levels				General Inspection Levels		
			S-1	S-2	S-3	S-4	I	II	III
2	to	8	A	A	A	A	A	A	B
9	to	15	A	A	A	A	A	B	C
16	to	25	A	A	B	B	B	C	D
26	to	50	A	B	B	C	C	D	E
51	to	90	B	B	C	C	C	E	F
91	to	150	B	B	C	D	D	F	G
151	to	280	B	C	D	E	E	G	H
281	to	500	B	C	D	E	F	H	J
501	to	1200	C	C	E	F	G	J	K
1,201	to	3,200	C	D	E	G	H	K	L
3,201	to	10,000	C	D	F	G	J	L	M
10,001	to	35,000	C	D	F	H	K	M	N
35,001	to	150,000	D	E	G	J	L	N	P
150,001	to	500,000	D	E	G	J	M	P	Q
500,001	and	over	D	E	H	K	N	Q	R

The AQLs are expressed as percentages so that, for example, the plans under 1.0 correspond to an AQL of $p = .01$ (or $\lambda = .01$). Notice also that the plans for AQLs larger than 100% refer only to mean nonconformities per unit applications, where it can make sense to have an AQL of $\lambda = 1.5, 2.5, 4.0, 6.5$, or 10.0. Where one encounters an arrow in the table, the intention is that one move either up or down as indicated, until encountering the first plan in the table. (One uses the sample size on the row of the first plan encountered, and should that sample size be more than the lot size, 100% inspection is done.) As already indicated, Table II-C for reduced inspection has in it rejection numbers that are larger than the acceptance numbers plus 1, leaving open the possibility that a lot be accepted but the system return to normal inspection.

Finally, in those cases where reduced inspection is to be employed, Military Standard 105 Table VIII, reproduced here as Table 8.12, provides **limit numbers** specifying what is a small enough count of total nonconforming (or total nonconformities) in 10 (or more) consecutive samples to allow a switch to reduced inspection. Where one encounters dots in the table, there is no possibility of switching to reduced inspection after only 10 consecutive accepted lots. Instead, one reads down the AQL column to find the first numerical entry. Enough lots must be sampled and accepted on normal inspection to accumulate a total sample size indicated by the row label for this entry, with a total number nonconforming (or total nonconformities) less than or equal to this first limit number in the column.

Table 8.9 Table II-A from Mil. Std. 105, Single Sampling Plans for Normal Inspection (Master Table)

Sample size code letter	Sample size	Acceptable Quality Levels (normal inspection) 0.010	0.015	0.025	0.040	0.065	0.10	0.15	0.25	0.40	0.65	1.0	1.5	2.5	4.0	6.5	10	15	25	40	65	100	150	250	400	650	1,000
		Ac Re	Ac Re	Ac Re	Ac Re	Ac Re	Ac Re	Ac Re	Ac Re	Ac Re	Ac Re	Ac Re	Ac Re	Ac Re	Ac Re	Ac Re	Ac Re	Ac Re	Ac Re	Ac Re	Ac Re	Ac Re	Ac Re	Ac Re	Ac Re	Ac Re	Ac Re
A	2	↓	↓	↓	↓	↓	↓	↓	↓	↓	↓	↓	↓	↓	↓	0 1	↓	↓	1 2	2 3	3 4	5 6	7 8	10 11	14 15	21 22	30 31
B	3	↓	↓	↓	↓	↓	↓	↓	↓	↓	↓	↓	↓	↓	0 1	↑	↓	1 2	2 3	3 4	5 6	7 8	10 11	14 15	21 22	30 31	44 45
C	5	↓	↓	↓	↓	↓	↓	↓	↓	↓	↓	↓	↓	0 1	↑	↓	1 2	2 3	3 4	5 6	7 8	10 11	14 15	21 22	30 31	44 45	↑
D	8	↓	↓	↓	↓	↓	↓	↓	↓	↓	↓	↓	0 1	↑	↓	1 2	2 3	3 4	5 6	7 8	10 11	14 15	21 22	30 31	44 45	↑	↑
E	13	↓	↓	↓	↓	↓	↓	↓	↓	↓	↓	0 1	↑	↓	1 2	2 3	3 4	5 6	7 8	10 11	14 15	21 22	30 31	44 45	↑	↑	↑
F	20	↓	↓	↓	↓	↓	↓	↓	↓	↓	0 1	↑	↓	1 2	2 3	3 4	5 6	7 8	10 11	14 15	21 22	↑	↑	↑	↑	↑	↑
G	32	↓	↓	↓	↓	↓	↓	↓	↓	0 1	↑	↓	1 2	2 3	3 4	5 6	7 8	10 11	14 15	21 22	↑	↑	↑	↑	↑	↑	↑
H	50	↓	↓	↓	↓	↓	↓	↓	0 1	↑	↓	1 2	2 3	3 4	5 6	7 8	10 11	14 15	21 22	↑	↑	↑	↑	↑	↑	↑	↑
J	80	↓	↓	↓	↓	↓	↓	0 1	↑	↓	1 2	2 3	3 4	5 6	7 8	10 11	14 15	21 22	↑	↑	↑	↑	↑	↑	↑	↑	↑
K	125	↓	↓	↓	↓	↓	0 1	↑	↓	1 2	2 3	3 4	5 6	7 8	10 11	14 15	21 22	↑	↑	↑	↑	↑	↑	↑	↑	↑	↑
L	200	↓	↓	↓	↓	0 1	↑	↓	1 2	2 3	3 4	5 6	7 8	10 11	14 15	21 22	↑	↑	↑	↑	↑	↑	↑	↑	↑	↑	↑
M	315	↓	↓	↓	0 1	↑	↓	1 2	2 3	3 4	5 6	7 8	10 11	14 15	21 22	↑	↑	↑	↑	↑	↑	↑	↑	↑	↑	↑	↑
N	500	↓	↓	0 1	↑	↓	1 2	2 3	3 4	5 6	7 8	10 11	14 15	21 22	↑	↑	↑	↑	↑	↑	↑	↑	↑	↑	↑	↑	↑
P	800	↓	0 1	↑	↓	1 2	2 3	3 4	5 6	7 8	10 11	14 15	21 22	↑	↑	↑	↑	↑	↑	↑	↑	↑	↑	↑	↑	↑	↑
Q	1,250	0 1	↑	↓	1 2	2 3	3 4	5 6	7 8	10 11	14 15	21 22	↑	↑	↑	↑	↑	↑	↑	↑	↑	↑	↑	↑	↑	↑	↑
R	2,000	↑ ↑ ↑	↑ ↑ ↑	1 2	2 3	3 4	5 6	7 8	10 11	14 15	21 22	↑ ↑ ↑	↑ ↑ ↑	↑ ↑ ↑	↑ ↑ ↑	↑ ↑ ↑	↑ ↑ ↑	↑ ↑ ↑	↑ ↑ ↑	↑ ↑ ↑	↑ ↑ ↑	↑ ↑ ↑	↑ ↑ ↑	↑ ↑ ↑	↑ ↑ ↑	↑ ↑ ↑	↑ ↑ ↑

↓ = Use first sampling plan below arrow. If sample size equals or exceeds lot or batch size, do 100% inspection.
↑ = Use first sampling plan above arrow.
Ac = Acceptance number.
Re = Rejection number.

Table 8.10 Table II-B from Mil. Std. 105, Single Sampling Plans for Tightened Inspection (Master Table)

Sample size code letter	Sample size	Acceptable Quality Levels (tightened inspection)																									
		0.010	0.015	0.025	0.040	0.065	0.10	0.15	0.25	0.40	0.65	1.0	1.5	2.5	4.0	6.5	10	15	25	40	65	100	150	250	400	650	1,000
letter	size	Ac Re	Ac Re	Ac Re	Ac Re	Ac Re	Ac Re	Ac Re	Ac Re	Ac Re	Ac Re	Ac Re	Ac Re	Ac Re	Ac Re	Ac Re	Ac Re	Ac Re	Ac Re	Ac Re	Ac Re	Ac Re	Ac Re	Ac Re	Ac Re	Ac Re	Ac Re
A	2	↓	↓	↓	↓	↓	↓	↓	↓	↓	↓	↓	↓	↓	↓	↓	↓	↓	↓	1 2	2 3	3 4	5 6	8 9	12 13	18 19	27 28
B	3	↓	↓	↓	↓	↓	↓	↓	↓	↓	↓	↓	↓	↓	↓	0 1	↓	↓	1 2	2 3	3 4	5 6	8 9	12 13	18 19	27 28	41 42
C	5	↓	↓	↓	↓	↓	↓	↓	↓	↓	↓	↓	↓	↓	0 1	↓	↓	1 2	2 3	3 4	5 6	8 9	12 13	18 19	27 28	41 42	↑
D	8	↓	↓	↓	↓	↓	↓	↓	↓	↓	↓	↓	↓	0 1	↓	↓	1 2	2 3	3 4	5 6	8 9	12 13	18 19	27 28	41 42	↑	↑
E	13	↓	↓	↓	↓	↓	↓	↓	↓	↓	↓	↓	0 1	↓	↓	1 2	2 3	3 4	5 6	8 9	12 13	18 19	27 28	41 42	↑	↑	↑
F	20	↓	↓	↓	↓	↓	↓	↓	↓	↓	↓	0 1	↓	↓	1 2	2 3	3 4	5 6	8 9	12 13	18 19	↑	↑	↑	↑	↑	↑
G	32	↓	↓	↓	↓	↓	↓	↓	↓	↓	0 1	↓	↓	1 2	2 3	3 4	5 6	8 9	12 13	18 19	↑	↑	↑	↑	↑	↑	↑
H	50	↓	↓	↓	↓	↓	↓	↓	↓	0 1	↓	↓	1 2	2 3	3 4	5 6	8 9	12 13	18 19	↑	↑	↑	↑	↑	↑	↑	↑
J	80	↓	↓	↓	↓	↓	↓	↓	0 1	↓	↓	1 2	2 3	3 4	5 6	8 9	12 13	18 19	↑	↑	↑	↑	↑	↑	↑	↑	↑
K	125	↓	↓	↓	↓	↓	↓	0 1	↓	↓	1 2	2 3	3 4	5 6	8 9	12 13	18 19	↑	↑	↑	↑	↑	↑	↑	↑	↑	↑
L	200	↓	↓	↓	↓	↓	0 1	↓	↓	1 2	2 3	3 4	5 6	8 9	12 13	18 19	↑	↑	↑	↑	↑	↑	↑	↑	↑	↑	↑
M	315	↓	↓	↓	↓	0 1	↓	↓	1 2	2 3	3 4	5 6	8 9	12 13	18 19	↑	↑	↑	↑	↑	↑	↑	↑	↑	↑	↑	↑
N	500	↓	↓	↓	0 1	↓	↓	1 2	2 3	3 4	5 6	8 9	12 13	18 19	↑	↑	↑	↑	↑	↑	↑	↑	↑	↑	↑	↑	↑
P	800	↓	↓	0 1	↓	↓	1 2	2 3	3 4	5 6	8 9	12 13	18 19	↑	↑	↑	↑	↑	↑	↑	↑	↑	↑	↑	↑	↑	↑
Q	1,250	↓	0 1	↓	↓	1 2	2 3	3 4	5 6	8 9	12 13	18 19	↑	↑	↑	↑	↑	↑	↑	↑	↑	↑	↑	↑	↑	↑	↑
R	2,000	0 1	↑	↓	1 2	2 3	3 4	5 6	8 9	12 13	18 19	↑	↑	↑	↑	↑	↑	↑	↑	↑	↑	↑	↑	↑	↑	↑	↑
S	3,150			1 2																							

↓ = Use first sampling plan below arrow. If sample size equals or exceeds lot or batch size, do 100% inspection.
↑ = Use first sampling plan above arrow.
Ac = Acceptance number.
Re = Rejection number.

Table 8.11 Table II-C from Mil. Std. 105, Single Sampling Plans for Reduced Inspection (Master Table)

Sample size code letter	Sample size	Acceptable Quality Levels (reduced inspection)†																									
		0.010	0.015	0.025	0.040	0.065	0.10	0.15	0.25	0.40	0.65	1.0	1.5	2.5	4.0	6.5	10	15	25	40	65	100	150	250	400	650	1,000
		Ac Re	Ac Re	Ac Re	Ac Re	Ac Re	Ac Re	Ac Re	Ac Re	Ac Re	Ac Re	Ac Re	Ac Re	Ac Re	Ac Re	Ac Re	Ac Re	Ac Re	Ac Re	Ac Re	Ac Re	Ac Re	Ac Re	Ac Re	Ac Re	Ac Re	Ac Re
A	2	↓	↓	↓	↓	↓	↓	↓	↓	↓	↓	↓	↓	↓	↓	0 1	↓	↓	1 2	2 3	3 4	5 6	7 8	10 11	14 15	21 22	30 31
B	2	↓	↓	↓	↓	↓	↓	↓	↓	↓	↓	↓	↓	↓	0 1	↑	↓	0 2	1 3	2 4	3 5	5 6	7 8	10 11	14 15	21 22	30 31
C	2	↓	↓	↓	↓	↓	↓	↓	↓	↓	↓	↓	↓	0 1	↑	↓	0 2	1 3	1 4	2 5	3 6	5 8	7 10	10 13	14 17	21 24	↑
D	3	↓	↓	↓	↓	↓	↓	↓	↓	↓	↓	↓	0 1	↑	↓	0 2	1 3	1 4	2 5	3 6	5 8	7 10	10 13	14 17	21 24	↑	↑
E	5	↓	↓	↓	↓	↓	↓	↓	↓	↓	↓	0 1	↑	↓	0 2	1 3	1 4	2 5	3 6	5 8	7 10	10 13	14 17	21 24	↑	↑	↑
F	8	↓	↓	↓	↓	↓	↓	↓	↓	↓	0 1	↑	↓	0 2	1 3	1 4	2 5	3 6	5 8	7 10	10 13	↑	↑	↑	↑	↑	↑
G	13	↓	↓	↓	↓	↓	↓	↓	↓	0 1	↑	↓	0 2	1 3	1 4	2 5	3 6	5 8	7 10	10 13	↑	↑	↑	↑	↑	↑	↑
H	20	↓	↓	↓	↓	↓	↓	↓	0 1	↑	↓	0 2	1 3	1 4	2 5	3 6	5 8	7 10	10 13	↑	↑	↑	↑	↑	↑	↑	↑
J	32	↓	↓	↓	↓	↓	↓	0 1	↑	↓	0 2	1 3	1 4	2 5	3 6	5 8	7 10	10 13	↑	↑	↑	↑	↑	↑	↑	↑	↑
K	50	↓	↓	↓	↓	↓	0 1	↑	↓	0 2	1 3	1 4	2 5	3 6	5 8	7 10	10 13	↑	↑	↑	↑	↑	↑	↑	↑	↑	↑
L	80	↓	↓	↓	↓	0 1	↑	↓	0 2	1 3	1 4	2 5	3 6	5 8	7 10	10 13	↑	↑	↑	↑	↑	↑	↑	↑	↑	↑	↑
M	125	↓	↓	↓	0 1	↑	↓	0 2	1 3	1 4	2 5	3 6	5 8	7 10	10 13	↑	↑	↑	↑	↑	↑	↑	↑	↑	↑	↑	↑
N	200	↓	↓	0 1	↑	↓	0 2	1 3	1 4	2 5	3 6	5 8	7 10	10 13	↑	↑	↑	↑	↑	↑	↑	↑	↑	↑	↑	↑	↑
P	315	↓	0 1	↑	↓	0 2	1 3	1 4	2 5	3 6	5 8	7 10	10 13	↑	↑	↑	↑	↑	↑	↑	↑	↑	↑	↑	↑	↑	↑
Q	500	0 1	↑	↓	0 2	1 3	1 4	2 5	3 6	5 8	7 10	10 13	↑	↑	↑	↑	↑	↑	↑	↑	↑	↑	↑	↑	↑	↑	↑
R	800	↑	↑	0 2	1 3	1 4	2 5	3 6	5 8	7 10	10 13	↑	↑	↑	↑	↑	↑	↑	↑	↑	↑	↑	↑	↑	↑	↑	↑

↓ = Use first sampling plan below arrow. If sample size equals or exceeds lot or batch size, do 100% inspection.
↑ = Use first sampling plan above arrow.
Ac = Acceptance number.
Re = Rejection number.
† = If the acceptance number has been exceeded but the rejection number has not been reached, accept the lot but reinstate normal inspection.

Table 8.12 Table VIII from Mil. Std. 105, Limit Numbers for Reduced Inspection

Number of sample units from last 10 lots or batches	Acceptable Quality Level																									
	0.010	0.015	0.025	0.040	0.065	0.10	0.15	0.25	0.40	0.65	1.0	1.5	2.5	4.0	6.5	10	15	25	40	65	100	150	250	400	650	1,000
20–29	•	•	•	•	•	•	•	•	•	•	•	•	•	•	•	0	0	2	4	8	14	22	40	68	115	181
30–49	•	•	•	•	•	•	•	•	•	•	•	•	•	•	0	0	1	3	7	13	22	36	63	105	178	277
50–79	•	•	•	•	•	•	•	•	•	•	•	•	•	0	0	2	3	7	14	25	40	63	110	181	301	
80–129	•	•	•	•	•	•	•	•	•	•	•	•	0	0	2	4	7	14	24	42	68	105	181	297		
130–199	•	•	•	•	•	•	•	•	•	•	•	0	0	2	4	7	13	25	42	72	115	177	301	490		
200–319	•	•	•	•	•	•	•	•	•	•	0	0	2	4	8	14	22	40	68	115	181	277	471			
320–499	•	•	•	•	•	•	•	•	•	0	0	1	4	8	14	24	39	68	113	189						
500–799	•	•	•	•	•	•	•	•	0	0	2	3	7	14	25	40	63	110	181							
800–1,249	•	•	•	•	•	•	•	0	0	2	4	7	14	24	42	68	105	181								
1,250–1,999	•	•	•	•	•	•	0	0	2	4	7	13	24	40	69	110	169									
2,000–3,149	•	•	•	•	•	0	0	2	4	8	14	22	40	68	115	181										
3,150–4,999	•	•	•	•	0	0	1	4	8	14	24	38	67	111	186											
5,000–7,999	•	•	•	0	0	2	3	7	14	25	40	63	110	181												
8,000–12,499	•	•	0	0	2	4	7	14	24	42	68	105	181													
12,500–19,999	•	0	0	2	4	7	13	24	40	69	110	169														
20,000–31,499	0	0	2	4	8	14	22	40	68	115	181															
31,500–49,999	0	1	4	8	14	24	38	67	111	186																
50,000 & Over	2	3	7	14	25	40	63	110	181	301																

• Denotes that the number of sample units from the last 10 lots or batches is not sufficient for reduced inspection for this AQL. In this instance more than 10 lots or batches may be used for the calculation, provided that the lots or batches used are the most recent ones in sequence, that they have all been on normal inspection, and that none has been rejected while on original inspection.

Example 8.8 **Using the Mil. Std. 105 Single Sampling Tables.** As an example of using the Military Standard 105 tables to set up an acceptance sampling system, suppose that two parties agree upon an AQL of .001 (i.e., .1%), a lot size of 200,000, and use of General Inspection Level II. Then the code letter P is prescribed by Table 8.8. Following that code letter and $AQL = .1$ (%) into Tables 8.9 through 8.11, one finds that:

1. on normal inspection, one should sample $n = 800$ items and accept a lot if $X \leq 2$ and reject it if $X \geq 3$,
2. on tightened inspection, one should sample $n = 800$ items and accept a lot if $X \leq 1$ and reject it if $X \geq 2$, and
3. on reduced inspection, one should take a sample of $n = 315$ items and accept a lot if $X \leq 1$, reject it if $X \geq 3$, and if $X = 2$ accept the lot but return to normal inspection.

Finally, since the sample size for normal inspection is 800, 10 lots accepted on normal inspection will represent a total sample size of $10(800) = 8{,}000$. In the 8,000–12,499 row and $AQL = .1$ column of Table 8.12 is the limit number 4. So, if reduced inspection is to be used, a switch can be made from normal to reduced inspection only when four or less nonconforming items (or nonconformities) are observed in 10 consecutive samples taken on normal inspection.

To illustrate the operation of the switching rules, two hypothetical sequences of sample results under this set of plans are indicated in Tables 8.13 and 8.14.

8.3.2 Comments on the Use of Mil. Std. 105

Military Standard 105 is a most famous document. Indeed, all that some people who think they want to do acceptance sampling know about the subject is that the standard prescribes some sample sizes and acceptance numbers. It is thus important to reflect briefly on the limitations and proper use of the standard (and its derivatives).

In the first place, it cannot be emphasized strongly enough that what the standard prescribes is an acceptance sampling *system*. It does not prescribe that one should use a sample size and acceptance number somehow drawn from it on an isolated lot. The AQL protection supposedly offered by the Mil. Std. 105 system refers only to what it can be expected to do when the whole system is applied to a stream of lots. It is a common and unfortunate abuse to use the normal inspection tables in the standard as a kind of catalog of sampling plans, wrongly expecting that the AQL headings in the table have something to do with how the individual plans in the columns will perform used by themselves alone. If one wants to choose a single sampling plan to use by itself (not in a system including switching between several plans) the rational way to choose it is to apply the OC and design material presented in Section 8.1, not to unthinkingly pull something out of Mil. Std. 105 Table II-A.

Table 8.13 A Hypothetical Sequence of Sample Results Under Mil. Std. 105 (AQL = .1%, Lot Size 200,000, General Inspection Level II)

Sample	Sample Size	X	Action/Comments
1	800	0	accept lot
2	800	2	accept lot
3	800	1	accept lot
4	800	1	accept lot
5	800	2	accept lot
6	800	0	accept lot
7	800	3	reject lot
8	800	0	accept lot
9	800	2	accept lot
10	800	4	reject lot/switch to tightened (2 out of 5 rejected)
11	800	2	reject lot
12	800	1	accept lot
13	800	0	accept lot
14	800	0	accept lot
15	800	0	accept lot
16	800	1	accept lot/switch to normal (5 in a row accepted)
17	800	1	accept lot
18	800	0	accept lot
19	800	3	reject lot
20	800	3	reject lot/switch to tightened (2 out of 4 rejected)
21	800	4	reject lot
22	800	1	accept lot
23	800	1	accept lot
24	800	0	accept lot
25	800	2	reject lot
26	800	1	accept lot
27	800	0	accept lot
28	800	0	accept lot
29	800	2	reject lot
30	800	1	accept lot/cease and take corrective action (10 in a row on tightened)

Second, exactly how to describe the properties of the 105 system when applied to a sequence of lots is not clear. The system is complicated and analyses of its mathematical properties have been only partially satisfactory. In particular, nothing so simple as an "OC" will capture the dynamic behavior of the system, even assuming that lots of constant quality are submitted. The Duncan and Schilling books mentioned earlier in this section have some references in this area, and other attempts to quantify the properties of the switching system appear from

Table 8.14 Another Hypothetical Sequence of Sample Results Under Mil. Std. 105 (AQL = .1%, Lot Size 200,000, General Inspection Level II)

Sample	Sample Size	X	Action/Comments
1	800	1	accept lot
2	800	1	accept lot
3	800	0	accept lot
4	800	1	accept lot
5	800	0	accept lot
6	800	1	accept lot
7	800	0	accept lot
8	800	0	accept lot
9	800	1	accept lot
10	800	0	accept lot (no switch since total nonconforming is 5)
11	800	0	accept lot/switch to reduced (total nonconforming in last 10 samples is 4)
12	315	0	accept lot
13	315	0	accept lot
14	315	1	accept lot
15	315	0	accept lot
16	315	1	accept lot
17	315	1	accept lot
18	315	2	accept lot/switch to normal
19	800	1	accept lot
20	800	0	accept lot
21	800	0	accept lot
22	800	0	accept lot
23	800	0	accept lot
24	800	0	accept lot
25	800	0	accept lot
26	800	1	accept lot
27	800	0	accept lot
28	800	0	accept lot/switch to reduced
29	315	0	accept lot
30	800	3	reject lot/switch to normal

time to time in publications like the *Journal of Quality Technology*. But since the problem is fairly complicated, a real discussion of the matter is not in keeping with the general goals of this book and not appropriate here.

The whole notion of using a sampling inspection system based on Mil. Std. 105 outside of its original realm of application also raises significant questions. Some have pointed out that the standard both implicitly assumes and also tends to promote an adversarial relationship between a supplier and a party taking receipt of its lots. It is the pressure of inspection of the lots that is implicitly counted on to keep p (or λ) below the AQL. This inspection amounts to a kind of checking up on the supplier, who presumably has little or no other motivation to supply high-quality lots. This may or may not be an appropriate view in the context of military procurement. But it is surely misguided in a modern business context where "suppliers" and "customers" share the common interest of making good end products that will be well received by end customers and keep both in business. A modern auto company and the subcontractors that supply parts to it are well advised to *cooperate* to make the best autos possible, not try to "get away with as much as possible."

Related to this last point is modern criticism of the whole notion of an AQL. It flies in the face of modern quality philosophy and its emphasis on constant improvement, to talk in terms of an "acceptable" level of p or λ. It is true that the originators of the AQL concept were careful to define it in terms of what is acceptable "for purposes of acceptance sampling," but the criticism still has force. Some critics have gone so far as to call an AQL "a license to steal" (again emphasizing the adversarial mindset typically implicit in the application of the standard).

In all, it is clear that Military Standard 105 can be abused and misused. It certainly should not be thought of as a primary tool of modern quality control (or even of attributes acceptance sampling). But historically, it has at least provided some visibility to the notion of acceptance sampling and some awareness of what is possible in the realm of sampling inspection.

8.4 CONTINUOUS INSPECTION PLANS

The sampling inspection methods of Sections 8.1 through 8.3 operate in a framework where items of interest are broken up into lots. It is also possible to apply product-oriented sampling inspection to a production stream of individual items. This section discusses one method of **continuous inspection**, Dodge's CSP-1 (continuous sampling plan 1).

It is important to state at the outset that although the attributes control charting methods of Section 3.3 and the present continuous inspection material might both find application in the production of a single stream of items, the goals of continuous inspection and process monitoring are quite different. The intention here is to produce a kind of on-line rectification (or removal) system that takes advantage of sampling (as opposed to 100% inspection) when quality seems to be good. There is no explicit provision or expectation that information about process behavior gained from continuous inspection will be fed back and used in process-monitoring or -improvement efforts. Thus, classical continuous inspection is focused *on the product, not on the process* producing it.

Dodge's CSP-1 operates as follows on a stream of items that can be individually conforming or nonconforming. One begins by inspecting all items until i

consecutive items have been found to be conforming. Then, for a sampling fraction f (usually taken to be the reciprocal of an integer), one begins to sample and inspect only a fraction f of the product, continuing in this sampling mode until a nonconforming item is encountered. When this happens, one begins again to inspect all items, looking for i conforming items in a row, and so on.

The value i is usually called the **clearing interval** for CSP-1, and there are at least three different ways in which the sampling might be implemented. It is possible to

1. select one item for inspection at random from every group of $1/f$ consecutive items produced (this was Dodge's original suggestion),
2. systematically select every $1/f$th item (this is an inferior, but administratively cleaner possibility), or
3. select each consecutive item for inspection with probability f (this is perhaps the most effective of the three possibilities, but also administratively the most unpleasant).

Those nonconforming items located both during 100% inspection phases and also at the ends of sampling phases can be either **rectified** (repaired or replaced with items somehow known to be conforming) or **removed** (culled from the product stream).

It is hopefully evident to the reader that the second method of operating a sampling phase of CSP-1 is extremely dangerous if there is any possibility of a cyclical trend in product quality that might unfortunately have a period that is a divisor of $1/f$. (If every 10th item is nonconforming while all others are conforming, $i = 5$ and $f = .1$, the plan will quickly enter a sampling phase from which it will never emerge.) The random sampling included in the first and third possibilities, while an administrative headache, acts as insurance against being trapped unaware in this way. It is hopefully also evident that while the product stream exiting CSP-1 will look better than the source stream, this is achieved essentially through diluting or culling some of the nonconforming items, not through fundamental improvement of a production mechanism.

While CSP-1 is perhaps intuitively appealing, quantifications of how it is likely to behave in practice and what it can be expected to accomplish are less than totally satisfactory. Intuitively, the method is intended to address situations where quality varies, doing 100% inspection when quality is poor and sampling inspection when it is good. But the only completely tractable mathematics for predicting CSP-1 behavior is "perspective B constant p" mathematics (mathematics that describes quality that *doesn't vary*). Despite the fact that this is less than what might be hoped for, it is useful to at least have the constant quality operating properties of CSP-1 as a baseline, so these will be presented next.

Suppose then, that consecutive items submitted to CSP-1 are (independently) nonconforming with some probability p. It is useful to think of the plan operating in terms of **cycles**, each consisting of a single **100% inspection phase** together with the corresponding subsequent **sampling phase**. Let

T_1 = the number of items inspected on a particular 100% inspection phase

and

T_2 = the number of items represented by the subsequent sampling phase.

To begin with, it is possible to determine the mean values of T_1 and T_2 under the constant quality assumption. That is, it turns out that

Mean Length of a 100% Inspection Phase

$$u = \mathrm{E}T_1 = \frac{1-(1-p)^i}{p(1-p)^i}. \tag{8.42}$$

Further (using the simple formula for the mean of the geometric distribution), it is easy enough to argue that

Mean Length of a Sampling Phase

$$v = \mathrm{E}T_2 = \frac{1}{fp}. \tag{8.43}$$

Continuing this line of development, the mean number of items in a cycle is clearly $u + v$, while the mean number inspected in a cycle is $u + fv$. (The presumption here is that if method 2 of implementing the sampling is used, the 100% inspection begins after the group of $1/f$ from which the first nonconforming item on the sampling phase is selected.) So if one considers a large number of cycles, a sensible measure of inspection load generated by CSP-1 is an average fraction inspected (AFI) defined by

Average Fraction Inspected for CSP-1

$$AFI(p) = \frac{u+fv}{u+v}, \tag{8.44}$$

where u and v are the means (and functions of p) defined in displays (8.42) and (8.43).

It is even possible to find an AOQ for CSP-1 under constant quality assumptions, a measure of the average fraction of items leaving the scheme that are nonconforming. In a given cycle, only those items that are uninspected in the sampling phase have a chance of both being nonconforming and continuing in the product stream undetected. On average, there are $p(1-f)v$ of those in a cycle. Since the mean cycle length is $u+v$, for the case of rectifying inspection (where any nonconforming item identified is repaired or replaced with a conforming one), a sensible AOQ figure is

$$AOQ(p) = \frac{p(1-f)v}{u+v} = p(1 - AFI(p)), \tag{8.45}$$

Average Outgoing Quality for CSP-1 with Rectification

where again u and v are defined in displays (8.42) and (8.43). Removal inspection calls for a modification of formula (8.45) to take account of the fact that while the mean cycle length is $u + v$, less than $T_1 + T_2$ items from a given cycle will end up in the product stream. The appropriate correction to formula (8.45) for the case of removal inspection is

$$AOQ(p) = \frac{p(1-f)v}{u + v - (1-p)^{-i}}. \tag{8.46}$$

Average Outgoing Quality for CSP-1 with Removal

Example 8.9 **Constant Quality Properties of a CSP-1 Plan.** To illustrate the use of the formulas (8.42) through (8.46), consider a case where a clearing interval of $i = 100$ is employed along with a sampling fraction of $f = .02$ in CSP-1. Then, for example, supposing that $p = .01$, from equation (8.42), the mean length of a 100% inspection part of a cycle is

$$u = \mathrm{E}T_1 = \frac{1 - (1 - .01)^{100}}{.01(1 - .01)^{100}} = 173.2,$$

and from equation (8.43) the mean number of items represented in a sampling phase is

$$v = \mathrm{E}T_2 = \frac{1}{.02(.01)} = 5{,}000.$$

These in turn imply (using equation (8.44)) that when $p = .01$ the average fraction of the product stream that will be inspected is

$$AFI(.01) = \frac{173.2 + .02(5{,}000)}{173.2 + 5{,}000} = .0528,$$

and the corresponding AOQ values are (from equations (8.45) and (8.46)) for rectifying inspection

$$AOQ(.01) = .01(1 - .0528) = .00947$$

and for removal inspection

Table 8.15 Some Properties of CSP-1 with $i = 100$ and $f = .02$

p	u	v	$AFI(p)$	**Rectification** $AOQ(p)$	**Removal** $AOQ(p)$
.00	100	∞	0	0	0
.01	173.2	5,000	.0528	.00947	.00948
.02	327.0	2,500	.1334	.01733	.01738
.03	667.6	1,667	.3003	.02010	.02118
.04	1,457	1,250	.5475	.01810	.01851
.05	3,358	1,000	.7751	.01124	.01170
.06	8,094	833.3	.9085	.00549	.00581

$$AOQ(.01) = \frac{.01(1-.02)5{,}000}{173.2 + 5{,}000 - (1-.01)^{-100}} = .00948\,.$$

As expected (on the grounds that removal inspection does not add back in conforming items for the nonconforming items identified, and therefore leaves the product stream relatively poorer in terms of conforming items) the removal inspection AOQ is slightly more than that for rectifying inspection.

Table 8.15 shows a few additional values of p and corresponding values of u, v, $AFI(p)$, and both $AOQ(p)$'s. Notice that as was the case for rectifying inspection associated with single sampling of lots, the AOQ's are small for both small and large p, peaking for a somewhat "moderate" value of p.

It is an interesting fact that the maximum of $AOQ(p)$ defined in equation (8.45) can be found in more or less explicit terms. (That is, one can find a kind of AOQL for rectifying continuous inspection.) Differentiating $AOQ(p)$ as a function of p and setting the derivative equal to 0, it follows that the maximum AOQ for the rectifying case of continuous inspection occurs for a value of p solving the equation

Equation for p Maximizing AOQ Under Rectifying CSP-1

$$(i+1)p - 1 = \left(\frac{1}{f} - 1\right)(1-p)^{i+1}\,. \qquad (8.47)$$

This equation is easily solved (for example, on a programmable calculator with a root-finding algorithm) to yield a value, say, p^*. Then the fact that p^* solves equation (8.47) can be used to show that

$$AOQL = \max_{p} AOQ(p) = \frac{(i+1)p^{*} - 1}{i}. \tag{8.48}$$

Average Outgoing Quality Limit for Rectifying CSP-1

So by solving equation (8.47) for p and plugging the solution into the right side of equation (8.48), one can arrive at an upper bound on the fraction nonconforming in the product stream after rectifying continuous inspection, *assuming that quality is constant.*

The fact that the upper bound on the CSP-1 AOQ just discussed holds only under constant quality conditions makes it less than completely satisfying. But the problem is that what CSP-1 will do when quality varies depends upon exactly *how* quality varies. About the best one can do in terms of a minimal guarantee of effectiveness regardless of how p might vary comes from a worst-case analysis supposing that the third method of selecting items to inspect (when on a sampling phase) is used. It turns out that regardless of how p might change, if one selects each consecutive item for inspection with probability f, the long-run fraction nonconforming surviving in the product stream is bounded above by

$$\frac{1-f}{if+1}. \tag{8.49}$$

Universal AOQL for Rectifying CSP-1 Without Constant p Assumption (Method 3 Sampling)

Unfortunately, while this bound is universal, it is generally so large as to be useless for small f and moderate i. Either the sampling fraction or the clearing interval must be very large to guarantee a small fraction nonconforming in the final product stream if p can vary without limits.

Example 8.9 continued. Consider both the constant quality and universal bounds on the average fraction nonconforming in the final product stream after rectifying continuous inspection with $i = 100$ and $f = .02$. The reader can verify that $p^{*} = .0307$ is (at least approximately) a solution to

$$(100+1)p - 1 = \left(\frac{1}{.02} - 1\right)(1-p)^{100+1}$$

suggested in display (8.47). Then plugging this value of p into equation (8.48) one has that under constant quality conditions,

$$AOQL = \max_{p} AOQ(p) = \frac{(100+1)(.0307) - 1}{100} = .021 .$$

By virtue of the on-line rectification, no more than 2.1% of the final product stream will be nonconforming.

If, however, one allows the possibility that quality varies in the product stream in the worst conceivable way, display (8.49) can only provide an upper bound of

$$\frac{1 - .02}{100(.02) + 1} = .327$$

for the fraction of nonconforming items in the rectified product stream exiting CSP-1. (And even this bound requires that sampling be done by selecting consecutive items with probability .02 for inspection.)

This discussion has dealt specifically with CSP-1. Many other plans of this general type have been suggested (some, for example, have more than just the two inspection rates 1 and f). But they all share the flavor and limitations of CSP-1. The bottom line on this topic seems to your authors to be that continuous inspection plans

1. promise substantial administrative headaches in terms of (randomly) fluctuating inspection load,
2. cannot really be counted on to effectively screen out nonconforming items in all circumstances unless f and/or i are quite large, and
3. are not tools for directly addressing root causes of poor quality and improving processes.

It therefore seems that while there may be places where their application makes sense, they cannot be recommended as standard tools for modern quality assurance. The fact that they are more or less standard topics of discussion in quality control texts seems more a matter of historical precedent and the need to prepare students to communicate with those already in the quality control field than a serious mandate for widespread application.

8.5 THE LEGITIMATE ROLE OF INSPECTION IN MODERN STATISTICAL QUALITY ASSURANCE

This chapter has far more than its share of caveats, warnings, and disclaimers. To some readers, this may seem so disconcerting that the question "So does this material have *any* relevance?" could arise. And indeed, there are some in the modern quality movement who claim that the general topic of product-oriented inspection is one that has outlived its usefulness and ought to disappear from both teaching and practice. That view is, however, much too extreme. Limited though it may be, the material of this chapter (and more broadly, product-oriented inspection) has its legitimate place in modern quality assurance. The purpose of this last section of this chapter is thus to lay out what product-oriented inspection can and should be used to do, and in particular to define the proper role of sampling inspection.

To begin with, while efforts to improve processes and thereby influence the quality of products they produce must be primary, it is simply an error in logic to infer that product-oriented inspection will never be necessary. Time and engineering resources are finite. Particularly in leading-edge technologies, it will never be possible to have all processes producing only perfect products. For example, 60% yields are considered excellent for some new-generation electronics production processes. In such contexts, it is clear that before installing IC chips on circuit boards, there *must be* product-oriented sorting of good chips from bad ones!

Product-oriented inspection provides means of grading items, culling/removing ones that are fatally flawed, and/or rectifying those that can be redeemed through an economically feasible amount of effort. These are legitimate activities. They don't directly support process-improvement efforts and should not be used as a substitute for them. But they can be both necessary and economically wise.

After coming to terms with the fact that product-oriented inspection can make sense, the question arises as to the extent that it can be accomplished on a **sampling** basis. It is extremely common for engineers aware of statistical methods to have some vague notion that inspection costs might be reduced through the use of sampling. But here again, careful thought is required. Whether or not *sampling* inspection makes sense depends upon what one knows about the quality of a lot or stream of items. In many contexts, if quality is somehow known and constant, there is no sensible role for further sampling. The paper "A Discussion of All-or-None Inspection Policies," by Vander Wiel and Vardeman which appeared in *Technometrics* in February 1994 has a thorough treatment of this point, and the following is a brief summary of some of the material in that paper.

Suppose a lot or stream of items is such that each item is (independently) conforming or nonconforming with respective probabilities $(1-p)$ and p. Suppose further that inspection errors are possible, and a conforming item fails inspection with probability w_G and passes with probability $(1-w_G)$, while a nonconforming item passes with probability w_D and fails with probability $(1-w_D)$. (Taking w_G and w_D to both be 0 covers the situation where inspection errors are not thought to be possible.) Finally, suppose that costs accrue *on a per-item basis* according to the structure outlined in Table 8.16.

Vander Wiel and Vardeman show that under this kind of model one should either inspect *all* items or *no* items (not part of the items). If one uses the abbreviation

$$K = (1-w_D)(k_{DU}-k_{DF}) + w_D(k_{DU}-k_{DP}) + w_G k_{GF}, \tag{8.50}$$

it turns out that if

$$K \le 0 \quad \text{or} \quad p \le \frac{k_I + w_G k_{GF}}{K}, \tag{8.51}$$

Table 8.16 Per-Item Costs in a Generic Inspection Problem

	Item Conforming	Item Nonconforming
Item Uninspected	0	k_{DU}
Item Passes Inspection	k_I	$k_I + k_{DP}$
Item Fails Inspection	$k_I + k_{GF}$	$k_I + k_{DF}$

one should inspect *no* items, while in other cases *all* items should be inspected. This is intuitively sensible. If quality is known to be good (p is small), inspection is a waste of resources. If it is known to be bad (p is large), it is better to suffer inspection costs than later take the consequences of passing nonconforming items. But there is no reason to inspect a portion of the (*a priori* indistinguishable as far as this analysis is concerned) items, and it should indeed be the economic consequences of the various eventualities that dictate which of "inspect all" or "inspect none" is most reasonable. In broad terms, this is exactly what the criterion specified in displays (8.50) and (8.51) prescribes.

So then, what is the proper role of the acceptance *sampling* material of this chapter? Where does it fit? The key to understanding this is to take a step back and remember what one must know in order to apply the kind of criterion specified in displays (8.50) and (8.51). Quality must be assumed to be constant and one has to know the relevant p. If quality varies and/or is unknown (p is unknown), the criterion cannot be implemented. And it is precisely then that acceptance sampling comes into its own. Sampling inspection (including acceptance sampling) is not fundamentally about taking remedial measures on items of fixed and known quality. It is, instead, about gathering information and making (admittedly fuzzy, due to sampling variation) economical determinations about the likely state of quality, in contexts where that quality varies/is unknown.

Indeed, where the costs represented in Table 8.16 can be evaluated (and, if relevant, the probabilities of inspection errors are also known), the economic criterion specified by displays (8.50) and (8.51) gives a rational way of choosing where the "drop" in an OC curve should occur. The value $(k_I + w_G k_{GF})/K$ is a kind of breakeven point between no inspection and 100% inspection in terms of the quality parameter p. Thus where p is not known, it makes sense to choose an acceptance sampling plan whose OC drops near this value. As discussed in Section 8.1, for the fraction nonconforming attributes sampling problem, this means that in rough terms one wants n and c such that

$$\frac{c}{n} \approx \frac{k_{\mathrm{I}} + w_{\mathrm{G}} k_{\mathrm{GF}}}{K} \,.$$

So it seems that product-oriented inspection has its place, and doing it on a sampling basis for information-gathering purposes is potentially useful when quality varies/is unknown. But a final assault on the usefulness of sampling inspection could possibly be mounted on the basis that in a modern process-oriented quality climate, one should *always* "know" the quality of items one is receiving (on the basis of process monitoring information collected upstream and passed along a production line).

There are two strong points to be made counter to this last argument against sampling inspection. In the first place, the argument is to some degree a matter only of semantics. If one "knows" lot quality from information gathered upstream and passed along with a lot of product, one may as well admit that the upstream information is serving two functions. It is simultaneously being used for upstream process-monitoring purposes and for acceptance-sampling purposes downstream. And the fact that it is being used for two purposes does not make the considerations discussed in this chapter any less relevant.

In the second place, saying that one "knows" current lot quality from upstream process-monitoring records is a fairly risky position to take. For one thing, it ignores the possibility that something unpleasant might happen to product after an upstream data collection point. Because a lot of fragile items is perfect when it leaves a supplier's facility on the East Coast does not necessarily imply that it is perfect when it arrives on a loading dock on the West Coast. And besides, the position ignores the possibility of human error. Sometimes one is well advised to do a small amount of sampling inspection simply to guard against gross blunders and the unexpected.

So finally, it can be said again (as it was at the beginning of this chapter) that it is appropriate that this book majors on process-oriented tools. But the product-oriented sampling inspection discussed in this chapter cannot be forgotten completely. In its proper place, it too is a valuable tool of modern quality assurance. More discussion of this point and some important references can be found in the article "The Legitimate Role of Inspection in Modern SQC" by Vardeman, which appeared in *The American Statistician* in 1986.

8.6 CHAPTER SUMMARY

This chapter has considered product-oriented inspection done on a sampling basis. The first section of the chapter introduced attributes single sampling plans, their properties, and their design/choice. Section 8.2 discussed the variables single sampling plans based on normal distributions that are sometimes recommended as alternatives to fraction nonconforming attributes plans. The third section treated the Military Standard 105 attributes sampling system, concentrating on the scheme's switching rules and the use of its basic single sampling tables. Dodge's so-called CSP-1 continuous inspection method was presented in Section 8.4 and some of its properties were discussed. The final section of the chapter discussed the proper place of product-oriented inspection in general and sampling inspection in particular in the modern quality assurance world.

8.7 CHAPTER 8 EXERCISES

8.1. Briefly discuss the roles of (1) acceptance sampling, (2) process control/monitoring, and (3) experimental design and analysis in a modern quality assurance program.

8.2. The AQL concept is sometimes criticized as amounting to "a license to steal." Give an argument for this interpretation.

8.3. It is often said that to simply take a sample size and acceptance number out of Mil. Std. 105 for use on a single isolated lot is an abuse of the standard. Give an argument for this point of view.

8.4. What are some standard modern arguments against the use of 100% inspection as the primary tool of guaranteeing product quality?

8.5. What is the "all or none" argument in product inspection circles? What are its practical limitations?

8.6. In what context is acceptance sampling an appropriate tool in modern quality assurance?

8.7. Another engineer tells your (common) boss, "Company X is consistently shipping us product that is 5% defective. We could live with 1%, but the 5% figure is causing serious problems on our assembly line." Your boss says, "Well, start using acceptance sampling on shipments from company X!" What would you say and why?

8.8. What are legitimate advantages and disadvantages of variables acceptance sampling in comparison to attributes acceptance sampling?

8.9. What do you have to say to a fellow engineer who says "We ought to look into acceptance sampling as a way of reducing our inspection costs around here"?

8.10. What do you say to a colleague who upon return (quite excited) from an SPC seminar says "We must cease dependence on mass inspection!!! Let's abandon final inspection on the $40,000 disk drives we're manufacturing!"?

8.11. Standard "unknown σ" variables acceptance sampling based on normal distributions essentially attempts to treat equally all (μ, σ) pairs with the same fraction outside of specifications. Describe a situation where this is not sensible, and among (μ, σ) pairs with a given fraction below some lower specification, large σ is to be preferred by a consumer.

8.12. On what basis is it sensible to criticize the relevance of the calculations usually employed to characterize the performance of continuous sampling plans?

8.13. An engineer and manager are convinced of the need for an attributes acceptance sampling scheme for shipments they have contracted to receive. They want an OC curve that drops in the vicinity of 6% nonconforming and want to be very sure (they want at least .95 probability) of rejecting lots with 8% or more nonconforming. The lot sizes are very large.

a. Find a sample size n and acceptance number c that meet the OC goals.

b. For the plan found in (a), find AOQ values corresponding to 4%, 6%, and 8% nonconforming. (Acceptance sampling is followed by 100% inspection of rejected lots, and if a nonconforming item is found it is rectified.)

8.14. Suppose that Mil. Std. 105 is to be employed in a situation where lots of $N = 4{,}000$ will be submitted, and a producer and consumer have agreed upon an AQL of .015 (1.5%) and use of general inspection level II.

a. Find single sampling sample sizes and acceptance numbers for this situation under normal, tightened, and reduced inspection.

b. Let the (single digit) values below represent the numbers of nonconforming items found in successive samples while using the standard in the situation described above. Circle values corresponding to rejected lots and indicate by slash marks and English notes switches between normal, tightened, and reduced plans.

6 8 9 4 3 8 2 1 0 3 4 5 6 4 8 3 2 4 1 5 6 5 0 1 2 0 1 0 2 1 3 4 4 8 9 6 5 7 4 3 8 5 3 7 2 1

c. Suppose that in fact "rejection" of a lot means that the unsampled portion is subjected to 100% inspection. If lot quality is at the AQL ($p = .015$, i.e., 1.5%), about how many items are inspected per lot (on average) when on normal inspection?

8.15. On what grounds might the routine civilian use of Mil. Std. 105 be criticized as leading to counterproductive supplier–consumer relationships?

8.16. Suppose a company purchases sheets of sandpaper in large lots. The lengths of sheets in a given lot are approximately normally distributed with standard deviation $\sigma = .06$ inch but the lot mean lengths vary. Sheets with length less than 11.0 inches are considered nonconforming.

a. Choose a sample size n and a number M so that rejecting a lot if a random sample of n sheets produces $\bar{x} < M$ is a variables acceptance sampling procedure with $Pa_1 = .9$ for $p_1 = .01$ and with $Pa_2 = .1$ for $p_2 = .05$. (p is the fraction of nonconforming sheets in the lot.)

b. What sample size is required for an attributes single sampling plan comparable to the variables plan above (i.e., having $Pa_1 = .9$ for $p_1 = .01$ and with $Pa_2 = .1$ for $p_2 = .05$)? (Use equation (8.21).)

c. The correct answer to (a) is substantially smaller than the correct answer to (b). At what price is this large reduction purchased? (What are possible weaknesses of the plan in (a)? How might the plan in (a) be more expensive to use and "fragile"?)

d. If the lot size is $N = 10,000$ find the AOQ and ATI for $p = .05$ if the plan from (b) is employed in rectifying inspection.

e. Suppose that (for better or worse) henceforth incoming inspection of these lots will be done according to Mil. Std. 105 and that $AQL = .01$ (1%) and general inspection level II have been negotiated with the supplier. The option for reduced inspection will be employed. Describe (including sample sizes and acceptance numbers) exactly how incoming inspection will proceed.

8.17. In an engineering research project conducted in cooperation with a heavy equipment manufacturer, the inside diameters (x) of a tubular steel part produced on a company lathe were monitored. Engineering specifications on the inside diameter were 1.1795 inches $\pm$ .0005 inch. Suppose that as far as a consumer is concerned, it is sensible to think about production runs of these parts as having inside diameters that are normally distributed with some (unknown) mean μ and standard deviation $\sigma = .0002$. If the consumer does variables acceptance sampling by sampling $n = 4$ items, measuring the inside diameters and rejecting a lot of parts if $\bar{x} < 1.1792$ or if $\bar{x} > 1.1798$, find the proportion nonconforming and the probability of lot acceptance, first if $\mu = 1.1795$ and then if $\mu = 1.1798$.

8.18. Consider an attributes single sampling scheme for fraction nonconforming. Let the sample size be $n = 100$ and the acceptance number be $c = 1$. Adopt perspective B.

a. Find *Pa* for such a plan, first if $p = .005$ and then if $p = .02$.

b. What are the AOQ and ATI of the plan if in fact $p = .005$ and lots of $N = 1{,}000$ items are subjected to rectifying inspection?

8.19. Refer to problem (8.18). For known σ variables sampling contexts, some texts resort to the use of a graphical device called a nomograph to set up acceptance sampling plans. In fact, as equations (8.38) and (8.39) show, there is no need to do this. To illustrate, consider a situation where there is a single upper specification $U = 450$, a normal distribution with $\sigma = 10$ is a reasonable description of measurements in a lot, and one wishes to find a sample size n and critical number M, such that if $\bar{x} > M$, the lot is rejected.

a. Find n and M such that the resulting plan has OC points matching those in part (a) of problem (8.18).

b. Is sample size here smaller or larger than the corresponding sample size from problem (8.18)?

8.20. Consider an attributes single sampling scenario.

a. Find type A *Pa* values for $p = .05, .1, .5, .9$, and .95, lot size $N = 20$, sample size $n = 5$, and $c = 0$ for a fraction nonconforming situation.

b. Find type A *Pa* values for $\lambda = .05, .1, .5, .9$, and .95, lot size $N = 20$, sample size $k = 5$, and $c = 0$ for a mean nonconformities per unit situation.

c. Redo (a) with $c = 1$.
d. Redo (b) with $c = 1$.

8.21. Consider an attributes single sampling scenario.
a. Find type B probability of acceptance values for $n = 5, c = 0$, and $p =$ .001, .01, .5, .9, and .95 for a fraction nonconforming situation.
b. Find type B probability of acceptance values for $k = 5, c = 0$, and $\lambda =$ .001, .01, .5, .9, and .95 for a mean nonconformities per unit situation.
c. Redo (a) with $c = 1$ and then $c = 2$.
d. Redo (b) with $c = 1$ and then $c = 2$.

8.22. Consider an attributes single sampling plan for fraction nonconforming with $n = 200$ and $c = 2$. Suppose the lot size is $N = 10,000$.
a. Find OC values for $p = .01, .1, .3, .5, .7$, and .95.
b. Find AOQ and ATI values for the values of p in (a) for a rectifying inspection scheme.
c. What is the AOQL?

8.23. Consider variables single sampling based on a normally distributed variable with a single (lower) specification for individual values, $L = 1.00$. Further, suppose that it is desirable to have $Pa_1 = .95$ for $p_1 = .03$ and $Pa_2 = .10$ for $p_2 = .10$.
a. Suppose $\sigma = .015$. Use formulas (8.38) and (8.39) and find the appropriate sample size, Δ_1, and a number M, such that if $\bar{x} < M$, the lot should be rejected.
b. For your plan from (a), find OC points (p, Pa) for $\mu = 1.025, 1.020, 1.015, 1.010$, and 1.005. (Use formulas (8.25) and (8.31).)
c. Suppose σ is unknown. Use formulas (8.40) and (8.41) and find the appropriate sample size and rejection criterion needed to meet the OC criteria. (Say explicitly for which combinations of $\bar{x}$ and s a lot should be rejected.)
d. For your plan from (c), use formulas (8.25) and (8.37) to find OC points (p, Pa) corresponding to values $-4.0, -3.5, -3.0, -2.5, -2.0, -1.5$, and -1.0 of the quantity $z_L = (L - \mu)/\sigma$.

8.24. Alternator-Voltage Regulator. Gitlow (in the 1984 volume *Proceedings—Case Study Seminar—Dr. Deming's Management Methods: How Are They Being Implemented in the U.S. and Abroad*) discussed a scenario like the following. The alternator-voltage regulator combination in an auto must provide a voltage of 13.2 volts to keep the battery's charge at 12 volts. Suppose engineers set a lower specification of 13.19 volts and an upper specification of 13.21 volts for the output of the alternator-voltage regulator combination, and that experience has shown that output voltages for a lot of assemblies of this type may be described using a normal distribution with $\sigma = .003$ volts. Find a variables acceptance sampling plan such that the acceptance probability is 95% for a lot with 3% of the assemblies outside specifications and the acceptance probability is 10% when 10% of the assemblies are outside specifications.

8.25. Refer to the **Alternator-Voltage Regulator** case in problem (8.24).

a. Find n and c for a fraction nonconforming attributes single sampling plan that meets the OC requirements set in problem (8.24). (Use the method of equations (8.19) through (8.21).)

b. Based on the information given in problem (8.24), why is perspective B appropriate?

c. Is the n in (a) smaller or larger than the one in problem (8.24)? Explain why, despite its apparent advantage, the variables plan needs to be used only with caution.

8.26. This problem concerns acceptance sampling for very small fractions nonconforming.

a. Find an attributes single sampling plan for fraction nonconforming (find n and c) that for $p_1 = 10^{-6}$ has $Pa_1 = .95$ and for $p_2 = 3 \times 10^{-6}$ has $Pa_2 = .10$. (Use the method of equations (8.19) through (8.21).)

b. In contrast to the attributes plan from (a), make use of the fact that the upper 10^{-6} point of the standard normal distribution is about 4.753, while the upper 3×10^{-6} point is about 4.526, to find the sample size required for a known σ single limit variables acceptance sampling plan meeting the OC criteria set in (a).

c. What is a practical weakness of the calculation in (b)?

8.27. A variables acceptance sampling scheme is used in the screening of incoming lots of 1000 insecticide containers for net contents. Any container with less than 32.0 ounces of insecticide is nonconforming. Suppose that fill levels in any lot are roughly normally distributed with standard deviation $\sigma = .1$ ounce, samples of $n = 4$ containers are checked and a lot is rejected if $\bar{x} < 32.1$ ounces.

a. If the mean content for a lot (μ) is such that $p = .05$ (5% of containers have less than 32.0 ounces of insecticide), what is Pa?

b. Suppose that rejection of a lot will be followed by 100% inspection and rectification of any underfilled containers. Find the AOQ and ATI for $p = .05$. (Equation (8.9) can be used and assume that the approximation following equation (8.11) is valid.)

c. It is possible to verify that if $p = .10$, Pa is about .29. Use this fact and your answer in (a) to find an attributes acceptance sampling plan for fraction nonconforming (find n and c) that has the same OC as the variables plan for $p = .05$ and $p = .10$. (Use the method of equations (8.19) through (8.21).)

8.28. Suppose an engineer wants a high probability (say, $Pa_1 = .90$) of accepting lots generated by a process with a small mean number (λ_1) of nonconformities per unit, and a low probability (say $Pa_2 = .05$) of accepting lots generated by a process with a large mean number (λ_2) of nonconformities per unit.

a. If $\lambda_1 = 10^{-4}$ and $\lambda_2 = 2 \times 10^{-4}$, use expression (8.19), Table A.7 and expression (8.21) to find c and k satisfying the engineer's criteria.

b. If $\lambda_1 = 10^{-4}$ and $\lambda_2 = 1.5\times10^{-4}$, use the approximation (8.20) and equation (8.21) to find c and k satisfying the engineer's criteria.

8.29. If the mean number of nonconformities per unit is .01, management and engineering decision-makers want an acceptance probability of .99. However, if the mean number of nonconformities is .1, it is desirable to have an acceptance probability of only 5%. Find an attributes acceptance sampling plan (find k and c) for this scenario.

8.30. Consider Dodge's Continuous Sampling Plan 1 (CSP-1). Suppose that in the sampling portion of a cycle, 1.5% of the items are sampled and when in the 100% inspection phase of a cycle, 50 conforming items in a row are required before 100% inspection is stopped.

a. Make AFI and AOQ plots for this plan if nonconforming items are rectified. (Compute for at least the values $p = 0, .01, .05, .1, .2, .3, .4, .5, .6, .7, .8, .9, .95, .99$, and 1.0.)

b. Find the AOQL. Is this "AOQL" appropriate/relevant if the true proportion nonconforming varies? Why or why not? If not, what is a legitimate upper bound on the average outgoing quality?

8.31. For the CSP-1 plan in problem (8.30), suppose that inspected nonconforming items are culled/removed from the product stream.

a. Plot the AOQ curve. (Compute for the same values of p as in problem (8.30).)

b. Find the AOQL. Compare this number to the value found in part (b) of problem (8.30). Which AOQL is better? In practical terms, what does it cost to produce this better value?

8.32. Suppose an engineer has in mind an AOQL for a CSP-1 plan. He also knows how often he will sample during a sampling phase (he has chosen f). The clearing interval (i) is then to be chosen.

a. Solve equation (8.48) for p^* to give the fraction nonconforming producing the AOQL (for a given clearing interval). Substitute this expression for p^* into equation (8.47) in place of p. (This gives a relationship between $AOQL, f$, and i.)

b. Suppose the desired AOQL is .02 and the sampling fraction f is to be .015. Use the equation in (a) (and trial and error if necessary) to find an appropriate clearing interval, i.

c. Suppose that using your plan in (b), the proportion nonconforming varies. What is a credible upper bound on the real AOQL?

8.33. Refer to problem (8.32), particularly part (a).

a. Set up a CSP-1 plan if it is desirable to sample 1 out of every 25 items in a sampling phase and have a resulting AOQL (assuming constant proportion of nonconforming items) of 1%. (Find i.)

b. Set up another CSP-1 plan if it is desirable to sample one out of every six items in a sampling phase and have a resulting AOQL (assuming constant proportion of nonconforming items) of 1%. (Find i.)

c. Find $AFI(p)$ for the plans of (a) and (b). (Compute for at least the values $p = 0, .005, .01, .015, .02, .05, .1, .5, .9, .95, .98, .985, .99, .995$, and 1.0.)

d. Contrast the predicted behaviors of the two plans in (a) and (b) (that share a common AOQL).

8.34. Suppose a clearing interval of $i = 50$ items and a sampling fraction of $f = .015$ are chosen for a CSP-1 continuous inspection scheme.

a. Find the p^* that solves equation (8.47) and the corresponding AOQL.

b. If the proportion nonconforming in a stream of lots is the value p^* that solves equation (8.47), find $E(T_1)$, $E(T_2)$, and AFI.

c. Find a legitimate upper bound for the long-run fraction nonconforming surviving in the product stream. What is practically appealing about this bound and what is not so appealing?

8.35. Answer the questions posed in problem (8.34) for $i = 100$ and $f = .03$. Comment on the predicted differences in operation of the two plans from this problem and problem (8.34).

8.36. Let $k_I = .5$, $k_{DU} = k_{DP} = 10.0$, $k_{DF} = 2.0$, and $k_{GF} = 3.0$ be per-item dollar costs in a generic inspection problem as specified in Table 8.16. Suppose that inspection methods are such that a conforming item passes inspection with probability .95 and a nonconforming item fails inspection with probability .99. If the proportion of conforming items in a product stream is known to be .99, should all items be inspected or no items be inspected? Why?

8.37. Refer to problem (8.36). Suppose the proportion nonconforming for a product stream generating a series of lots is unknown and/or varies unpredictably, and that the cost figures from problem (8.36) are appropriate.

a. Is the "all-or-none" conclusion in force? Why or why not?

b. If one were inclined to develop an acceptance sampling plan for this scenario, near what value of p should one try to place the drop of the OC curve? Explain.

8.38. In some acceptance sampling circles, the terms "consumer's risk" and "producer's risk" are used. The producer's risk is the probability of rejecting a lot that is at an appealing quality level (p or λ). The consumer's risk is the probability of accepting a lot that is at an unappealing quality level (p or λ). Consider problem (8.16). If the producer wanted to be sure lots with at most 1% nonconforming were accepted and consumers desired to have a small probability of accepting lots with 5% or more nonconforming, what are the consumer's risk and producer's risk in this problem?

CHAPTER 9

The Total Quality Management Environment

Statistical methods for quality assurance are not applied in a vacuum. Real people working from particular philosophical bases, in organizations with their own peculiar cultures, make use of tools like those presented in this book. Most often, the environment in which quality tools are used has as much (or more) to do with the success or failure of quality assurance efforts as the effectiveness of the particular methods employed. This book is primarily about statistical tools. Nevertheless, an effort has been made to interweave comments about the best of modern quality philosophy and culture with discussion of the statistical methods, and so to place the tools into their proper contexts. This last brief chapter is meant as a kind of philosophical summary, reviewing in one place the elements of popular modern quality culture, and providing some comments on the positives and potential negatives of the current climate.

9.1 TOTAL QUALITY MANAGEMENT

The most popular and common terminology associated with modern quality philosophy and culture is **Total Quality Management (TQM)**. This name is meant to convey the notion that in a world economy, successful organizations will *manage* the *total*ity of what they do with a view toward producing *quality* work. While TQM is actively promoted these days as appropriate for application in areas as diverse as manufacturing, education, and government, exactly what ideas and practices make it up is presently not well defined. It seems that in the views of some, essentially every potentially useful management and engineering paradigm is part of TQM. But that kind of very broad view is not really useful for thinking critically about the current quality scene. So here the discussion will be limited to the matters listed in Table 9.1, that seem to your authors to come up most frequently when people discuss TQM. Our method of exposition will be to first simply expand briefly upon the items listed in the table and to then offer some critique of their unenlightened application.

9.1.1 What Are the Elements of TQM?

TQM takes to heart the old saying "the customer is king." The end customer for a physical product or a service is viewed as the final consideration in corporate decisions. A variety of communication tools and techniques (including "quality function deployment") are called upon to ensure that the ultimately important "voice of the customer" is heard everywhere in a business, and throughout all phases of the production of a product or service, from the earliest stages of product development through the in-field servicing of goods previously manufactured.

But TQM's **customer orientation** also goes beyond this obvious meaning. In theory, it extends to every facet of an organization's activity. People are taught to view themselves as having internal "vendors" that pass work to them and "customers" in their own organizations to whom they pass work. Effective communication both up and down the workflow network is emphasized, and every member of an organization is enjoined to do his or her work in a way that makes the jobs of those who must "consume" his or her internal "products" as easy as possible.

Table 9.1 Elements of TQM Emphasis

1. Customer Focus
2. Process/System Orientation
3. Continuous Improvement
4. Self-Assessment and Benchmarking
5. Change to Flat Organizations "Without Barriers"
6. "Empowered" People/Teams and Employee Involvement
7. Management (and Others') Commitment (to TQM)
8. Appreciation/Understanding of Variability
9. Quality "Gurus"

The TQM emphasis on satisfying internal customers is strongly related to its focus on **processes and systems**. The theory is that only by concentrating on understanding and improving the processes by which things are done is there any hope of improving an organization's efficiency and ultimate product(s). Much of the corporate TQM training that goes on these days concerns methods for the analysis and improvement of work processes. These tools find application not only to rather concrete physical processes that, for example, are used to assemble an automobile, but also to less obvious processes like that used to bill a corporate customer for goods delivered.

TQMers carry out their process analysis in an intellectual framework that sees an organization's many processes fitting together in a large system. The billing process needs to mesh with various production processes, which need to mesh with the product-development process, which needs to mesh with the sales process, and so on. There is much planning and communication that needs to go on to see that these work together in harmony within an organization. But there is also recognition that *other* organizations, external suppliers and customers, need to be seen as part of "the system." So, the reasoning goes, a company's products can be only as good as the raw materials with which it works. And there is thus an emphasis in TQM on involving a broader and broader "superorganization" (our terminology) in process- and system-improvement efforts.

In the TQM framework, improvement is a job that is never finished. This work of **continual improvement** is typically aided by statistical and logical problem-solving tools. The dominant view is that whatever is the best possible corporate performance today will be woefully inadequate tomorrow and that an organization must constantly be about improving every facet of everything it does. People are now used to seeing order-of-magnitude improvements in computer performance produced every few years. To the TQM mind, such habitual quantum improvement in effectiveness "ought" to be realized in all dimensions of corporate life in all business sectors. Some of the motivation for this view is pragmatic, related to the notion that "our" competition isn't standing still, and if "we" don't improve, "they" will grab our market share. But in many cases, the view is more philosophic, amounting to a kind of modern corporate moral or ethic, an idea of what is "right" in a kind of evolutionary, progressive environment.

In the effort to be continually better, it is held to be important to know what the "best-in-class" practices are for a given business sector or activity. Hence, TQM circles also have their strong proponents of **benchmarking activities.** The emphasis here is on finding out how an organization's techniques compare to the best in the world. Where an organization is behind, every effort is made to quickly emulate the leader's performance, and where an organization's methodology is "state of the art," opportunities for yet another quantum improvement are to be considered.

It is rather standard TQM doctrine that the approach can really only be effective in organizations that are appropriately structured and properly unified in their acceptance of the viewpoint. Hence, there is a strong emphasis in the movement on **changing corporate cultures and structures** to enable this effectiveness. Proponents of TQM simultaneously emphasize the importance of involving all corporate citizens in TQM activities, beginning with the highest levels of management, and at the same time reducing the number of layers between the "top" and "bottom" of an organization, making it more egalitarian. Cross-functional project teams composed of employees from various levels of an organization (operating in consensus-building modes, with real authority not only to suggest changes but to see that they are implemented, and drawing on the various kinds of wisdom resident in the organization about how to make progress) are standard TQM fare. And one of the corporate evils most loudly condemned is the human tendency to create "little empires" inside an organization that in fact compete with each other, rather than cooperate in ways that are good for the organization as a whole.

In a dimension most closely related to the subject of statistics, the TQM movement places emphasis on understanding and appreciating the consequences of **variability.** In fact, providing training in elementary statistics (including the basics of describing variation through numerical and graphical means, and often some basic Shewhart control charting) is a typical early step in TQM programs. (Incidentally, that TQM respect for the basics of statistics provides a natural jumping-off point for the application of the more advanced tools discussed in this text.)

And finally, it is a fact of life that the TQM landscape is dotted with many (competing) **quality consultants** and their bands of loyal followers. There are "big names" like the late W.E. Deming, J.M. Juran, A.V. Feigenbaum, and P. Crosby and literally thousands of less famous individuals, who will in some cases provide guidance in implementing the ideas of more famous quality leaders, and in others provide instruction in their own modifications of the systems of others. The sets of terminology and action items promoted by this diverse set of individuals vary consultant to consultant, in keeping with the need for them to have unique products to sell. (This diversity is at least part of the reason that defining the precise boundaries of TQM is presently so difficult.)

9.1.2 Some Limitations

As just described, TQM is attractive enough as an approach to focusing an organization's work that it may seem difficult to find any real basis upon which to critique it. Professor G. Box, for example, has referred to TQM in such positive terms as "the democratization of science." And when kept in perspective, your authors are generally supportive of the emphases of TQM *in the realm of commerce*. But it is possible to lose perspective and, by pushing the TQM emphases too far or applying them where they are not really appropriate, to create unintended and harmful consequences. So this final subsection is dedicated to a plea for balance in the application of the TQM paradigm.

To begin with what is very close to the root of the matter, consider first TQM's "customer focus." To become completely absorbed with what some customers want amounts to embracing them as the final arbiters of what is to be done. And that is a basically amoral (or ultimately immoral) final position. This point holds in the realm of commerce, but is even more obvious when the TQM customer-focus paradigm is applied in areas other than business.

For example, it is laudable to try to make government or educational systems more efficient. But these institutions deal in fundamentally moral arenas. We should want our governments to operate morally, whether or not that is currently in vogue with the majority of (customer) voters. People should want their children to go to schools where serious content is taught, real academic achievement is required, and depth of character and intellect are developed, whether or not it is a "feel-good" experience and as popular as MTV with the (customer) students, or satisfies the job-training desires of (customer) business concerns making demands on the educational system. And ultimately, we should fear for a country whose people expect other individuals and all public institutions to immediately gratify their most trivial whims (as deserving customers). Big words and concepts like "self-sacrifice," "duty," "principle," "integrity," and so on seem to have little relevance in a "customer-driven" world. But what "the customer" wants is not always even consistent, let alone moral or wise.

The TQM preoccupation with the analysis and improvement of processes and systems has already received some criticism even in business circles, as often taking on a life of its own and becoming an end in itself, independent of the fundamental purposes of a company. Rationality is an important part of the human "standard equipment" and it is only good stewardship to be moderately organized about how things are done. But enough is enough. The effort and volume of paperwork connected with planning (and documentation of that planning) and auditing (what has been done in every conceivable matter) has increased exponentially in the past few years in American business, government, and academia. What is happening in many cases amounts to a monumental triumph of form over substance. In a sane environment, smart and dedicated people will naturally do reasonable things. Occasionally, TQM-like tools are useful in helping them think through a problem. But slavish preoccupation with the details of how things are done and endless generation of vision and mission statements, strategic plans, process analyses, outcome assessments, and so forth can turn a relatively small task for one person into a big one for a group, with an accompanying huge loss of productivity.

There are other aspects of the TQM emphases on the analysis of processes, continuous improvement, and the benchmarking notion that deserve mention. A preoccupation with formal benchmarking has the natural tendency to produce homogenization and the stifling of genuine creativity and innovation. When an organization invests a large effort in determining what others are doing, it is very hard to then turn around and say "So be it. That's not what we're about. That doesn't suit our strengths and interests. We'll go a different way." Instead, the natural tendency is to conform, to "make use" of the carefully gathered data and strive to be like others. And frankly, the process-analysis tools of flow diagrams, flip charts, and nominal group technique applied in endless TQM meetings are not the stuff of which first-order innovations are born. Rather, those almost always come from really bright and motivated people working hard on a problem *individually* and perhaps occasionally coming together for free-form discussions of what they've been doing and what might be possible.

The TQM notions that there is wisdom at all levels of an organization that should be tapped, and that it is best to deemphasize distinctions between people at those various stations in the organization (and encourage all to actively participate in improvement efforts) perhaps come the closest of any of the TQM doctrines and practices to squaring with reality and deserving universal application. But even so (at least as they seem to be increasingly practiced), they are not quite on target. To begin with, treating people decently is not simply good organizational practice, it is *morally right.* When such treatment becomes a "method" of enhancing organizational effectiveness, it amounts only to a means of manipulation whose continuation is as fragile as the motivation to provide it is shallow. We should not need TQM to tell us to be kind, and if we do, others have no firm basis upon which to expect us to continue to be so when it no longer suits our purposes.

It has further been your authors' observation that when treated as a method or technique (rather than being a manifestation of good character), "respect" for the input of others can produce quite misguided results and implications. While it is true that people who might typically be overlooked often have important insights into how problems might be solved and progress might be made, it is *not* true that all opinions on a given matter need be equally valid or that resident "within" a group is necessarily the wherewithal to make progress.

But, for example, much nonsense about "group/cooperative learning" is currently being put forth in educational circles under such assumptions. That multidisciplinary teams (of individual experts) are needed in industry to design and produce complicated products does *not* imply that unstructured group discussion is a workable (let alone effective) means of teaching most subjects. "Group work" has its place in education in the realm of synthesis and reinforcement of content already on the table, but it is not a sensible means of presentation of technical material. Too much misguided "respect" for the opinions of the uninformed produces a kind of effective paralysis and inability to discriminate between the important and the mundane, an inability to truly move forward.

And finally, the almost cultish, guru-laden nature of the current TQM scene should cause sensible individuals considerable consternation. It is important that people be given credit for good work that they do and ideas that originate with them. And some TQM leaders are genuinely distinguished individuals who are worthy of great respect. But not worship. Good ideas about quality assurance are good independent of whether they can be associated with a leading figure. And bad ideas and emphases are bad even if they come from the pen of a great person. Modern quality engineers and the organizations they serve will be much better off if they think clearly, work hard, and learn to use good tools, than if they spend their energy acting as devotees of a particular quality guru.

In the end, one has in TQM a sensible set of emphases, provided they are used in limited ways, in appropriate arenas, by ethical and thinking people. And in their best incarnation, they provide fertile ground for the effective use of the statistical methods of quality assurance that have been the main concern in this book.

APPENDIX A

Tables

Table A.1 Control Chart Constants

n	d_2	d_3	c_4	A_2	A_3	B_3	B_4	B_5	B_6	D_1	D_2	D_3	D_4
2	1.128	0.853	0.7979	1.880	2.659		3.267		2.606		3.686		3.267
3	1.693	0.888	0.8862	1.023	1.954		2.568		2.276		4.358		2.574
4	2.059	0.880	0.9213	0.729	1.628		2.266		2.088		4.698		2.282
5	2.326	0.864	0.9400	0.577	1.427		2.089		1.964		4.918		2.115
6	2.534	0.848	0.9515	0.483	1.287	0.030	1.970	0.029	1.874		5.078		2.004
7	2.704	0.833	0.9594	0.419	1.182	0.118	1.882	0.113	1.806	0.205	5.204	0.076	1.924
8	2.847	0.820	0.9650	0.373	1.099	0.185	1.815	0.179	1.751	0.387	5.307	0.136	1.864
9	2.970	0.808	0.9693	0.337	1.032	0.239	1.761	0.232	1.707	0.546	5.394	0.184	1.816
10	3.078	0.797	0.9727	0.308	0.975	0.284	1.716	0.276	1.669	0.687	5.469	0.223	1.777
11	3.173	0.787	0.9754	0.285	0.927	0.321	1.679	0.313	1.637	0.812	5.534	0.256	1.744
12	3.258	0.778	0.9776	0.266	0.886	0.354	1.646	0.346	1.610	0.924	5.592	0.284	1.716
13	3.336	0.770	0.9794	0.249	0.850	0.382	1.618	0.374	1.585	1.026	5.646	0.308	1.692
14	3.407	0.762	0.9810	0.235	0.817	0.406	1.594	0.399	1.563	1.121	5.693	0.329	1.671
15	3.472	0.755	0.9823	0.223	0.789	0.428	1.572	0.421	1.544	1.207	5.737	0.348	1.652
20	3.735	0.729	0.9869	0.180	0.680	0.510	1.490	0.504	1.470	1.548	5.922	0.414	1.586
25	3.931	0.709	0.9896	0.153	0.606	0.565	1.435	0.559	1.420	1.804	6.058	0.459	1.541

This table is from *Quality Control and Industrial Statistics,* by A. J. Duncan, and was originally adapted from the *A.S.T.M. Manual on Quality Control of Materials,* Table B2 and the *ASQC Standard A1*, Table 1.

Table A.2 Standard Normal Cumulative Probabilities

$$\Phi(z) = \int_{-\infty}^{z} \frac{1}{\sqrt{2\pi}} \exp(-\frac{t^2}{2})dt$$

z	.00	.01	.02	.03	.04	.05	.06	.07	.08	.09
−3.4	.0003	.0003	.0003	.0003	.0003	.0003	.0003	.0003	.0003	.0002
−3.3	.0005	.0005	.0005	.0004	.0004	.0004	.0004	.0004	.0004	.0003
−3.2	.0007	.0007	.0006	.0006	.0006	.0006	.0006	.0005	.0005	.0005
−3.1	.0010	.0009	.0009	.0009	.0008	.0008	.0008	.0008	.0007	.0007
−3.0	.0013	.0013	.0013	.0012	.0012	.0011	.0011	.0011	.0010	.0010
−2.9	.0019	.0018	.0018	.0017	.0016	.0016	.0015	.0015	.0014	.0014
−2.8	.0026	.0025	.0024	.0023	.0023	.0022	.0021	.0021	.0020	.0019
−2.7	.0035	.0034	.0033	.0032	.0031	.0030	.0029	.0028	.0027	.0026
−2.6	.0047	.0045	.0044	.0043	.0041	.0040	.0039	.0038	.0037	.0036
−2.5	.0062	.0060	.0059	.0057	.0055	.0054	.0052	.0051	.0049	.0048
−2.4	.0082	.0080	.0078	.0075	.0073	.0071	.0069	.0068	.0066	.0064
−2.3	.0107	.0104	.0102	.0099	.0096	.0094	.0091	.0089	.0087	.0084
−2.2	.0139	.0136	.0132	.0129	.0125	.0122	.0119	.0116	.0113	.0110
−2.1	.0179	.0174	.0170	.0166	.0162	.0158	.0154	.0150	.0146	.0143
−2.0	.0228	.0222	.0217	.0212	.0207	.0202	.0197	.0192	.0188	.0183
−1.9	.0287	.0281	.0274	.0268	.0262	.0256	.0250	.0244	.0239	.0233
−1.8	.0359	.0351	.0344	.0336	.0329	.0322	.0314	.0307	.0301	.0294
−1.7	.0446	.0436	.0427	.0418	.0409	.0401	.0392	.0384	.0375	.0367
−1.6	.0548	.0537	.0526	.0516	.0505	.0495	.0485	.0475	.0465	.0455
−1.5	.0668	.0655	.0643	.0630	.0618	.0606	.0594	.0582	.0571	.0559
−1.4	.0808	.0793	.0778	.0764	.0749	.0735	.0721	.0708	.0694	.0681
−1.3	.0968	.0951	.0934	.0918	.0901	.0885	.0869	.0853	.0838	.0823
−1.2	.1151	.1131	.1112	.1093	.1075	.1056	.1038	.1020	.1003	.0985
−1.1	.1357	.1335	.1314	.1292	.1271	.1251	.1230	.1210	.1190	.1170
−1.0	.1587	.1562	.1539	.1515	.1492	.1469	.1446	.1423	.1401	.1379
−0.9	.1841	.1814	.1788	.1762	.1736	.1711	.1685	.1660	.1635	.1611
−0.8	.2119	.2090	.2061	.2033	.2005	.1977	.1949	.1922	.1894	.1867
−0.7	.2420	.2389	.2358	.2327	.2297	.2266	.2236	.2206	.2177	.2148
−0.6	.2743	.2709	.2676	.2643	.2611	.2578	.2546	.2514	.2483	.2451
−0.5	.3085	.3050	.3015	.2981	.2946	.2912	.2877	.2843	.2810	.2776
−0.4	.3446	.3409	.3372	.3336	.3300	.3264	.3228	.3192	.3156	.3121
−0.3	.3821	.3783	.3745	.3707	.3669	.3632	.3594	.3557	.3520	.3483
−0.2	.4207	.4168	.4129	.4090	.4052	.4013	.3974	.3936	.3897	.3859
−0.1	.4602	.4562	.4522	.4483	.4443	.4404	.4364	.4325	.4286	.4247
−0.0	.5000	.4960	.4920	.4880	.4840	.4801	.4761	.4721	.4681	.4641

Table A.2 Standard Normal Cumulative Probabilities (continued)

z	.00	.01	.02	.03	.04	.05	.06	.07	.08	.09
0.0	.5000	.5040	.5080	.5120	.5160	.5199	.5239	.5279	.5319	.5359
0.1	.5398	.5438	.5478	.5517	.5557	.5596	.5636	.5675	.5714	.5753
0.2	.5793	.5832	.5871	.5910	.5948	.5987	.6026	.6064	.6103	.6141
0.3	.6179	.6217	.6255	.6293	.6331	.6368	.6406	.6443	.6480	.6517
0.4	.6554	.6591	.6628	.6664	.6700	.6736	.6772	.6808	.6844	.6879
0.5	.6915	.6950	.6985	.7019	.7054	.7088	.7123	.7157	.7190	.7224
0.6	.7257	.7291	.7324	.7357	.7389	.7422	.7454	.7486	.7517	.7549
0.7	.7580	.7611	.7642	.7673	.7704	.7734	.7764	.7794	.7823	.7852
0.8	.7881	.7910	.7939	.7967	.7995	.8023	.8051	.8078	.8106	.8133
0.9	.8159	.8186	.8212	.8238	.8264	.8289	.8315	.8340	.8365	.8389
1.0	.8413	.8438	.8461	.8485	.8508	.8531	.8554	.8577	.8599	.8621
1.1	.8643	.8665	.8686	.8708	.8729	.8749	.8770	.8790	.8810	.8830
1.2	.8849	.8869	.8888	.8907	.8925	.8944	.8962	.8980	.8997	.9015
1.3	.9032	.9049	.9066	.9082	.9099	.9115	.9131	.9147	.9162	.9177
1.4	.9192	.9207	.9222	.9236	.9251	.9265	.9279	.9292	.9306	.9319
1.5	.9332	.9345	.9357	.9370	.9382	.9394	.9406	.9418	.9429	.9441
1.6	.9452	.9463	.9474	.9484	.9495	.9505	.9515	.9525	.9535	.9545
1.7	.9554	.9564	.9573	.9582	.9591	.9599	.9608	.9616	.9625	.9633
1.8	.9641	.9649	.9656	.9664	.9671	.9678	.9686	.9693	.9699	.9706
1.9	.9713	.9719	.9726	.9732	.9738	.9744	.9750	.9756	.9761	.9767
2.0	.9773	.9778	.9783	.9788	.9793	.9798	.9803	.9808	.9812	.9817
2.1	.9821	.9826	.9830	.9834	.9838	.9842	.9846	.9850	.9854	.9857
2.2	.9861	.9864	.9868	.9871	.9875	.9878	.9881	.9884	.9887	.9890
2.3	.9893	.9896	.9898	.9901	.9904	.9906	.9909	.9911	.9913	.9916
2.4	.9918	.9920	.9922	.9925	.9927	.9929	.9931	.9932	.9934	.9936
2.5	.9938	.9940	.9941	.9943	.9945	.9946	.9948	.9949	.9951	.9952
2.6	.9953	.9955	.9956	.9957	.9959	.9960	.9961	.9962	.9963	.9964
2.7	.9965	.9966	.9967	.9968	.9969	.9970	.9971	.9972	.9973	.9974
2.8	.9974	.9975	.9976	.9977	.9977	.9978	.9979	.9979	.9980	.9981
2.9	.9981	.9982	.9983	.9983	.9984	.9984	.9985	.9985	.9986	.9986
3.0	.9987	.9987	.9987	.9988	.9988	.9989	.9989	.9989	.9990	.9990
3.1	.9990	.9991	.9991	.9991	.9992	.9992	.9992	.9992	.9993	.9993
3.2	.9993	.9993	.9994	.9994	.9994	.9994	.9994	.9995	.9995	.9995
3.3	.9995	.9995	.9996	.9996	.9996	.9996	.9996	.9996	.9996	.9997
3.4	.9997	.9997	.9997	.9997	.9997	.9997	.9997	.9997	.9997	.9998

This table was generated using the "CDF" command in Minitab.

Table A.3 ARLs for EWMA Charts

	λ															
$\mathcal{D}^*$	.05	.1	.2	.3	.4	.5	.75	1.0	.05	.1	.2	.3	.4	.5	.75	1.0
	$\mathcal{K}^* = 1.00$								$\mathcal{K}^* = 1.50$							
0	18	10	6.4	4.9	4.2	3.8	3.3	3.2	48	27	16	13	11	9.3	7.9	7.5
.25	13	8.7	5.7	4.5	3.9	3.5	3.1	3.0	26	18	13	10	8.9	8.1	7.0	6.9
.50	8.2	6.0	4.4	3.7	3.3	3.0	2.7	2.7	13	10	7.9	6.9	6.2	5.9	5.4	5.5
.75	5.6	4.3	3.3	2.9	2.6	2.5	2.3	2.3	8.4	6.6	5.3	4.7	4.4	4.2	4.0	4.2
1.00	4.3	3.3	2.6	2.4	2.1	2.0	1.9	1.9	6.2	4.9	3.9	3.5	3.2	3.1	3.0	3.2
1.25	3.5	2.7	2.2	1.9	1.8	1.7	1.6	1.6	4.9	3.8	3.1	2.7	2.6	2.4	2.4	2.5
1.50	2.9	2.3	1.8	1.7	1.6	1.5	1.4	1.4	4.1	3.2	2.5	2.3	2.1	2.0	1.9	2.0
1.75	2.5	2.0	1.6	1.5	1.4	1.3	1.3	1.3	3.5	2.8	2.2	1.9	1.8	1.7	1.6	1.7
2.00	2.3	1.8	1.4	1.3	1.3	1.2	1.2	1.2	3.1	2.4	1.9	1.7	1.6	1.5	1.4	1.4
2.25	2.1	1.6	1.3	1.2	1.2	1.1	1.1	1.1	2.8	2.2	1.7	1.5	1.4	1.4	1.3	1.3
2.50	1.9	1.5	1.2	1.1	1.1				2.6	2.0	1.6	1.4	1.3	1.2	1.2	1.2
2.75	1.7	1.3	1.1						2.4	1.9	1.4	1.3	1.2	1.2	1.1	1.1
	$\mathcal{K}^* = 1.75$								$\mathcal{K}^* = 2.00$							
0	78	44	27	20	17	15	13	12	128	73	45	35	29	26	23	22
.25	34	25	18	15	14	13	11	11	44	34	27	23	21	20	19	19
.50	16	13	10	9.2	8.6	8.3	8.0	8.5	19	16	13	12	12	12	12	14
.75	10	7.9	6.5	5.9	5.6	5.5	5.5	6.1	12	9.4	8.0	7.4	7.3	7.3	7.9	9.2
1.00	7.3	5.7	4.6	4.2	4.0	3.9	3.9	4.4	8.4	6.6	5.4	5.1	4.9	4.9	5.3	6.3
1.25	5.7	4.5	3.6	3.2	3.1	3.0	2.9	3.2	6.6	5.1	4.2	3.8	3.7	3.6	3.8	4.4
1.50	4.8	3.7	3.0	2.6	2.5	2.4	2.3	2.5	5.4	4.2	3.4	3.1	2.9	2.8	2.8	3.2
1.75	4.1	3.2	2.5	2.2	2.1	2.0	1.9	2.0	4.6	3.6	2.9	2.6	2.4	2.3	2.3	2.5
2.00	3.6	2.8	2.2	1.9	1.8	1.7	1.6	1.7	4.0	3.1	2.5	2.2	2.0	1.9	1.9	2.0
2.25	3.2	2.5	2.0	1.7	1.6	1.5	1.4	1.4	3.6	2.8	2.2	1.9	1.8	1.7	1.6	1.7
2.50	2.9	2.3	1.8	1.6	1.4	1.4	1.3	1.3	3.3	2.5	2.0	1.7	1.6	1.5	1.4	1.4
2.75	2.7	2.1	1.6	1.4	1.3	1.3	1.2	1.2	3.0	2.3	1.8	1.6	1.4	1.4	1.3	1.3
3.00	2.5	2.0	1.5	1.3	1.2	1.2	1.1	1.1	2.8	2.2	1.7	1.4	1.3	1.3	1.2	1.2

Table A.3 ARLs for EWMA Charts (continued)

$\mathcal{D}^*$	λ = .05	.1	.2	.3	.4	.5	.75	1.0	.05	.1	.2	.3	.4	.5	.75	1.0
	$\mathcal{K}^* = 2.25$								$\mathcal{K}^* = 2.50$							
0	215	125	77	61	53	48	42	41	379	223	141	113	99	91	82	81
.25	57	48	40	36	34	33	33	35	74	67	61	59	58	58	61	66
.50	23	19	17	17	17	18	20	23	27	24	23	24	25	27	33	41
.75	13	11	9.7	9.4	9.6	9.9	11	15	15	13	12	12	13	14	18	25
1.00	9.5	7.6	6.5	6.2	6.1	6.3	7.3	9.4	11	8.7	7.7	7.5	7.8	8.3	11	15
1.25	7.4	5.8	4.8	4.5	4.4	4.4	4.9	6.3	8.3	6.6	5.6	5.3	5.3	5.5	6.7	9.5
1.50	6.1	4.7	3.9	3.5	3.4	3.3	3.6	4.4	6.8	5.3	4.4	4.1	4.0	4.0	4.7	6.3
1.75	5.2	4.0	3.2	2.9	2.8	2.7	2.8	3.2	5.7	4.5	3.6	3.3	3.2	3.1	3.4	4.4
2.00	4.5	3.5	2.8	2.5	2.3	2.2	2.2	2.5	5.0	3.9	3.1	2.8	2.6	2.6	2.7	3.2
2.25	4.0	3.1	2.5	2.2	2.0	1.9	1.9	2.0	4.4	3.4	2.7	2.4	2.3	2.2	2.2	2.5
2.50	3.6	2.8	2.2	2.0	1.8	1.7	1.6	1.7	4.0	3.1	2.4	2.2	2.0	1.9	1.8	2.0
2.75	3.3	2.6	2.0	1.8	1.6	1.5	1.4	1.4	3.6	2.8	2.2	2.0	1.8	1.7	1.6	1.7
3.00	3.1	2.4	1.9	1.6	1.5	1.4	1.3	1.3	3.4	2.6	2.1	1.8	1.6	1.5	1.4	1.4
	$\mathcal{K}^* = 2.75$								$\mathcal{K}^* = 3.00$							
0	703	421	272	222	198	185	171	168	1384	842	560	466	421	397	375	370
.25	98	96	97	100	103	107	119	132	134	145	163	179	194	209	246	281
.50	31	30	31	35	39	44	59	78	37	37	44	53	64	75	111	155
.75	18	15	15	16	18	20	29	44	20	18	19	22	26	31	51	81
1.00	12	10	9.1	9.3	10	11	16	25	14	11	11	12	13	16	26	44
1.25	9.3	7.4	6.4	6.3	6.5	7.0	9.6	15	10	8.3	7.4	7.5	8.2	9.2	14	25
1.50	7.5	5.9	5.0	4.7	4.7	4.9	6.2	9.5	8.3	6.6	5.6	5.4	5.6	6.1	8.7	15
1.75	6.3	4.9	4.0	3.7	3.7	3.7	4.4	6.3	6.9	5.4	4.6	4.3	4.3	4.5	5.8	9.5
2.00	5.5	4.3	3.4	3.1	3.0	3.0	3.3	4.4	6.0	4.7	3.8	3.5	3.4	3.5	4.2	6.3
2.25	4.9	3.8	3.0	2.7	2.6	2.5	2.6	3.2	5.3	4.1	3.3	3.0	2.9	2.8	3.2	4.4
2.50	4.4	3.4	2.7	3.4	2.2	2.1	2.1	2.5	4.8	3.7	2.9	2.6	2.5	2.4	2.5	3.2
2.75	4.0	3.1	2.4	2.2	2.0	1.9	1.8	2.0	4.3	3.3	2.6	2.4	2.2	2.1	2.1	2.5
3.00	3.7	2.8	2.2	2.0	1.8	1.7	1.6	1.7	4.0	3.0	2.4	2.1	2.0	1.9	1.8	2.0

Table A.3 ARLs for EWMA Charts (continued)

$\mathcal{D}^*$	λ: .05	.1	.2	.3	.4	.5	.75	1.0	.05	.1	.2	.3	.4	.5	.75	1.0
	$\mathcal{K}^* = 3.50$								$\mathcal{K}^* = 4.00$							
0	12851	4106	2880	2487	2313	2227	2158	2149								
.25	281	385	553	695	825	951	1246	1503	1059	1330	2462	3525	4560	5577	7984	10090
.50	54	65	101	148	203	267	469	724	82	103	295	535	848	1234	2520	4237
.75	26	25	33	46	64	89	182	334	33	38	67	120	204	324	833	1730
1.00	17	15	16	20	27	36	78	161	20	19	26	40	65	104	304	741
1.25	12	10	10	11	14	18	37	82	15	13	14	18	27	41	123	336
1.50	9.9	8.0	7.2	7.5	8.4	10	20	44	12	9.7	9.4	11	14	20	55	161
1.75	8.2	6.5	5.6	5.6	5.9	6.7	11	25	9.6	7.8	7.0	7.4	8.7	11	28	82
2.00	7.1	5.5	4.6	4.4	4.5	4.9	7.3	15	8.2	6.5	5.6	5.6	6.2	7.3	15	44
2.25	6.2	4.8	4.0	3.7	3.6	3.8	5.1	9.4	7.2	5.6	4.7	4.5	4.7	5.3	9.4	25
2.50	5.6	4.3	3.5	3.2	3.1	3.1	3.8	6.3	6.4	5.0	4.1	3.8	3.8	4.1	6.3	15
2.75	5.0	3.9	3.1	2.8	2.7	2.6	2.9	4.4	5.8	4.5	3.6	3.3	3.2	3.3	4.5	9.5
3.00	4.6	3.5	2.8	2.5	2.4	2.3	2.4	3.2	5.3	4.0	3.2	2.9	2.8	2.8	3.4	6.3
3.25	4.2	3.3	2.6	2.3	2.1	2.0	2.0	2.5	4.8	3.7	3.0	2.7	2.5	2.5	2.8	4.4
3.50	4.0	3.0	2.4	2.1	2.0	1.9	1.8	2.0	4.5	3.4	2.7	2.4	2.3	2.2	2.3	3.2
3.75	3.7	2.8	2.2	2.0	1.8	1.7	1.6	1.7	4.2	3.2	2.5	2.3	2.1	2.0	2.0	2.5
4.00	3.5	2.7	2.1	1.9	1.7	1.6	1.4	1.4	3.9	3.0	2.4	2.1	2.0	1.8	1.7	2.0

This table was prepared using a program written by Stephen V. Crowder.

Table A.4 One-Sided CUSUM ARLs

$\mathcal{S}^*$	$\mathcal{H}^*$ .25	.50	.75	1.0	1.5	2.0	2.5	3.0	3.5	4.0	5.0	6.0	8.0	10.0
−3.00	1732	4293												
−2.75	740	1728	4275											
−2.50	335	737	1715	4222										
−2.25	161	333	727	1678										
−2.00	82	159	325	702	3791									
−1.75	44	80	154	308	1419									
−1.50	25	43	76	142	550	2377								
−1.25	15	24	40	69	221	766	2722							
−1.00	9.4	14	22	35	94	259	716	1963	5341					
−.75	6.2	8.9	13	19	43	94	206	443	944	2004	9008			
−.50	4.4	5.9	8.1	11	21	39	68	118	200	335	931	2553	18966	
−.25	3.2	4.2	5.4	7.0	12	18	27	39	56	77	142	251	737	2072
0	2.5	3.1	3.8	4.8	7.1	10	13	17	22	27	38	51	84	125
.25	2.0	2.4	2.9	3.4	4.8	6.3	8.0	9.7	11	13	17	21	29	37
.50	1.7	1.9	2.3	2.6	3.5	4.5	5.4	6.4	7.4	8.4	10	12	16	20
.75	1.4	1.6	1.9	2.1	2.7	3.4	4.1	4.7	5.4	6.1	7.4	8.7	11	14
1.00	1.3	1.4	1.6	1.8	2.2	2.7	3.3	3.8	4.3	4.8	5.8	6.8	8.8	11
1.25	1.2	1.3	1.4	1.5	1.9	2.3	2.7	3.1	3.5	3.9	4.7	5.5	7.1	8.7
1.50	1.1	1.2	1.3	1.4	1.7	2.0	2.3	2.7	3.0	3.3	4.0	4.7	6.0	7.3
1.75		1.1	1.2	1.3	1.5	1.8	2.1	2.4	2.7	2.9	3.5	4.1	5.2	6.4
2.00			1.1	1.2	1.3	1.6	1.9	2.1	2.4	2.6	3.1	3.6	4.6	5.6
2.25				1.1	1.2	1.4	1.7	1.9	2.2	2.4	2.8	3.3	4.2	5.0
2.50					1.2	1.3	1.5	1.8	2.0	2.2	2.6	3.0	3.8	4.6
2.75					1.1	1.2	1.4	1.6	1.8	2.0	2.4	2.7	3.5	4.2
3.00						1.2	1.3	1.5	1.7	1.9	2.2	2.5	3.2	3.9
3.25						1.1	1.2	1.4	1.6	1.8	2.1	2.4	3.0	3.6
3.50							1.2	1.3	1.5	1.7	2.0	2.2	2.8	3.4
3.75							1.1	1.2	1.4	1.6	1.9	2.1	2.7	3.2
4.00								1.2	1.3	1.5	1.9	2.1	2.5	3.1

This table was prepared using a program written by Fan Fatt Gan.

Table A.5 ARLs for Combined High and Low Side CUSUM Schemes (For values of $\mathscr{D}^*$ larger than those in this table, use Table A.4 with $\mathscr{S}^* = \mathscr{D}^* - \mathscr{H}^*$)

		$\mathscr{H}^*$													
$\mathscr{K}^*$	$\mathscr{D}^*$	.25	.50	.75	1.0	1.5	2.0	2.5	3.0	3.5	4.0	5.0	6.0	8.0	10.0
.25	0	1.6	2.1	2.7	3.5	5.8	9.1	14	20	28	39	71	125	368	1036
	.25	1.6	2.0	2.6	3.3	5.3	7.9	11	15	20	25	37	50	84	125
	.50	1.5	1.9	2.3	2.9	4.3	5.9	7.7	9.5	11	13	17	21	29	37
	.75	1.4	1.7	2.0	2.4	3.4	4.4	5.4	6.4	7.4	8.4	10	12	16	20
	1.00	1.3	1.5	1.8	2.1	2.7	3.4	4.1	4.7	5.4	6.1	7.4	8.7	11	14
	1.25	1.2	1.4	1.6	1.8	2.2	2.7	3.2	3.7	4.3	4.8	5.8	6.8	8.8	11
	1.50	1.2	1.3	1.4	1.5	1.9	2.3	2.7	3.1	3.5	3.9	4.7	5.5	7.1	8.7
.50	0	2.2	3.0	4.1	5.6	11	19	34	59	100	168	465	1277	9483	
	.25	2.1	2.8	3.8	5.1	9.1	15	24	36	53	74	139	249	737	2072
	.50	1.9	2.5	3.3	4.2	6.6	9.6	13	17	22	27	38	51	84	125
	.75	1.7	2.2	2.7	3.3	4.7	6.3	7.9	9.7	11	13	17	21	29	37
	1.00	1.6	1.8	2.2	2.6	3.5	4.4	5.4	6.4	7.4	8.4	10	12	16	20
	1.25	1.4	1.6	1.8	2.1	2.7	3.4	4.1	4.7	5.4	6.1	7.4	8.7	11	14
	1.50	1.3	1.4	1.6	1.8	2.2	2.7	3.3	3.8	4.3	4.8	5.8	6.8	8.8	11
.75	0	3.1	4.5	6.5	9.6	21	47	103	221	472	1002	4504			
	.25	3.0	4.2	6.0	8.5	17	34	62	111	192	328	931	2553		
	.50	2.6	3.5	4.8	6.4	11	18	27	39	56	77	142	251	737	2072
	.75	2.2	2.9	3.6	4.6	7.0	10	13	17	22	27	38	51	84	125
	1.00	1.9	2.3	2.8	3.4	4.8	6.3	8.0	9.7	11	13	17	21	29	37
	1.25	1.6	1.9	2.2	2.6	3.5	4.3	5.4	6.4	7.4	8.4	10	12	16	20
	1.50	1.4	2.6	1.9	2.1	2.7	3.4	4.1	4.7	5.4	6.1	7.4	8.7	11	14
1.00	0	4.7	7.1	11	18	47	129	358	981	2671	7256				
	.25	4.4	6.5	9.8	15	36	84	191	423	918	2004				
	.50	3.7	5.2	7.4	10	20	38	68	117	200	335	931	2553		
	.75	3.0	3.9	5.2	6.9	11	18	27	39	56	77	142	251	737	2072
	1.00	2.4	3.0	3.8	4.7	7.1	10	13	17	22	27	38	51	84	125
	1.25	2.0	2.4	2.8	3.4	4.8	6.3	8.0	9.7	11	13	17	21	29	37
	1.50	1.7	1.9	2.2	2.6	3.5	4.5	5.4	6.4	7.4	8.4	10	12	16	20

Table A.5 ARLs for Combined High and Low Side CUSUM Schemes (continued)

		$\mathcal{H}^*$													
$\mathcal{K}^*$	$\mathcal{D}^*$	**.25**	**.50**	**.75**	**1.0**	**1.5**	**2.0**	**2.5**	**3.0**	**3.5**	**4.0**	**5.0**	**6.0**	**8.0**	**10.0**
1.25	0	7.4	12	20	34	111	383	1361	4807						
	.25	6.8	11	17	28	80	233	672	1888	5341					
	.50	5.4	8.0	12	18	41	93	205	443	944	2004	9008			
	.75	4.1	5.7	7.9	11	21	38	68	118	200	335	931	2553		
	1.00	3.1	4.1	5.4	7.0	12	18	27	39	56	77	142	251	737	2072
	1.25	2.4	3.1	3.8	4.7	7.1	10	13	17	22	27	38	51	84	125
	1.50	2.0	2.4	2.9	3.4	4.8	6.3	8.0	9.7	11	13	17	21	29	37
1.50	0	12	21	38	71	275	1188	5431							
	.25	11	18	32	56	192	696	1685							
	.50	8.4	13	21	34	92	256	716	1963	5341					
	.75	6.0	8.7	13	19	42	94	206	443	994	2004	9008			
	1.00	4.3	5.9	8.1	11	21	39	68	118	200	335	931	2553		
	1.25	3.2	4.1	5.4	7.0	12	18	27	39	56	77	142	251	737	2072
	1.50	2.5	3.1	3.8	4.8	7.1	10	13	17	22	27	38	51	84	125
2.00	0	41	79	163	351	1895									
	.25	34	65	127	260	1250	6915								
	.50	23	40	73	138	540	2377								
	.75	15	24	40	68	221	776	2722							
	1.00	9.3	14	22	35	94	259	716	1963	5341					
	1.25	6.2	8.9	13	19	43	94	206	443	944	2004	9008			
	1.50	4.4	5.9	8.1	11	21	39	68	118	200	335	931	2553		

Blanks in this table indicate large ARLs.
This table was prepared using Table A.4 and formula (4.20).

Table A.6 ARLs for Combined Individuals and Moving Range Charts

$\mathcal{D}^*$	$\mathcal{M}^*$									
	2.0	2.5	3.0	3.5	4.0	2.0	2.5	3.0	3.5	4.0
	$\mathcal{R}^* = 3.0$					$\mathcal{R}^* = 3.5$				
0	18	30	34	34	35	21	52	77	83	84
.25	16	28	34	34	35	19	46	74	83	84
.50	12	24	32	34	35	13	34	65	81	84
.75	8.6	18	29	33	35	9.1	22	51	75	83
1.00	6.1	13	24	31	34	6.2	14	35	65	81
1.25	4.4	8.8	18	27	34	4.4	9.3	23	50	75
1.50	3.2	6.1	13	22	32	3.2	6.3	15	34	65
1.75	2.5	4.4	8.8	16	29	2.5	4.4	9.6	22	50
2.00	2.0	3.2	6.1	11	24	2.0	3.2	6.4	14	34
2.25	1.7	2.5	4.4	7.8	18	1.7	2.5	4.5	9.3	22
2.50	1.4	2.0	3.2	5.4	13	1.4	2.0	3.3	6.3	14
2.75	1.3	1.7	2.5	3.9	8.8	1.3	1.7	2.5	4.4	9.3
3.00	1.2	1.4	2.0	2.9	6.1	1.2	1.4	2.0	3.3	6.3
3.25	1.1	1.3	1.7	2.3	4.4	1.1	1.3	1.7	2.5	4.4
	$\mathcal{R}^* = 4.0$					$\mathcal{R}^* = 4.5$				
0	22	72	166	218	230	22	79	285	576	705
.25	19	60	149	212	229	19	65	231	526	694
.50	14	40	108	193	226	14	41	141	398	652
.75	9.2	24	69	158	217	9.2	25	78	253	557
1.00	6.2	15	41	112	195	6.2	15	43	144	408
1.25	4.4	9.4	25	71	157	4.4	9.5	25	79	255
1.50	3.2	6.3	15	43	110	3.2	6.3	15	44	144
1.75	2.5	4.4	9.6	25	69	2.5	4.4	9.6	25	78
2.00	2.0	3.2	6.4	15	41	2.0	3.2	6.4	15	43
2.25	1.7	2.5	4.5	9.8	24	1.7	2.5	4.5	9.8	25
2.50	1.4	2.0	3.3	6.5	15	1.4	2.0	3.3	6.5	15
2.75	1.3	1.7	2.5	4.5	9.3	1.3	1.7	2.5	4.5	9.5
3.00	1.2	1.4	2.0	3.3	6.3	1.2	1.4	2.0	3.3	6.3
3.25	1.1	1.3	1.7	2.5	4.4	1.1	1.3	1.7	2.5	4.4

Table A.6 ARLs for Combined Individuals and Moving Range Charts (continued)

$\mathscr{D}^*$	$\mathscr{M}^*$ 2.0	2.5	3.0	3.5	4.0	2.0	2.5	3.0	3.5	4.0
	$\mathscr{R}^* = 5.0$					$\mathscr{R}^* = 5.5$				
0	22	80	351	1301	2305	22	80	368	1894	6673
.25	19	66	271	1050	2158	19	66	280	1380	5452
.50	14	41	153	616	1721	14	41	155	699	3184
.75	9.2	25	81	318	1122	9.2	25	81	331	1547
1.00	6.2	15	44	161	617	6.2	15	44	161	711
1.25	4.4	9.5	25	84	313	4.4	9.5	25	84	331
1.50	3.2	6.3	15	45	157	3.2	6.3	15	45	160
1.75	2.5	4.4	9.6	26	81	2.5	4.4	9.6	26	82
2.00	2.0	3.2	6.4	16	44	2.0	3.2	6.4	16	44
2.25	1.7	2.5	4.5	9.8	25	1.7	2.5	4.5	9.8	25
2.50	1.4	2.0	3.3	6.5	15	1.4	2.0	3.3	6.5	15
2.75	1.3	1.7	2.5	4.5	9.5	1.3	1.7	2.5	4.5	9.5
3.00	1.2	1.4	2.0	3.3	6.3	1.2	1.4	2.0	3.3	6.3
3.25	1.1	1.3	1.7	2.5	4.4	1.1	1.3	1.7	2.5	4.4
	$\mathscr{R}^* = 6.0$					$\mathscr{R}^* = 6.5$				
0	22	80	370	2108	12453	22	80	370	2142	15115
.25	19	66	281	1484	8654	19	66	281	1499	9813
.50	14	41	155	721	3983	14	41	155	723	4187
.75	9.2	25	81	335	1691	9.2	25	81	335	1721
1.00	6.2	15	44	161	734	6.2	15	44	161	738
1.25	4.4	9.5	25	83	334	4.4	9.5	25	83	335
1.50	3.2	6.3	15	44	161	3.2	6.3	15	44	161
1.75	2.5	4.4	9.6	25	82	2.5	4.4	9.6	25	82
2.00	2.0	3.2	6.4	15	44	2.0	3.2	6.4	15	44
2.25	1.7	2.5	4.5	9.8	25	1.7	2.5	4.5	9.8	25
2.50	1.4	2.0	3.3	6.5	15	1.4	2.0	3.3	6.5	15
2.75	1.3	1.7	2.5	4.5	9.5	1.3	1.7	2.5	4.5	9.5
3.00	1.2	1.4	2.0	3.3	6.3	1.2	1.4	2.0	3.3	6.3
3.25	1.1	1.3	1.7	2.5	4.4	1.1	1.3	1.7	2.5	4.4

This table was prepared using a program written by Stephen V. Crowder.

Table A.7 Chi-Square Distribution Quantiles

ν	$Q(.005)$	$Q(.01)$	$Q(.025)$	$Q(.05)$	$Q(.1)$	$Q(.9)$	$Q(.95)$	$Q(.975)$	$Q(.99)$	$Q(.995)$
1	0.000	0.000	0.001	0.004	0.016	2.706	3.841	5.024	6.635	7.879
2	0.010	0.020	0.051	0.103	0.211	4.605	5.991	7.378	9.210	10.597
3	0.072	0.115	0.216	0.352	0.584	6.251	7.815	9.348	11.345	12.838
4	0.207	0.297	0.484	0.711	1.064	7.779	9.488	11.143	13.277	14.860
5	0.412	0.554	0.831	1.145	1.610	9.236	11.070	12.833	15.086	16.750
6	0.676	0.872	1.237	1.635	2.204	10.645	12.592	14.449	16.812	18.548
7	0.989	1.239	1.690	2.167	2.833	12.017	14.067	16.013	18.475	20.278
8	1.344	1.646	2.180	2.733	3.490	13.362	15.507	17.535	20.090	21.955
9	1.735	2.088	2.700	3.325	4.168	14.684	16.919	19.023	21.666	23.589
10	2.156	2.558	3.247	3.940	4.865	15.987	18.307	20.483	23.209	25.188
11	2.603	3.053	3.816	4.575	5.578	17.275	19.675	21.920	24.725	26.757
12	3.074	3.571	4.404	5.226	6.304	18.549	21.026	23.337	26.217	28.300
13	3.565	4.107	5.009	5.892	7.042	19.812	22.362	24.736	27.688	29.819
14	4.075	4.660	5.629	6.571	7.790	21.064	23.685	26.119	29.141	31.319
15	4.601	5.229	6.262	7.261	8.547	22.307	24.996	27.488	30.578	32.801
16	5.142	5.812	6.908	7.962	9.312	23.542	26.296	28.845	32.000	34.267
17	5.697	6.408	7.564	8.672	10.085	24.769	27.587	30.191	33.409	35.718
18	6.265	7.015	8.231	9.390	10.865	25.989	28.869	31.526	34.805	37.156
19	6.844	7.633	8.907	10.117	11.651	27.204	30.143	32.852	36.191	38.582
20	7.434	8.260	9.591	10.851	12.443	28.412	31.410	34.170	37.566	39.997
21	8.034	8.897	10.283	11.591	13.240	29.615	32.671	35.479	38.932	41.401
22	8.643	9.542	10.982	12.338	14.041	30.813	33.924	36.781	40.290	42.796
23	9.260	10.196	11.689	13.091	14.848	32.007	35.172	38.076	41.638	44.181
24	9.886	10.856	12.401	13.848	15.659	33.196	36.415	39.364	42.980	45.559
25	10.520	11.524	13.120	14.611	16.473	34.382	37.653	40.647	44.314	46.928
26	11.160	12.198	13.844	15.379	17.292	35.563	38.885	41.923	45.642	48.290
27	11.808	12.879	14.573	16.151	18.114	36.741	40.113	43.195	46.963	49.645
28	12.461	13.565	15.308	16.928	18.939	37.916	41.337	44.461	48.278	50.994
29	13.121	14.256	16.047	17.708	19.768	39.087	42.557	45.722	49.588	52.336
30	13.787	14.953	16.791	18.493	20.599	40.256	43.773	46.979	50.892	53.672
31	14.458	15.655	17.539	19.281	21.434	41.422	44.985	48.232	52.192	55.003
32	15.134	16.362	18.291	20.072	22.271	42.585	46.194	49.480	53.486	56.328
33	15.815	17.074	19.047	20.867	23.110	43.745	47.400	50.725	54.775	57.648
34	16.501	17.789	19.806	21.664	23.952	44.903	48.602	51.966	56.061	58.964
35	17.192	18.509	20.569	22.465	24.797	46.059	49.802	53.204	57.342	60.275
36	17.887	19.233	21.336	23.269	25.643	47.212	50.998	54.437	58.619	61.581
37	18.586	19.960	22.106	24.075	26.492	48.364	52.192	55.668	59.893	62.885
38	19.289	20.691	22.878	24.884	27.343	49.513	53.384	56.896	61.163	64.183
39	19.996	21.426	23.654	25.695	28.196	50.660	54.572	58.120	62.429	65.477
40	20.707	22.164	24.433	26.509	29.051	51.805	55.759	59.342	63.691	66.767

This table was generated using the "INVCDF" command in Minitab.

For $\nu > 40$ the approximation $Q(p) \approx \nu\left(1 - \frac{2}{9\nu} + Q_z(p)\sqrt{\frac{2}{9\nu}}\right)^3$ can be used.

Table A.8 *t*-Distribution Quantiles

ν	$Q(.9)$	$Q(.95)$	$Q(.975)$	$Q(.99)$	$Q(.995)$	$Q(.999)$	$Q(.9995)$
1	3.078	6.314	12.706	31.821	63.657	318.317	636.607
2	1.886	2.920	4.303	6.965	9.925	22.327	31.598
3	1.638	2.353	3.182	4.541	5.841	10.215	12.924
4	1.533	2.132	2.776	3.747	4.604	7.173	8.610
5	1.476	2.015	2.571	3.365	4.032	5.893	6.869
6	1.440	1.943	2.447	3.143	3.707	5.208	5.959
7	1.415	1.895	2.365	2.998	3.499	4.785	5.408
8	1.397	1.860	2.306	2.896	3.355	4.501	5.041
9	1.383	1.833	2.262	2.821	3.250	4.297	4.781
10	1.372	1.812	2.228	2.764	3.169	4.144	4.587
11	1.363	1.796	2.201	2.718	3.106	4.025	4.437
12	1.356	1.782	2.179	2.681	3.055	3.930	4.318
13	1.350	1.771	2.160	2.650	3.012	3.852	4.221
14	1.345	1.761	2.145	2.624	2.977	3.787	4.140
15	1.341	1.753	2.131	2.602	2.947	3.733	4.073
16	1.337	1.746	2.120	2.583	2.921	3.686	4.015
17	1.333	1.740	2.110	2.567	2.898	3.646	3.965
18	1.330	1.734	2.101	2.552	2.878	3.610	3.922
19	1.328	1.729	2.093	2.539	2.861	3.579	3.883
20	1.325	1.725	2.086	2.528	2.845	3.552	3.849
21	1.323	1.721	2.080	2.518	2.831	3.527	3.819
22	1.321	1.717	2.074	2.508	2.819	3.505	3.792
23	1.319	1.714	2.069	2.500	2.807	3.485	3.768
24	1.318	1.711	2.064	2.492	2.797	3.467	3.745
25	1.316	1.708	2.060	2.485	2.787	3.450	3.725
26	1.315	1.706	2.056	2.479	2.779	3.435	3.707
27	1.314	1.703	2.052	2.473	2.771	3.421	3.690
28	1.313	1.701	2.048	2.467	2.763	3.408	3.674
29	1.311	1.699	2.045	2.462	2.756	3.396	3.659
30	1.310	1.697	2.042	2.457	2.750	3.385	3.646
40	1.303	1.684	2.021	2.423	2.704	3.307	3.551
60	1.296	1.671	2.000	2.390	2.660	3.232	3.460
120	1.289	1.658	1.980	2.358	2.617	3.160	3.373
∞	1.282	1.645	1.960	2.326	2.576	3.090	3.291

This table was generated using the "INVCDF" command in Minitab.

Table A.9a Factors for Two-Sided Tolerance Intervals for Normal Distributions

n	95% Confidence			99% Confidence		
	$p = .90$	$p = .95$	$p = .99$	$p = .90$	$p = .95$	$p = .99$
2	31.092	36.519	46.944	155.569	182.720	234.877
3	8.306	9.789	12.647	18.782	22.131	28.586
4	5.368	6.341	8.221	9.416	11.118	14.405
5	4.291	5.077	6.598	6.655	7.870	10.220
6	3.733	4.422	5.758	5.383	6.373	8.292
7	3.390	4.020	5.241	4.658	5.520	7.191
8	3.156	3.746	4.889	4.189	4.968	6.479
9	2.986	3.546	4.633	3.860	4.581	5.980
10	2.856	3.393	4.437	3.617	4.294	5.610
11	2.754	3.273	4.282	3.429	4.073	5.324
12	2.670	3.175	4.156	3.279	3.896	5.096
13	2.601	3.093	4.051	3.156	3.751	4.909
14	2.542	3.024	3.962	3.054	3.631	4.753
15	2.492	2.965	3.885	2.967	3.529	4.621
16	2.449	2.913	3.819	2.893	3.441	4.507
17	2.410	2.868	3.761	2.828	3.364	4.408
18	2.376	2.828	3.709	2.771	3.297	4.321
19	2.346	2.793	3.663	2.720	3.237	4.244
20	2.319	2.760	3.621	2.675	3.184	4.175
25	2.215	2.638	3.462	2.506	2.984	3.915
30	2.145	2.555	3.355	2.394	2.851	3.742
35	2.094	2.495	3.276	2.314	2.756	3.618
40	2.055	2.448	3.216	2.253	2.684	3.524
50	1.999	2.382	3.129	2.166	2.580	3.390
60	1.960	2.335	3.068	2.106	2.509	3.297
80	1.908	2.274	2.988	2.028	2.416	3.175
100	1.875	2.234	2.936	1.978	2.357	3.098
150	1.826	2.176	2.859	1.906	2.271	2.985
200	1.798	2.143	2.816	1.866	2.223	2.921
500	1.737	2.070	2.721	1.777	2.117	2.783
1000	1.709	2.036	2.676	1.736	2.068	2.718
∞	1.645	1.960	2.576	1.645	1.960	2.576

This table was adapted from *Tables for Normal Tolerance Limits, Sampling Plans and Screening,* by R. E. Odeh and D. B. Owen, published by Marcel Dekker, Inc.

Table A.9b Factors for One-Sided Tolerance Intervals for Normal Distributions

n	95% Confidence			99% Confidence		
	$p = .90$	$p = .95$	$p = .99$	$p = .90$	$p = .95$	$p = .99$
2	20.581	26.260	37.094	103.029	131.426	185.617
3	6.155	7.656	10.553	13.995	17.370	23.896
4	4.162	5.144	7.042	7.380	9.083	12.387
5	3.407	4.203	5.741	5.362	6.578	8.939
6	3.006	3.708	5.062	4.411	5.406	7.335
7	2.755	3.399	4.642	3.859	4.728	6.412
8	2.582	3.187	4.354	3.497	4.285	5.812
9	2.454	3.031	4.143	3.240	3.972	5.389
10	2.355	2.911	3.981	3.048	3.738	5.074
11	2.275	2.815	3.852	2.898	3.556	4.829
12	2.210	2.736	3.747	2.777	3.410	4.633
13	2.155	2.671	3.659	2.677	3.290	4.472
14	2.109	2.614	3.585	2.593	3.189	4.337
15	2.068	2.566	3.520	2.521	3.102	4.222
16	2.033	2.524	3.464	2.459	3.028	4.123
17	2.002	2.486	3.414	2.405	2.963	4.037
18	1.974	2.453	3.370	2.357	2.905	3.960
19	1.949	2.423	3.331	2.314	2.854	3.892
20	1.926	2.396	3.295	2.276	2.808	3.832
25	1.838	2.292	3.158	2.129	2.633	3.601
30	1.777	2.220	3.064	2.030	2.515	3.447
35	1.732	2.167	2.995	1.957	2.430	3.334
40	1.697	2.125	2.941	1.902	2.364	3.249
50	1.646	2.065	2.862	1.821	2.269	3.125
60	1.609	2.022	2.807	1.764	2.202	3.038
80	1.559	1.964	2.733	1.688	2.114	2.924
100	1.527	1.927	2.684	1.639	2.056	2.850
150	1.478	1.870	2.611	1.566	1.971	2.740
200	1.450	1.837	2.570	1.524	1.923	2.679
500	1.385	1.763	2.475	1.430	1.814	2.540
1000	1.354	1.727	2.430	1.385	1.762	2.475
∞	1.282	1.645	2.326	1.282	1.645	2.326

This table was adapted from *Tables for Normal Tolerance Limits, Sampling Plans and Screening,* by R. E. Odeh and D. B. Owen, published by Marcel Dekker, Inc.

Table A.10a Factors for Simultaneous 95% Two-Sided Confidence Limits for Several Means

	Number of Means													
ν	1	2	3	4	5	6	7	8	9	10	12	14	16	32
2	4.303	5.571	6.340	6.886	7.306	7.645	7.929	8.172	8.385	8.573	8.894	9.162	9.390	10.529
3	3.182	3.960	4.430	4.764	5.023	5.233	5.410	5.562	5.694	5.812	6.015	6.184	6.328	7.055
4	2.776	3.382	3.745	4.003	4.203	4.366	4.503	4.621	4.725	4.817	4.975	5.107	5.221	5.794
5	2.571	3.091	3.399	3.619	3.789	3.928	4.044	4.145	4.233	4.312	4.447	4.560	4.657	5.150
6	2.447	2.916	3.193	3.389	3.541	3.664	3.769	3.858	3.937	4.008	4.129	4.230	4.317	4.760
7	2.365	2.800	3.055	3.236	3.376	3.489	3.585	3.668	3.740	3.805	3.916	4.009	4.090	4.498
8	2.306	2.718	2.958	3.127	3.258	3.365	3.454	3.532	3.600	3.660	3.764	3.852	3.927	4.310
9	2.262	2.657	2.885	3.046	3.171	3.272	3.357	3.430	3.494	3.552	3.650	3.733	3.805	4.169
10	2.228	2.609	2.829	2.983	3.103	3.199	3.281	3.351	3.412	3.467	3.562	3.641	3.710	4.058
11	2.201	2.571	2.784	2.933	3.048	3.142	3.220	3.288	3.347	3.400	3.491	3.568	3.634	3.969
12	2.179	2.540	2.747	2.892	3.004	3.095	3.171	3.236	3.294	3.345	3.433	3.507	3.571	3.897
13	2.160	2.514	2.717	2.858	2.967	3.055	3.129	3.193	3.249	3.299	3.385	3.457	3.519	3.836
14	2.145	2.493	2.691	2.830	2.936	3.022	3.095	3.157	3.212	3.260	3.344	3.415	3.475	3.784
15	2.131	2.474	2.669	2.805	2.909	2.994	3.065	3.126	3.180	3.227	3.309	3.378	3.438	3.740
16	2.120	2.458	2.650	2.784	2.886	2.969	3.039	3.099	3.152	3.199	3.279	3.347	3.405	3.701
17	2.110	2.444	2.633	2.765	2.866	2.948	3.017	3.076	3.127	3.173	3.253	3.319	3.376	3.668
18	2.101	2.432	2.619	2.749	2.849	2.929	2.997	3.055	3.106	3.151	3.229	3.295	3.351	3.638
19	2.093	2.421	2.606	2.734	2.833	2.912	2.979	3.037	3.087	3.132	3.209	3.273	3.329	3.611
20	2.086	2.411	2.594	2.721	2.819	2.897	2.963	3.020	3.070	3.114	3.190	3.254	3.308	3.587
24	2.064	2.380	2.558	2.681	2.775	2.851	2.914	2.969	3.016	3.059	3.132	3.193	3.246	3.513
30	2.042	2.350	2.522	2.641	2.732	2.805	2.866	2.918	2.964	3.005	3.075	3.133	3.184	3.439
36	2.028	2.331	2.499	2.615	2.704	2.775	2.834	2.885	2.930	2.970	3.038	3.094	3.143	3.391
40	2.021	2.321	2.488	2.602	2.690	2.760	2.819	2.869	2.913	2.952	3.019	3.075	3.123	3.367
60	2.000	2.292	2.454	2.564	2.649	2.716	2.772	2.821	2.863	2.900	2.964	3.018	3.064	3.295
120	1.980	2.264	2.420	2.527	2.608	2.673	2.727	2.773	2.814	2.849	2.910	2.961	3.005	3.225
144	1.977	2.259	2.415	2.521	2.602	2.666	2.720	2.766	2.806	2.841	2.902	2.952	2.996	3.214
∞	1.960	2.237	2.388	2.491	2.569	2.631	2.683	2.727	2.766	2.800	2.858	2.906	2.948	3.156

This table was prepared using a program for the calculation of quantiles of the studentized maximum modulus distribution, written by Daniel L. Rose.

Table A.10b Factors for Simultaneous 95% One-Sided Confidence Limits for Several Means

	Number of Means													
ν	1	2	3	4	5	6	7	8	9	10	12	14	16	32
2	2.920	4.075	4.834	5.397	5.842	6.208	6.516	6.781	7.014	7.220	7.573	7.867	8.118	9.364
3	2.353	3.090	3.551	3.888	4.154	4.372	4.557	4.717	4.858	4.983	5.199	5.380	5.535	6.315
4	2.132	2.722	3.080	3.340	3.544	3.711	3.852	3.974	4.082	4.179	4.345	4.484	4.604	5.212
5	2.015	2.532	2.840	3.062	3.234	3.376	3.495	3.599	3.690	3.772	3.912	4.031	4.132	4.650
6	1.943	2.417	2.696	2.894	3.049	3.175	3.282	3.374	3.455	3.528	3.653	3.758	3.849	4.312
7	1.895	2.340	2.599	2.783	2.925	3.041	3.139	3.224	3.299	3.365	3.480	3.577	3.660	4.085
8	1.860	2.285	2.530	2.703	2.837	2.946	3.038	3.117	3.187	3.250	3.357	3.447	3.525	3.923
9	1.833	2.243	2.479	2.644	2.772	2.875	2.962	3.038	3.104	3.163	3.265	3.351	3.424	3.801
10	1.812	2.211	2.439	2.598	2.720	2.820	2.904	2.976	3.039	3.096	3.193	3.275	3.346	3.707
11	1.796	2.186	2.407	2.561	2.680	2.776	2.857	2.927	2.988	3.042	3.136	3.215	3.283	3.631
12	1.782	2.164	2.380	2.531	2.647	2.740	2.819	2.886	2.946	2.999	3.090	3.166	3.232	3.569
13	1.771	2.147	2.359	2.506	2.619	2.710	2.787	2.853	2.911	2.962	3.051	3.126	3.190	3.517
14	1.761	2.132	2.340	2.485	2.596	2.685	2.760	2.825	2.881	2.932	3.018	3.091	3.154	3.473
15	1.753	2.119	2.324	2.467	2.576	2.663	2.737	2.800	2.856	2.905	2.990	3.062	3.123	3.436
16	1.746	2.108	2.311	2.451	2.558	2.645	2.717	2.779	2.834	2.883	2.966	3.036	3.096	3.403
17	1.740	2.099	2.299	2.437	2.543	2.628	2.700	2.761	2.815	2.863	2.945	3.014	3.073	3.375
18	1.734	2.090	2.288	2.425	2.530	2.614	2.684	2.745	2.798	2.845	2.926	2.994	3.052	3.349
19	1.729	2.083	2.279	2.415	2.518	2.601	2.671	2.731	2.783	2.830	2.910	2.977	3.034	3.327
20	1.725	2.076	2.271	2.405	2.507	2.590	2.659	2.718	2.770	2.816	2.895	2.961	3.018	3.307
24	1.711	2.055	2.245	2.375	2.474	2.554	2.621	2.678	2.728	2.772	2.848	2.912	2.967	3.244
30	1.697	2.034	2.219	2.346	2.442	2.519	2.584	2.639	2.687	2.730	2.803	2.864	2.917	3.183
36	1.688	2.020	2.202	2.327	2.421	2.496	2.559	2.613	2.660	2.702	2.773	2.833	2.884	3.142
40	1.684	2.014	2.194	2.317	2.410	2.485	2.547	2.600	2.647	2.688	2.758	2.817	2.868	3.122
60	1.671	1.993	2.169	2.289	2.379	2.451	2.511	2.563	2.607	2.647	2.715	2.771	2.820	3.063
120	1.658	1.974	2.145	2.261	2.349	2.418	2.476	2.526	2.569	2.607	2.672	2.726	2.773	3.005
144	1.656	1.971	2.141	2.257	2.344	2.413	2.471	2.520	2.563	2.601	2.665	2.719	2.765	2.995
∞	1.645	1.955	2.121	2.234	2.319	2.386	2.442	2.490	2.531	2.568	2.630	2.682	2.727	2.948

This table was prepared using a program for the calculation of quantiles of the studentized extreme deviate distribution, written by Daniel L. Rose.

Table A.11a .95 Quantiles of the Studentized Range Distribution

	Number of Means to Be Compared													
ν	**2**	**3**	**4**	**5**	**6**	**7**	**8**	**9**	**10**	**11**	**12**	**13**	**15**	**20**
5	3.64	4.60	5.22	5.67	6.03	6.33	6.58	6.80	6.99	7.17	7.32	7.47	7.72	8.21
6	3.46	4.34	4.90	5.30	5.63	5.90	6.12	6.32	6.49	6.65	6.79	6.92	7.14	7.59
7	3.34	4.16	4.68	5.06	5.36	5.61	5.82	6.00	6.16	6.30	6.43	6.55	6.76	7.17
8	3.26	4.04	4.53	4.89	5.17	5.40	5.60	5.77	5.92	6.05	6.18	6.29	6.48	6.87
9	3.20	3.95	4.41	4.76	5.02	5.24	5.43	5.59	5.74	5.87	5.98	6.09	6.28	6.64
10	3.15	3.88	4.33	4.65	4.91	5.12	5.30	5.46	5.60	5.72	5.83	5.93	6.11	6.47
11	3.11	3.82	4.26	4.57	4.82	5.03	5.20	5.35	5.49	5.61	5.71	5.81	5.98	6.33
12	3.08	3.77	4.20	4.51	4.75	4.95	5.12	5.27	5.39	5.51	5.61	5.71	5.88	6.21
13	3.06	3.73	4.15	4.45	4.69	4.88	5.05	5.19	5.32	5.43	5.53	5.63	5.79	6.11
14	3.03	3.70	4.11	4.41	4.64	4.83	4.99	5.13	5.25	5.36	5.46	5.55	5.71	6.03
15	3.01	3.67	4.08	4.37	4.59	4.78	4.94	5.08	5.20	5.31	5.40	5.49	5.65	5.96
16	3.00	3.65	4.05	4.33	4.56	4.74	4.90	5.03	5.15	5.26	5.35	5.44	5.59	5.90
17	2.98	3.63	4.02	4.30	4.52	4.70	4.86	4.99	5.11	5.21	5.31	5.39	5.54	5.84
18	2.97	3.61	4.00	4.28	4.49	4.67	4.82	4.96	5.07	5.17	5.27	5.35	5.50	5.79
19	2.96	3.59	3.98	4.25	4.47	4.65	4.79	4.92	5.04	5.14	5.23	5.31	5.46	5.75
20	2.95	3.58	3.96	4.23	4.45	4.62	4.77	4.90	5.01	5.11	5.20	5.28	5.43	5.71
24	2.92	3.53	3.90	4.17	4.37	4.54	4.68	4.81	4.92	5.01	5.10	5.18	5.32	5.59
30	2.89	3.49	3.85	4.10	4.30	4.46	4.60	4.72	4.82	4.92	5.00	5.08	5.21	5.47
40	2.86	3.44	3.79	4.04	4.23	4.39	4.52	4.63	4.73	4.82	4.90	4.98	5.11	5.36
60	2.83	3.40	3.74	3.98	4.16	4.31	4.44	4.55	4.65	4.73	4.81	4.88	5.00	5.24
120	2.80	3.36	3.68	3.92	4.10	4.24	4.36	4.47	4.56	4.64	4.71	4.78	4.90	5.13
∞	2.77	3.31	3.63	3.86	4.03	4.17	4.29	4.39	4.47	4.55	4.62	4.68	4.80	5.01

This table is taken from *Probability and Statistics for Engineering and the Sciences,* by J. L. Devore, and originally abridged from *Biometrika Tables for Statisticians,* E. S. Pearson and H. O. Hartley, editors.

Table A.11b .99 Quantiles of the Studentized Range Distribution

ν	Number of Means to Be Compared													
	2	3	4	5	6	7	8	9	10	11	12	13	15	20
5	5.70	6.98	7.80	8.42	8.91	9.32	9.67	9.97	10.24	10.48	10.70	10.89	11.24	11.93
6	5.24	6.33	7.03	7.56	7.97	8.32	8.61	8.87	9.10	9.30	9.48	9.65	9.95	10.54
7	4.95	5.92	6.54	7.01	7.37	7.68	7.94	8.17	8.37	8.55	8.71	8.86	9.12	9.65
8	4.75	5.64	6.20	6.62	6.96	7.24	7.47	7.68	7.86	8.03	8.18	8.31	8.55	9.03
9	4.60	5.43	5.96	6.35	6.66	6.91	7.13	7.33	7.49	7.65	7.78	7.91	8.13	8.57
10	4.48	5.27	5.77	6.14	6.43	6.67	6.87	7.05	7.21	7.36	7.49	7.60	7.81	8.23
11	4.39	5.15	5.62	5.97	6.25	6.48	6.67	6.84	6.99	7.13	7.25	7.36	7.56	7.95
12	4.32	5.05	5.50	5.84	6.10	6.32	6.51	6.67	6.81	6.94	7.06	7.17	7.36	7.73
13	4.26	4.96	5.40	5.73	5.98	6.19	6.37	6.53	6.67	6.79	6.90	7.01	7.19	7.55
14	4.21	4.89	5.32	5.63	5.88	6.08	6.26	6.41	6.54	6.66	6.77	6.87	7.05	7.39
15	4.17	4.84	5.25	5.56	5.80	5.99	6.16	6.31	6.44	6.55	6.66	6.76	6.93	7.26
16	4.13	4.79	5.19	5.49	5.72	5.92	6.08	6.22	6.35	6.46	6.56	6.66	6.82	7.15
17	4.10	4.74	5.14	5.43	5.66	5.85	6.01	6.15	6.27	6.38	6.48	6.57	6.73	7.05
18	4.07	4.70	5.09	5.38	5.60	5.79	5.94	6.08	6.20	6.31	6.41	6.50	6.65	6.97
19	4.05	4.67	5.05	5.33	5.55	5.73	5.89	6.02	6.14	6.25	6.34	6.43	6.58	6.89
20	4.02	4.64	5.02	5.29	5.51	5.69	5.84	5.97	6.09	6.19	6.28	6.37	6.52	6.82
24	3.96	4.55	4.91	5.17	5.37	5.54	5.69	5.81	5.92	6.02	6.11	6.19	6.33	6.61
30	3.89	4.45	4.80	5.05	5.24	5.40	5.54	5.65	5.76	5.85	5.93	6.01	6.14	6.41
40	3.82	4.37	4.70	4.93	5.11	5.26	5.39	5.50	5.60	5.69	5.76	5.83	5.96	6.21
60	3.76	4.28	4.59	4.82	4.99	5.13	5.25	5.36	5.45	5.53	5.60	5.67	5.78	6.01
120	3.70	4.20	4.50	4.71	4.87	5.01	5.12	5.21	5.30	5.37	5.44	5.50	5.61	5.83
∞	3.64	4.12	4.40	4.60	4.76	4.88	4.99	5.08	5.16	5.23	5.29	5.35	5.45	5.65

This table is taken from *Probability and Statistics for Engineering and the Sciences,* by J. L. Devore, and originally abridged from *Biometrika Tables for Statisticians,* E. S. Pearson and H. O. Hartley, editors.

APPENDIX B

Answers to Selected Exercises

CHAPTER 1

1.1. It is possible that many items measured have responses that deviate greatly from the target value on both the high and low sides. Averaging cancels or masks these extremes. Thus, it is possible that many items measured are not fit for use because the quality characteristic of interest is far from that necessary to produce acceptable performance.

1.3. Determining whether or not brake systems have measured features consistent with design requirements reveals quality of conformance. If measurements of a brake system's performance are compared to established safety standards, this is assessment of performance quality.

1.5. **a.** The number of good chips produced and the total number of chips produced were recorded over a long period and the corresponding ratio was about 14%.

b. The data collection activity would consist of sorting good from bad chips and recording counts of good and bad chips, along with documentation of the day, time, plant, etc. Summarization would involve finding the long-term ratio of good chips to total chips sorted, and possibly making a plot of ratios of good chips to total chips versus time. The conclusions that the process yield is an unsatisfactory 14% and that a bottleneck has been created are data-based inferences.

1.7. Vendor 1 price is (\$20.00/tube) × (1 million tubes/999,999 conforming) = \$20.00002/conforming tube. Vendor 2 price is (\$19.00/tube) × (1 million tubes/900,000 conforming) = \$21.11111/conforming tube. Vendor 3 price is (\$18.00/tube) × (?) = \$?/conforming tube.

1.10. **a.** The immediate customer is the operator that places the bags over the rods and into the box.

b. Yes. The hole locations could have a very small variability but all be off-center by about the same amount. If this amount is too large, the holes will not fit over the rods.

c. No. The thinking is not correct. Perhaps the operator has to stretch or crimp many of the bags to get them over the rods. A better method of evaluating quality of production is to measure hole locations (relative to bag edges) and distances between holes.

d. No. The process of hole punching and sealing must be stable or consistent before experimentation. Otherwise, the results of the experimentation cannot reliably be attributed to different experimental conditions. (Changes in response could be due more to instability in the process than to different experimental conditions.)

1.13. **a.** Final Inspection/Production Test = 1.3.
GE Rotor Assembly/Production Test = 100.
GE Assembly Teardown/Production Test = 300.
In Customer's Airplane/Production Test = 1000.
At Unscheduled Engine Removal/Production Test = 6000.

b. The ratios in (a) use the "Production Test" cost as the denominator, so the comparison to the schematic should also use "Production" as the denominator. Thus, the schematic produces two comparable ratios, 100/10 = 10 (for Assembly/Test) and 1000/10 = 100 (for Field). The two company ratios related to "Assembly" were 100 and 300, much larger than 10. The two company "Field" ratios were 1000 and 6000, again much larger than 100. The Final Inspection/Production Test ratio identified by GE does not have a comparable ratio in the schematic.

c. If consistently implemented, Step 3 in the Six-Step-Cycle of Table 1.1 can produce significant financial benefits and prevent poor perceptions of the company traceable to nonconformances. (These can produce loss of customer confidence and reduced future orders.)

1.18. **a.** Identify, say, two levels of feed rate and two levels of stop delay. For each of the four different combinations of feed rate and stop delay, make a large number of cuts, each of equal length, on the same type of material. Repeat this with, say, at least two cutting inserts per combination. Obtain the average tool wear (averaged over the different inserts for a particular feed rate/stop delay combination). Compare the average tool wears.

b. The important variable is tool or insert wear.

c. Measure insert wear.

d. Increasing tool life will allow customers to purchase a smaller number of inserts to make a given number of cuts (i.e., get more value for a given purchase price).

CHAPTER 2

2.1. a. The "firing" order was not recorded for the nine pellets.

b. The plot of y vs x does not appear to be linear. The plot of $\ln(y)$ vs $\ln(x)$ does appear to be linear. Thus, y appears to be related to x by $y = ax^b$.

2.2. a. The response variable is the depth of cut (in standard units).

b. $\overline{x}_{100} = 7.4, \overline{x}_{500} = 26.0, \overline{x}_{1000} = 35.4$.

c. $R_{100} = 3.0, R_{500} = 5.7, R_{1000} = 4.1$.

$R_{100}/d_2 = 3/2.059 = 1.4570$.

$R_{500}/d_2 = 5.7/2.059 = 2.7683$.

$R_{1000}/d_2 = 4.1/2.059 = 1.9912$.

$s_{100} = 1.296, s_{500} = 2.620, s_{1000} = 1.734$.

d. $\overline{R}/2.059 = 2.0722$.

e. Averaging improves precision. The concept of calibration is most closely associated with accuracy.

2.5. b. Estimated $\sigma_{\text{repeatability}}$ is .0000633333/1.128 = .000056146.

c. Estimated $\sigma_{\text{reproducibility}}$ is 3.3433×10^{-4} (no repeatability component).

d. Estimated percent of total measurement variance due to repeatability is $(.000056146)^2/((.000056146)^2 + (3.3433)^2 \times 10^{-8}) = .027428$ or 2.74%.

e. The estimated percent of total measurement variance due to reproducibility becomes $1 - .027428$ or 97.26%.

f. The plot illustrates the high proportion of variability due to reproducibility because Operator B is much different from A and C for every part.

g. $6\times(3.39013)\times10^{-4}/.002 = 1.017$. This is very large compared to .1. Thus, the measurement system is not good.

2.8. a. The rating for repeatability is unacceptable because "% Gage" for repeatability is 6(.00072695/.004)(100%) = 109%. The "% Gage" for reproducibility is 6(.0004625244/.004)(100%) = 69.379%. This rating is also unacceptable.

b. $\overline{y}$ for operator 1 is $\mu + \Sigma\alpha_i/25 + \beta_1 + \Sigma\alpha\beta_{i1}/25 + \overline{\epsilon}_{.1.}$.

$\overline{y}$ for operator 2 is $\mu + \Sigma\alpha_i/25 + \beta_2 + \Sigma\alpha\beta_{i2}/25 + \overline{\epsilon}_{.2.}$.

The subscript i runs from 1 to 25.

c. $(\sigma_\beta^2 + \sigma_{\alpha\beta}^2/25 + \sigma^2/50)^{1/2}$.

2.9. a. $\alpha_i + \alpha\beta_{i\,\text{Lourits}} + \overline{\epsilon}_{i\,\text{Lourits}}$.

b. $\sigma_\alpha^2 + \sigma_{\alpha\beta}^2 + \sigma^2/2$.

c. $R/d_2(25)$ estimates the square root of the answer to (b).

d. $R/d_2(25) = .000763165$.

e. $\sigma_\alpha^2 + \sigma_{\alpha\beta}^2$.

f. $\sigma_\alpha^2 + \sigma_{\alpha\beta}^2 + \sigma^2$.

g. $(.000763165)^2 - (1/2)(.00056/1.128)^2 = 4.594208 \times 10^{-7}$.

h. $4.594208 \times 10^{-7} + (.00056/1.128)^2 = 7.054208 \times 10^{-7}$.

2.12. **a.** The purpose is eliminating psychological pressure to produce the same measurement in repeated evaluations by the same operator.

b. Operator/piece ranges R_{ij} are

	Hartong	Hart	Spears	Evering	Jobe
Part 1	.011	.024	.021	.003	.004
Part 2	.004	.008	.012	.011	.013

c. Operator/piece means $\bar{x}_{ij}$ are

	Hartong	Hart	Spears	Evering	Jobe
Part 1	3.476	3.4633	3.47533	3.47333	3.472
Part 2	3.25667	3.2436	3.25267	3.24233	3.24833

d. Ranges of operator/piece means are $\Delta_1 = .0126667$ and $\Delta_2 = .0143333$.

e. $R_{23} = .012 \times (1/400) \times 10^4 = .3\text{g/m}^2$.

f. Estimated $\sigma_{\text{repeatability}}$ is $.0111/1.693 = .0065564$. Thus the gage repeatability rating is $6(.0065564) \times (1/400) \times 10^4/4 = .2459$ or 24.59%, which is somewhere between marginal and unacceptable.

g. Estimated $\sigma_{\text{reproducibility}}$ is $((.0135/2.326)^2 - (1/3)(.0065564)^2)^{1/2} = .004399683$. Thus, $6(.004399683) \times (1/400) \times 10^4/4 = .16499 = 16.499\%$, which is acceptable to marginal.

h. The estimated gage capability ratio is $6(.19357 + .42986)^{1/2} \times 10^{-2} \times (1/400) \times 10^4/4 = .29609$ or 29.6%, which is somewhere between marginal and unacceptable.

i. $11.8436\ \text{g/m}^2$.

2.14. **a.** The set of operators from which the five operators were selected is the group to which conclusions can legitimately be extended.

b. The set of paper pieces from which the two pieces were randomly drawn is the universe to which conclusions can legitimately be extended.

c. $I = 2, J = 5, m = 2$.

d. Using ANOVA, $\hat{\sigma}_{\text{repeatability}} = (2.0955)^{1/2} = 1.447584$.

e. Using the range approach, $\hat{\sigma}_{\text{repeatability}} = 1.2855$.

f. Using ANOVA, $\hat{\sigma}_{\text{reproducibility}} = (.651625)^{1/2} = .80723$.

g. Using the range approach, $\hat{\sigma}_{\text{reproducibility}} = .772708$.

h. Using ANOVA, $\hat{\sigma}_{\text{overall}} = (2.747119)^{1/2} = 1.65744$.

i. Using the range approach, $\hat{\sigma}_{\text{overall}} = (2.2496)^{1/2} = 1.49986$.

2.16. **a.** $L_c = z_1\sigma_{\text{measurement}}(2)^{1/2} = 1.645(.002)(1.4142135) = .00465276$. Since $y_{\text{new}} - y_{\text{blank}} = .006$ exceeds L_c, there is evidence of Thorium beyond that in the blank.

b. $L_c = z_1\sigma_{\text{measurement}}(2)^{1/2} = 2.33(.002)(2)^{1/2} = .0065902$. Since $y_{\text{new}} - y_{\text{blank}} = .006$ does not exceed L_c, there is not sufficient evidence to establish the presence of Thorium beyond that in the blank.

c. $L_c = z_1\sigma_{\text{measurement}}(2)^{1/2} = .006$ implies $z_1 = 2.12132$ or a risk level α between .0166 and .017.

d. Measurements must come from normal distributions with common variance for the blank and field measurements.

e. $A = L_c - z_2\sigma_{\text{measurement}}(2)^{1/2} = .00465276-(-1.645(.002)(1.4142135)) = 2(.00465276) = .00930552$.

f. A is the Lower Limit of Detection (L_d).

2.17. a. $A = 2(L_c) = 2(z_1\sigma_{\text{measurement}}(2)^{1/2}) = 2(2.33(.002)(2)^{1/2}) = 2(.0065902) = .0131804$.

b. (.00930552/58.2) = .000159889 or 159.889 μg/l.

c. (.0131804/58.2) = .000226467 or 226.467 μg/l.

d. $10(.002)(2)^{1/2} = .02828427$. Thus .02828427/58.2 = .000485984 or 485.984 μg/l.

2.19. a. $10(30.847544)(2)^{1/2} = 436.2502$ μg.

b. $10(30.847544)(1) = 308.47544$ μg.

c. Yes, there is only one source of variability (measurement error on the field sample) when the blank mean is known. The lower limit of detection (L_d) when both risk levels are .05 and the blank is unknown was shown to be 143.526 μg in part (e) of problem 2.18. If the blank value is known, the L_d becomes $2(1.645)(30.847544)(1) = 101.4884$ μg, i.e., knowing the true blank value decreases the L_d.

2.21. (In each of the entries below, σ abbreviates $\sigma_{\text{measurement}}$.)

	L_c	L_d	$10\,\sigma_{\bar{y}_f-\bar{y}_b}$
$n_b = 1,\ n_f = 1$	2.3264σ	4.6528σ	14.142σ
$n_b = \infty,\ n_f = 1$	1.645σ	3.29σ	10σ

CHAPTER 3

3.1. The point of Shewhart control charting is to provide a detection tool for process changes. If an unstable situation is identified the charts suggest when the instability began and (indirectly) what is needed to improve process consistency.

3.3. Since c charts and u charts are attributes-data charts, their application requires determination of conformance/nonconformance. If different criteria are used for determining conformance/nonconformance, one plotted X or $\hat{u}$ equal to, say, 3, may reflect something different than another plotted X or $\hat{u}$ of 3.

3.5. "Specification limits" are boundaries for the quality variable of interest delineating product functionality or acceptability. On the other hand, for Q a function of values of the quality variable of interest (this could be a sample average, sample standard deviation, sample range, etc.), control limits are boundaries (usually $3\sigma_Q$ above and below an agreed-upon value for μ_Q)

such that if Q is outside these boundaries, the process is thought to be unstable. The "specification limits" are applied to individual values of the quality variable (X), whereas the "control limits" are applied to Q. Specification limits have to do with product acceptability while control limits have to do with process stability.

3.7. ARL stands for "average run length." If a process changes, it is desirable to stop it and find out why this has occurred. The time interval between samples is usually called a period. The average number of periods before the control chart "flags" a change is the ARL.

3.11. Engineering specification limits are appropriate for judging single values of the quality variable of interest. Control limits on the $\overline{x}$ chart are appropriate for judging subgroup averages of n observations. Having specification limits on an $\overline{x}$ chart (except possibly when $n = 1$) is comparing apples with oranges. Further, a control chart is used to evaluate process stability. The specifications are meant to define product acceptability or functionality. It is possible for all items (x's) to be within specifications while one has an unstable process ($\overline{x}$ outside control limits). Also, the control limits are determined from either current process data or known values of the process parameters. Specification limits should be determined independent of the process, to guarantee (1) competitive advantage, (2) safety, and (3) performance for the customer.

3.12. The statement is not accurate. The job of control charting is to "flag" or "detect" an unstable or inconsistent process. An unstable process is not desirable, although it may produce all items within specifications. Process changes can on occasion produce fortuitous product improvement, though quality degradation is more common.

3.15. No, the notion of a chart guiding continuous regulatory efforts is inappropriate. The name "monitoring chart" is better than "control chart." Control limits are boundaries used to flag data that would be unusual under stable conditions. After observing an unusual plotted statistic, analysts should stop the process and look for physical causes. If causes are found, they should be addressed before restarting the process and the associated monitoring.

3.18. **a.** $LCL_{\overline{x}} = 8.679$, $UCL_{\overline{x}} = 13.721$, no LCL_s, $UCL_s = 3.690$. s chart and $\overline{x}$ chart reveal no instabilities.

b. $\hat{\sigma} = 1.8793$.

c. $P(Z < -5.959) \approx 0$.

d. Reduce the length from 11.2 units to 5.6377 units above nominal. This is a $(5.5622)(1/64) = .0869$-inch reduction. The yearly savings will be $(.0869 \text{ inch})(1/.25)(100,000) = \$34,764$.

e. $\overline{\overline{x}} = 5.6377$, $\hat{\sigma} = 1.8793$, $LCL_{\overline{x}} = 5.6377 - 3(1.8793)/\sqrt{3} = 2.38266$, $UCL_{\overline{x}} = 5.6377 + 3(1.8793)/\sqrt{3} = 8.89274$, no LCL_s, $UCL_s = B_6\hat{\sigma} = 4.277$.

f. $\mu = 4.37877$, $P(\overline{x} < LCL_{\overline{x}}) = .0329$, $ARL = 1/.0329 = 30.4 \approx 30$ or 31. For $p = .01$, $P(X \geq 1) = 1 - P(X = 0) = .0297$, $ARL = 1/.0297$, $ARL \approx 33$ or 34.

3.19. **a.** $\overline{x}_{\text{pooled}} = 11.2938$, from the equation in problem 3.17, $\hat{\sigma}_1 = 1.89606$. Using the equation in problem 3.17 $s_{\text{pooled}} = 1.8693 = \hat{\sigma}_2$.

d. Centerline for the $\overline{x}$ chart is constant. The centerline for the s chart, $c_4(n_i)\hat{\sigma}$, is a function of n_i and is not constant when n_i changes.

e. With $\overline{x}_{\text{pooled}} = 11.29375$, $\hat{\sigma}_2 = 1.8693$, $P(Z < -6.0418) \approx 0$.

3.21. **a.** $\overline{x}_{\text{pooled}} = 11.1381$; $\hat{\sigma}_1 = 1.88373$ from the formula in problem 3.17 using the s_i and $c_4(n_i)$; $\hat{\sigma}_2 = 1.8791$ using the formula in problem 3.17 for s_{pooled}; $\hat{\sigma}_3 = 1.83265$ from the formula in problem 3.21 using the R_i and $d_2(n_i)$.

d. Using $\hat{\sigma}_3 = 1.83265$ one gets the table

$\hat{\sigma}_R$	1.5632	1.6274	1.6127	1.5834	1.5541	1.5266	1.5028
n	2	3	4	5	6	7	8

f. No, $\hat{\sigma}_{R_i}$ and $\hat{\sigma}_{\overline{x}_i}$ are based on n_i (the n_is are not all equal).

g. The $\overline{x}$ chart has a constant centerline, $\overline{x}_{\text{pooled}}$. The R chart has different centerlines because $ER_i = \mu_{R_i}$ is a function of n_i.

3.27. **a.** This is an attribute data setting because counts are being recorded.

b. The final assembly of a single jet engine makes up a "subgroup."

c. The Poisson distribution. The minimum count is 0 with no upper bound. The probability of a nonconformance on a very small portion of the final assembly is very small.

d. $\hat{\lambda} = 21.20$, $\hat{\sigma} = \sqrt{21.20} = 4.6043$.

e. $UCL_c = 35.013$, $LCL_c = 7.3871$, $centerline = 21.20$. The process does not appear stable over the period of study. Counts for July 16 and 25 are below the LCL and above the UCL for August 4, 6, and 9. Further, the count on August 10 violates the rule concerning nine points in a row on one side of the centerline.

f. Both inspectors must use the same criteria for defining an imperfection/nonconformance.

3.31. **a.** $P(X = 5) = .00564$.

b. $P(X > 5) = .00091$.

c. $EX = 1.32$.

d. $\sigma^2 = 1.1748$.

e. $P(X \geq 1) = .75301$ = probability at least 1 (in 12) is cracked. Let Y equal the number of sets of 12 castings that have at least one cracked casting. $P(Y \geq 1) = 1 - P(Y = 0) = 1 - \binom{10}{0}(.75301)^0(.24699)^{10} = .9999992$.

3.37. **a.** $\Delta X(t)$ represents how much the variable X should be changed after observing Y at time t (with the goal that the next observed Y will be pushed toward the target value T). The adjustment $\Delta X(3)$ is made after observing $Y(3) = 0$.

b. If $\kappa_2 = 4$, then $\kappa_1 = 2$, $\kappa_3 = 1$.

c.

t	3	4	5	6	7	8	9	10	11	12
$\Delta X(t)$	18	1	10	1	12	0	−2	−6	−3	−4

d.

t	4	5	6	7	8	9	10
Number of periods considered	9	8	7	6	5	4	3
MSE	1.888	1.625	1.286	1.333	.8	.75	1

e. It seems the controlled process has stabilized. The MSE for the "last n periods" seems to have stabilized near 1.

3.39. a. Since $\Delta X(t) = 5$ produces 1.5 g/m^2 increase in paper dry weight, let $E(t) = 1.5$ and 5 ticks $= 0 \cdot \Delta E(t) + \kappa_2(1.5) + 0 \cdot \Delta^2 E(t)$ or $\kappa_2 = 3.33$.

b.

t	1	2	3	4	5	6	7	8
$E(t)$	−2.1	−.6	−1.3	2.9	−1.5	−.3	1.6	−1.7
$\Delta X(t)$	−6.99	−1.998	−4.329	9.657	−4.995	−.999	5.328	−5.661

e.

t	2	3	4	5	6
Number of periods considered	7	6	5	4	3
MSE	2.6071	2.9817	3.24	1.9475	1.8466

f. With only 5 points to plot it is difficult to authoritatively conclude the controlled process has stabilized. But since the last two MSEs are much smaller than the initial MSEs it seems possible the controlling equation is effective and that a limiting MSE near 2 is possible.

3.43. c.

t	4	5	6	7	8	9
Number of periods considered	8	7	6	5	4	3
MSE	1.335	.9543	.3067	.24	.21	.1967

d. The plot of MSE vs. number of periods considered does suggest the controlled process has stabilized with a limiting MSE near .2.

CHAPTER 4

4.2. **a.** $UCL_x = 6.482$, no LCL_x. $UCL_{MR} = 7.7$, no LCL_{MR}.
b. ARL is close to 155.
c. MRs are 4, 2.5, 0, 1, 0, 1, 2.5, 1.5, 3.
d. No change in variability. No change in location.
e. $UCL_x = 7.77$, no LCL_x. $UCL_{MR} = 5.67$, no LCL_{MR}.
f. $ARL = 226$. The chart in (e) is designed to pick up shifts in σ at the expense of degraded ability to pick up a shift in μ.

4.4. **a.** $UCL_x = 8.8627$, no LCL_x, *centerline* $= 3.384$.
$UCL_{MR} = 6.732$, no LCL_{MR}, *centerline* $= 2.06$.
b. Nine points on one side of the centerline on the moving range chart suggests a possible change in process short-term variability.
c. The MR chart should be considered first. The x-chart is based on a constant σ assumption.
d. $\bar{x} = 3.384$, $\hat{\sigma} = 1.826$.

4.11. **a.** $UCL_{EWMA} = 4.9990427$, $LCL_{EWMA} = 4.9989573$, *centerline* $= 4.999$.
b. $23 \leq ARL \leq 27$.
c. $9 \leq ARL \leq 11$.
d. $ARL \approx 14$ or 15.

4.14. **a.** $UCL_{EWMA} = .14595$, *centerline* $= .13$ and $LCL_{EWMA} = .11406$. The CUSUM chart has $k_1 = .14$, $k_2 = .12$, $h = .3204$, $-h = -.3204$.

b. The EWMA scheme produces:

$\Delta\mu$	.02	.04	.06	.08
ARL	27	11	6.8	5

The CUSUM scheme produces:

$\Delta\mu$	.02	.04	.06	.08
ARL	29	11	7.1	5.2

c. $UCL_x = .254$, $LCL_x = .006$, *centerline* $= .13$. $UCL_{MR} = .1828$, no LCL_{MR}.

d. The individuals chart produces:

$\Delta\mu$	.02	.04	.06	.08
ARL	141	43	15	6.4

e. It is clear the EWMA and CUSUM schemes are uniformly better than the individuals chart for $\Delta\mu \leq .08$. The EWMA chart is perhaps slightly better than the CUSUM scheme. (Statistical theory suggests that this apparent small advantage is illusory.)

4.15. **a.** $EWMA_0 = .025$, $\lambda = .11$, $UCL_{EWMA} = .025377$, $LCL_{EWMA} = .0246227$.
b. $\mathcal{H}^* = 2.71$, $\lambda = .11$, $\mathcal{D}^* = .3464$, $ARL \approx 60$.
c. $UCL_{\bar{x}} = .02673$, $LCL_{\bar{x}} = .02327$, *centerline* $= .025$, all-OK $ARL = 370$.

d. $ARL \approx 221$ by linear interpolation.
e. The EWMA chart will tend to detect the small change in process mean more quickly than the $\bar{x}$ chart.

4.16. a. $UCL_{EWMA} = 1.936537$, $LCL_{EWMA} = 1.936463$, $\mathcal{D}^* = 1$, $\mathcal{K}^* = 2.318$, $\lambda = .18$, $6.7 \leq ARL \leq 7.5$.
b. $k_1 = 1.936525$, $k_2 = 1.936475$, $h = .000175$, $\mathcal{H}^* = 3.5$, $\mathcal{D}^* = 1$, $\mathcal{K}^* = .5$, $ARL \approx 7.4$.
c. EWMA and CUSUM schemes have similar ARL properties. But EWMA schemes have inferior "worst-case" behavior.

4.25. a. $k_1 = 17.6137$, $k_2 = 16.386$, $h = 1.049315$, $U_0 = .52465$, $L_0 = -.52465$.
b. $U_i = 0$ for every i except $i = 7$ and 15, $U_7 = .0263$, $U_{15} = .0863$. No signs of a shift up. $L_1 = -1.171$, $L_4 = -1.85$, $L_5 = -1.6964$, $L_6 = -1.42275$ indicates a shift down.

4.31. a. Variable 1: $UCL_{\bar{x}} = 23.354$, $LCL_{\bar{x}} = 16.646$, $centerline = 20$. Only subgroup 4 has $\bar{x}$ outside control limits.
Variable 2: $UCL_{\bar{x}} = 23.354$, $LCL_{\bar{x}} = 16.646$, $centerline = 20$. Only subgroup 4 has $\bar{x}$ outside control limits.
b. $centerline = p = 2$, $UCL = 8$.

subgroup	1	2	3	4	5	6	7	8
X^2	0	3.56	8.89	10.89	0	8	32	3.56

Yes, we have points out-of-control (i.e., subgroups 3, 4, 6, and 7).
c. Subgroups 3, 6, and 7 have sample means that are related to μ_1 and μ_2 in a different fashion than that described by the positive correlation indicated in $\boldsymbol{V}$.

4.36. a.

$$\hat{\boldsymbol{V}} = \begin{pmatrix} .00000210 & -.00000004 & .00000041 \\ -.00000004 & .00000020 & .00000019 \\ .00000041 & .00000019 & .00000046 \end{pmatrix} \begin{matrix} \text{A} \\ \text{B} \\ \text{C} \end{matrix}$$

(columns A, B, C)

$\bar{x}_A = 1.4998$, $\bar{x}_B = .32413$, $\bar{x}_C = .07944$.

b.

$$\hat{\boldsymbol{V}} = \begin{pmatrix} .000001286 & .000000071 & .000000429 \\ .000000071 & .000000214 & .000000232 \\ .000000429 & .000000232 & .000000482 \end{pmatrix}$$

c.

part	1	2	3	4	5	6	7	8
X^2	4.5236	4.6018	1.1429	4.1076	5.4268	1.3673	6.8182	1.8154

$UCL_{X^2} = 10.348$. Since all $X^2 < 10.348$, conclude that the process was stable.

CHAPTER 5

5.5. **a.** $Q_z(.1) = -1.2815$, $Q_z(.5) = 0$, $Q_z(.9) = 1.2815$.
b. $\bar{x} = 56.006$, $s = 42.518$, $Q_x(.1) = 1.5192$, $Q_x(.5) = 56.006$, $Q_x(.90) = 110.4928$.
c. $Q_x(.1) = 14.77$, $Q_x(.5) = 45.65$, $Q_x(.9) = 103.64$.
d. $\bar{x} = 22.142$, $s = 22.007$, $Q_x(.1) = -6.0599$, $Q_x(.5) = 22.142$, $Q_x(.9) = 50.3439$.
e. $Q_x(.1) = 2.6999$, $Q_x(.5) = 15.05$, $Q_x(.9) = 50.7198$.
f. For the 52 wells, use the direct approach because it doesn't seem the data come from a normal distribution. Problem 5.4(c) suggests a serious departure from normality. Thus, use $Q_x(.1) = 14.77$, $Q_x(.5) = 45.67$, $Q_x(.9) = 103.6399$.
g. For the alternative (12 wells), use the direct approach because it doesn't seem the data come from a normal distribution. Problem 5.4(d) suggests a serious departure from normality. Thus, use $Q_x(.1) = 2.699$, $Q_x(.5) = 15.05$, $Q_x(.9) = 50.7198$.

5.7. **a.** No, the order doesn't change. 0, 1.3862, 1.6094, 2.30259, 2.39790.
b. $Q_x(.3) = 4$, $Q_x(.5) = 5$, $Q_x(.7) = 10$, $Q_{\ln(x)}(.3) = 1.3862$, $Q_{\ln(x)}(.5) = 1.6094$, $Q_{\ln(x)}(.7) = 2.3025$.
c. $\exp(1.6094) = 4.9998 \approx 5.0$.
d. $\exp(1.3862) = 3.9996 \approx 4.0$.
e. The p quantile of the x distribution corresponds to the p quantile of the $\ln(x)$ distribution, i.e., $\ln(Q_x(p)) = Q_{\ln(x)}(p)$ or $\exp(Q_{\ln(x)}(p)) = Q_x(p)$.

5.10. **a.** $(-43.911, 155.9233)$ has a 95% chance of containing productions from 95% of all wells like the 52 sampled. (Units are 1000 barrels.) ($\tau = 2.35$).
b. $(-30.273, 142.285)$ has a 95% chance of containing the production from any one well like the 52 sampled. (Units are 1000 barrels.) ($t_{51} = 2.01$).
c. $(-21.376, \infty)$ has a 99% chance of containing productions of 90% of all wells like the 52 sampled. (Units are 1000 barrels.) ($\tau = 1.82$) $B = -21.376$.

5.17. **a.** (2.8972, 3.178) has a 95% chance of including diameters from 90% of 3-inch saddles produced like the 20 sampled.
(3.9212, 4.526) has a 95% chance of including diameters from 90% of 4-inch saddles produced like the 20 sampled.
(6.2747, 6.6569) has a 95% chance of including diameters from 90% of 6.5-inch saddles produced like the 20 sampled.
b. (2.9078, 3.1676) has a 95% chance of including the diameter for any single 3-inch saddle produced like the 20 sampled.
(3.9439, 4.5033) has a 95% chance of including the diameter for any single 4-inch saddle produced like the 20 sampled.

(6.2891, 6.6425) has a 95% chance of including the diameter for any single 6.5-inch saddle produced like the 20 sampled.

c. There is a 99% chance that 90% of all saddles like those sampled have diameters of at least 2.8999.
There is a 99% chance that 90% of all saddles like those sampled have diameters of at least 3.9268.
There is a 99% chance that 90% of all saddles like those sampled have diameters of at least 6.2783.

d. There is a 95% chance the next 3-inch saddle selected will have a diameter of at least 2.9304.
There is a 95% chance the next 4-inch saddle selected will have a diameter of at least 3.9926.
There is a 95% chance the next 6.5-inch saddle selected will have a diameter of at least 6.3198.

5.18. a. 60.825%.
b. 90.48%.
c. 87.84%.
d. 95.238%.

5.28. a. .81398 is a 95% lower bound for C_p. .44314 is a 95% lower bound for C_{pk}.
b. The estimated C_p reflects potential performance. The estimated C_{pk} reflects current performance. $\overline{x}$ is not at the nominal value $(U + L)/2$.
c. Both estimated indices would increase.
d. To credibly compare a series of estimated C_p or C_{pk} values, the lower and upper specifications must be constant.
e. (44.9727, 44.9861) is a 99% prediction interval for the next journal diameter selected. The linear normal probability plot and stability of aim and short-term variability give this interval credibility.
f. (44.9713, 44.9874). This interval is 95% sure to include 99% of journal diameters.

5.30. a. $\mu_X + \mu_Y$.
b. X and Y must have $\text{cov}(X, Y) = 0$.
c. $\sigma_{X+Y} = (\sigma_X^2 + \sigma_Y^2)^{1/2}$.
d. $\overline{x} + \overline{y}$.
e. $(s_x^2 + s_y^2)^{1/2}$.
f. Let $w = x + y$. Thus, data values $w_1, w_2, \ldots, w_n$ are available. Calculate s_w^2. Then s_w is an estimate of σ_{X+Y}. In (e) the expression is a valid estimate of σ_{X+Y} only when $\text{cov}(X, Y) = 0$. If $\rho_{XY} \neq 0$, $\text{cov}(X, Y) \neq 0$.
g. No. $\overline{x} + \overline{y} = \overline{w}$.

5.31. a. $\text{E}XY = \mu_X \mu_Y = 300$.
b. $\sigma_{area}^2 \approx (\mu_Y)^2\sigma_X^2 + (\mu_X)^2\sigma_Y^2$, $\sigma_{area} \approx 9.01387$.
c. $\text{E}(volume) = 3600$.
d. $\sigma_{volume}^2 \approx (\mu_Y \mu_Z)^2\sigma_X^2 + (\mu_X \mu_Z)^2\sigma_Y^2 + (\mu_X \mu_Y)^2\sigma_Z^2$, $\sigma_{volume}^2 \approx 19{,}800$, $\sigma_{volume} \approx 140.7125$.

5.33. **a.** $I = 10$ = number of bundles, $J = 4$ = number of rods sampled from each bundle.

b. $y_{ij} = \mu + \alpha_i + \beta_{ij}$.

c. 4.661×10^{-6} in^2.

d. 8.231×10^{-5} in^2.

e. $\sigma_\beta^2 + \sigma_\alpha^2$ is estimated as 8.6972×10^{-5} in^2.

f. $\sigma_\alpha^2/(\sigma_\alpha^2 + \sigma_\beta^2) = .9464$ or 94.64%.

5.37. **a.** $\hat{\sigma}_\lambda^2 \approx 1.0927 \times 10^{-3}$, $\hat{\sigma}_\lambda = .033056$.

b. % for $C = 6.291$, % for $L = 36.24$, % for $T_1 = .3626$, % for $T_2 = .3624$, % for $\tau = 0$, % for $D = 56.77$. Bar diameter (D) contributes most to the variation in experimentally determined heat conductivity.

CHAPTER 6

6.1. **a.** Two-way factorial experiment.

b. Six treatment combinations.

c. Conclusions based on this experiment can legitimately be extended to synthetic material made with raw materials similar to those used in the experiment under one of the six treatment combinations implemented in the experiment.

d. $y_{ijk} = \mu + \alpha_i + \beta_j + \alpha\beta_{ij} + \epsilon_{ijk}$.

y_{ijk} = iron content measured from the kth sample at time level i and coating level j.

α_i = main effect of time at level i, i.e., the true average iron contamination at level i of time minus the overall average.

β_j = main effect of coating at level j, i.e., the true average iron content at level j of coating minus the overall average.

$\alpha\beta_{ij}$ = interaction effect of time at level i and coating at level j, i.e., the mean iron content for the ith time level and jth coating minus the sum of the ith time level mean and jth coating level mean, minus the overall mean.

ϵ_{ijk} = deviation of the kth measured iron content at the ith time level and jth coating level from the mean iron content at the ith time and jth coating levels.

e. Six degrees of freedom.

f. Two factors. Three levels of Curing Time. Two levels of Coating.

g. 3×2.

6.7. **a.** One factor. The two levels are Avimid-N and VCAP-75 polymer resins.

b. $y_{ij} = \mu + \alpha_i + \epsilon_{ij}$. $i = 1, 2, j = 1, \ldots, 4$.

y_{ij} is the percent weight loss from the jth piece of polymer i.

μ is the grand mean percent weight loss for all pieces from all polymers considered.

α_i is the percent weight loss deviation (from the grand average percent weight loss) attributable to polymer i.

ϵ_{ij} is the percent weight loss deviation (from the average percent weight loss of the i th polymer) attributable to the j th piece of the i th polymer.

c. Two treatments. Avimid-N, VCAP-75.

d. $\bar{y}_{\text{Avimid-N}} = 9.7699, \bar{y}_{\text{VCAP-75}} = 26.6238.$
$\hat{\epsilon}_{11} = .8877, \hat{\epsilon}_{12} = -.6685, \hat{\epsilon}_{13} = -.6790, \hat{\epsilon}_{14} = .46,$
$\hat{\epsilon}_{21} = -.7433, \hat{\epsilon}_{22} = -3.0362, \hat{\epsilon}_{23} = -.1365, \hat{\epsilon}_{24} = 3.916.$

e. $s_P = 2.1225.$

f. Some departure from normality is apparent in this experiment.

6.10. a. One factor. Five levels. Avimid-N, VCAP-75, N-CYCAP, PMR-II-50, AFR700B are the five levels, each level being a polymer type.
$y_{ij} = \mu + \alpha_i + \epsilon_{ij}$. y_{ij}, μ, α_i and ϵ_{ij} are interpreted as in exercise 6.7(b). $i = 1, \ldots, 5, j = 1, \ldots, 4.$
Five treatments. The five polymer types Avimid-N, VCAP-75, N-CYCAP, PMR-II-50, AFR700B are the five treatments.

	Avimid-N	VCAP-75	N-CYCAP	PMR-II-50	AFR700B
Mean	Avg. 9.7699	26.6238	25.1871	28.7791	26.7489
Residuals	.88765	−.7433	.03675	−1.2574	1.6838
	−.6685	−3.0362	.18155	.35913	2.2059
	−.67905	−.1365	−.60195	1.106	−1.96257
	.45995	3.916	.38365	−.2076	−1.92717

$s_{\text{pooled}} = 1.748.$
No strong departure from normality is indicated on the normal probability plot.

b. $.95 = 1 - (5 - 5\gamma)$. $\gamma = .99$ is the confidence level for each of five intervals in order to assure the set of five intervals has a simultaneous confidence level of at least 95%. The resulting five intervals are:

Avimid-N	VCAP-75	N-CYCAP
(7.1943, 12.3456)	(24.0481, 29.1995)	(22.6114, 27.7628)

PMR-II-50	AFR700B
(26.2034, 31.3548)	(24.1732, 29.3246)

The P-R method using a joint 95% confidence level produces the following five intervals:

Avimid-N	VCAP-75	N-CYCAP
(7.2274, 12.3126)	(24.0813, 29.1663)	(22.6446, 27.7296)

PMR-II-50	AFR700B
(26.2366, 31.3216)	(24.2064, 29.2914)

The P-R intervals are better. With the same family-wise confidence level, the corresponding intervals are uniformly shorter.

c. Ten different pairs.

d.

Avimid-N	minus	VCAP-75	$(-20.6734, -13.0346)$
Avimid-N	minus	N-CYCAP	$(-19.2366, -11.5978)$
Avimid-N	minus	PMR-II-50	$(-22.8286, -15.1898)$
Avimid-N	minus	AFR700B	$(-20.7984, -13.1596)$
VCAP-75	minus	N-CYCAP	$(-2.3827, 5.2561)$
VCAP-75	minus	PMR-II-50	$(-5.9747, 1.6641)$
VCAP-75	minus	AFR700B	$(-3.9445, 3.6943)$
N-CYCAP	minus	PMR-II-50	$(-7.4114, .2274)$
N-CYCAP	minus	AFR700B	$(-5.3812, 2.2576)$
PMR-II-50	minus	AFR700B	$(-1.7892, 5.8496)$

e. Avimid-N is best.

6.16. a. There are methods available to analyze 2^2 factorial with unbalanced data.

b. Two factors. Two levels for each factor.

c. $y_{ijk} = \mu + \alpha_i + \beta_j + \alpha\beta_{ij} + \epsilon_{ijk}$.
$i = 1, 2 \qquad n_{11} = 1, n_{12} = 4.$
$j = 1, 2 \qquad n_{21} = 2, n_{22} = 3.$
$k = 1, \ldots, n_{ij}$.

d. Let $i = 1$ correspond to Avimid-N and $i = 2$ correspond to PMR-II-50.
$\hat{\alpha}_2 = 9.46875, \hat{\alpha}_1 = -9.46875.$
Let $j = 1$ correspond to position 1 and $j = 2$ correspond to position 2.
$\hat{\beta}_2 = -.81875, \hat{\beta}_1 = .81875.$

e. $\widehat{\alpha\beta}_{11} = -.98125, \widehat{\alpha\beta}_{12} = .98125, \widehat{\alpha\beta}_{21} = .98125, \widehat{\alpha\beta}_{22} = -.98125.$

f. $(-2.0299, .067404)$. Other intervals are not needed because $\alpha\beta_{12} = -\alpha\beta_{11}, \alpha\beta_{21} = -\alpha\beta_{11}, \alpha\beta_{22} = -\alpha\beta_{21} = \alpha\beta_{11}$.

g. $(-10.1609, -8.77653)$ is a 95% confidence interval for α_1. $(8.77653, 10.1609)$ is a 95% confidence interval for α_2. $(.126532, 1.510968)$ is a 95% confidence interval for β_1. $(-1.510968, -.126532)$ is a 95% confidence interval for β_2.

h. (−2.7367, −.53821) includes the main effect difference (for position 2 minus position 1) with 90% confidence.

i. (17.8382,20.03679) includes the main effect difference (for PMR-II-50 minus Avimid-N) with 90% confidence.

6.19. a. Four factors. Factor A is oven position, two levels: position 1 and position 2. Factor B is kapton, two levels: with and without. Factor C is preprocessing time, two levels: 15 minutes and two hours. Factor D is dianhydride type, two levels: polymer grade and electronic grade.

b. 2^4.

c. Sixteen treatment combinations.

d. No replication. Each treatment combination occurred only once.

e. $y_{ijklm} = \mu + \alpha_i + \beta_j + \alpha\beta_{ij} + \gamma_k + \alpha\gamma_{ik} + \beta\gamma_{jk} + \alpha\beta\gamma_{ijk} + \delta_l + \alpha\delta_{il} + \beta\delta_{jl} + \alpha\beta\delta_{ijl} + \gamma\delta_{kl} + \alpha\gamma\delta_{ikl} + \beta\gamma\delta_{jkl} + \alpha\beta\gamma\delta_{ijkl} + \epsilon_{ijklm}$. $m = 1$ for every i, j, k, l.

f. $\hat{\mu} = 4.3375$, $a_2 = \hat{\alpha}_2 = .1375$, $b_2 = \hat{\beta}_2 = .0375$, $ab_{22} = \widehat{\alpha\beta}_{22} = .0375$, $c_2 = \hat{\gamma}_2 = -.025$, $ac_{22} = \widehat{\alpha\gamma}_{22} = .025$, $bc_{22} = \widehat{\beta\gamma}_{22} = -.025$, $abc_{222} = \widehat{\alpha\beta\gamma}_{222} = .025$, $d_2 = \hat{\delta}_2 = -.525$, $ad_{22} = \widehat{\alpha\delta}_{22} = -.125$, $bd_{22} = \widehat{\beta\delta}_{22} = -.1$, $abd_{222} = \widehat{\alpha\beta\delta}_{222} = .05$, $cd_{22} = \widehat{\gamma\delta}_{22} = -.0125$, $acd_{222} = \widehat{\alpha\gamma\delta}_{222} = .0375$, $bcd_{222} = \widehat{\beta\gamma\delta}_{222} = -.0375$, $abcd_{2222} = \widehat{\alpha\beta\gamma\delta}_{2222} = .0625$.

g. and **h.** Let $p = (i - 1/2)/15$.

i	$Q_z(p)$	$Q_z((1+p)/2)$
1	−1.8339	.04179
2	−1.28155	.12566
3	−.96742	.21043
4	−.72791	.29674
5	−.5244	.38532
6	−.34069	.47704
7	−.16789	.57297
8	0	.67449
9	.16789	.7835
10	.34069	.90273
11	.5244	1.03643
12	.72791	1.19182
13	.96742	1.38299
14	1.28155	1.64485
15	1.83391	2.12805

j. Yes, there appears to be a statistically detectable effect on mean percent weight loss. It seems a smaller percent weight loss occurs with electronic grade dianhydride (i.e., Factor D appears to be statistically significant).

6.21. **a.** 28.

b. $s_{\text{pooled}} = .3$, $d.f. = 8$.

c. $\Delta = 1.187939$.

d.

Treatment	Kapton	Preprocessing Time	Dianhydride	95% P-R Simultaneous
1	no	15 min	polymer	(4.00075,5.49925)
2	yes	15 min	polymer	(4.25075,5.74925)
3	no	2 hours	polymer	(3.95075,5.44925)
4	yes	2 hours	polymer	(4.25075,5.74925)
5	no	15 min	electronic	(3.10075,4.59925)
6	yes	15 min	electronic	(3.10075,4.59925)
7	no	2 hours	electronic	(3.15075,4.64925)
8	yes	2 hours	electronic	(2.90075,4.39925)

e. Let the 2nd or high levels of factors A, B and C be kapton, two hours, and electronic grade respectively, $(-.19795, .14795)$ is a 95% interval estimate for $\alpha\beta_{22}$, $(-.27295,.07295)$ is a 95% interval estimate for $\alpha\gamma_{22}$, $(-.18545,.16045)$ is a 95% interval estimate for $\beta\gamma_{22}$, $(-.21045,.13545)$ is a 95% interval estimate for $\alpha\beta\gamma_{222}$. The group confidence is $\gamma \geq 1 - 4(1 - .95) = .80$.

f. $(\alpha_2 - \alpha_1) + (\alpha\gamma_{22} - \alpha\gamma_{12})$ is the kapton mean minus no-kapton mean for electronic grade resin averaged over time. Let $E = \alpha_2 - \alpha_1 + \alpha\gamma_{22} - \alpha\gamma_{12} = 2\alpha_2 + 2\alpha\gamma_{22}$. The estimate of $2\alpha_2 + 2\alpha\gamma_{22}$ is $\hat{E} = 2\hat{\alpha}_2 + 2\widehat{\alpha\gamma}_{22}$. Since each treatment combination mean is based on (constant) $n = 2$, $\text{Var}(\hat{E}) = 4\ \text{Var}(\hat{\alpha}_2) + 4\ \text{Var}(\widehat{\alpha\gamma}_{22})$. From display (6.31), the estimated variance of $\hat{E}$ is $8s_p^2(1/8)^2(8/2) = s_p^2(1/2)$. Thus, the confidence limits for E become $(2\hat{\alpha}_2 + 2\widehat{\alpha\gamma}_{22}) \pm t_8 s_p (1/2)^{1/2}$, i.e., $(-.6141, .36417)$. $(\alpha_2 - \alpha_1) + (\alpha\beta_{21} - \alpha\beta_{11}) + (\alpha\gamma_{22} - \alpha\gamma_{12}) + (\alpha\beta\gamma_{212} - \alpha\beta\gamma_{112})$ is the kapton mean minus the no-kapton mean for electronic grade resin at 15 min preprocessing time. Further, $(\alpha_2 - \alpha_1) + (\alpha\beta_{22} - \alpha\beta_{12}) + (\alpha\gamma_{22} - \alpha\gamma_{12}) + (\alpha\beta\gamma_{222} - \alpha\beta\gamma_{122})$ is the kapton mean minus the no-kapton mean for electronic grade at two hours preprocessing time. The 95% interval estimate for $\alpha_2 - \alpha_1 = 2\alpha_2$ is $(-.2709, .4209)$. Since in part (e), all intervals for two- and three-factor interactions included 0, the interval $(-.2709, .4209)$ is best (it is shorter and is estimating the quantity of interest).

g. $(\alpha_2 - \alpha_1) + (\alpha\gamma_{21} - \alpha\gamma_{11})$ is the kapton mean minus the no-kapton mean for polymer grade resin averaged over time. Let $E = (\alpha_2 - \alpha_1) + (\alpha\gamma_{21} - \alpha\gamma_{11}) = 2\alpha_2 + 2\alpha\gamma_{21}$. The estimate of E is $\hat{E} = 2\hat{\alpha}_2 + 2\widehat{\alpha\gamma}_{21}$. Since each treatment combination mean is based on (constant) $n = 2$, $\text{Var}(\hat{E}) = 4\ \text{Var}(\hat{\alpha}_2) + 4\ \text{Var}(\widehat{\alpha\gamma}_{21})$. From display (6.31), the estimated variance of

$\hat{E}$ is $8s_p^2(1/8)^2(8/2) = s_p^2(1/2)$. Thus, the confidence limits for E become $(2\hat{\alpha}_2 + 2\widehat{\alpha\gamma}_{21}) \pm t_8 s_p(1/2)^{1/2}$, i.e., $(-.21417, .76417)$. $(\alpha_2 - \alpha_1) + (\alpha\beta_{21} - \alpha\beta_{11}) + (\alpha\gamma_{21} - \alpha\gamma_{11}) + (\alpha\beta\gamma_{211} - \alpha\beta\gamma_{111})$ is the kapton mean minus the no-kapton mean for polymer grade at 15 min. preprocessing time. Further, $(\alpha_2 - \alpha_1) + (\alpha\beta_{22} - \alpha\beta_{12}) + (\alpha\gamma_{21} - \alpha\gamma_{11}) + (\alpha\beta\gamma_{221} - \alpha\beta\gamma_{121})$ is the kapton mean minus the no-kapton mean for polymer grade at 2 hours preprocessing time. Using the same argument as in (f), the best 95% interval estimate for the kapton mean minus the no-kapton mean is $(-.2709, .4209)$, i.e., the 95% interval estimate for $\alpha_2 - \alpha_1$.

h. A 95% interval estimate for the polymer grade mean minus electronic grade mean, averaged over time, with kapton is (.761,1.739). A 95% interval estimate for the difference in the two dianhydride main effects ($\gamma_1 - \gamma_2 = 2\gamma_1$), polymer minus electronic, is (.7041,1.3959). Since in part (e) all two-factor and three-factor interactions can be judged to be 0, $\gamma_1 - \gamma_2$ is equal to polymer grade mean minus electronic grade mean for any time/kapton combination. The shorter interval is "best", i.e., (.7041,1.3959).

CHAPTER 7

7.1. a. 128 runs, 12,800 tiles.

b. Eight treatment combinations, 800 tiles.

c. Four generators.

e. 15 effects. I ↔ ABD ↔ ACE ↔ BCF ↔ CDG ↔ BEG ↔ AFG ↔ DEF ↔ ABCG ↔ BCDE ↔ ACDF ↔ ABEF ↔ ADEG ↔ BDFG ↔ CEFG ↔ ABCDEFG.

7.2. a. C ↔ AB, E ↔ AD, F ↔ BD, G ↔ ABD.

b. I ↔ ABC ↔ ADE ↔ BDF ↔ CDG ↔ BEG ↔ AFG ↔ CEF ↔ ABDG ↔ BCDE ↔ ACDF ↔ ABEF ↔ ACEG ↔ BCFG ↔ DEFG ↔ ABCDEFG.

c. BC, DE, FG, BDG, CDF, BEF, CEG, ABDF, ACDG, ABEG, ACEF, ABCDE, ABCFG, ADEFG, BCDEFG are all aliased with the A effect.

e. $\hat{\mu} = .24125$, $\hat{\alpha}_2 = -.11375$, $\hat{\beta}_2 = .02625$, $\widehat{\alpha\beta}_{22} = .01125$, $\hat{\delta}_2 = -.05125$, $\widehat{\alpha\delta}_{22} = .06375$, $\widehat{\beta\delta}_{22} = -.10625$, $\widehat{\alpha\beta\delta}_{22} = .08875$.

h. From the two plots, it appears the A main effect plus aliases is detectably larger than the other effects plus aliases.

7.10. a. $2^{4-1} = 8, p = 4, q = 1$.

c. I ↔ ABCD.

d. A ↔ BCD, B ↔ ACD, C ↔ ABD, D ↔ ABC.

e. $\mu + \alpha\beta\gamma\delta_{2222}, \alpha_2 + \beta\gamma\delta_{222}, \beta_2 + \alpha\gamma\delta_{222}, \alpha\beta_{22} + \gamma\delta_{22}, \gamma_2 + \alpha\beta\delta_{222}, \alpha\gamma_{22} + \beta\delta_{22}, \beta\gamma_{22} + \alpha\delta_{22}, \delta_2 + \alpha\beta\gamma_{222}$.

f. $\widehat{\mu + \alpha\beta\gamma\delta_{2222}} = 81.355$, $\widehat{\alpha_2 + \beta\gamma\delta_{222}} = .98$, $\widehat{\beta_2 + \alpha\gamma\delta_{222}} = .365$, $\widehat{\alpha\beta_{22} + \gamma\delta_{22}} = .26$, $\widehat{\gamma_2 + \alpha\beta\delta_{222}} = 2.72$, $\widehat{\alpha\gamma_{22} + \beta\delta_{22}} = -.125$, $\widehat{\beta\gamma_{22} + \alpha\delta_{22}} = .41$, $\widehat{\delta_2 + \alpha\beta\gamma_{222}} = 4.355$.

g. Individual 95% interval estimates for the sums of effects in (f) are of the form *estimate* $\pm .7988$. Thus, $\mu + \alpha\beta\gamma\delta_{2222}$ is estimated with the 95% interval (80.5562,82.1538).

h. $\mu + \alpha\beta\gamma\delta_{2222}$, $\alpha_2 + \beta\gamma\delta_{222}$, $\gamma_2 + \alpha\beta\delta_{222}$ and $\delta_2 + \alpha\beta\gamma_{222}$ are statistically detectable sums of effects, i.e., the corresponding 95% intervals do not include 0.

7.22. c. Yes, there was replication, i.e., more than one response was available for points (.29,.44,.27), (.13,.67,.20), (.45,.35,.20), (.45,.21,.34), (.13,.53,.34).

d. No, all variables have a lower bound greater than 0 and an upper bound less than 1.

e. $\hat{y} = .685 + .33x_1 - .023x_2$.

f. $\hat{y} = -2.41 + 9.39x_1 + 8.28x_2 - 7.1x_1^2 - 5.74x_2^2 - 11.23x_1x_2$. The quadratic is a better fit.

g. No-intercept form of (e) is $\hat{y} = 1.01x_1 + .662x_2 + .685x_3$. No-intercept form of (f) is $\hat{y} = -.12x_1 + .13x_2 - 2.41x_3 + 1.61x_1x_2 + 7.1x_1x_3 + 5.74x_2x_3$.

7.23. a. ($x_1 = .4008, x_2 = .33021$) is the stationary point. It is in the experimental region. $x_3 = .26899$ and ($x_1 = .4008, x_2 = .33021, x_3 = .26899$) is in the experimental region. This point produces a maximum because both eigenvalues (-12.0611 and $-.7789$) are negative.

7.26. a. 153g NH_3, 51g NH_3, 270°C, 230°C, 500g H_2O, 100g H_2O, 1200 psi, 500 psi.

b. (102g NH_3, 250°C, 300g H_2O, 850 psi) is the center point. The eight uncoded star points are: (173.4,250,300,850), (30.6,250,300,850), (102,278,300,850), (102,222,300,850), (102,250,580,850), (102,250,20,850), (102,250,300,360), (102,250,300,1340).

c. 25 design points.

d. The stationary point (coded) is ($x_1 = .265, x_2 = 1.033, x_3 = .287, x_4 = 1.675$). This point is not in the experimental region. Since $\lambda_1 = -7.55$, $\lambda_2 = -6.01$, $\lambda_3 = -2.16$ and $\lambda_4 = 2.60$, the stationary point is a saddle point. (There are both positive and negative eigenvalues.)

e. For $x_3 = -1.4$ and $x_4 = -1.4$, $\hat{y} = 41.1433 - 3.5228x_1 - 13.8528x_2 + 2.194x_1x_2 - 4.292x_2^2 - 6.332x_1^2$. The stationary point has $x_1 = -.58359$ and $x_2 = -1.7629$. Since $\lambda_1 = -6.8$, $\lambda_2 = -3.814$, the stationary point is a maximum $\hat{y} = 54.3822$.

For $x_3 = -1.4$ and $x_4 = 0$, $\hat{y} = 52.4718 - 1.309x_1 - 9.9244x_2 + 2.194x_1x_2 - 6.332x_1^2 - 4.292x_2^2$. The stationary point has $x_1 = -.31776$ and $x_2 = -1.23737$. Since $\lambda_1 = -6.8$, $\lambda_2 = -3.81$, the stationary point is a maximum $\hat{y} = 58.82$.

For $x_3 = -1.4$ and $x_4 = 1.4$, $\hat{y} = 53.9768 + .904x_1 - 5.996x_2 - 6.332x_1^2 - 4.292x_2^2 + 2.194x_1x_2$. The stationary point has $x_1 = -.0519$ and $x_2 = -.7118$. Since $\lambda_1 = -6.8$, $\lambda_2 = -3.81$, the stationary point is a maximum $\hat{y} = 56.0872$.

For $x_3 = -1.3$ and $x_4 = -1.4$, $\hat{y} = 40.2228 - 3.5372x_1 - 13.0522x_2 - 6.332x_1^2 - 4.292x_2^2 + 2.194x_1x_2$. The stationary point has $x_1 = -.56788$ and $x_2 = -1.665675$. Since the eigenvalues are both negative, the stationary point is a maximum $\hat{y} = 52.0975$.
For $x_3 = -1.3$ and $x_4 = 0$, $\hat{y} = 51.5925 - 1.3238x_1 - 9.1238x_2 - 6.332x_1^2 - 4.292x_2^2 + 2.194x_1x_2$. The stationary point has $x_1 = -.30205$ and $x_2 = -1.14$. Since both eigenvalues are negative, the stationary point is a maximum $\hat{y} = 56.9934$.
For $x_3 = -1.3$ and $x_4 = 1.4$, $\hat{y} = 53.13866 + .8896x_1 - 5.1954x_2 - 6.332x_1^2 - 4.292x_2^2 + 2.194x_1x_2$. The stationary point has $x_1 = -.036212$ and $x_2 = -.61449$. Since both eigenvalues are negative, the stationary point is a maximum $\hat{y} = 54.7188$.

f. The point $x_1 = -.31776$, $x_2 = -1.237$, $x_3 = -1.4$, $x_4 = 0$ is recommended, $\hat{y} = 58.82$. In uncoded or raw values, (85.794g NH_3, 225.26°C, 20g H_2O, 850 psi) is the recommended point to maximize response.

CHAPTER 8

8.13. a. $n \approx 498$, $c \approx 29$.
b. $AOQ(.04) \approx .0392$, $AOQ(.06) \approx .0289$, $AOQ(.08) \approx .0031$.

8.14. a. Normal Inspection, $n = 200$, $c = 7$. Tightened Inspection, $n = 200$, $c = 5$. Reduced Inspection, $n = 80$, $c = 3$, $X = 4, 5$ accept lot and return to normal inspection, $X = 6$ reject lot. Go to reduced only if cumulative number of nonconforming is ≤ 22 for 10 consecutive accepted normal inspection lots.
c. $ATI(.015) = 243$, using Type B formulas because $n = 200$ is much less than $N = 4000$.

8.16. a. $n = 15$, $m = 11.119$.
b. $107 \leq n \leq 111$, choose $n = 109$, $c = 2$.
c. The assumption of a normal distribution is vital to the calculation of n in part (a). The accuracy of the normal distribution in its tails is critical, but if the normal assumption is to be violated, it is very likely departure will occur in the tails.
d. $AOQ(.05) = .004295$, $ATI(.05) = 9151$.
e. Normal inspection with code letter L, $n = 200$, $X \leq 5$ accept lot, $X \geq 6$ reject lot. Tightened inspection with code letter L, $n = 200$, $X \leq 3$ accept lot, $X \geq 4$ reject lot. Reduced inspection, code letter L, $n = 80$, $X \leq 2$ accept the lot, $X = 3$ or 4 accept the lot and return to normal inspection, $X \geq 5$ reject the lot and return to normal inspection. Go to reduced inspection from normal inspection only if the cumulative nonconforming items from the last 10 sets of $n = 200$ (from 10 accepted lots) is at most 14.

8.17. With $\mu = 1.1795$, $p = .0124$. With $\mu = 1.1795$, $Pa = .9974$. With $\mu = 1.1798$, $p = .1587$. With $\mu = 1.1798$, $Pa = .5$.

8.27. **a.** $\mu = 32.1645$, $Pa = .9015$.

b. $AOQ = .045075$, $ATI \approx 103$.

c. $n \approx 93$, $c \approx 7$.

8.32. **a.** $p^* = [(AOQL)i + 1]/(i + 1)$

$(AOQL)i = [(1 - f)/f][1 - [(AOQL)i + 1]/(i + 1)]^{i+1}$.

b. $i = 115$.

c. .3614

8.34. **a.** $p^* = .0639$, $AOQL = .04518$.

b. $ET_1 = 410$, $ET_2 = 1044$, $AFI = .2928$.

c. .56286 without the assumption that product quality is constant. .04518 when product quality is constant.

Bibliography

American National Standards Committee (1981). *ANSI/ASQC Z1.4-1981,* American Society for Quality Control, Milwaukee.

Anand, K.N., S.M. Bhadkamkar, and R. Moghe (1994–95). "Wet Method of Chemical Analysis of Cast Iron: Upgrading Accuracy and Precision Through Experimental Design," *Quality Engineering,* Vol. 7, No. 1, pp. 187–208.

Bisgaard, S. (1994). "Blocking Generators for Small 2^{k-p} Designs," *Journal of Quality Technology,* Vol. 26, No. 4, pp. 288–296.

Bisgaard, S. and H.T. Fuller (1995–1996). "Reducing Variation With Two-Level Factorial Experiments," *Quality Engineering,* Vol. 8, No. 2, pp. 373–377.

Box, G.E.P. and N.R. Draper (1969). *Evolutionary Operation,* Wiley, New York.

Box, G.E.P. and N.R. Draper (1986). *Empirical Model Building and Response Surfaces,* Wiley, New York.

Box, G.E.P., W.G. Hunter, and J.S. Hunter (1978). *Statistics for Experimenters,* Wiley, New York.

Brassard, M. (editor) (1984). *Proceedings–Case Study Seminar–Dr. Deming's Management Methods: How They Are Being Implemented in the U.S. and Abroad,* Growth Opportunity Alliance of Lawrence, Andover, MA.

Brezler, P. (1986). "Statistical Analysis: Mack Truck Gear Heat Treat Experiments," *Heat Treating,* Vol. 18, No. 11, pp. 26–29.

Brown, D.S., W.R. Turner, and A.C. Smith (1958). "Sealing Strength of Wax-Polyethylene Blends," *Tappi,* Vol. 41, No. 6, pp. 295–300.

Burdick, R.K. and F.A. Graybill (1992). *Confidence Intervals on Variance Components,* Marcel Dekker, New York.

Burns, W.L. (1989–90). "Quality Control Proves Itself in Assembly," *Quality Engineering,* Vol. 2, No. 1, pp. 91–101.

Burr, I.W. (1953). *Engineering Statistics and Quality Control,* McGraw-Hill, New York.

Burr, I.W. (1979). *Elementary Statistical Quality Control,* Marcel Dekker, New York.

Champ, C.W. and W.H. Woodall (1987). "Exact Results for Shewhart Control Charts With Supplementary Runs Rules," *Technometrics,* Vol. 29, No. 4, pp. 393–399.

Champ, C.W. and W.H. Woodall (1990). "A Program to Evaluate the Run Length Distribution of a Shewhart Control Chart with Supplementary Runs Rules," *Journal of Quality Technology,* Vol. 22, No. 1, pp. 68–73.

Chau, K.W. and W.R. Kelley (1993). "Formulating Printable Coatings via D-Optimality," *Journal of Coatings Technology,* Vol. 65, No. 821, pp. 71–78.

Collins, W.H. and C.B. Collins (1994). "Including Residual Analysis in Designed Experiments," *Quality Engineering,* Vol. 6, No. 4, pp. 547–565.

Cornell, J.A. (1990). *Experiments With Mixtures: Designs, Models and the Analysis of Mixture Data,* 2nd edition, Wiley, New York.

Cornell, J.A. (1995). "Fitting Models to Data From Mixture Experiments Containing Other Factors," *Journal of Quality Technology,* Vol. 27, No. 1, pp. 13–33.

Currie, L.A. (1968). "Limits for Qualitative Detection and Quantitative Determination," *Analytical Chemistry,* Vol. 40, No. 3, pp. 586–593.

Crowder, S.V. (1987). "Computation of ARL for Combined Individual Measurement and Moving Range Charts," *Journal of Quality Technology,* Vol. 19, No. 2, pp. 98–102.

Crowder, S.V. (1987). "A Program for the Computation of ARL for Combined Individual Measurement and Moving Range Charts," *Journal of Quality Technology,* Vol. 19, No. 2, pp. 103–106.

Crowder, S.V. (1987). "Average Run Lengths of Exponentially Weighted Moving Average Control Charts," *Journal of Quality Technology,* Vol. 19, No. 3, pp. 161–164.

Crowder, S.V. (1987). "A Simple Method for Studying Run-Length Distributions of Exponentially Weighted Moving Average Charts," *Technometrics,* Vol. 29, No. 4, pp. 401–407.

Crowder, S.V. (1989). "Design of Exponentially Weighted Moving Average Schemes," *Journal of Quality Technology,* Vol. 21, No. 3, pp. 155–162.

Crowder, S.V., K.L. Jensen, W.R. Stephenson, and S.B. Vardeman (1988). "An Interactive Program for the Analysis of Data from Two-Level Factorial Experiments via Probability Plotting," *Journal of Quality Technology,* Vol. 20, No. 2, pp. 140–148.

Duncan, A.J. (1986). *Quality Control and Industrial Statistics,* 5th edition, Irwin, Homewood, IL.

Eibl, S., U. Kess, and F. Pukelsheim (1992). "Achieving a Target Value for a Manufacturing Process: A Case Study," *Journal of Quality Technology,* Vol. 24, No. 1, pp. 22–26.

Ermer, D.S. and G.M. Hurtis (1995–1996). "Advanced SPC for Higher-Quality Electronic Card Manufacturing," *Quality Engineering,* Vol. 8, No. 2, pp. 283–299.

Gan, F.F. (1993). "The Run Length Distribution of a Cumulative Sum Control Chart," *Journal of Quality Technology,* Vol. 25, No. 3, pp. 205–215.

Grego, J.M. (1993). "Generalized Linear Models and Process Variation," *Journal of Quality Technology,* Vol. 25, No. 4, pp. 288–295.

Hahn, J.G. and W.Q. Meeker (1991). *Statistical Intervals: A Guide for Practitioners,* Wiley, New York.

Hendrix, C.D. (1979). "What Every Technologist Should Know About Experimental Design," *Chemical Technology,* Vol. 9, No. 3, pp.167–174.

Hill, W.J. and W.R. Demler (1970). "More on Planning Experiments to Increase Research Efficiency," *Industrial and Engineering Chemistry,* Vol. 62, No. 10, pp. 60–65.

Holmes, D.S. and E.A. Mergen (1993). "Improving the Performance of the T^2 Control Chart," *Quality Engineering,* Vol. 5, No. 4, pp. 619–625.

Kolarik, W.J. (1995). *Creating Quality: Concepts, Systems, Strategies and Tools,* McGraw-Hill, New York.

Koons, G.F. and M.H. Wilt (1985). "Design and Analysis of an ABS Pipe Compound Experiment," in *Experiments in Industry: Design, Analysis and Interpretation of Results,* American Society for Quality Control, Milwaukee, pp. 113–117.

Kurotori, I.S. (1966). "Experiments With Mixtures of Components Having Lower Bounds," *Industrial Quality Control,* Vol. 22, No. 11, pp. 592–596.

Lawson, J.S. (1990–1991). "Improving a Chemical Process Through Use of a Designed Experiment," *Quality Engineering,* Vol. 3, No. 2, pp. 215–235.

Lawson, J.S. and J.L. Madrigal (1994). "Robust Design Through Optimization Techniques," *Quality Engineering,* Vol. 6, No. 4, pp. 593–608.

Leigh, H.D. and T.D. Taylor (1990). "Computer-Generated Experimental Designs," *Ceramic Bulletin,* Vol. 69, No. 1, pp. 100–106.

Lochner, R.H. and J.E. Matar (1990). *Designing for Quality: An Introduction to the Best of Taguchi and Western Methods of Statistical Experimental Design,* Chapman and Hall, London and New York.

Lucas, J.M. (1982). "Combined Shewhart-CUSUM Quality Control Schemes," *Journal of Quality Technology,* Vol. 14, No. 2, pp. 51–59.

Lucas, J.M. and R.B. Crosier (1982). "Fast Initial Response for CUSUM Quality-Control Schemes: Give Your CUSUM a Head Start," *Technometrics,* Vol. 24, No. 3, pp. 199–205.

Mielnik, E.M. (1993–1994). "Design of a Metal-Cutting Drilling Experiment: A Discrete Two-Variable Problem," *Quality Engineering,* Vol. 6, No. 1, pp. 71–98.

Miller, A., R.R. Sitter, C.F.J. Wu, and D. Long (1993–1994). "Are Large Taguchi-Style Experiments Necessary? A Reanalysis of Gear and Pinion Data," *Quality Engineering,* Vol. 6, No. 1, pp. 21–37.

Moen, R.D., T.W. Nolan, and L.P. Provost (1991). *Improving Quality Through Planned Experimentation,* McGraw-Hill, New York.

Myers, R.H. (1976). *Response Surface Methodology,* Edwards Brothers, Ann Arbor.

Nair, V. N. (editor) (1992). "Taguchi's Parameter Design: A Panel Discussion," *Technometrics,* Vol. 34, No. 2, pp. 127–161.

Nelson, L. S. (1984). "The Shewhart Control Chart-Tests for Special Causes," *Journal of Quality Technology,* Vol. 16, No. 4, pp. 237–239.

Neter, J., M.H. Kutner, C.J. Nachtsheim, and W. Wasserman (1996). *Applied Linear Statistical Models,* 4th edition, Irwin, Chicago.

Ophir, S., U. El-Gad, and M. Snyder (1988). "A Case Study of the Use of an Experimental Design in Preventing Shorts in Nickel-Cadmium Cells," *Journal of Quality Technology,* Vol. 20, No. 1, pp. 44–50.

Quinlan, J. (1985). "Product Improvement by Application of Taguchi Methods," *American Supplier Institute News* (special symposium edition), American Supplier Institute, Dearborn, MI, pp. 11–16.

Ranganathan, R., K.K. Chowdhury, and A. Seksaria (1992). "Design Evaluation for Reduction in Performance Variation of TV Electron Guns," *Quality Engineering,* Vol. 4, No. 3, pp. 357–369.

Schilling, E.G. (1982). *Acceptance Sampling in Quality Control,* Marcel Dekker, New York.

Schneider, H., W.J. Kasperski, and L. Weissfeld (1993). "Finding Significant Effects for Unreplicated Fractional Factorials Using the n Smallest Contrasts," *Journal of Quality Technology,* Vol. 25, No. 1, pp. 18–27.

Sirvanci, M.B. and M. Durmaz (1993). "Variation Reduction by the Use of Designed Experiments," *Quality Engineering,* Vol. 5, No. 4, pp. 611–618.

Snee, R.D. (1981). "Developing Blending Models for Gasoline and Other Mixtures," *Technometrics,* Vol. 23, No. 2, pp. 119–130.

Snee, R.D. (1985). "Computer-Aided Design of Experiments: Some Practical Experiences," *Journal of Quality Technology,* Vol. 17, No. 4, pp. 222–236.

Snee, R.D. (1985). "Experimenting With a Large Number of Variables," in *Experiments in Industry: Design, Analysis and Interpretation of Results,* American Society for Quality Control, Milwaukee, pp. 25–35.

Snee, R.D., L.B. Hare, and J.R. Trout (editors) (1985). *Experiments in Industry: Design, Analysis and Interpretation of Results,* American Society for Quality Control, Milwaukee.

Sutter, J.K., J.M. Jobe, and E. Crane (1995). "Isothermal Aging of Polyimide Resins," *Journal of Applied Polymer Science,* Vol. 57, No. 12, pp. 1491–1499.

Taguchi, G. and Y. Wu (1980). *Introduction to Off-Line Quality Control,* Japan Quality Control Organization, Nagoya.

Tomlinson, W.J. and G.A. Cooper (1986). "Fracture Mechanism of Brass/Sn-Pb-Sb Solder Joints and the Effect of Production Variables on the Joint Strength," *Journal of Materials Science,* Vol. 21, No. 5, pp. 1730–1734.

Tracy, N.D., J.C. Young, and R.L. Mason (1995). "A Bivariate Control Chart for Paired Measurements," *Journal of Quality Technology,* Vol. 27, No. 4, pp. 370–376.

Vander Wiel, S.A. and S.B. Vardeman (1994). "A Discussion of All-or-None Inspection Policies," *Technometrics,* Vol. 36, No. 1, pp. 102–109.

Vardeman, S.B. (1986). "The Legitimate Role of Inspection in Modern SQC," *The American Statistician,* Vol. 40, No. 4, pp. 325–328.

Vardeman, S.B. (1994). *Statistics for Engineering Problem Solving,* PWS Publishing, Boston.

Walpole, R.E. and R.H. Myers (1993). *Probability and Statistics for Engineers and Scientists,* 5th edition, Macmillan, New York.

Wernimont, G. (1989–1990). "Statistical Quality Control in the Chemical Laboratory," *Quality Engineering,* Vol. 2, No. 1, pp. 59–72.

Western Electric Company (1984). *Statistical Quality Control Handbook,* 2nd edition, Western Electric Company, New York.

Wetherill, G.B. and D.W. Brown (1991). *Statistical Process Control: Theory and Practice,* Chapman and Hall, London and New York.

Zwickl, R.D. (1985). "An Example of Analysis of Means for Attribute Data Applied to a 2^4 Factorial Design," *ASQC Electronics Division Technical Supplement,* Issue 4, American Society for Quality Control, Milwaukee.

Index

Quality assurance applications from industry discussed in the text and in the problem sections are included in the index, i.e., "Plastics packaging". Only the page number where the example is first introduced in a given chapter is included. Terms containing numbers are placed in the index as if spelled out. Thus "2^p factorials" would be located as if spelled "two", "80-20 rule" would be located as if spelled "eighty", etc.

Also note: an italic "t" after a page number refers to a table on that page.